Quantum Uncertainties

Recent and Future Experiments and Interpretations

NATO ASI Series

Advanced Science Institutes Series

A series presenting the results of activities sponsored by the NATO Science Committee, which aims at the dissemination of advanced scientific and technological knowledge, with a view to strengthening links between scientific communities.

The series is published by an international board of publishers in conjunction with the NATO Scientific Affairs Division

A	**Life Sciences**	Plenum Publishing Corporation
B	**Physics**	New York and London
C	**Mathematical and Physical Sciences**	D. Reidel Publishing Company Dordrecht, Boston, and Lancaster
D	**Behavioral and Social Sciences**	Martinus Nijhoff Publishers
E	**Engineering and Materials Sciences**	The Hague, Boston, Dordrecht, and Lancaster
F	**Computer and Systems Sciences**	Springer-Verlag
G	**Ecological Sciences**	Berlin, Heidelberg, New York, London,
H	**Cell Biology**	Paris, and Tokyo

Recent Volumes in this Series

Series B: Physics

Quantum Uncertainties

Recent and Future Experiments and Interpretations

Edited by

William M. Honig

Curtin University
Perth, Western Australia

David W. Kraft

University of Bridgeport
Bridgeport, Connecticut

and

Emilio Panarella

National Research Council of Canada
Ottawa, Canada

Plenum Press
New York and London
Published in cooperation with NATO Scientific Affairs Division

PREFACE

    In the past 15 years we have witnessed an increase in efforts to test the Quantum Mechanics (QM) paradigm. This is evidenced by the increased number of workers attracted to this field and by several international conferences, such as those in Perugia (1982), Bari (1983) and New York (1986). Much of the recent work in this field has concerned experimental tests which derive from the Einstein-Podolsky-Rosen paradox, the Bell Theorem and other considerations. Although these studies have shed much light on the fundamentals of QM, other experiments and alternative interpretations may have been neglected. It seemed, therefore, both necessary and timely for a concentrated and integrated examination of these experiments and interpretations. This volume is a compilation of the invited papers and ensuing discussions at a workshop held at the University of Bridgeport, Bridgeport, Connecticut, U.S.A., June 23-27, 1986.

    The events which led to this conference began with correspondence between two of us, WMH and EP, in late 1984. It was felt that an ideal location for such a conference might be in the U.S., near a large metropolitan center; DWK the joined the organizing committee and offered the Bridgeport site. When NATO's support for an Advanced Research Workshop was secured, invitations for papers were extended to individuals active in this field; others were invited to participate as observers. The conference attained an international character with the participation of attendees from twelve nations.

    The choice of a suitable name for the workshop proved difficult. After much discussion, the organizers selected "Quantum Violations: Recent and Future Experiments and Interpretations." Once the conference was in session, however, it became apparent that "uncertainties" was more appropriate than "violations." Hence this volume bears the title "Quantum Uncertainties: Recent and Future Experiments and Interpretations."

    The structure of the sessions departed from the usual format in that all discussion was reserved for an extended period at the end of each day. This format allowed for both extensive discussion of the papers presented on that day and for additional discussion of papers presented on previous days. Approximately nine hours of discussions were held, most of which are contained herein. No one familiar with the controversial nature of the workshop will be surprised that the discussions proved spirited.

    The conference organizers wish to thank the individuals and organizations listed in the Acknowledgements, as well as the attendees for their

Proceedings of a NATO Advanced Research Workshop on
Quantum Violations: Recent and Future Experiments and Interpretations,
held June 23–27, 1986,
in Bridgeport, Connecticut

Library of Congress Cataloging in Publication Data

NATO Advanced Research Workshop on Quantum Violations: Recent and Future
    Experiments and Interpretations (1986: Bridgeport, Conn.)
    Quantum uncertainties.

    (NATO ASI series. Series B, Physics; vol. 162)
    "Proceedings of a NATO Advanced Research Workshop on Quantum Viola-
tions: Recent and Future Experiments and Interpretations, held June 23–27,
1986, Bridgeport, Connecticut"—Verso of t.p.
    Includes bibliographical references and index.
    1. Quantum theory—Congresses. I. Honig, William M. II. Kraft, David W. III.
Panarella, Emilio. IV. Title. V. Series: NATO ASI series. Series B, Physics; v. 162.
QC173.96.N38   1986   530.1′2   87-20223
ISBN-13: 978-1-4684-5388-1   e-ISBN-13: 978-1-4684-5386-7
DOI: 10.1007/978-1-4684-5386-7

© 1987 Plenum Press, New York
Softcover reprint of the hardcover 1st edition 1987
A Division of Plenum Publishing Corporation
233 Spring Street, New York, N.Y. 10013

participation and timely submission of manuscripts. We thank Dr. M. De Haan
of the Université Libre de Bruxelles for his oral presentation entitled
Quantum Zeno Effects and Quantum Irreversibility.

William M. Honig, Director
David W. Kraft, Associate Director
Emilio Panarella, Associate Director

ACKNOWLEDGEMENTS

The Conference Directors are indebted to the following individuals and
institutions for their support and assistance.

Dr. M. di Lullo and the North Atlantic Treaty Organization (NATO).
Prof. John de Laeter and Western Australian Institute of Technology.*
Mr. Don Manzer and WAITEC.
Dr. Edwin Eigel, Provost, and the Faculty and Staff of the University
    of Bridgeport.
National Research Council of Canada, and, in particular, H.S. Cuccaro
    and L. Charbonneau of the Stenographic Services, and C.C. Eamer,
    R. Vallieres and D. Saumure of the Drafting and Photography Section.
Dr. Bern Dibner and the Burndy Library of Science.
Dr. Anita R. Kraft and Jill L. Kraft for editorial assistance.

_____

* Recently renamed Curtin University

CONTENTS

SECTION III

SECTION IV

SECTION V

OPENING REMARKS-REPLACING THE QUANTUM PARADIGM

William M. Honig, Conference Director

Curtin University, (formerly known as the
Western Australian Institute of Technology)
Perth, Bentley, 6102, Western Australia

I take great pleasure in opening this conference. I have a number of
remarks to make about what I think may be our common attitudes with
respect to the Quantum Paradigm. I hope they will occasion your agreement,
disagreement, or modification and that your response will help initiate the
oral discussions in the final period of today's and each day's program.

This is the place and now is the time where alternates to Quantum
Mechanics (QM), can be discussed. Here such discussions can be carried out to
the bitter end free from preemptory and angry criticism, and the proponents
will not be under sufferance from those who favor QM. Those who would
slightly extend, reinterpret, or reexplain QM will carry here only an equal
status with other discussants.

It should be acknowledged that this conference title: Quantum Violations,
etc., is somewhat peculiar and in many people's minds is quite premature. It
should, however, aid in making the point that some or even many
experiments in the past are susceptible to alternate interpretations to that of
QM.

The major purpose of this conference is, I think, to discuss both theoretical
and experimental matters related to testing, modifying, or even replacing the
Quantum Paradigm.

These activities, how they started, and how they might proceed in order to
attract the serious attention of those uncommitted to these issues, also
deserve comment.

They are based on the remark which Einstein made to Born in the early 1950s just after David Bohm's first papers appeared discussing the viability of alternate concepts to QM and 'hidden variables'. The remark (paraphrased) was, "Bohm's ideas are somehow too cheap". I both disagree and agree with this remark. At that time when Bohm started his efforts for the reincarnation of anti-QM it was vitally necessary for someone to enhearten those whose thoughts were in such directions. I remember that the subject of anti-QM was then anathema because of the prevailing climate of opinion and it was impossible to even complete a sentence on this subject without it being banished from the conversation.

It was Bohm's bravery that brought this subject back into serious contention. His competence was in showing that an alternate crypto-deterministic or hidden variable approach was capable of remaining consistent with the methods and results of QM and that von Neumann's strictures were more limited than prevailing opinion believed.

These early steps have established today's more liberal climate of opinion and indeed have provided the conditions from which this conference springs.

Having said this, I now give the sense in which I can agree with Einstein's remark. I have, in the past, for more than six years been the Editor of a journal devoted to frank speculations in science. In that time I read hundreds of papers all devoted to heavily modifying or replacing one or another paradigm in science. Almost without exception all these efforts were devoted to showing that an alternate set of conceptions would deliver the same experimental results or practical equations which the original paradigm provides.

I have come to feel very strongly that a new theory must deliver experimental predictions which are confirmed and which are not derivable from the ruling paradigm and that otherwise such ideas are not deserving of attention by professional workers in that field. Thus only after the fertility and heuristic qualities of an idea are literally demonstrated does it deserve to be proposed as an occupant of the serious working time of those not directly involved in these activities.

I suppose the Einstein was thinking of matters of this kind when he made his remark. Since he was concerned with alternates to QM I think he had already accepted the points that Bohm made and was looking for something further.

The general acceptance of QM and of its slightly older brother Special Relativity (STR) occurred only after their obvious usefulness for the prediction of new measurements was demonstrated. Particularly for QM, its acceptance was propagated from the laboratories which confirmed the dazzling measurement accuracies which over the past 60 years have strengthened the position of QM.

I must emphasize my conclusion that it is an unfair imposition upon the time and attention of even unbiased workers in the field to expect them to automatically commit time to anti-QM ideas.

The establishment would welcome or at least be neutral to an experimental anti-QM attitude. This is because those committed to QM would expect that experiments which are performed to test QM will even more deeply confirm it, just as the Aspect experiment and its predecessors seem to have done. The current consensus on this subject is that no local hidden variable can duplicate recent experimental results and non-local hidden variables, whatever they may mean, are still largely unexplored.

This anti-QM attitude appears to be an open approach that is willing and anxious to take its chances in the laboratory with frankly falsifiable experimentally testable suppositions. This approach carries with it, its own formula for success. The performance of an experiment whose character and whose results are underivable from the QM paradigm will thus carry with it automatic acceptance of the ideas involved. This is, of course, provided that such experiments are confirmed and more important are simple and dramatic enough to properly impress.

This would counter the charge that anti-QM is the realm of the arm-chair philosopher with nothing to lose because nothing is being risked.

I think it has been accepted that there might be alternates to QM that are more philosophically satisfying but which deliver no experimental results differing from QM. I think that the lack of interest in these approaches is reasonable and without the literal demonstration of the fertility and heuristic qualities of a new approach no one should give up on what we have inherited. Although there have been some temporary exceptions the glories of our science have all had to pass through the same kinds of trials in the laboratory.

Turning now to our common sympathies: Many of us believe that QM has erected procedural and psychological barriers to the deeper study of nature. Many of us believe that the search and inquiry into microscopic physical reality is not over. Many of us believe in the standing which Physics enjoyed during the Age of the Enlightenment. Of this time someone has said, "Physics has always been the intellectual fountainhead of the human race".

The standing of physics has declined during this century. The selection of careers in the Biological Sciences by many talented people who might have gone into Physics is due, I think, to the contemporary perception that that is where the intellectual action is. This is a reasonable development if QM has indeed specified the ultimate limits of efforts in the microscopic realm.

If our attempts to go deeper into the QM realm fail in the laboratory, I for one will be satisfied to accept QM rather than some equivalent theory offering only philosophical comfort.

We have not come here to bury QM and certainly not to praise it. Our aim here is, I think, to initiate actions, both theoretical and experimental, which could ultimately result in the digestion of QM by an as yet incomplete or unborn paradigm. Those who disagree with these aims embrace this approach because it also harbors the possibility for confirming the QM paradigm even more strongly then before.

I will parade my prejudices now in order to make it easier for you to do the same in the daily final discussions. The formal presentation of previously prepared papers is the usual practice at meetings like this. In a field such as this, however, the oral discussions provide the best opportunities for direct confrontation of ideas, from which corrections, modifications, and syntheses can arise. It is usually the part of any meeting which is remembered longest and may have the most important influence on the future evolution of these ideas.

I comment on six necessary characteristics which a post-QM paradigm might need to have which I think are important. I believe these requirements cannot be accomplished piecemeal because of their interdependence. We are all, I think, familiar with each of these requirements but I commend your consideration to how such difficult tasks can be accomplished simultaneously (or in a single theory).

Generally, a new post-QM paradigm ought to solve the photon point-wave conundrum and suggest experiments to test this. It ought to bring back pictures or images of physical reality. This would not be merely pictures of mathematical terms in equations such as we do have today. At the present time, such images for, say, the fundamental particles are ruled out if invariance is invoked or adhered to. Invariance is too useful and too important to be dumped, but it should be possible to retain it operationally while a Lorentzian approach is substituted for Special Relativity.

This appears to lead to a fluid plenum (an ether) requirement, but one which supports relativistic invariance. This will permit the circumvention of the epistemological problem of retaining all of the results of STR (invariance) while simultaneously also permitting unique fluid (non-invariant appearing) models for the fundamental particles.

I think that Einstein's greatest contribution is that he invented a new way of doing science. It consists of constructing a theory using axioms which may appear inconsistent, paradoxical, crazy, etc., but, whose deductions give confirmed experimental predictions while the axioms themselves are difficult or impossible to empirically demonstrate or understand. This method permits us to skip ahead of deficiencies in our determinate understandings. This method, of course, is the one also used in QM. Thus both STR and QM are in this way closely related so that any changes in the QM paradigm may need also to affect the STR paradigm. Even if a post-QM paradigm becomes established which satisfies the wishes of many of us, this kind of axiomatic structure possessed by QM and STR may still be useful for the even further task of skipping ahead in the presence of deficiencies of our future determinate understandings.

The six characteristics I remark on are:

1. Relativistic Invariance must be retained on an operational basis but the conundrum of the axioms of the Special Theory of Relativity (STR) should be solved with a crypto-deterministic formulation. I am personally committed to the ideas of Geoffrey Builder of the University of Sydney who in the late '50s and the '60s showed that a fluid plenum could explain and be used to derive all of the results of STR. It gave no results counter to or beyond STR and was meant, because of the clarity of its concepts, to be a teaching aid. To put it in a very small nutshell, this is a neo-Lorentzian approach which postulates a fluid plenum and declares that the velocity of light is equal to c only in the cosmological (or absolute) rest frame. In all other rest frames the

Lorentz contractions which are to be considered as literal, are such that the space and time distortions are unmeasurable inside these frames. The magnitudes of these variations occur to exactly the extent which makes the measurement of the velocity of light also appear to be c in all the non-cosmological rest frames. Thus the velocity of light as c is literally true in the cosmological rest frame and is operationally true in all the rest frames which are in uniform motion with respect to the cosmological rest frame.

2. Since invariance is retained on an operational basis one may utilize fluids or some sort of ether to recover and provide the previously forbidden literal picturizable microscopic models for the fundamental particles (electrons, protons, photons, etc.,) The manifold continuous and discrete phenomena from the field of hydrodynamics should be of great heuristic aid here.

3. The concept of non-locality ought to be utilized and turned into a practical concept by extending the meaning of metrics. Helmholtz, Poincare, and others have written on how space can be considered a subjective perception. If the metric of space is a relative matter then that which is local in one rest frame may appear to be non-local in another rest frame. Thus a photon may appear to be local only in its own (but not physically accessible) rest frame while it appears to be non-local in physical rest frames. In this way it will be discrete in both rest frames.

4. A meaning should be found for the Psi and the related electromagnetic wave functions. It ought to include a logical explanation and also a set-theoretic interpretation for the imaginary number, i, and for imaginary exponentials. This is necessary because even a post-QM paradigm would try to make use of most of the techniques of QM wherever possible because it will be mainly in the post-quantum physical realm that the new paradigm would be most useful. In this way, the post-QM paradigm will obey a Correspondence Principle and efficiently utilize our past knowledge. Thus such a post-QM paradigm will not eliminate QM but will continue to use those techniques with the explicit explanation that the QM methods are simple to use in their field of action and were developed as very good ways of temporarily leapfrogging defects in our determinate understandings. Only at the limits of the quantum approach would the post-quantum approach begin to become useful. The necessity for meanings for the imaginary exponentials resides also in the conceptual integration which must occur between the QM and post-QM concepts. The logical and set-theoretic meanings for the imaginary exponential functions should provide this.

5. The first uses for the post-QM paradigm would be for the design and prediction of experiments in this realm. It is the only activity which will make possible the acceptance of the new paradigm. It should, however, be possible with such a theory to calculate the mass ratio of the fundamental particles, provide a physical explanation for the fine structure constant, and permit fluid models for each of the fundamental particles to be constructed. These models would then contain all those physical features which characterize the particle. These tasks are secondary however; many would be concerned with these matters only after acceptance of the post-QM experiments has occurred.

6. Finally all the above would still have to be related to a physical mechanism for gravitation. Since, however, fluids would now characterize both space and the fundamental particles themselves, it should be possible to give gravitational attraction a fluidic mechanism.

I close with an endearing comment made by Simon Diner at the end of a paper of his which has recently appeared.

Don't worry, there is order behind all this chaos.

THEORETICAL IMPLICATIONS OF TIME-DEPENDENT

DOUBLE RESONANCE NEUTRON INTERFEROMETRY

Jean-Pierre Vigier

Institut Henri Poincaré
Laboratoire de Physique Théorique
11, rue P. et M. Curie, 75005 Paris

INTRODUCTION

The purpose of this communication is to discuss the implications of neutron interferometric experiments on the possible interpretation of the quantum formalism. The most recent one, which is a time-dependent double resonance experiment performed by the Vienna experimentalists[1] following a suggestion of our group[2], has farreaching implications, which, as we hope to show, establish the validity of the causal Stochastic Interpretation of Quantum Mechanics (SIQM) as the most adequate theoretical tool in grasping quantum "paradoxes". This approach which follows the views of Einstein and de Broglie in their controversy with Bohr and Heisenberg, develops the model of de Broglie's pilot wave theory[3] and Bohm's quantum potential concept[4]. This explains why, before we discuss neutron interferometry, we find it useful to expose the basic ingredients of this model for an ordinary double slit situation.

THE DOUBLE SLIT CONFIGURATIONS IN THE SIQM

In the usual double slit configuration one is confronted with the quantum puzzle, that quantum objects (photons, electrons, neutrons, etc.) appear as particles on the screen while they simultaneously possess wave-like properties manifested by the intensity interference pattern. Adherents of the Copenhagen Interpretation of Quantum Mechanics (CIQM) utilize this behaviour to argue in favor of a specific form of wave particle duality : the complementary concept of wave or particle entity. In the two slit experiment either we design an apparatus to observe interference, i.e. wave property (and thus forgo the description in terms of space-time coordination) or we design an incompatible arrangement to determine the space-time motion (i.e. the particle property) and consequently exclude the possibility of observing interference. The combination of the two is forbidden in principle and held impossible, by the complementarity "no-go" postulate. The quantum description thus concerns only the statistical prediction of results in well defined experiments. The wave function ψ is the most complete description of an individual state that can exist : the probability of finding a particle at position r is given by $|\psi(r)|^2$ if a measurement is made. Herewith is tied the concept of probability wave collapse in every measurement, in order to account for an instantaneous reduction of the extended ψ structure to a "point-particle" phenomenon when the measuring device intervenes.

In the SIQM, originally presented by de Broglie[3] and Bohm[4] and recently extended by Vigier[5] to include a subquantal random covariant Dirac aether[6], different premisses exist :

(a) the quantum mechanical description in terms of the wave ψ is incomplete in the sense of Einstein.

(b) the description may be supplemented with real physical motions, i.e. trajectories of particles.

(c) the wave particle duality is conceived as a concept which combines the wave and the particle aspect, the particle being a singularity which beats in phase with the wave[7] in agreement with de Broglie's pilot wave concept.

The formalism utilized by the SIQM is equivalent to the standard quantum formalism since the Schrödinger equation

$$\frac{\hbar}{i} \frac{\partial \psi}{\partial t} (\vec{x}, t) = \frac{\hbar^2}{2m} \nabla^2 \psi(\vec{x}, t) + V(\vec{x}, t) \tag{1}$$

is reinterpreted as follows. Writing $\psi = R\exp\left(i\frac{S}{\hbar}\right)$ and performing a real and imaginary part decomposition one obtains a Hamilton-Jacobi equation

$$\frac{\partial S}{\partial t} + \frac{(\nabla S)^2}{2m} - V - \frac{\hbar^2}{2m} \frac{\nabla^2 R}{R} = 0 \tag{2}$$

and a continuity equation

$$\frac{\partial \rho}{\partial t} + \nabla \left(\rho \frac{\nabla S}{m} \right) = 0 \tag{3}$$

We note directly that in the (H-J) equation a non-classical term appears, i.e. Bohm's quantum potential $Q = -\frac{\hbar^2}{2m} \frac{\nabla^2 R}{R}$ which influences the particle motion in the quantum domain in a non-classical way. As can be readily deduced from the continuity equation the real average motions of the particle can be represented by trajectories derived from $\vec{p} = \nabla S$.

The entirely new features introduced by the presence of the quantum potential can be summarized as follows :
- It depends on the form of the wave function (since $Q = -\frac{\hbar^2}{2m} \frac{\nabla^2 R}{R}$), not on the amplitude and does not in general decrease with distance. This introduces a type of non-locality since distant effects are mediated through Q to the particle behaviour.
- It depends (through the evolution of ψ according to the wave equation) on the whole environment; the distribution of trajectories is altered by changes in the environment (e.g. opening and closing slits).
- It depends on the quantum state. Particle trajectories are determined by their initial positions at t = 0, the structure of the environment and the initial quantum state.
- In the many body case the quantum potential acting on each particle is a function of the positions of all other particles in addition to the above factors. This introduces non-locality and allows a description and explanation of quantum statistics in terms of correlated distinguishable particle motions.

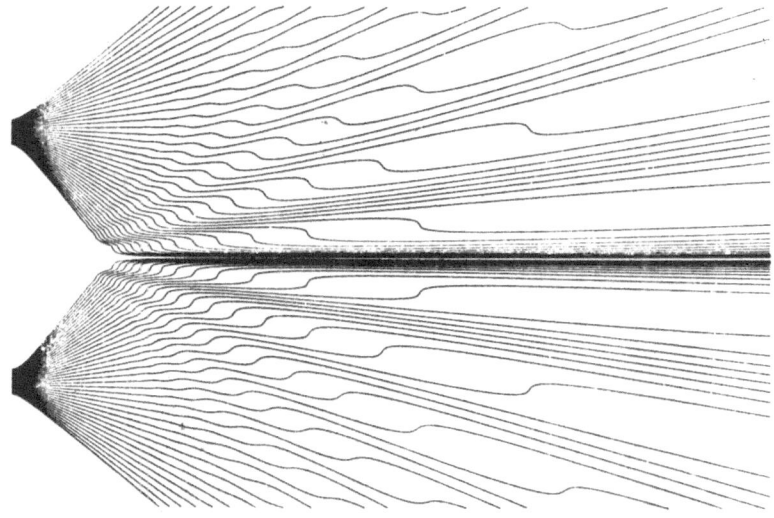

Fig. 1. Particle trajectories through two Gaussian slits.

A calculation for the double slit has been performed using the quantum potential approach[8]. The resulting trajectories are displayed in Fig.1. and the corresponding quantum potential in Fig.2.

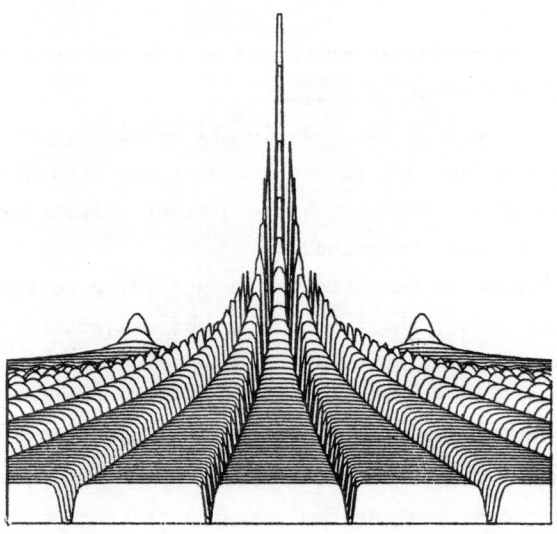

Fig.2 The Quantum Potential for the two Gaussian slits as viewed from the screen.

The matching of the trajectory lines on interference maxima corresponds to the quantum potential flat maxima while the "kinks" in the trajectories correspond to a particle traversing its deep peaked minima. The intensity distribution on the screen corresponds to the trajectory density distribution. We therefore get, via the quantum potential, a result where each particle travelling in the apparatus "knows" about or responds non-locally to the global structure of its environment. (e.g. the presence of two slits).

DOUBLE SLIT CONFIGURATIONS IN NEUTRON INTERFEROMETRY

Neutron interferometry has recently become a powerful tool in testing the predictions of quantum theory and revisits the double slit problem in a modified version, which comes very close to the core of the trajectory plus interference problem. The Vienna group has conducted experiments with attenuated neutron beams so that one records self inter-ference of neutrons[9-12]. In these experiments the quantum theoretical pre-dictions have been confirmed.

As we will now demonstrate, in analogy to double slit configurations, a detailed account of the individual microprocesses occurring within the interferometer is possible and is provided by the quantum potential interpretation. There is however a crucial novelty with respect to the double slit situations. Due to the presence of an additional degree of freedom, the neutron spin, we can namely show that if one accepts the energy-momentum conservation in individual microprocesses, then one is inevitably led to accept that individual neutrons travel through one slit only.

We shall show later that this point plays a crucial role in the discussion. For the Einstein-de Broglie school energy-conservation holds rigorously in each individual microprocess where the total particle energy ($E = - \partial S / \partial t$) is no longer equal to just the kinetic energy but is also determined by the quantum potential contribution. Equations of energy balance have to take this fact explicitly into account. Analogous reasoning leads to a generalized angular momentum conservation law as a result of the action of "quantum forces" i.e. quantum potential gradients and torques. This is not true for the CIQM. Heisenberg deduced from the uncertainty relations statistical conservation of energy-momentum but denies energy-momentum conservation in individual processes. To clarify this point, let us now treat the neutron interferometry experiments more extensively.

TIME DEPENDENT NEUTRON INTERFEROMETRY

A strongly polarized monokinetic neutron beam (i.e. with the magnetic moments oriented in the +z direction perpendicular to the plane of the figure by an external magnetic field) enters the interferometer. The beam intensity has been reduced by chopping to 2-3 neutrons/second with a velocity of ~ 2073 ms^{-1} (with an error bar $\pm$ 1.4%). Since the compton wave length is $\sim 10^{-13}$ cm this clearly establishes that only one wavepacket can be present at a time in a device (silicon monocrystal) ~ 10 cm long since its passage takes ~ 30 μsec. Any detected neutron (with an efficiency $>$ 99%) is thus observed long before the next has left its uranium atom within the pile. Observed interferences in such a situation clearly confirm that quantum particles (here neutrons) only interfere with themselves (since the neutrons are detected individually). For the sake of completeness we will present the three variations of interference experiments starting from the static, time independent case.

Experiment I (no spin-flippers operating)

Individual polarized neutron waves (i.e. the incoming wave-packets) $\psi = |\uparrow_z\rangle$ are split by the Bragg planes of a first slab, a, cut in the monocrystal. This separation gives rise to two partial beams $\psi_I = e^{i\chi}|\uparrow_z\rangle$ and $\psi_{II} = |\uparrow_z\rangle$, the term $e^{i\chi}$ corresponding to the phase-shift induced by a movable rotating crystal (denoted by χ in Fig.3)

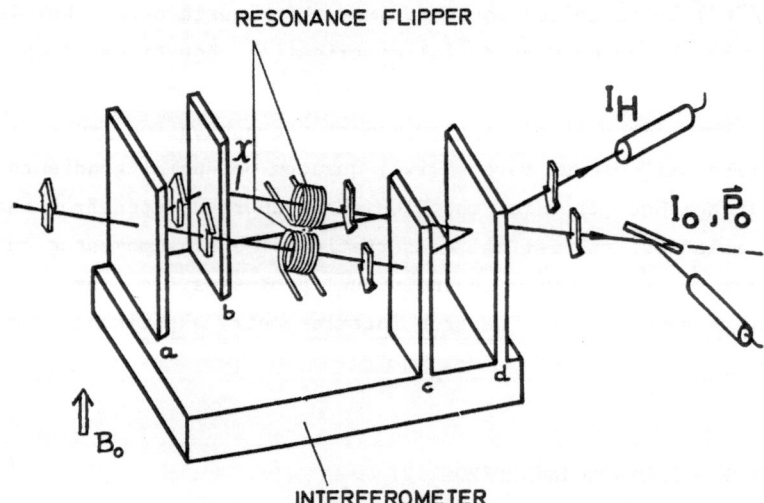

RESONANCE FLIPPER

INTERFEROMETER

Fig.3. Schematic arrangement of the radio frequency flip coils
within the skew-symmetric neutron interferometer in the
double resonance experiment. As indicated too, a Heusler
crystal allows to analyze the polarization of the O-beam
(with permission of H. Rauch).

introduced to produce a controllable phase difference between the paths
I and II which will yield modulations. A check is made to ensure that
the crystal is pure enough (i.e. its Bragg planes are sufficiently parallel) so that the waves ψ_I and ψ_{II} have no observable energy difference and suffer no modification of their $\uparrow_z$ polarization. With both
spin-flippers turned off the wave packets ψ_I and ψ_{II} travel along both
paths, are reflected on two slabs b and c, recombine on the slab d and

superpose (interfere) on two beams O and H (O containing a supplementary polarizer) where they are detected in detectors D_A and D_O. Independently of any interpretation, the quantum mechanical formalism has been shown to be entirely correct in this case, the superposition in O and H being modulated with the phase shift χ to give for the intensity

$$ I = (\psi_I + \psi_{II})^* \cdot (\psi_I + \psi_{II}) = 2\,(1 + \cos\chi) \qquad (4) $$

while the polarization remains in the +z direction i.e.

$$ \vec{P} = (P_x, P_y, P_z) = (0,0,1) \qquad (5) $$

Due to Bragg reflection-transmission in d one obtains two sinusoïdal intensity patterns for O and H, the maxima of O corresponding to minima of H and vice versa.

This first (successful) experiment by the Vienna group has been extended in two ways :

Experiment II (one spin flipper only)

The first is to add to the set-up one radio frequency spin flipper only (represented by the coil I in Fig.3) which operates at $\sim 100\%$ efficiency. As is well known (and as has been checked experimentally) practically every neutron ($> 96\%$) whose wave packet $|\uparrow_z\rangle$ goes through such a spin flipper loses one photon to the coil by resonance, i.e. flips its spin to $|\downarrow_z\rangle$ with a loss of energy $\Delta E = \hbar\omega_{rf} = 2\mu B_0$, in the present set-up. The quantum formalism says that the superposition of $\psi_I = e^{i\chi} \cdot |\uparrow_z\rangle$ and $\psi_{II} = |\downarrow_z\rangle$ yields a constant intensity along with a final polarization

$$ \vec{P} = \left(\cos(\omega_{rf}t - \chi),\ \sin(\omega_{rf}t - \chi),\ 0 \right) \qquad (6) $$

so that each neutron emerging from the interferometer is polarized in the X-Y plane and has lost an energy $\Delta E/2$. Here also quantum mechanics has been shown to be verified by experiment.

Experiment III (two spin flippers)

The second is to have two synchronized spin flippers working on both paths I and II simultaneously (Fig.3.). The quantum formalism implies that the superposition on beams 0 and H is $e^{i\chi}|\downarrow_z\rangle + |\downarrow_z\rangle$ which yields the intensity

$$I = 2(1+\cos\chi) \tag{7}$$

and the polarization

$$\vec{P} = (0,0,-1) \tag{8}$$

It is these predictions which have just been verified in Grenoble[1]. Each individual neutron has been shown experimentally to leave the interferometer with spin down (i.e. to have lost an energy ΔE to the spin flippers) and also to have taken part in interference.

THEORETICAL DISCUSSION

In order to account for these experiments the original causal interpretation of the Schrödinger equation may be generalized following Bohm et al.[13] to include the concept of spin which can be interpreted in terms of well defined motions determined by the Pauli equation. The extra degrees of freedom in the two-component Pauli spinor are given by the Euler angles θ, ψ, ϕ which define a general state of rotation given by the spinor

$$\Psi = R \begin{bmatrix} \cos\frac{\theta}{2} \ e^{i(\psi+\phi)/2} \\ i\sin\frac{\theta}{2} \ e^{i(\psi-\phi)/2} \end{bmatrix} \tag{9}$$

in terms of which the spin vector is given by $\vec{S} = \frac{\hbar}{2}\frac{\Psi^+\vec{\sigma}\Psi}{\Psi^+\Psi}$. The evolution of a Pauli spinor is then given by the Pauli equation

$$i\hbar\frac{\partial\Psi}{\partial t} = H\Psi \tag{10}$$

where $H = -\frac{\hbar^2}{2m}(\vec{P}-i\frac{e}{c}\vec{A})^2 + \mu(\vec{\sigma}\cdot\vec{H}) + V$ with $\mu = -\frac{e\hbar}{2mc}$, $\vec{H}$ the magn. field

and Ψ as above. The equations of motion can be derived from the Pauli-Hamiltonian as described by Bohm et al.[13], by introducing in addition to the pair $\rho = R^2$, $\psi/2$ the canonical variables $\rho\cos\theta$, $-\phi/2$. We then find

$$\frac{\partial\rho}{\partial t} + \nabla(\rho\cdot\vec{v}) = 0 \tag{11}$$

i.e. a continuity equation, and a Hamilton-Jacobi type equation

$$-\frac{\hbar}{2}\frac{\partial\psi}{\partial t} + \frac{\hbar}{2}\cos\Theta\cdot(\vec{v}\cdot\nabla)\phi + \frac{1}{2}m\vec{v}^2 + Q + H_s = 0 \tag{12}$$

where $\vec{v} = \frac{\hbar}{2m}(\nabla\psi + \cos\Theta\cdot\nabla\phi) - \frac{e}{c}\vec{A}$ is the particle velocity containing spin contributions, $Q = -\frac{\hbar^2}{2m}\frac{\nabla^2 R}{R}$ is the usual quantum potential and $H_s + \frac{\hbar}{2}\cos\Theta(\vec{v}\cdot\nabla)\phi$ is a spin dependent addition to the quantum potential (which introduces a new spin-orbit coupling), with

$$H_s = \frac{\hbar^2}{2m}\left[(\nabla\Theta)^2 + \sin^2\Theta(\nabla\phi)^2\right] - \frac{e}{mc}\vec{s}\cdot\vec{H} \tag{13}$$

The equation of motion of the spin vector $\vec{s}$ can also be written as

$$\frac{d\vec{s}}{dt} = -\frac{\vec{s}}{\rho} \times \sum_i \frac{\partial}{\partial x_i}\left(\rho\frac{\partial\vec{s}}{\partial x_i}\right) - \frac{e}{mc}\vec{s}\times\vec{H} \tag{14}$$

where the term $-\frac{\vec{s}}{\rho}\times\sum_i\frac{\partial}{\partial x_i}(\rho\frac{\partial\vec{s}}{\partial x_i})$ is an additional "quantum torque", which introduces a quantum mechanical precession of the spin vector. It is evident from these equations that even in the absence of magnetic fields free spinning particle trajectories will not coincide with Schrödinger trajectories, nor will the orientation of the spin vector remain constant if the particle is in a non-stationary spin state. The spin dependent quantum potential and quantum torque are non-zero under these conditions since the spin orientation has non-zero spatial gradients $\nabla\phi$, $\nabla\Theta$. With this general scheme based on the Pauli equation a detailed account of all three experiments can be given. In all cases we assume both beams to have a gaussian wave-packet profile.

It has been shown[14] that for experiment I both beams have the same spin orientation, e.g. $\binom{1}{0}$ and the combined wave function possesses the same spinor symmetry and exhibits a spatial interference pattern. As a consequence the Euler angle $\Theta = 0$, the spin vector has no spatial dependence and the spin dependent quantum potential and the quantum torque are zero. Every incoming neutron preserves its spin orientation in the constant external magnetic field and therefore the spin dependent part of the energy is conserved.

The quantum potential and associated ensembles of trajectories may be calculated numerically by adopting a simple model in which the action of the sets of crystal planes is simulated by a square potential barrier[15]. In Fig.4. we plot the effective potential (quantum potential plus classical square barrier) for the symmetric case (equal intensity in O and H beams).

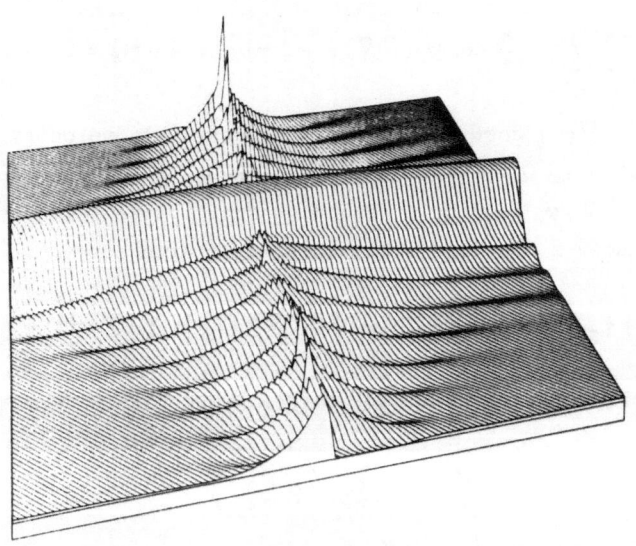

Fig.4. The effective potential at the last set of crystal planes,
symmetric case. The severely modified square potential can be
seen across the centre of the figure.

The effect of this potential is shown in the trajectories plotted in Fig.5;

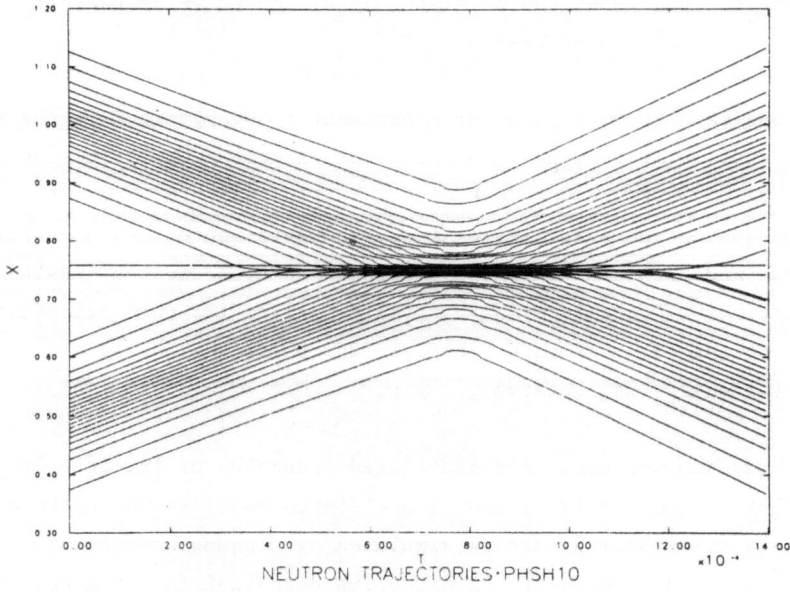

NEUTRON TRAJECTORIES·PHSH10

Fig.5. The trajectories associated with Fig.4. The horizontal lines
indicate the position of the square potential barrier.

no trajectories cross the potential barrier. Rotating the phase shifter plate changes the relative phase of the beams and the quantum potential develops in a markedly different manner as shown in Fig.6.

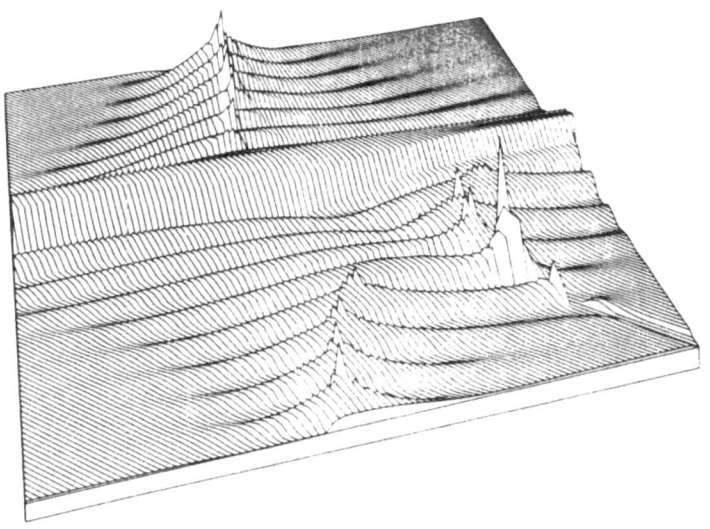

Fig.6. The effective potential which produces a maximum intensity
       in the H-beams.

This effective potential produces trajectories as shown in Fig.7. in which

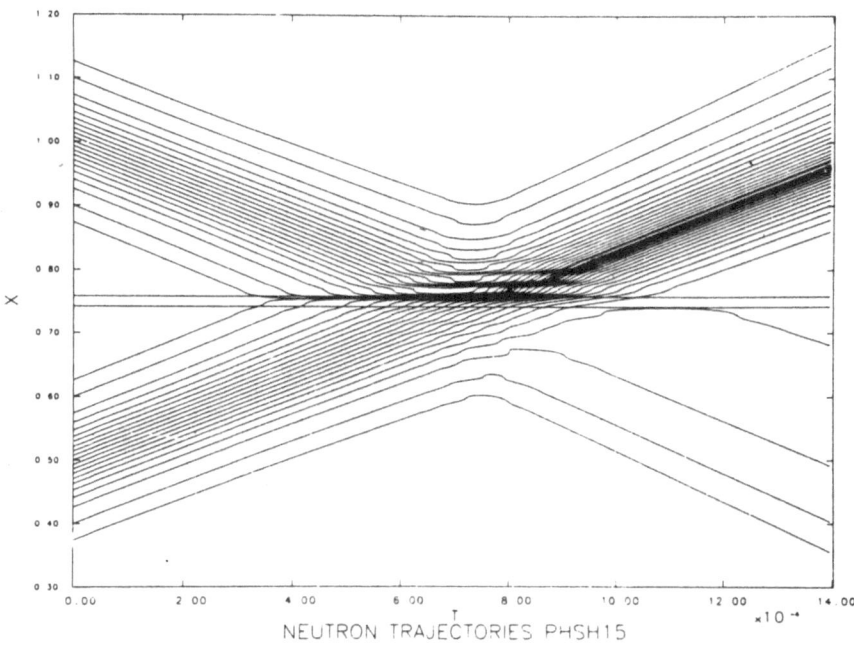

NEUTRON TRAJECTORIES PHSH15

Fig.7. The associated trajectories with Fig.6.

most initial positions with an appreciable probability of occupation lead
to trajectories which enter the H beam. Thus the fate of any individual
neutron is decided at the outset by its initial position in the beam and
by the configuration of the apparatus.

In Experiment II where a r.f. spin flipper operates in one of the
beams the situation is quite different. A detailed calculation shows[16]
that due to the different partial beam spinor symmetry the combined wave
function does not exhibit spatial interference despite the phase difference between the beams. The trajectories are similar to those of Fig.5.
But the Euler angles θ and ϕ have a spatial variation and therefore
the spin dependent addition to the quantum potential and the quantum
torque are non-zero. The beams emerging from the interferometer have defi-
nite spin vectors with $\theta = \pi/2$ and $\phi = -\left[\frac{\pi}{2} - \omega_{rf}t + \chi\right]$ in agreement with
quantum predictions and experimental results. Originally spin-up or spin-
down vectors in the partial beams are twisted to the horizontal position
(x-y plane) by the quantum torque, as shown in Fig.8 where we plot $\theta(z,t)$,

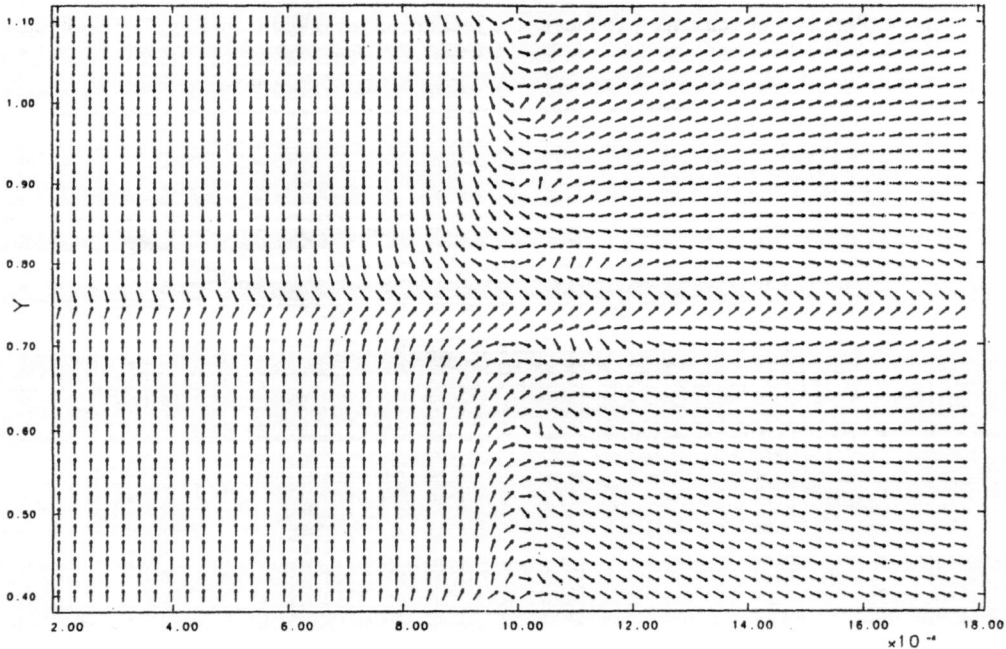

Fig.8. The polar orientation angle θ of the spin vector plotted
on a mesh of points (z,t) in the region of the last set of
crystal planes.

and have opposite directions depending on which partial emerging beam the neutrons enter when leaving the superposition region. We consider now the question of energy and angular momentum conservation in each individual microprocess. A statistical conservation of both quantities is always guaranteed : before superposition there is an average value of total spin zero ($\uparrow$, $\downarrow$) ; afterwards the spins in the forward and deviated beams lie in opposite directions in the xy plane. But in the causal interpretation, where definite trajectories exist for the spinning particles, the total angular momentum is also conserved in an individual passage of a neutron since the quantum torque rotates the spin vector from the +z or −z direction into the xy plane in the region of superposition. Similar reasoning applies to energy conservation. Every neutron entering the interferometer in the spin up state has a definite energy in the external magnetic field. If it goes through the partial beam with no spin-flipper an energy of ($-\hbar\omega_{rf}/2$) is supplied by the spin dependent quantum potential so that an energy balance is established in the horizontal position in the superposition region. If on the other hand the neutron passes through the spin flipper and loses an energy $\Delta E = \hbar\omega_{rf}$ it gains an energy $\hbar\omega_{rf}/2$ due to the spin dependent quantum potential so that an energy balance is again established. This picture thus goes beyond just an average energy conservation, which is of course also valid due to the equal relative frequencies of neutron passages through each of the two partial beams.

In Experiment III we have one rf spin flipper operating in each of the arms of the interferometer. A calculation using our general scheme shows that due to the equal partial-beam spinor symmetry (both spinor-waves are inverted) the combined wave function exhibits spatial interference modulated by χ and no interference in the spin-polarization. The trajectories and quantum potentials are shown in Experiment I. The Euler angles θ and ϕ have no spatial variation and therefore, exactly as in Experiment I, both the spin dependent quantum potential and the quantum torque vanish. Angular momentum is absolutely conserved individually and statistically in a "classical" way since quantum torque effects are absent. In fact every neutron that enters the interferometer with spin $\uparrow_z$ is found with spin $\downarrow_z$ in the superposition region due to the $\vec{s} \times \vec{H}$ term operating in the spin flipper.

The two states $|\uparrow_z\rangle$ and $|\downarrow_z\rangle$ differ by $\Delta E = \hbar\omega_{rf}$ which is exactly the resonance energy-exchange with the spin-flipper. Since this energy is exchanged with only one spin flipper, due to the resonance exchange condition, it seems to us that we have direct evidence of a neutron passage through one of the arms of the interferometer. Since no <u>actual</u> (irreversible) measurement takes place, we are of course not in a position to say through which spin-flipper the neutron has passed, but we can nevertheless state that the neutron has passed through one coil <u>or</u> the other. In order to understand what is happening inside the interferometer it is thus necessary to employ both the wave and particle aspects of matter in the same experiment, and the quantum potential approach allows one to consistently do this. Whilst spatial interference exists, energy conservation implies a particle passage through one of the interferometer arms. This is precisely the reason why we proposed[2] this experiment. We follow the neutron as a particle in the interferometer and observe interference, since the spinor wave propagates at the same time through both arms of the apparatus.

Our model accounts realistically for all three experiments, preserving a spacetime coordination of events as well as the strict validity of the conservation laws. Reality is non-separable in the sense that the quantum potential and quantum torque are context-dependent, but this same reality is analyzable and intuitively comprehensible, at least in principle.

A MACROSCOPIC QUANTUM BEAT PHENOMENON

An interesting variation of experiment III was performed[1] by driving asynchronously the two resonance coils, namely by introducing two slightly different frequencies ω_{r_1} and ω_{r_2}. The conditions for a spin flip operation are still fullfilled due the broad resonance curve, i.e. $\Delta\omega = |\omega_{r_1} - \omega_{r_2}| \ll \delta\omega_{1/2}$ where $\delta\omega_{1/2}$ is the halfwidth of the resonance curve. Hereby the wave functions of the two partial beams are respectively modified in approximately the following way (we neglect the constant phase shift $\sim \Delta\omega \cdot \Delta t$ due to the finite time of flight of the neutron in the coil)

$$\Psi_I + \Psi_{II} \propto e^{i(\omega-\omega_{r_1})t} |\uparrow_z\rangle + e^{i\chi} e^{i(\omega-\omega_{r_2})t} |\downarrow_z\rangle \qquad (15)$$

and the resulting intensity modulation has the following form

$$I \propto (1 + \cos[\chi + \Delta\omega t]) \qquad (16)$$

14

This time dependent pattern which describes a quantum beat phenomenon possesses a sensitivity up to 10^{-18}eV for macroscopic time scales, and offers itself as a measuring process of extremely small energy differences for diversified applications[1].

We will not insist on this remarkable verification of quantum laws, but we will briefly comment a second aspect which is manifest in these quantum beat phenomena and turns out to be very important for the interpretation of quantum mechanics although it has so far been neglected in the discussion. Let us follow explicitly the processes in the interferometer. The quantum formalism implies that the neutron obtains the additional phase factor corresponding to $\Delta E_1 = \hbar \omega_{r_1}$ if it passes through path I or $\Delta E_2 = \hbar \omega_{r_2}$ if it goes through path II. However since we measure a spin down at the exit of the interferometer we know that the neutron has lost an energy $\Delta E = \hbar \omega_r = 2\mu B_0$ due to spin inversion. Since we know that ω_r differs from ω_{r_1} and ω_{r_2} we can conclude that the missing energy $|\Delta E - \Delta E_1|$ or $|\Delta E - \Delta E_2|$ must have been converted to the kinetic component. The fact now is that this kinetic energy difference, which lies well below the monochromaticity $\Delta E_B \sim 10^{-4}$eV of the beam[1], is detectable by means of the beating frequency of the intensity. This has the following implications : (a) An energy value below the energy uncertainties of the beam is directly measurable showing thus implicitly the reality of definite energy values below the uncertainties. (b) The criticisms formulated on the basis of energy uncertainties against a possible measurement of the $\hbar \omega_{rf}$ in the synchronized experiment III are herewith invalidated. In fact this argument was based on the fact that $\hbar \omega_{rf} \ll \Delta E_B$, and here a value $\hbar(\omega_{r_1} - \omega_{r_2}) \ll \hbar \omega_{rf}$ has been experimentally detected. Furthermore another argument is invalidated according to which energy transfers below the uncertainty limit do not produce collapse because they do not constitute a quantum measurement[17]. In fact here an energy below ΔE_B is <u>detected</u> and no collapse occurs nevertheless.(c) Since the varying kinetic energy is at the origin of this beat effect, one can deduce herefrom the reality of the kinetic energy variation and implicitly the reality of the spin-flip energy transfer which both result from the passage of the neutron through one resonator only, establishing thus the existence of spacetime paths in the interferometer.

This qualitative discussion must be supplemented by a calculation of the quantum potential dependence and the trajectory formation in the interferometer. By doing so we can see how the time dependence of the quantum potential shifts the trajectories from the O to the H beam and back as a function of $\Delta\omega$ and the time and describe this effect in terms of causal space time trajectories and definite spin values. We leave this task for a future publication.

CONCLUSIONS

We conclude with a simple reasoning which can be summarized in the following points for the double resonance experiment.
1. This set-up never contains more than one neutron at a time. In fact each time a neutron is detected the next one "is still confined in the nucleus of the reactor fuel"[1].
2. Each neutron enters the interferometer with its spin (magnetic moment) oriented upwards i.e. parallel to the constant magnetic field in which the whole set-up is imbedded.
3. Each neutron leaves the set-up (and is detected by one of the detectors) with its spin oriented downward i.e. antiparallel to the constant magnetic field. As a consequence it has lost an identical quantum of energy $\Delta E = \hbar\omega_{rf}$.
4. Now let us assume
a) with Einstein and de Broglie that the neutron's energy and impulsion (i.e. mass multiplied by velocity) are always conserved in all micro-processes where there is an exchange of energy.
b) with Einstein and Bohr that all exchange of energy is tied with the "particle" aspect of matter.
c) that in an individual exchange of energy a resonator can absorb quanta in the resonance frequency only.

Then one must accept that this exchange of energy $K\omega_{rf}$ i.e. the manifestation of the neutron's particle aspect, has necessarily happened in one or the other spin-flipper (i.e. in one spin-flipper only) but not in both spin-flippers simultaneously because this would violate the resonance condition.
5. Since the third experiment has shown that each individual neutron localized itself on an interference pattern, i.e. according to point 1 that it interferes with itself in Dirac's sense one can conclude following Rauch's own terms : "At the place of superposition every neutron has the information that there have been two equivalent paths I and II through the interferometer, which have a certain phase difference causing the neutron to join the beam in the forward or deviated direction"[18].
6. From the combination of points 4 and 5 one deduces that between the

source and the detectors each neutron manifests itself as a wave (because of the interference) and as a particle (because of the loss of energy) simultaneously. In other terms each neutron is a wave and a particle simultaneously.

7. From point 6 one deduces that between the source and detector the wave representation of the neutron as ψ_{in}, then ψ_I and ψ_{II}, then ψ_o and ψ_H is correct but no complete since it does not contain the fact that each neutron has manifested itself in one of the spin-flippers. To quote Rauch again:"This experiment shows explicitly that the interference properties of ψ_I and ψ_{II} can be preserved even when real exchange of energy has occured : which is intuitively a measuring process"[18].

   If one recalls that for the Copenhagen Interpretation all measurements imply an instantaneous collapse of the wave packets one could suggest that in CIQM the apparition of the neutron in spin flipper I <u>or</u> II would instantaneously destroy ψ_I or ψ_{II} and thus destroy the observed interference. In SIQM, on the contrary, the neutron moves in one packet (ψ_I or ψ_{II}) only the other (ψ_{II} or ψ_I) being empty of the particle aspect of matter.

   This reasoning implies that if one accepts absolute energy momentum conservation in all cases (an assumption which evidently corresponds to a fundamental postulate of physics) then one is forced to admit that Einstein was right in the Bohr-Einstein controversy.

ACKNOWLEDGEMENT : The author is indebted to Prof. H. Rauch for many enlightening discussions as well as for his kind permission to use his yet unpublished material and to reproduce one of his figures.

REFERENCES

1. G. Badurek, H. Rauch and D. Tupinger, "Neutron Interferometric Double Resonance Experiment", to appear in Phys.Rev.A.

2. C. Dewdney, Ph. Gueret, A. Kyprianidis and J.P. Vigier, Phys.Lett. 102A : 291 (1984).

   C. Dewdney, A. Garuccio, A. Kyprianidis and J.P. Vigier Phys.Lett.104A : 325 (1984).

3. L. de Broglie, Non Linear Wave Mechanics, Elsevier, Amsterdam, 1960.

4. D. Bohm, Phys.Rev. 85 : 166 (1952).

5. J.P. Vigier, Astr.Nachr. 303 : 55 (1982).

6. P.A.M. Dirac, Nature, 168 : 906 (1951).

7. Ph. Gueret and J.P. Vigier, Lett. Nuov. Cim. 38 : 125 (1983).

8. C. Philippidis, C. Dewdney and B. Hiley, Nuov.Cim. B52 : 15 (1979).
   C. Dewdney, Ph.D.Thesis, London 1983.

9. J. Summhammer, G. Badurek and U. Kischko, Phys.Lett. 90A : 110 (1982).

10. G. Badurek, H. Rauch, J. Summhammer, U. Kischko and A. Zeilinger,
    J.Phys. A16 : 1133 (1983).

11. J. Summhammer, G. Badurek, H. Rauch, U. Kischko and A. Zeilinger,
    Phys.Rev. A27 : 2523 (1983).

12. G. Badurek, H. Rauch and J. Summhammer, Phys.Rev.Lett. 51 : 1015 (1983).

13. D. Bohm, R. Schiller and J. Tiomno, Suppl. al Nuov.Cim. Serie X. (1955).

14. C. Dewdney, P.R. Holland, A. Kyprianidis and J.P. Vigier, Trajectories
    and spin vector orientations in the causal interpretation of the Pauli
    equation, preprint Inst.H.Poincaré, submitted to Phys.Rev.D. (1986).

15. C. Dewdney, Phys.Lett. 109A : 377 (1985).

16. C. Dewdney, P.R. Holland, A. Kyprianidis, J.P. Vigier, Spin superposi-
    tion in neutron interferometry, preprint Inst.H.Poincaré (1985).

17. H. Rauch, "Polarized Neutron Interferometry", Proceedings of the Int.
    Conf. "New Techniques and Ideas in Quantum Measurement Theory", N.Y.
    Jan. 1986.

18. H. Rauch, Test of Quantum Mechanics by Matter Wave Interferometry,
    Int.Symp.Found.Qu.Mech. Tokyo, Aug. 1983, Proceed.J.Phys.Soc. Japan,
    1985.

CALCULATIONS IN THE CAUSAL INTERPRETATION

OF QUANTUM MECHANICS

Christopher Dewdney

Department of Applied Physics
and Physical Electronics
Portsmouth Polytechnic
Park Building, King Henry I St, Portsmouth, UK

INTRODUCTION

In this paper we present a series of computer calculations carried
out in order to demonstrate exactly how the causal interpretation works in
specific cases. In this way we show how the causal interpretation can
account for the essential features of single and two particle nonrelativistic quantum mechanics, including spin, in terms of well defined
individual particle motions.

Quantum mechanics only presents great difficulties for those who
believe that the task of physics is to describe the structure of the
material world. For quantum phenomena seem to defy the imagination and our
intuitive notions about how matter behaves, structured by classical physics,
do not appear to serve as useful guides when attempting to conceive the
structure of the quantum world. The fact that "classical" notions of the
world, instead of clarifying our experience, only lead to ambiguity, when
we attempt to conceive what may lie beyond the statistical predictions of
the theory, reflects the deep crisis that quantum mechanics has bought
about. Of course it is fair to say that many physicists, who have "learned
to stop worrying and love the statistics", would deny the existence of such
a crisis. If pressed with questions of interpretation these physicists
tend to resort to some variation of the Copenhagen interpretation. But how
many are really prepared to accept the consequences and to give up any
possibility of understanding the statistical predictions of quantum mechanics in terms of some underlying reality? How many are satisfied by Bohr's
resolution of the difficulties in terms of a particular restrictive epis-
temology; that is a particular opinion about how we come to know and what
it is possible to know; or Wigner's idea that consciousness must be introduced in order to make anything definite; or Everett's idea of multiple
splitting universes. Is quantum mechanics just an abstract formalism for
connecting the statistical results recorded at the presumably unproblematic
classical level; or is it indicative or a new order in the structure of
the material world; that is a new ontology.

THE CAUSAL INTERPRETATION OF QUANTUM MECHANICS

One way of exploring a possible underlying structure is through the
causal interpretation, proposed by de Broglie[1] and rediscovered by Bohm in
1952[2]. In this context the wave function is not held to exhaust the

possibilities of description of individual systems, but does, as usual, encompass the limits of prediction. This approach allows a description of quantum phenomena in terms of well defined individual particle motions, the statistics have no special status and neither does measurement. The disturbance caused by measurement can be analysed, but not avoided, and the Heisenberg uncertainty relations are interpreted as statistical scatter relations which arise in the repeated measurement, on similarly prepared systems, of well defined variables. It is assumed that a particle, an electron for example, has a well defined position, momentum and spin vector at all times. In addition the particle always has an associated ψ wave. The evolution of the particle coordinates depends on the form rather than the amplitude of the associated wave, as can be seen from the equations of motion derived from the appropriate wave equation. As we shall see these equations show that the particle is acted on by new forces and torques which account for the observed quantum phenomena and distinguish classical from quantum behaviour. The specific calculations presented here show clearly the new features that this description entails.

As an illustration of the causal interpretation of the Schrodinger equation we present detailed calculations of the individual particle motions associated with the scattering of Gaussian wave packets from square potentials. Since the many body case introduces some new features we then illustrate the interdependence of the motions of two particles contained within a harmonic oscillator potential.

In order to illustrate the causal interpretation of spin we present detailed calculations of particle trajectories and spin vector orientations during a simple spin superposition process and during the passage through a Stern-Gerlach inhomogeneous magnetic field. This latter interaction amounts to what is usually called a measurement of the spin. We then apply the interpretation in the two body case to describe the spin correlations predicted in Bohm's version of the Einstein-Podolsky-Rosen experiment[3].

THE CAUSAL INTERPRETATION IN THE CONTEXT OF THE ONE PARTICLE, ONE DIMENSIONAL, TIME DEPENDENT SCHRÖDINGER EQUATION

In order to illustrate the interpretation of the one particle Schrödinger equation, the process of scattering from square potentials has been chosen since it exhibits so many of the essentially quantum features of matter. In introductory texts this subject is usually treated in terms of plane waves using the time independent Schrödinger equation to calculate transmission and reflection coefficients. Thus it is, for example, that a finite transmission probability can be calculated even when the incident energy of a particle is less than that of the potential barrier from which it scatters. But how can this tunnelling be explained? What has happened when an individual particle appears on the other side of the barrier? (According to Bohr this is not a meaningful question, but let us put Bohr asied and consider it.)

Evidently any process requires a time dependent description and the solution of the time dependent Schrödinger equation in the presence of various scalar potentials is very important, since many essential quantum phenomena are exhibited in such processes, such as; interference in the formation of Weiner fringes in front of potential edges; particle reflection at potential edges; penetration into and tunnelling through classically forbidden regions, and so on. Further more, simple models may be constructed, within this context, of more complex phenomena such as alpha emission, propagation in crystal lattices and interferometers and the like.

Some years ago Goldberg et al[4] produced several computer generated motion pictures which displayed the development in time of the probability density during the scattering of Gaussian wave packets from various square potentials. However, the description of the phenomena in terms of the probability density, although illuminating, does not in fact aid the understanding of what may be happening on an individual basis. However, their numerical methods can be adapted for exactly this purpose within the framework of the causal interpretation and computer movies of particle tunnelling through potential barriers, for one example, have in fact been produced[5].

The evolution of the wave function is given by Schrödinger's equation

$$\frac{-\hbar^2}{2m} \nabla^2 \psi + V\psi = i\hbar \frac{\partial \psi}{\partial t}$$

which with the substitution $\psi = Re^{iS/\hbar}$ may be separated into two equations. The first

$$Q + \frac{(\nabla S)^2}{2m} + V = -\frac{\partial S}{\partial t}$$

can be interpreted as a Hamilton-Jacobi equation, in which the particle velocity V is given by $\nabla S/m$ and Q is an additional quantum potential where

$$Q = \frac{-\hbar^2}{2m} \frac{\nabla^2 R}{R}$$

The second may be interpreted as a continuity equation

$$\frac{\partial p}{\partial t} + \nabla \cdot (p\underline{v}) = 0$$

where $p = R^2$.

The algorithm developed by Goldberg et al and referred to above, can be adapted to provide a causal description of the quantum features of the processes associated with square potentials in terms of well defined individual particle motions. Thus a motion picture can be produced which shows exactly what is happening to an individual particle as it scatters from the potential.

In order to solve for the individual particle motions it is necessary to specify the initial position of the particle and the initial form of the wave function. Since the numerical method then furnishes R and S for all x and t a simple procedure allows the determination of the particle trajectory from

$$\frac{m d\underline{x}}{dt} = \nabla S$$

In the following the initial wave function is assumed to be of Gaussian form

$$\psi_0(x) = \exp\left[-(x-0.5)^2/2\sigma_0^2\right] \exp(ik_0 x) \qquad (1)$$

in arbitrary units where h=1, m=0.5. The initial momentum k and the width and magnitude of the square potential barrier are chosen to produce a transmission coefficient of one half. In this case the initial particle energy is 0.8V and so "classically" any particle with this energy would be reflected. The quantum potential approach as applied here gives a clear picture, in terms of individual particle motions, of how some particles become reflected and some "tunnel through", depending on their initial position. Here the results of the numerical integration are displayed at a number of time steps during the scattering in figure one. Each frame from the motion picture shows the probability density (dotted line), the square potential (dashed line), the effective potential ie the square potential plus the quantum potential, (solid line). Also displayed are the position and kinetic energy, $(\nabla S)^2/2m$, of a range of possible particle motions. The horizontal position representing motion along the X axis and the vertical position the particle energy. From this set of frames alone it can be seen that the process of tunnelling arises as a result of the severe modification of the classical potential brought about by the quantum potential. In particular the quantum potential is negative in the region of the barrier and so it may be possible for a particle to have sufficient energy to pass through this region. The outcome in an individual case (remember we are discussing an ensemble of single particles) depends on the (uncontrollable) initial position of the particle within the wave packet. It can be seen that all those particles that start from a position in the forward part of the wave packet enter the transmitted beam whereas those in the trailing part are reflected. As is clearly shown in figure one it is the development of a series of steep peaks and troughs in the quantum potential that accounts for the process of reflection and depending on its initial position a particle may be reflected some distance in front of the classical potential.

It should be emphasised here that the assumption of a particle trajectory avoids all the usual problems associated with "wave packet collapse". The particle is either transmitted or reflected in each individual scattering quite independently or whether we set up some detecting device to establish what happend or not. The causal interpretation thus provides an accessable model of what happens during an essentially quantum phenomenon, tunnelling. It does this simply by assuming that the particle exists with well defined coordinates and is associated with a ψ wave. It is not necessary to introduce any further assumptions such as wave packet collapse or many worlds.

The description given here can be extended[6] to give an account of the motion of neutrons in a single crystal interferometer such as has been used by Rauch[7] to test fundamental quantum phenomena. A simple model of such a device can be constructed by considering each set of crystal planes to be replaced by a semi-transparent surface (see figure two). Further, the important component of the motion is perpendicular to the surface. The description given above then represents what happens at the first two sets of crystal planes. Two wave packets then converge on the last set of crystal planes and in order to simulate what happens here we simply need to alter the initial wave function in the numerical integration programme and initiate two sets of initial particle positions on either side of the potential as representatives of the possible individual particle motions. Thus we replace ψ of equation one with

$$\psi_o(x) = \exp[-(x-0.5)^2/2\sigma_o^2]\exp(ik_ox) + \exp[-(x-1.0)^2/2\sigma_o^2]$$

$$x \, \exp[-i(k_ox+\phi)]\exp(i\chi)$$

where χ is the variable relative phase shift between the two packets and ϕ a constant symmeterising phase factor. Figure three shows a set of frames from the motion picture produced for this process in the symmetric case in which each emerging beam has the same intensity. Figure four shows a set of frames when the relative phase is adjusted by $\pi/2$. Clearly the development of the quantum potential is no longer symmetric, the assymetry leading to nearly all the representative particles emerging in one beam.

NON-LOCALITY AND THE CAUSAL INTERPRETATION OF TWO PARTICLE MOTION

The two particle Schrödinger equation can also be interpreted within

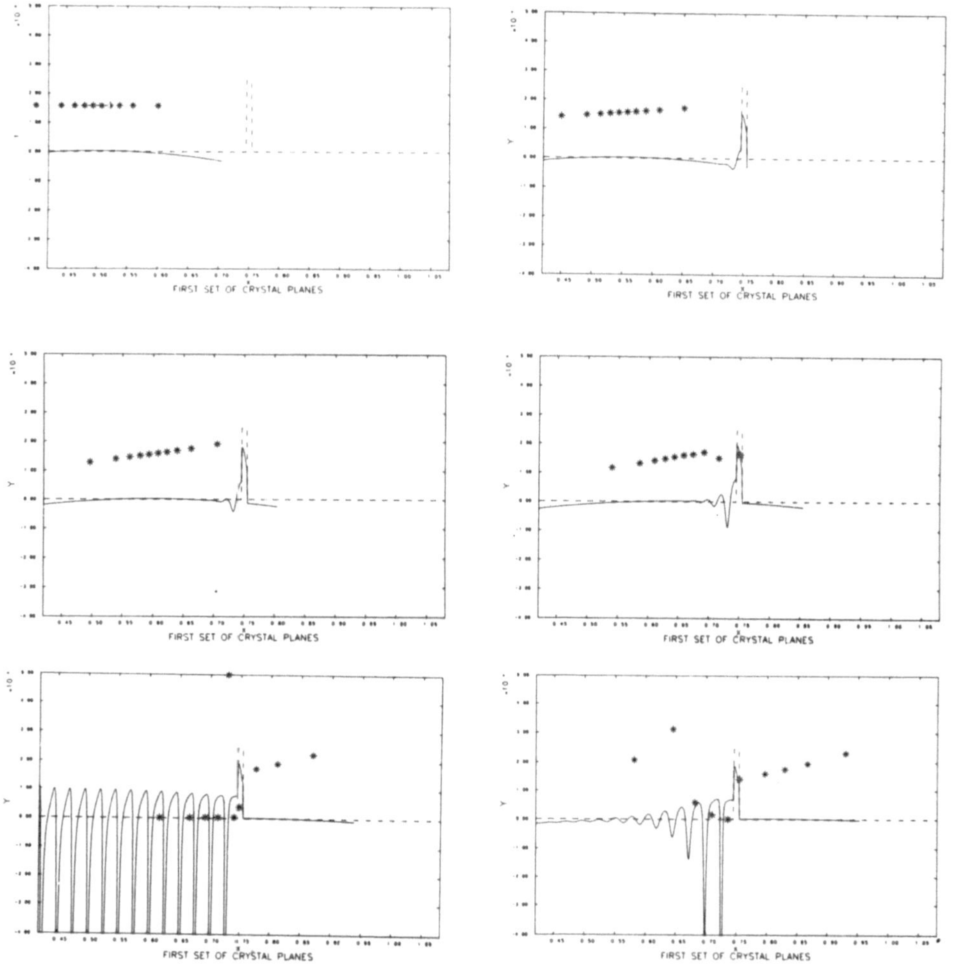

Fig.1 The effective potential (solid line), the classical potential (dotted line) and a set of representative single particle positions and energies during the scattering of a Gaussian wave packet from a square potential barrier. Vertical scale represents energy, horizontal position.

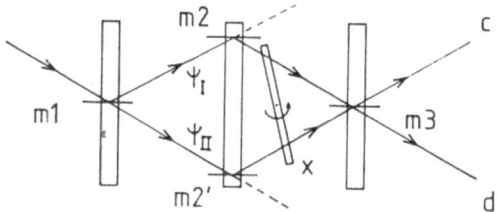

Fig.2 The neutron interferometer, crystal planes replaced by semi-transparent surfaces.

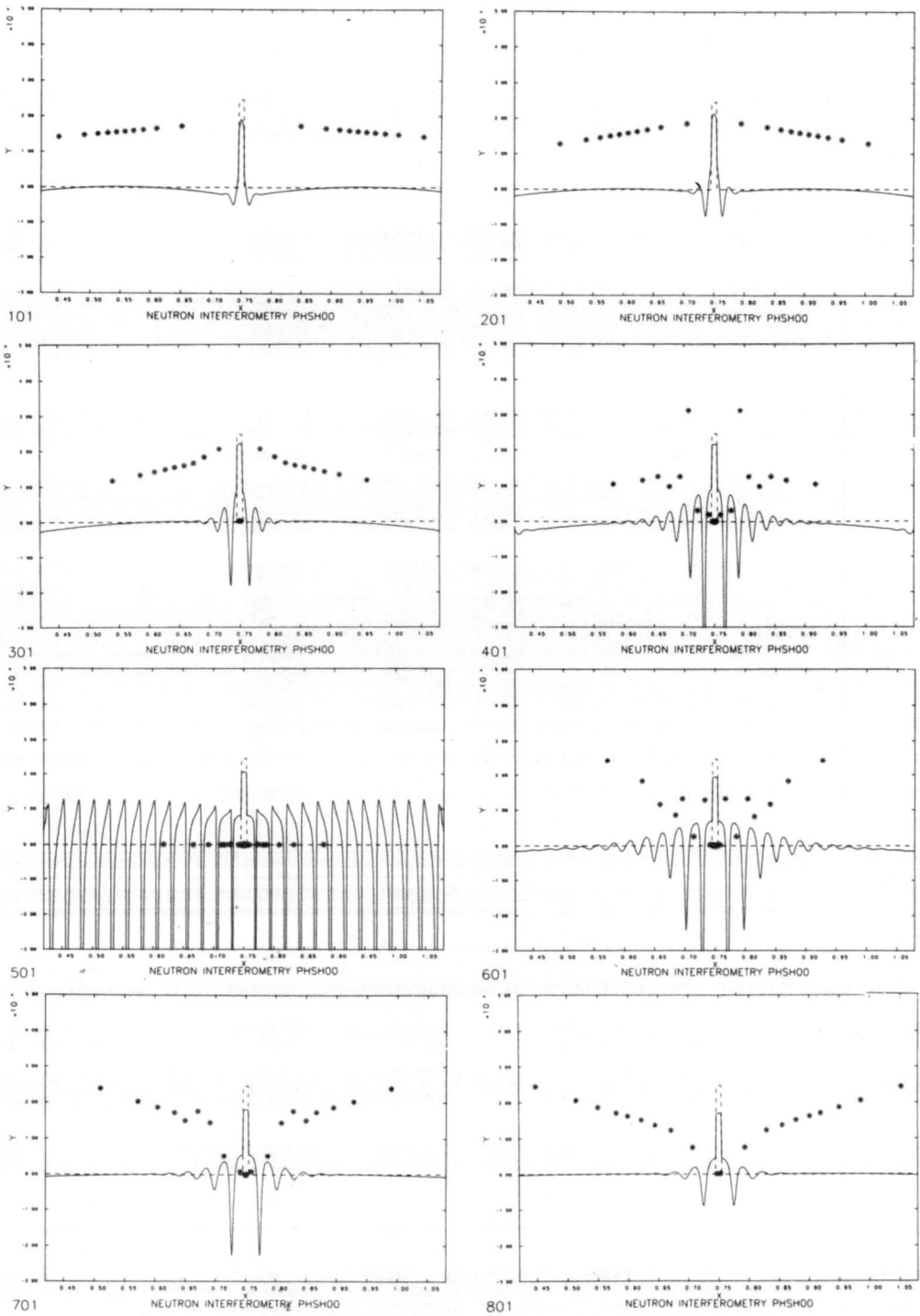

Fig.3 The effective potential (solid line), the classical potential
(dotted line), and a set of representative particle positions and
energies during the scattering of two Gaussian wave packets from a Square
potential barrier. Relative phase $\chi = \pi/2$ A simple model of the
neutron interferometer.

24

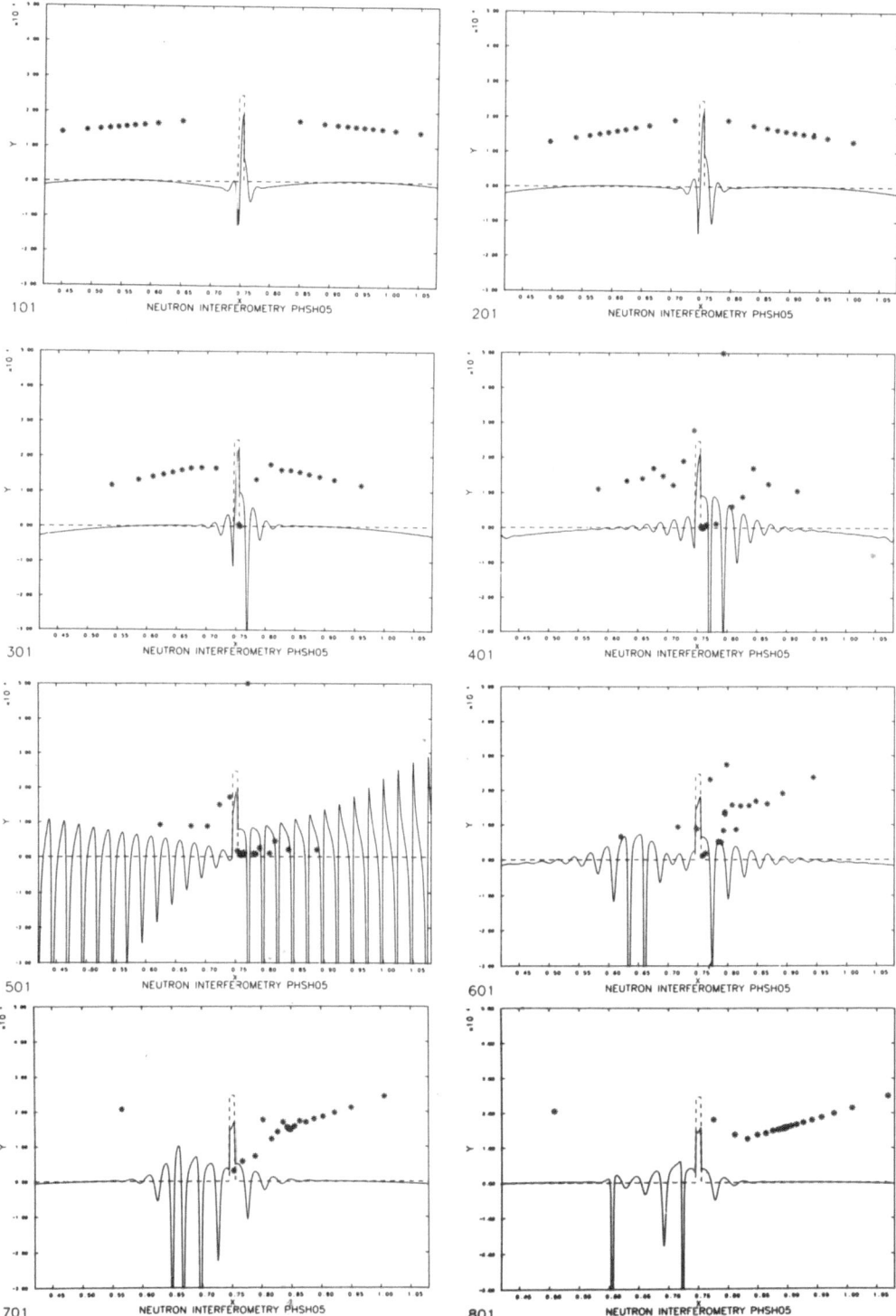

Fig.4 The effective potential (solid line), the classical potential (dotted line), and a set of representative particle positions and energies during the scattering of two Gaussian wave packets from **a Square** potential barrier. Relative phase $\gamma = \pi$.

the framework of the causal interpretation. Writing

$$(\underline{x}_1,\underline{x}_2,t) = R(\underline{x}_1,\underline{x}_2,t)e^{iS(\underline{x}_1,\underline{x}_2,t)}$$

we find

$$\frac{-h^2}{2m}\frac{\nabla_1{}^2R}{R} \frac{-h^2}{2m}\frac{\nabla_2{}^2R}{R} + \frac{(\nabla_1S)^2}{2m} + \frac{(\nabla_cS)^2}{2m} + V = \frac{-\partial S}{\partial t}$$

a two particle Hamilton-Jacobi equation, and

$$\frac{\partial p}{\partial t} + \nabla_1.(p\underline{v}_1) + \nabla_2.(p\underline{v}_2) = 0$$

a continuity equation, where

$$p = R^2$$

$$v_1 = \nabla_1S/m \quad v_2 = \nabla_2S/m$$

and ∇_1 and ∇_2 operate on coordinates one and two respectively. Evidently the quantum potential acting on particle one, say, depends not only on the coordinates of particle one but on those of particle two as well. The velocities also show this inter-relationship.

In order to illustrate the conditions under which this inter-relationship becomes significant consider the case of two particles in a harmonic oscillator potential[8]. A wave packet solution may be constructed for the motion of a single particle in such a potential which is, incidently, non-dispersive.

$$\psi(x,t) = \exp(-iwt)\exp\left[-\tfrac{1}{2}(x-x_o\cos wt)^2\right] x \exp[\tfrac{1}{2}i(\tfrac{1}{2}x_o{}^2\sin 2wt - 2xx_o\sin wt)]$$

Assuming that there are two particles, one in each of the packets ψ_a and ψ_b, centred initially at x_o and $-x_o$ respectively, then there are three possible wave functions that may be written. These are:

$$\phi_{MB} = \alpha_{MB}\,\psi_a\,(x_1,t)\,\psi_b\,(x_2,t)$$

$$\phi_{BE} = \alpha_{BE}\,[\psi_a(x_1,t)\,\psi_b(x_2,t) + \psi_b(x_1,t)\,\psi_a(x_2,t)]$$

$$\phi_{FD} = \alpha_{FD}\,[\psi_a(x_1,t)\,\psi_b(x_2,t) - \psi_b(x_1,t)\,\psi_a(x_2,t)]$$

where the α's are normalisation coefficients.

Now in the first case the wave function is a simple product and it is fairly evident that the factorisable wave function enables the two particle Schrödinger equation to be itself factored into two separate one particle equations. The same is true of the equations of motion derived in the causal interpretation and hence each particle exhibits, independently, the motion of a single particle in a harmonic oscillator potential. The trajectories are shown in figure five.

The other two wave functions correspond to those written when the two particles are said to be indistinguishable, being symmetric (the Bose-Einstein case) and anti-symmetric (the Fermi-Dirac case). Of course the particles can only be considered to be indistinguishable when the two constituent wave packets overlap. When they do not the particles are in principle distinguishable by their histories. In the causal interpretation, of course, the particles are always distinguishable, in analysis if not in practice, and hence the different statistics they obey cannot be accounted

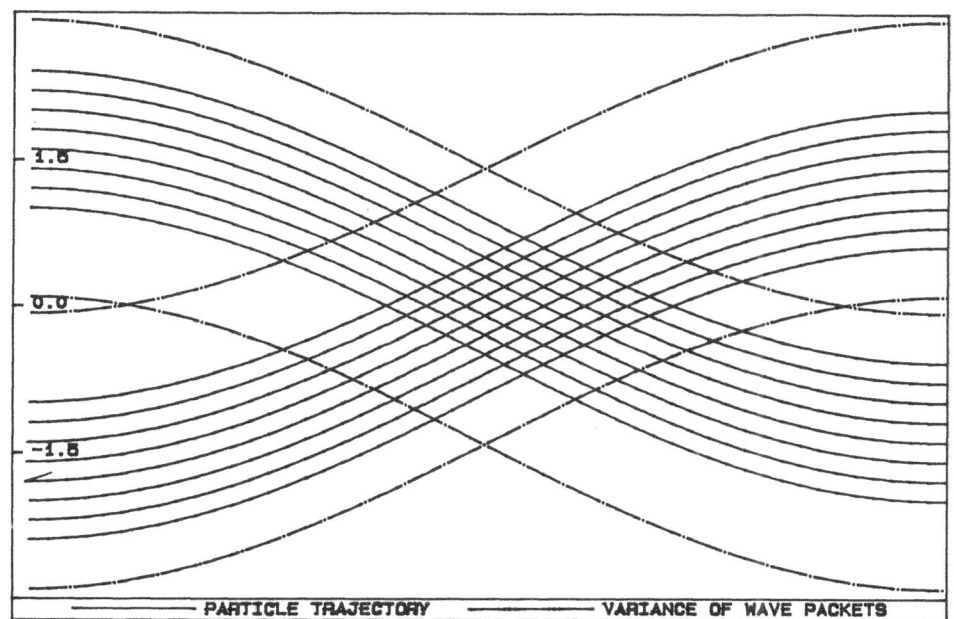

Fig.5 Particle trajectories for two particles in a harmonic oscillator potential with a factorisable wave function. Maxwell-Boltzmann statistics.

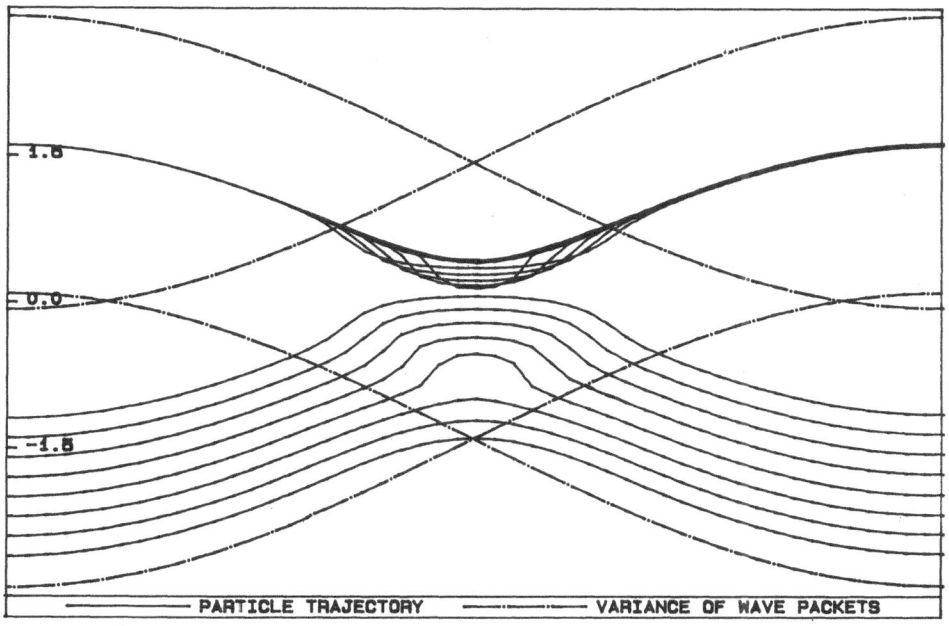

Fig.6 Correlated pairs of particle trajectories for two particles in a harmonic oscillator potential symmetric wave function. Initial position of particle one is the same for a range of initial positions of particle two. Bose-Einstein statistics.

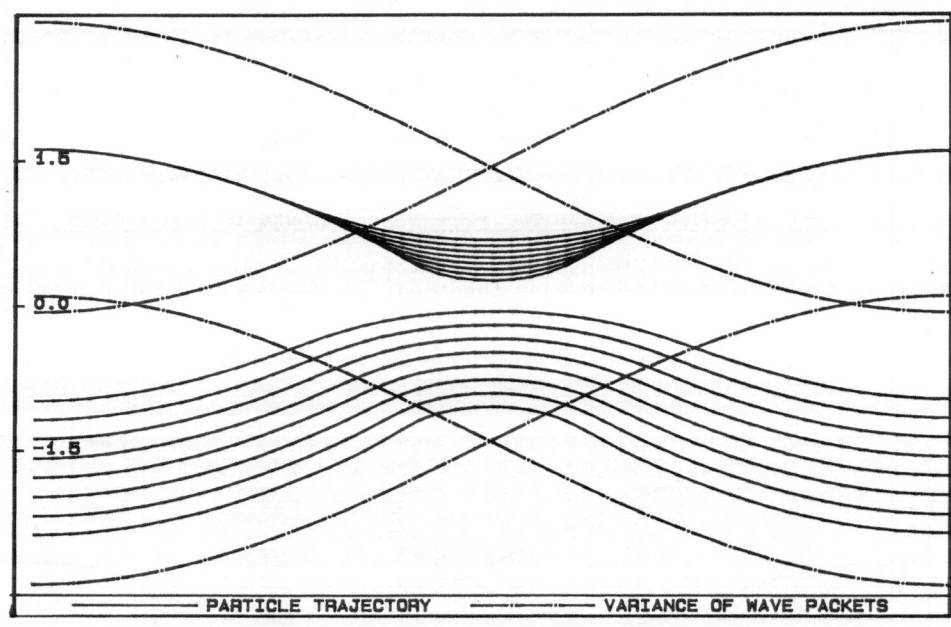

Fig.7 Correlated pairs of particle trajectories for two particles in a
harmonic oscillator potential, anti-symmetric wave function. Initial
position of particle one is chosen to be the same for a range of initial
positions of particle two.

for by reference to indistinguishability. Instead, as is demonstrated here, the different statistics arise as a result of the development of the two-body quantum potential which depends on the wave function assumed.

Figure 6 shows a set of correlated pairs of trajectories, for the symmetric wave function ϕ_{BE}, in which the initial position of particle one, $x_1(0)$, is chosen to be the same for each pair of trajectories whilst the initial position of particle two, $x_2(0)$, is different in each case. It is very clearly shown in the form of the correlated trajectories that the motion of each particle depends on the position of the other. In figure seven we plot the correlated trajectories that arise from the anti-symmetric wave function ϕ_{FD}. The initial positions are chosen as for the symmetric case and clearly the different wave function introduces a different form of correlation, in which, on the average, the particles tend to stay further apart. It should also be noticed that the magnitude of the effect is only appreciable in the region of overlap between the two wave packets ψ_a and ψ_b. This is understandable since the different statistics arise from the interference terms. In this case then the strength of the non-local interaction depends on the amount of overlap between the two wave packets. This is interesting since it tells us that the fate of particle one, for example, will not be appreciably altered by what happens to particle two when we can be sure that they are separated. In the example studied here then we may say that when the particles can be distinguished, according to the definitions of the usual approach, we can expect no non-local correlation of their trajectories. Whereas when they become indistinguishable, by virtue of the overlap of the two wave packets, non-local correlation will arise. According to this discussion non-local interaction between the particles, in the causal interpretation, only exists in those case in which the particles cannot be said to be definitely separated in space in the usual interpretation.

In the example of correlated particle motion suggested by Einstein, Podolsky and Rosen (EPR)[9] the wave function that they proposed is

$$\psi = \int dk_1 \int dk_2 \, \delta(k_1+k_2) e^{2\pi i(k_1 x_1 + k_2 x_2)} e^{-2\pi i k_2 d} = \int dk_1 e^{2\pi i k_1 (x_1 - x_2 + d)}$$

This wave function implies that measurement of the momentum of one of the particles allows us to deduce that of the other: $k_1 = -k_2$. Also measurement of the position of one of the particles allows us to deduce the position of the other: $x_1 = x_2 - d$. The essential point of EPR's argument, (at the time), was that we may choose whether to measure the momentum or the position of particle one, say and hence infer the value of the momentum or position of particle two, without interacting with it at all. EPR then argued that if we can, in this way, predict with certainty the values of these quantities of particle two then they must be pre-existent. That is the particles must possess a well defined position and momentum, even before the measurements are carried out. Furthermore the above reasoning is independent of how far apart the particles actually are, the correlation should exist to infinity. Their conclusion was that the quantum mechanical description of reality through the wave function is incomplete.

From the point of view of our discussion of the two particles in the harmonic oscillator potential we can see that, in the EPR case, the correlation between the particles exists to infinity, as a result of the particular wave function assumed. In particular the particles, in the state defined by them, cannot be considered to be separated in space in the usual approach, and hence a non-local connection can be expected to exist in the approach of the causal interpretation. It can be seen then that the use of spatially infinite waves by EPR is responsible for the distance independence of the correlation. As has been pointed out by de Broglie[1], it is this feature that enabled Bohr[10] to reply to the problem posed by the EPR example. Further on we examine the same problem from the point of view of the version of the EPR argument proposed by Bohm, dealing with a correlation between the particles' spins, we conclude that even when the particles' spatial packets do not overlap, a non-local correlation may still exist between the spins. In the next section we discuss the interpretation of the spin in the single particle case.

THE CAUSAL INTERPRETATION IN THE CONTEXT OF THE PAULI EQUATION, THE TREATMENT OF SPIN

In the usual interpretation spin is simply treated as an empirically required addition to the angular momentum. It is argued that no intuitive model is possible, which can provide an understanding of quantum phenomena associated with the spin. (This is the case for all quantum phenomena in the usual approach.) Since the operators for the components of the spin along three mutually perpendicular directions ($\hat{s}_x$, $\hat{s}_y$, $\hat{s}_z$) do not commute, it is argued that they can not be simultaneously well defined. As is usually the case, the value of a quantity, in this case the spin component, does not become definite until a measurement "throws" the system into an eigenstate of the observable being measured.

The causal interpretation can be extended to include the description of spin, as has been demonstrated by Bohm et al[11] and also by Takabayasi[12]. More recently it has been shown how the causal interpretation of spin actually works in a series of specific calculations, which show explicitly the continuous motion of the well defined spin vector during a spin superposition experiment[13] and during the passage through a Stern-Gerlach measuring device[14].

In the causal interpretation the Pauli spinor is interpreted as defining the state of rotation of a body in terms of the Eulerian rotation angles θ, ϕ, χ, relative to a standard spinor ($\begin{smallmatrix}1\\0\end{smallmatrix}$) defining a z direction, according to

$$\psi = Re^{i\chi/2} \begin{pmatrix} \cos\frac{\theta}{2} e^{i\phi/2} \\ i\sin\frac{\theta}{2}e^{-i\phi/2} \end{pmatrix} \tag{2}$$

The spin vector is defined to be

$$\underline{s} = \frac{h}{2} \frac{\psi^{+}\underline{\sigma}\psi}{\psi^{+}\psi}$$

The Pauli spinor evolves according to the equation

$$ih \frac{\partial\psi}{\partial t} = \left\{ \frac{-h^2}{2m} (\nabla - \frac{ie}{hc} \underline{A})^2 + \mu\underline{B}\cdot\underline{\sigma} + eA_o + V \right\} \psi \tag{3}$$

where $A(z,t)$ is the external vector potential, $\underline{B} = \nabla x\underline{A}$, A_o is the external electric potential and V any other scalar pot, and the velocity of the particles is given by

$$\underline{v} = \frac{h}{m} (\frac{\nabla\psi}{2} + \cos\theta\frac{\nabla\phi}{2} - \frac{e}{hc} \underline{A})$$

The equations of motion can be derived by direct substitution of 2 in 3 and are

$$\frac{h}{2} (\frac{\partial\chi}{\partial t} + \cos\theta\frac{\partial\phi}{\partial t}) + \frac{1}{2}mv^2 + Q + Q_s + \frac{2\mu}{h} \underline{B}\cdot\underline{s} + eA_o + V = 0 \tag{4}$$

a Hamilton-Jacobi equation.

30

Where

$$Q = \frac{-h^2}{2m} \frac{\nabla^2 R}{R}$$

is the quantum potential and

$$Qs = \frac{h^2}{8m} [(\nabla\theta)^2 + \sin^2\theta(\nabla\phi)^2]$$

is a spin dependent addition. The total energy is given by

$$\frac{h}{2} \left(\frac{\partial\chi}{\partial t} + \cos\theta\frac{\partial\phi}{\partial t}\right)$$

The equation of motion of the spin vector can be written

$$\frac{ds}{dt} = \frac{1}{mp} \underline{s} \times \partial_i(p\partial_i\underline{s}) + \frac{2\mu}{h} \underline{B} \times \underline{s} \tag{5}$$

where the first term on the r.h.s. is an additional quantum torque and $p=R^2$. the continuity equation is

$$\frac{\partial p}{\partial t} + \nabla\cdot(p\underline{v}) = 0$$

From the equations of motion (4), 5 and 6 it can be seen that even in the absence of magnetic fields, free spinning particle trajectories will not be the same as Schrödinger trajectories nor will the spin vector orientation remain constant if the particle is in a non-stationary spin state.

PASSAGE THROUGH AN INHOMOGENEOUS MAGNETIC FIELD: THE STERN-GERLACH APPARATUS

In the usual approach it is said that when a component of the spin is measured, along z say, the system is "thrown" into an eigenstate of the corresponding operator, $\hat{S}_z = h/2\sigma_z$. However the effect of such a measurement is to disturb the other x and y components which become indefinite, and hence the spin cannot be considered to be well defined even after the measurement has been carried out. In the following we demonstrate how the causal interpretation can account for these features of the quantum description in terms of a well defined but continuously variable spin vector.

We represent the initial state, of an atom with an angular momentum of h/2, by the spinor wave function:

$$\psi_0(z) = f_0(z)\{c_+ u_+ + c_- u_-\}$$

where c_+ and c_- are unknown real coefficients, $f_0(z)$ a localised packet, $u_+ = \binom{1}{0}$ and $u_- = \binom{0}{1}$ which we assume to be of Gaussian form.

Writing

$$c_+ = |c_+| e^{iS_+/h}$$

$$c_- = |c_-| e^{iS-/h}$$

the initial orientation is

$$\theta_0 \cos^{-1}(|c+|^2 - |c-|^2)$$

$$\phi_0 = \frac{S_+-S_-}{h} - \frac{\pi}{2}$$

$$\chi_o = \frac{S_+ + S_-}{\hbar} - \frac{\pi}{2}$$

Solution of the Pauli equation with the appropriate interaction Hamiltonian yields $\psi(z,t)$. The velocities and orientations of a set of representative single particle motions can then be calculated for various choices of the parameters c_+ and c_-. Figures 8 and 9 show the spin dependent trajectories, the field of orientations $\theta(z,t)$ and $\phi(z,t)$ respectively for the choice $c_+ = c_- = (0.5)^{\frac{1}{2}}$. This illustration demonstrates explicitly that it is possible to describe the process in terms of well defined particle motions. Clearly the quantum torque aligns the particle's spin vector parallel or anti-parallel to the field. Which of the two alternatives is actually realised in a particular case depends on the actual initial values of the hidden variables of both the system (here the spin vector direction) and the apparatus (here the particle position, since by observing this we can deduce the value of the spin). In the particular example described here the spin dependent quantum potential splits the packet at its centre and as the two packets separate the quantum torque rotates the spin vector to align parallel or anti-parallel to the analysing field. Thus in the causal interpretation $\underline{s} = \frac{\hbar}{2} (0, 0, 1)$.

Evidently, in this description the outcome of the measurement is related deterministically to the actual (uncontrollable) initial values of the hidden variables but measurement does not simply reveal them. Rather the evolution of the spin variable is correlated with the evolution of the apparatus variable (the particle position), the correlation being introduced by the inhomogeneous field. It is the existence of well defined variables in the system and in the apparatus, evolving according to the causal equations of motion, that ensures that unique initial conditions lead to unique and well defined outcomes. In this way it can be seen that wave packet collapse is a redundant hypothesis.

Evidently there is nothing special or extraordinary about measurement, it is simply a particular case of the correlated evolution of the variables of the two systems according to the laws of quantum mechanics. During this evolution the apparatus variable enters the space of one of a series of unambiguously distinguishable states each of which is correlated with different state of the system under observation.

NON-LOCAL SPIN CORRELATIONS IN THE TWO PARTICLE CASE

The description given above of spin half particles can be extended to the two particle case. This enables a description to be given of the EPR experiment in the form proposed by Bohm, in which the correlations are between spin measurements carried out on each of two particles of intrinsic angular momentum $\hbar/2$ in a singlet state. Bohm's version of the experiment has been of great historical importance and the recent results obtained with photon polarization can be easily discussed in terms of spin rather than polarization. Although it is still possible to question whether EPR-type experiments have finally demonstrated the existence of non-local phenomena in physics this has become increasingly difficult. Indeed it only seems to be possible by denying the validity of many body quantum theory.

We have already seen in the foregoing that the particle motions calculated in the causal interpretation of many body phenomena exhibit non-local correlations under certain circumstances. In particular we saw that when the wave function can be written as a sum of products of individual spatial wave functions that non-local correlation will only produce observable results when these functions overlap. We now demonstrate that the causal interpretation of the two body Pauli equation naturally implies that non-local correlations will indeed exist between two spin half

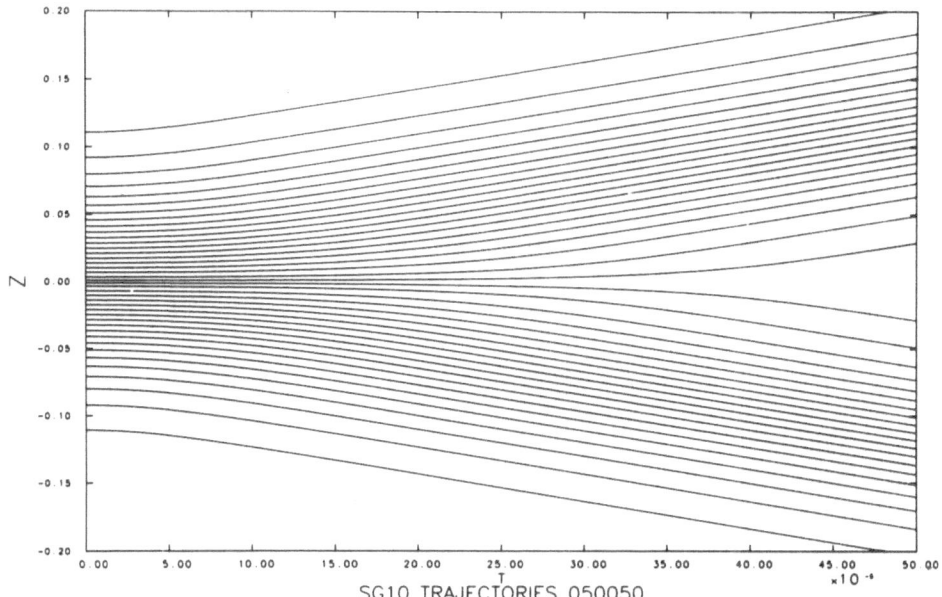

SG10 TRAJECTORIES 050050

Fig.8 Spin dependent trajectories, from a Gaussian distribution of initial positions, at the exit of a Stern-Gerlach field orientated in the z direction. Initial spin vector orientation perpendicular to the field.

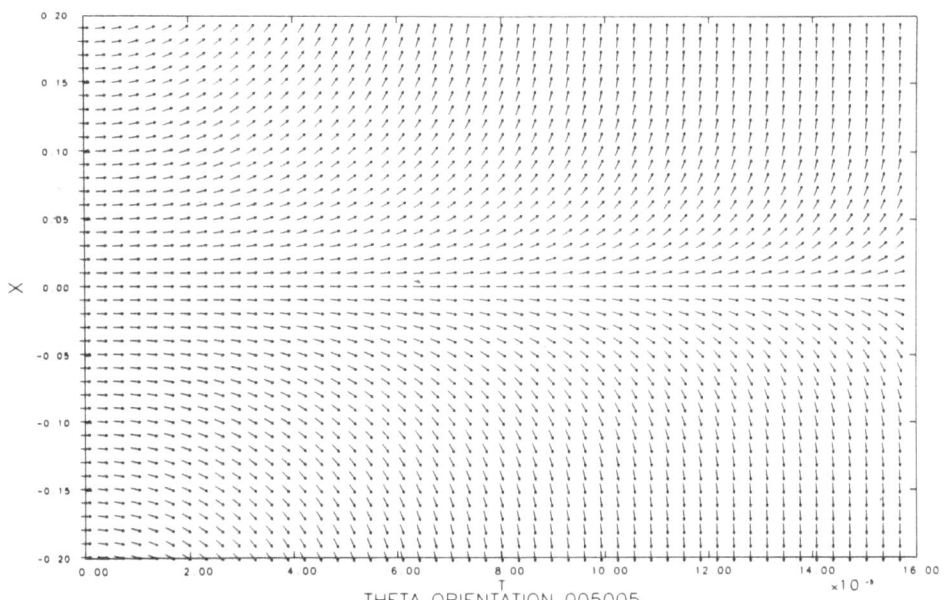

THETA ORIENTATION 005005

Fig.9 The field of orientations Θ(z,t) corresponding with the trajectories of figure 8.

particles in the singlet state. We present for the first time plots of the correlated trajectories of the particles and the evolution of their spin vectors as a result of the measurement of the spin components of each. Our conclusion is that although the model presented here is an idealised one, it provides an insight into the meaning of non-locality in terms of an underlying causal process, in a way that no other interpretation of quantum mechanics has managed to do.

Consider a system of two spin one half particles of masses m_1, m_2 and charges e_1, e_2 respectively, which are placed in external electromagnetic fields and possibly interact. The two body Pauli equation is:

$$i\hbar\frac{\partial\psi}{\partial t} = \left\{\frac{-h^2}{2m_1}(\nabla_1-\frac{ie}{hc}\underline{A}_1(x_1,x_2))^2 \frac{-h^2}{2m}(\nabla_2-\frac{ie_2}{hc}\underline{A}_2(\underline{x}_1,\underline{x}_2))^2+W_1+W_2+V\right\}\psi \quad (7)$$

where x_1, x_2 are the coordinates of particles 1 and 2 and

$$\psi = \psi_{ab}(\underline{x}_1,\underline{x}_2,t)$$

is the wave function of the system $V=V(\underline{x}_1,\underline{x}_2,t)$ is the total external plus internal scalar potential.

$$W_1 = W_1(\underline{x}_1,\underline{x}_2,t) = \mu_1 \underline{H}_1(\underline{x}_2,\underline{x}_1)\cdot\underline{\sigma}_1$$

$$W_2 = W_2(\underline{x}_1,\underline{x}_2,t) = \mu_2 \underline{H}_2(\underline{x}_1,\underline{x}_2)\cdot\underline{\sigma}_2$$

where μ_1, μ_2 are the particles magnetic moments with

$$\underline{H}_1 = \nabla_1\times\underline{A}_1 \qquad \underline{H}_2 = \nabla_2\times\underline{A}_2$$

and σ_1, σ_2 are two sets of Pauli matrices which commute and operate independently.

Writing

$$\psi_{ab}(\underline{x}_1,\underline{x}_2,t) = R(\underline{x}_1,\underline{x}_2,t)e^{iS(\underline{x}_1,\underline{x}_2,t)/h}\phi_{ab}(\underline{x}_1,\underline{x}_2,t)$$

where R and S are real amplitude and phase functions respectively and $\phi^+.\phi.=1$, we may deduce a Hamilton-Jacobi equation

$$\frac{\partial S}{\partial t} - ih\phi^+\frac{\partial\phi}{\partial t} + \tfrac{1}{2}m_1\underline{v}_1^2 + \tfrac{1}{2}m_2\underline{v}_2^2 + Q_1+Q_2+H_{1s}+H_{2s}+\frac{2\mu_1\underline{H}_1\cdot\underline{s}_1}{h}+\frac{2\mu_2\underline{H}_2,\underline{s}_2}{h}+v = 0$$

and a continuity equation

$$\frac{\partial p}{\partial t} + \nabla_1.(p\underline{v}_1) + \nabla_2.(p\underline{v}_2) = 0$$

where $p = \psi^+\psi = R^2$ is the configuration space probability density,

$$v_i = \frac{ih}{2m_i p} \psi^+\overleftrightarrow{\nabla}_i\psi - \frac{e_i}{m_ic}\underline{A}_i = \frac{1}{m_i}(\nabla_iS - ih\phi^+\nabla_i\phi - \frac{e_i}{c}A_i), \quad i = 1,2$$

are the velocities of the particles which contain spin dependent contributions,

$$Q_i = \frac{h^2}{2m_i}\frac{\nabla_i^2R}{R}$$

are the usual quantum potentials which arise in the two body case and

$$H_{is} = \frac{-h^2}{2m_i}(\nabla_i\phi^+.\nabla_i\phi+(\phi^+\nabla_i\phi)^2), \quad i = 1,2$$

34

are spin dependent additions to the quantum potentials, and

$$\underline{s}_i = \frac{h}{2} \phi^+ \underline{\sigma}_i \phi = \frac{h}{2} (\psi^+ \underline{\sigma}_i \psi)/p \qquad i = 1,2$$

are the spin vectors which we shall adopt here as describing the local spin orientation of each particle. The total energy of the system

$$\frac{\partial S}{\partial t} - ih\phi^+ \frac{\partial \phi}{\partial t}$$

is clearly spin dependent. Each of the above functions depends on the coordinates of both particles, and it is simple to show that the trajectories and spin vectors of the two particles will only evolve independently when the wave function factorizes:

$$\psi_{ab} (\underline{x}_1, \underline{x}_2, t) = \psi_{1a}(\underline{x}_1, t) \psi_{2b}(\underline{x}_2, t)$$

THE EPR EXPERIMENT

The basic set up is as shown in Figure 10. A pair of spin half particles of mass m and magnetic moment μ are formed at 0 in a simultaneous eigenstate of the spin operator in the z-direction $(h/2(\sigma_{z1}+\sigma_{z2})$ and the total spin operator $h^2/4(\sigma_1+\sigma_2)^2$ of eigenvalue zero. The particles separate in the y direction and pass through Gaussian slits orientated so as to produce packets in the directions of the analysing fields of two identical Stern-Gerlach devices. The magnet 2 is set to measure spin in the z-direction, and magnet 1 has been rotated couterclockwise through an angle δ about the y axis so that it has a gradient in the z^1 direction.

At the entrance to the fields the wavefunction is

$$\psi_0 = \frac{1}{\sqrt{2}} f_1(z_1') f_2(z_2) (u_+v_- - u_-v_+)$$

where $f_1(z_1')f_2(z_2)$ are normalised packets, z_1, z_2 are the coordinates of the particles 1 and 2 in the z and z' directions respectively and $\sigma_{z1}u_\pm = \pm u_\pm$, $\sigma_{z2}v_\pm = \pm v_\pm$. The state (8) predicts the following expectation value for the correlations of the spins measured in the z, z^1 directions:

$$\langle \sigma_{z1}', \sigma_{z2} \rangle = -\cos \delta$$

In treating this problem we have suppressed the motion in the y direction since this is not relevant to the measuring process. We only assume that the particles are sufficiently far apart on the y axis so that they do not interact and the measuring devices cannot influence one another. As we saw in the single particle case discussed above the state before the measurement takes place is one in which the spin is independent of position. The Stern-Gerlach devices introduce couplings between the spins (the variables measured) and the particle positions (the apparatus coordinates). As in the single particle case at the exit of each magnet two superposed packets are formed which separate with time along the direction of the analysing fields. The calculation of the motions when the fields are aligned along different directions are given in our paper[16]. Here we demonstrate explicitly the correlated particle motions which arise in the causal interpretation by presenting the results of these calculations in terms of correlated particle motions and spin vector orientations.

If we write

$$f_1(z_1') = R_1(z_1') e^{iS_1(z_1')/h}, \quad f_2(z_2) = R_2(z_2) e^{iS_2(z_2)/h}$$

35

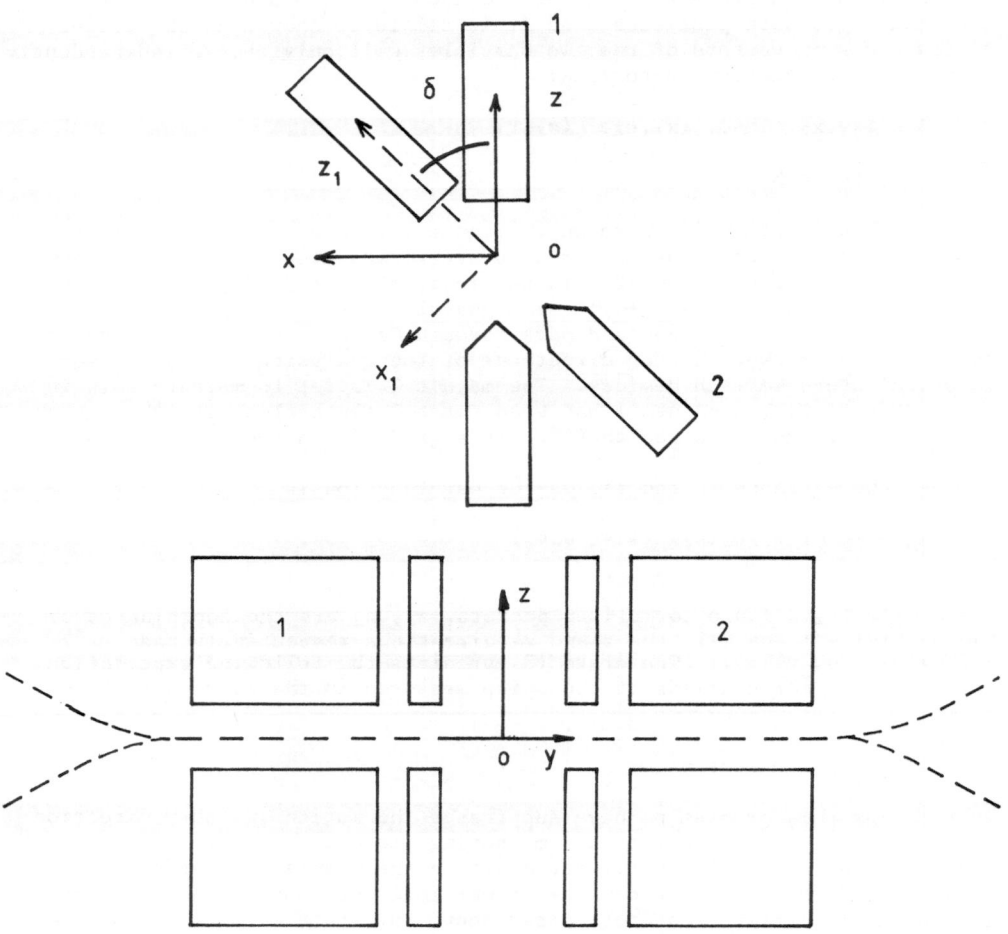

Fig.10 The experimental apparatus of the EPR-Bohm experiment.

then we find that the velocities are simply given by

$$v_1 = \frac{\nabla_1 S_1}{m} \qquad v_2 = \frac{\nabla_2 S_2}{m}$$

The spin vectors are given by $\underline{s_1} = \underline{s_2} = 0$. This clearly demonstrates the context dependence of particle properties in the causal interpretation since the individual spin vectors are here zero in the singlet state, something that cannot occur in the single particle case.

In order to understand what happens during the measuring process we consider first the simplest case which is arrived at by setting $H_2, A_2 = 0$ in equation 7, i.e. we make a measurement on one side only. The calculation yields the result that at the exit of the field the particle undergoing measurement enters one of the separating packets, depending on its initial position on the z axis, under the influence of the total spin dependent quantum potential, and its spin vector component in the direction of the analysing field changes continuously from 0 to $+h/2$ or $-h/2$ as the packets become separated in the z direction. Simultaneously the spin vector component in the z direction of the second particle not undergoing measurement changes in the opposite sense from 0 to $-h/2$ or $+h/2$ as a result of the operation of the non-local quantum torque. That is the spin of particle two depends on the position and hence spin of particle one. The velocities however in this case remain independent.

$$v_2 = \frac{\nabla_2 S_2}{m} \qquad v_1 = \frac{|R_{1+}|^2 \, \nabla_1 S_{1+} + |R_{1-}|^2 \, \nabla_1 S_{1-}}{m(|R_{+1}|^2 + |R_{-1}|^2)}$$

where the + and − subscripts refer to the two separating packets formed in the field. The trajectory of particle two is not affected by the measurement on particle one and the trajectory of particle one depends simply on local factors. If we were subsequently to measure the spin of particle two in the z direction we would of course find the opposite result to that found for particle one. The trajectories and spin vector magnitudes (indicated by the length of the arrow which always lies in the z direction) are shown in figure 11.

Consider now the case in which both Stern-Gerlach devices are operational and set to measure the spin component in the z direction on both particles simultaneously. The motion of each particle for any pair of trajectories depends sensitively on the choice of both initial positions at the entrance slits to the Stern-Gerlach devices. The calculation in this case yields the results plotted in figure 12. These results were calculated by taking the initial position of particle one to be the same in each case and then calculating the correlated trajectories which develop for a representative range of initial positions of particle two. The situation when the initial position of particle one $z_{1,0}$ is chosen to be equal to that of particle two $z_{2,0}$ represents a bifurcation point (actually a bifurcation line in configuration space). If $z_{2,0} < z_{1,0}$ then particle two has a negative velocity and its z spin component decreases from 0 to $-h/2$ whilst the correlated particle one has a positive velocity and its z spin component increases from 0 to $+h/2$, with a corresponding result if $z_{2,0} > z_{1,0}$. Clearly the fate of each particle depends sensitively on what happens to the other. In figure 13 we illustrate the same phenomenon with a different choice of the constant $z_{2,0}$.

The illustrations presented here demonstrate clearly how the results of experiments in quantum mechanics, including EPR-type experiments, can be accounted for in terms of a reality in which well defined and continuously variable quantities evolve in a deterministic manner according to the equations of motion of the cuasal interpretation. They also illustrate

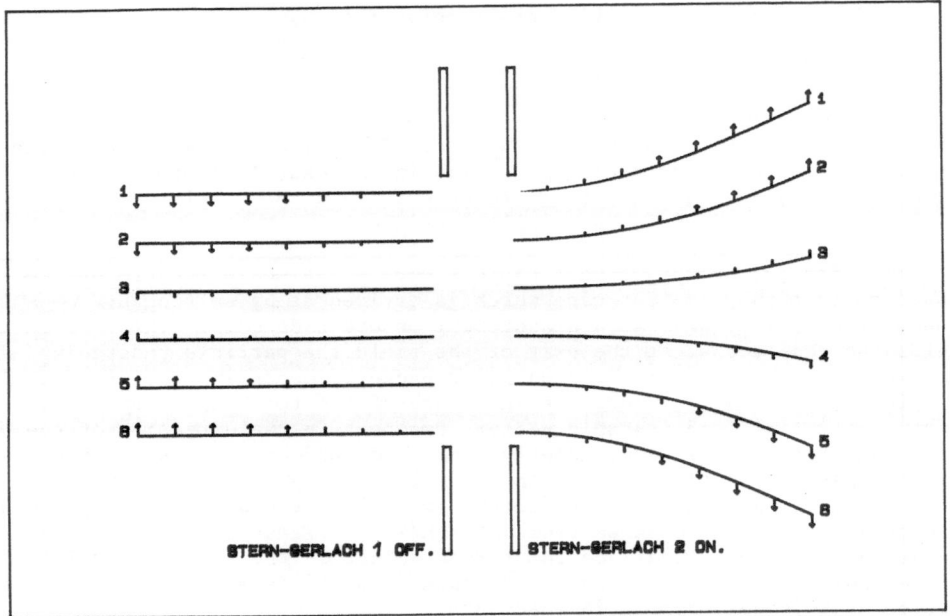

Fig.11 Trajectories and correlated spin vector orientations for two
particles initially in a singlet state after the impulsive measurement
of the z component of the spin of particle two only.

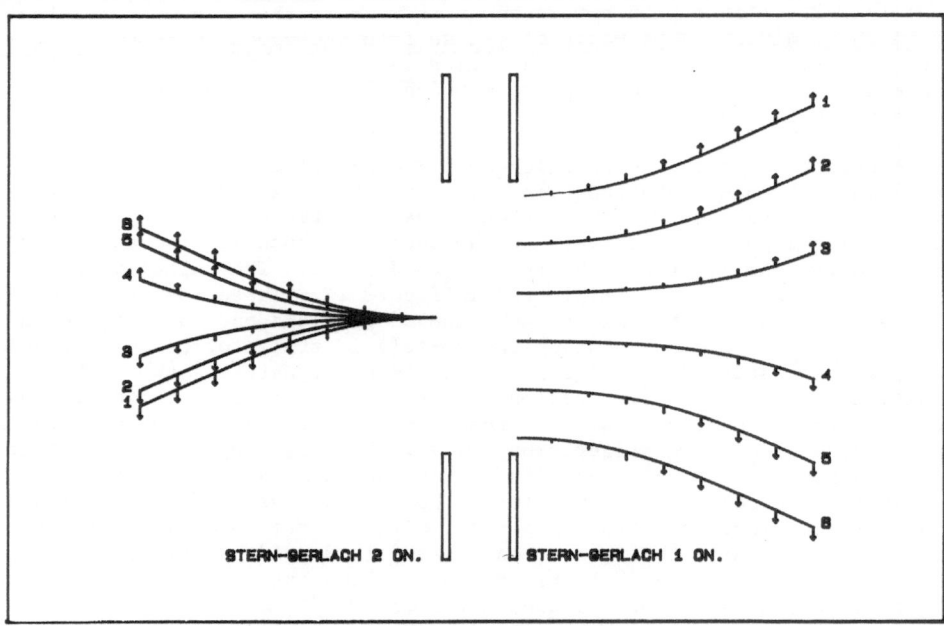

Fig.12 Correlated pairs of trajectories and correlated spin vector orientations for two particles initially in a singlet state after the impulsive
measurement of the spin in the z direction on both particles. $z_1,0 =$
constant, $z_2,0 =$ variable.

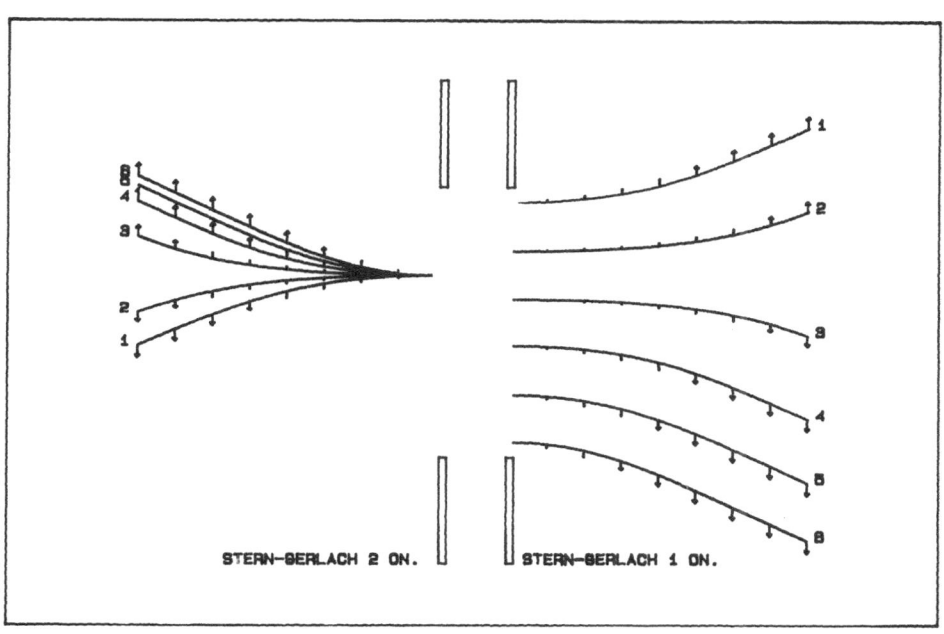

Fig.13 Correlated pairs of trajectories and correlated spin vectors for two particles initially in a singlet state after the impulsive measurement of the spin the the z direction on both particles. $z_1,0$ = a different constant to that of figure 12, $z_2,0$ variable.

that the fundamentally new feature of matter introduced in the quantum
theory is a kind of wholeness in which the behaviour of an individual particle is irreducibly connected with its context, expressed through the
wave function, evidenced in the two particle case by the existence of non-
local connection. Elements of reality exist, but they are context
dependent, they are not described by quantum theory which deals with eigenvalues of operators and in general are not simply revealed in measurement
interactions. As we have seen the interactions regarded as measurements
are in fact those in which a particular variable of a "measured" system
becomes correlated with a particular apparatus coordinate according to
deterministic laws of evolution of the whole undivided system-plus-
apparatus. In this sense the elements of reality in quantum theory are
essentially different to the elements of reality in Newtonian physics. In
the history of science each new epoch-making discovery has indicated a new
feature matter, one of which was the introduction of the idea of the field.
In our opinion assuming reality has this fundamentally new feature of
wholeness is preferable to assuming that it does not exist except when we
are looking at it!

REFERENCES

1.   L de Broglie,"Non linear wave mechanics", Elsevier, Amsterdam (1969)
2.   D Bohm, Phys Rev 85: 166 and 180 (1952)
3.   D Bohm, "Quantum Theory", Prentice Hall, New York (1951)
4.   A Goldberg, H M Schey and J L Schwarz, Am J Phys 35: 177 (1967)
5.   C Dewdney and B J Hiley, Found Phys 12: 27 (1982)
6.   C Dewdney, Phys Lett 109A: 377 (1985)
7.   H Rauch, Proceedings of the International Symposium on the Foundations
      of Quantum Mechanics, Tokyo (1983): 277 (see also: S A Werner and
      A G Klein in: "Neutron Scattering", D H Price and K Skold eds,
      Academic Press (1984))
8.   C Dewdney, A Kyprianidis and J P Vigier, J Phys A 17: L741 (1984)
9.   A Einstein, N Rosen, B Podolsky, Phys Rev 48: 777 (1935)
10.   N Bohr, in: "Albert Einstein: Philosopher-Scientist", P A Schlipped,
      p200-241. Evanston (1949)
11.   D Bohm, R Schiller, J Tiomno, Supp Nuovo Cim 1: 48 (1955)
12.   T Takabayasi, Prog Theor Phys 14: 283 (1955)
13.   C Dewdney, P R Holland, A Kyprianidis and J P Vigier, "Trajectories
      and spin vector orientations in the causal interpretation of the
      Pauli equation": Institut Henri Poincare preprint (1985)
14.   C Dewdney, P R Holland, A Kyprianidis, J P Vigier, "What happens in
      a spin measurement", Phys Lett, to appear, (1986)

CALCULATION OF AMPLIFICATION PROCESSES GENERATED BY EMPTY QUANTUM WAVES

F. Selleri

Dipartimento di Fisica dell Università
INFN - Sezione di Bari
via Amendola 173
I-70126 - Bari, Italy

I. INTRODUCTION

When Einstein wrote in 1949 his "Reply to Criticisms"[1], he stated that the problem of wave-particle dualism was "probably the most interesting subject" to discuss:

> I now come to what is probably the most interesting subject which absolutely must be discussed in connection with the detailed arguments of my highly esteemed colleagues Born, Pauli, Heitler, Bohr, and Margenau. They are all firmly convinced that the riddle of the double nature of all corpuscles (corpuscular and undulatory character) has in essence found its final solution in the statistical quantum theory ... In what follows I wish to adduce reasons which keep me from falling in line... .

In the same book, Bohr[2] refers to Einstein's picture of the electromagnetic field by recalling some discussions they had on such problems:

> The discussions, to which I have often reverted in my thoughts, added to all my admiration for Einstein a deep impression of his detached attitude. Certainly, his favored use of such picturesque phrases as "ghost waves "Gespensterfelder) guiding the photons" implied no tendency to mysticism, but illuminated rather a profound humour behind his piercing remarks.

In two previous papers[3] it was shown that the available evidence on the nature of dualism, both for photons and for massive particles, can be interpreted very naturally, at least at the physical-qualitative leve, in terms of an empty wave accompanying the propagation of energized and localized particles. This natural interpretation holds, for instance, for the neutron interferometric experiments and for the well known Janossy-Narag experiment.

If the empty wave exists really it should be detectable, one way or another. How can one hope to reveal the presence of a wave which does not carry energy or momentum? This problem can have an answer if it is noticed that one does not only measure energy changing processes but

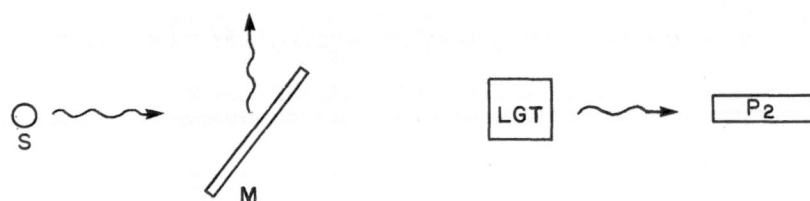

Fig. 1 Apparatus for the detection of empty quantum waves. S
is the source, M a semitransparent mirror, P_1 and P_2
are photon detectors, and LGT is the amplifier (Laser
Gain Tube).

probabilities as well: The wave could reveal its presence by modifying
decay probabilities for an unstable system.

Consider the following experimental situation.

An atomic source emits a well-defined spectral line of wavelength λ
selected with an optical filter. The source is designed to randomly emit
the photons, one by one, in such a way that there is never more than one
photon at a time in the whole apparatus. Each photon is made to fall on
a semitransparent mirror M, which "splits" it into a transmitted part and
a reflected one. As is well known, no coincidences above the casual
background are observed if a phototube is placed on each of the
transmission and reflection paths. This means that the property enabling
a photon to discharge a photomultiplier is not split by M but is made to
choose between the two paths. More accurate analyses reveal that the
entire energy $E = hc/\lambda$ is always found on either one or the other of the
two paths. Thus, whatever the carrier of energy is, we see then that <u>it</u>
is not subdivided by the mirror. Something <u>is</u> however, as shown by the
low-intensity permanence of the interference pattern.

We wish to study the possibility that what is split is just
Einstein's <u>Gespensterfelder</u> (or Bohr's virtual waves). It is therefore
assumed that the mirror diverts the entire energy in one direction even
as it splits the wave into transmitted and reflected components. Suppose
we interpose a phototube P_1 (Fig. 1) in the reflection path and
concentrate our attention on the case where this counter is triggered:
According to our postulated picture, one must then conclude that only a
virtual wave is present in the transmitted beam.

The transmitted beam is made to pass through a laser gain tube
(LGT), where the associated wave packet, though devoid of energy and
momentum, has a chance to reveal its existence by generating a
zero-energy-transfer stimulated emission. Indeed, one may assume that
the molecules filling the LGT are maintained in an excited energy band
which, on de-excitation, gives rise to radiation that includes the

wavelength λ of the incoming wave. The emitted photon could then be revealed by the phototube P_2 that follows the LGT. In this way, P_1P_2 coincidences would reveal the existence of a zero-energy undulatory phenomenon transmitted by M. The space-time propagation of this entity could be demonstrated by checking that P_1P_2 coincidences disappear whenever an obstacle is inserted before LGT in the transmitted beam.

A positive outcome of this experiment would definitely indicate that something propagates in space and time which does not carry energy- momentum but which can, all the same, induce transitions. This would practically amount to the discovery of a new level of reality and would tend to put on a firmer basis the old controversy concerning the nature of the wave-particle dualism.

II. DETECTION PROBABILITIES

The coincidence rate for the experiment of Fig. 1 will now be considered in the case of laser pulses. Coherent light pulses of finite and known duration are emitted from a laser and are known to impinge on the semitransparent mirror M in the time intervals:

$$(t_o, t_o + \Delta t); \quad (2t_o, 2t_o + \Delta t); \quad \ldots$$

with $\Delta t \ll t_o$. The times of entrance of these pulses in the detectors P_1 and P_2 are calculated simply by adding to the previous times the ratio between the dector-mirror distances and the speed of light. Coherent pulses are characterized by a Poisson distribution, so that the probability $p(n)$ of finding n photons in a pulse is:

$$p(n) = e^{-<n>} \frac{<n>^n}{n!}, \tag{1}$$

if $<n>$ is the average photon number.

The mirror M will split each pulse in a transmitted part (with t photons) and in a reflected part (with $n - t$ photons).

The probability $\pi(n,t)$ that t photons will be transmitted through M, if n arrive at its surface, is assumed to be:

$$\pi(n,t) = \left(\frac{1}{2}\right)^n \binom{n}{t}. \tag{2}$$

Let $D_1(n-t)$ $[D_2(t+\ell)]$ be the probability that the photomultiplier $P_1[P_2]$ will detect at least one of the $n - t$ reflected photons [at least one of the $t+\ell$ photons coming out from the LGT]. If η_1 and η_2 are the quantum efficiencies of P_1 and P_2, respectively, one can write:

$$\begin{aligned} D_1(n-t) &= 1 - (1-\eta_1)^{n-t} \\ D_2(t+\ell) &= 1 - (1-\eta_2)^{t+\ell} \end{aligned} \tag{3}$$

In fact, photons are detected (or non detected) independently; η_1 is the probability that P_1 detects a given photon; $1-\eta_1$ is the probability that P_1 does not detect it; $(1-\eta_1)^{n-t}$ is the probability that P_1 does not detect any one of the $n - t$ incoming photons, so that

$$1 - (1-\eta_1)^{n-t}$$

is the probability that P_1 detects something. The argument for the case of D_1 is similar.

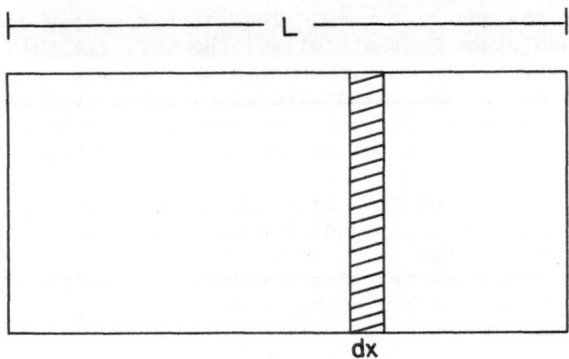

Fig. 2   Laser Gain Tube of rectangular shape with length L for
the experiment of Fig. 1. Amplifying waves impinge
from left.

One can introduce two further probabilities:

$q(t+\ell, x)$ will be the probability that $t+\ell$ photons are found at a
depth x inside the LGT if t enters from the left [see
Fig. 2]. Only photon <u>emission</u> is assumed possible, so
that $\ell \geqslant 0$.

$C_{12}(n)$ will be the probability of a P_1P_2 coincident count if n
photons are impinging on LGT. Coincident counts are
defined as taking place within appropriate time intervals
of width not larger than Δt.

One can obviously write:

$$C_{12}(n) = \sum_{t=o}^{n} \pi(n,t) \; D_1(n-t) \sum_{\ell=o}^{\infty} q(t+\ell,x) \; D_2(t+\ell) \qquad (4)$$

The inclusion of the term with t = n is possible because $D_1(o) = 0$,
consistently with (3). Note also that $C_{12}(o) = 0$.

In the statistical interpretation of quantum theory the so called
"coherent pulse" is a statistical ensemble of pulses, with average photon
number <n>, in which every given photon number n is present with a frequency given by (1). The probability of a P_1P_2 coincident count for an
incoming "coherent pulse" is therefore given by:

$$C_{12} = \sum_{n=o}^{\infty} p(n) \; C_{12}(n) \qquad (5)$$

Equations (4) and (5) will allow concrete empirical predictions to be
deduced from the models for stimulated emission which will be introduced
in the following section.

III. STIMULATED EMISSION PROCESSES

The only unknown probabilities in (4) and (5) are the

q(t+ℓ, x).

In order to calculate them one must introduce the following four probabilities:

Pdx   is the probability that a photon be generated by stimulated emission in the length - interval dx from a "normal" photon (namely, from a photon which has not crossed M).

P'dx   is the probability of the same event from a photon which has crossed M and whose associated wave has been attenuated by M.

P_0dx   is the probability of the same event for an empty wave (quantum hν reflected, wave attenuated by M).

Sdx   is the probability for the spontaneous emission of a photon in the length - interval dx.

The parameters P, P', P_0, S are all positive. They are referred to a LGT of rectangular shape [see Fig. 2] and homogeneous, so that these parameters are all x - independent.

Elementary thermodynamical considerations lead to the conclusion that

$$P_0 + P' = P. \tag{6}$$

In fact, as can be easily shown, if the previous relation were not satisfied, Planck's formula would not hold when a semitransparent mirror is inserted inside a cavity where matter is in equilibrium with radiation. As a consequence, if a second cavity were connected to the first one through a small hole with a suitable frequency filter, well known paradoxical situations would be obtained. The only way out is to assume the validity of (6).

Two opposite assumptions about stimulated emission will now be examined:

CASE 1:   Stimulated emission is due exclusively to the energetic quantum hν, and has nothing to do with the wave. Therefore:

$$P_0 = 0 \; ; \; P' = P \tag{7}$$

CASE 2:   Stimulated emission is only due to the wave and the presence/absence of the quantum hν is totally irrelevant. Therefore:

$$P_0 = P' = \frac{1}{2} P \tag{8}$$

These two physically interesting possibilities will be obtained as particular cases of our general calculations. For the time being we will call Qdx the total probability for the stimulated emission of a photon in the length-interval dx, where

$$Q = t \, P' + (n-t) \, P_0 \tag{9}$$

since t photons and (n-t) empty waves are transmitted.

The adoption of (9) allows us to deal with CASE 1 and CASE 2 above in an unified way. In fact this relation is fully justified from a physical point of view in either situation. In CASE 1, one must use Eq. (7) which leads to $Q = t P$, exactly as expected, since the t energy quanta should stimulate independently from one another and give additive effects. In CASE 2, one must use Eq. (8) which leads to:

$$Q = n P/2,$$

independent of t. Of course, in a realistic approach to the electromagnetic (empty) waves, one expects their amplitudes to be fixed in the source (and to be reduced by a factor of $1/\sqrt{2}$ in M) independently of the (random) number t of transmitted photons. An undulatory approach to stimulated emission should reasonably assume that the emission probability Q is proportional to the squared amplitude of the wave. If E_n is the electromagnetic wave amplitude associated with n energetic quanta $h\nu$, one expects, from a semiclassical analogy:

$$|E_n|^2 = f\, n\, h\nu,$$

where f is a (n-independent) proportionality factor. It follows from the previous relation that:

$$|E_n|^2 = n\, |E_1|^2,$$

so that if P is due to the action of $|E_1|^2$ and Q to the action of $|E_n|^2$, one can understand $Q = n P/2$, if one also keeps into account the factor of 2 reduction of $|E_n|^2$, owing to the action of the semitransparent mirror M. Therefore it can be concluded that the proposed treatment of stimulated emission [Eq (9) above] is fully adequate for dealing with the two physically interesting situations called CASE 1 and CASE 2 above.

Turning now to the spontaneous emission terms S, which adds its effects systematically to those of Q [see Eq (12) below], it can be noted that there exist two extreme possibilities:

$S = 0$ corresponds to a complete elimination of spontaneous emission by means of filters, geometrical selections, and so on.

$S = P$ corresponds to a full competition of spontaneous with stimulated emission. This is the case found in the textbook treatment of matter in equilibrium with radiation, where the emission probability in a given mode k is proportional to $(n_k + 1)$, n_k being the number of quanta in that mode and the "1" term being the contribution of spontaneous emission.

In the studied experiment of Fig. 1 one is not dealing with matter in equilibrium with radiation, and the effects of spontaneous emission can be somewhat reduced with respect to the extreme case $S = P$. Something like

$$S = \frac{1}{2} P \tag{10}$$

should give a good idea of spontaneous emission for the experiment under consideration.

IV. EQUATIONS AND SOLUTIONS

Referring once more to Fig. 2 one can say that the probability of finding $t+\ell$ photons at depth $x+dx$ inside the LGT is given by the sum of two terms: the first one equals the probability of having $t+\ell$ photons at depth x, times the probability that no photon be emitted between x and $x+dx$; the second one equals the probability of having $t+\ell-1$ photons at depth x, times the probability that a single photon be emitted between x and $x+dx$. Therefore:

$$q(t+\ell,\ x+dx) = q(t+\ell,x)\left[1-(\ell P+Q+S)dx\right] + q(t+\ell-1,x)\left[(\ell-1)P+Q+S\right]dx \qquad (11)$$

From this equation one can easily obtain:

$$\frac{d\ q(t+\ell,x)}{dx} = -(\ell P+Q+S)\ q(t+\ell,x) + \left[(\ell-1)P+Q+S\right]q(t+\ell-1,x) \qquad (12)$$

This is a set of coupled differential equations of the first order which allow one to calculate $q(t+\ell,x)$, with $\ell \geqslant 1$, given $q(t,x)$. It is not difficult to show that the latter quantity satisfies an equation of the type (12) with $\ell=0$ and with the second term missing in the r.h.s. Coupled with the initial condition $q(t,o)=1$, this equation can be solved and gives

$$q(t,x) = e^{-(Q+S)x} \qquad (13)$$

The latter result allows one to solve all the equations of the type (12) with $\ell \geqslant 1$, which must be coupled with the initial conditions

$$q(t+\ell,o) = 0, \quad \text{if } \ell \geqslant 1. \qquad (14)$$

The general solution is given by

$$q(t+\ell,x) = \frac{\alpha(\alpha+1)\ldots(\alpha+\ell-1)}{\ell!}\ e^{-(Q+S)x}\ (1-e^{-Px})^{\ell} \qquad (15)$$

where

$$\alpha = \frac{Q+S}{P} = t+(n-2t)\ \beta+\gamma \qquad (16)$$

and

$$\beta = \frac{P_0}{P}; \quad \gamma = \frac{S}{P} \qquad (17)$$

It has already been agreed that a good guess for the spontaneous emission parameter γ is:

$$\gamma = \frac{1}{2}$$

[see Eq (10)]. The parameter β expresses the strength of empty-wave stimulation. $\beta = 0$ corresponds to the CASE 1 above and $\beta = \frac{1}{2}$ to the CASE 2.

Since the total depth of the LGT is obtained from $x = L$, the probability $\Gamma(t)$ that P_2 detects at least one photon is given by:

$$\Gamma(t) = \sum_{\ell=0}^{\infty} q(t+\ell,L)\ D_2(t+\ell). \qquad (18)$$

47

TABLE 1

Comparison of coincidence rates with stimulated emission only due to energy $[C_{12}^{(1)}]$ and only due to waves $[C_{12}^{(2)}]$.

$<n>$	γ	$C_{12}^{(1)}$	$C_{12}^{(2)}$	ratio (2)/(1)
.5	0	.00150	.00358	2.39
.5	1/4	.00253	.00452	1.79
.5	1/2	.00352	.00541	1.54
.5	3/4	.00446	.00627	1.41
.5	1	.00537	.00709	1.32
1	0	.00573	.00970	1.69
1	1/4	.00765	.01144	1.50
1	1/2	.00948	.01310	1.38
1	3/4	.01123	.01469	1.31
1	1	0.01290	0.01621	1.26

By substituting (15) and (3) in (18) and carrying out the summation, one obtains:

$$\Gamma(t) = 1 - (1-\eta_2)^t (A_o^\beta)^n (A_o^{1-2\beta})^t A_o^\gamma \tag{19}$$

where

$$A_o = (1-\eta_2+\eta_2\ e^{PL})^{-1} \tag{20}$$

An independent calculation shows that e^{PL} is the average photon output from the LGT, when one photon enters into this instrument from the left. Therefore e^{PL} is exactly what is usually called "the amplification" of the LGT.

Substitution of (19), (2) and (3) into the expression (4) for $C_{12}(n)$ leads to:

$$C_{12}(n) = 1 - (\tfrac{1}{2})^n \left\{ (2-\eta_1)^n - [A_o^\beta + (1-\eta_2)\ A_o^{1-\beta}]^n\ A_o^\gamma \right.$$

$$\left. + [(1-\eta_1)\ A_o^\beta + (1-\eta_2)\ A_o^{1-\beta}]^n\ A_o^\gamma \right\} \tag{21}$$

A final substitution of (21) in (5) leads to:

$$C_{12} = 1 - \exp\left\{-<n>\ \eta_1/2\right\}$$

$$- A_o^\gamma \exp\left\{<n>\ [A_o^\beta + (1-\eta_2)\ A_o^{1-\beta} - 2]/2\right\}$$

$$+ A_o^\gamma \exp\left\{<n>\ [(1-\eta_1)\ A_o^\beta + (1-\eta_2)\ A_o^{1-\beta} - 2]/2\right\} \tag{22}$$

The previous general expression can be applied to two important cases. If the stimulated emission is due to the action of the energetic quanta, one should take $\beta = 0$, as was shown above, and (22) becomes:

$$C_{12}^{(1)} = 1 - \exp\{-<n> \eta_1/2\}$$
$$- A_o^\gamma \exp\{<n>[A_o(1-\eta_2) - 1]/2\}$$
$$+ A_o^\gamma \exp\{<n>[A_o(1-\eta_2) - 1 - \eta_1]/2\} \tag{23}$$

If instead stimulated emission is due to the action of the wave, as here proposed, one should adopt $\beta = \frac{1}{2}$ and (22) then becomes:

$$C_{12}^{(2)} = 1 - \exp\{-<n> \eta_1/2\}$$
$$- A_o^\gamma \exp\{<n>[\overline{\sqrt{A_o}}(1 - \frac{\eta_2}{2}) - 1]\}$$
$$+ A_o^\gamma \exp\{<n>[\overline{\sqrt{A_o}}(1 - \frac{\eta_1+\eta_2}{2}) - 1]\} \tag{24}$$

The comparison of (23) and (24) is shown in Table 1 for the fixed values

$$\eta_1 = \eta_2 = 0.1$$

of the quantum efficiencies of the photomultipliers P_1 and P_2, and for an amplification

$$e^{PL} = 3.$$

These values imply

$$A_o = .83333.$$

As it can be seen, the difference between (23) and (24) is easily observable in all cases.

V. CONCLUSIONS

The previous calculation shows that the proposed experiment can easily be carried out with incoming laser pulses: Particle and Wave stimulation gives rise to measurably different predictions.

It is amusing to notice that our formulae, for the case of particle stimulation, are identical to those "deduced" in textbooks from quantum electrodynamics. We therefore propose that the usual treatment of stimulated emission is wrong and that it should be modified by taking into account our corrected equations. This is not very surprising, since the usual textbook approach is not based on a straightforward quantum − electrodynamical calculation but requires the introduction of independent approaches, like for instance the master equation.

The formulae of the present paper still contain a difficulty, since the probabilities for different incoming photon numbers are added incoherently.

This difficulty can however be eliminated, as it will be shown in a forthcoming publication.

REFERENCES

1. A. Einstein; Reply to Criticisms, in P.A. Schilpp (ed.); <u>Albert Einstein: Philosopher-Scientist</u>, 3rd edn., Open Court, LaSalle, Illinois (1970). The quoted sentence is at p. 666.
2. N. Bohr, in the same book, p. 206.
3. F. Selleri; Found. Phys. <u>12</u>, 1087 (1982).
   F. Selleri; <u>Gespensterfelder</u> in <u>The Wave-Particle Dualism</u>; S. Diner et al. Eds., Reidel, Dordrecht (1984).

RELATIVE METRICS AND

PHYSICAL MODELS FOR NON-LOCAL PARTICLES

William M. Honig
Curtin University, (formerly known as the
Western Australian Institute of Technology)
Perth, Bentley, 6102, Western Australia

ABSTRACT

A presentation is made that attempts to find a meaning and a physical representation for non-local particles. This is based on how the space metric can be considered as a relative concept so that its form and specification may be different for different observers of the same spatial arena. The non-local particles are limiting forms for toroids and are thus one-dimensional rings representing the toroidal half wavelength elements of dipole radiation which are considered as discrete entities. They are candidates for the role of hidden variables. In the so called electromagnetic rest frame these rings can according to relative metrics be local particles. This may resolve some of the QM and EPR problems.

I. INTRODUCTION

The problem of finding a meaning and a physical model for a non-local particle turns on what we mean by the word 'particle'. Mathematically this has been taken as a three-dimensional delta function in 3-space. This fits with the meaning for the word discrete, which may be taken as an extended local object which has spatial boundaries which are finite.

A discrete object such as a sphere can be linked with a particle if one defines a sphere in a space and then shrinks the sphere about its center so that in the limit it approaches its center point. In this case, that center point can be considered as a particle which represents the sphere for those conditions where distances are always large compared to the dimensions of the sphere. This, in fact, is the kind of use to which this concept is put in QM. Thus, for dimensions large with respect to the classical electron radius, the electron is considered as a particle. This representation of the electron is further bolstered with the QM rules and strictures on uncertainty, etc.

There would appear to be no way of extending these concepts to a 'non-local particle' without extending our understanding of what we mean by space or the metric of a space. In Section II a discussion based on previous work[1] is given where it is shown how the metric of space can be considered as a relative and thus a subjective concept. Although Reimann and Gauss have developed the concept of the metric as an essentially invariant measure on space, many others, especially Helmholtz[2] and Poincare[3] have at least discussed how space can be considered as non-invariant and thus how it can be a quite subjective concept. Both of the last named have discussed how the appearance of space and spatial objects can vary radically in the opinion or perception of one observer as compared to the perceptions of another simultaneous observer of the same space and objects.

This is based on the fact that rest frame distortions universal within that rest frame will not be measurable, perceivable, or even conceivable within that rest frame to its inhabitants (or observers whose dimensions and senses are likewise distorted and are unmeasurable to themselves). Simultaneously, observers outside such rest frames who do not suffer these distortions will indeed observe them in those rest frames. This results in the conclusion that arbitrary coordinate systems such as, for instance, that with the spherical coordinates R, θ, ϕ can first be considered in the usual manner, as consisting of coordinates which are radial, angular, etc.

On the other hand, to the appropriate observer, these same designations can be considered as the symbols for a Euclidean flat space where the above coordinates will have an appearance identical to the Cartesian coordinates x, y, z. The equations and mappings relating both sets of coordinate symbols to each other and the appearance of spatial phenomena will depend on each observer's point of view.

The usefulness for this is for the major example to be discussed: that a mapping can be provided between two different observers in such a way that a 'particle' , (a 3-dimensional delta function, a point) in one observer's view becomes a 'non-local particle', (a one dimensional ring and thus a mathematical but still an extended object) to another observer of the same object.

In Section III, a physical example of the above case is discussed. We try to show how, say, the internal one-dimensional ring axis of a 'smoke ring' together with the toroidal surface of the 'smoke ring' itself, can be mapped and transformed. This toroid could, after a sequence of linked mappings and transformations appear to the appropriate observer, to be a sphere. The point center of the sphere is the final form for the ring axis of the original 'smoke ring' and the surface of the sphere is the final form of the surface of the original 'smoke ring'. Both 'smoke ring' and sphere are considered in a physical context. The mapping between them, which in the mathematical sense is not isomorphic, can be given a useful physical and fluidic meaning and the singularities of the mappings can be shown to be physically negligible.

In Section IV, finally, we can take the point center of the sphere as an example of the 'canonical particle' representation and the one dimensional ring axis of a 'smoke ring' as the model for a 'non-local particle' representation. The reason for the introduction of the 'smoke ring' is that it is a conceptual model relevant to the generation of the toroidal half wavelength dipole field distributions emitted by a fluid model droplet electron[1] in rectilinear acceleration and deceleration.

This object, first sketched by Heinrich Hertz, the half wavelength dipole field distribution, is our model for a discrete electromagnetic object which we have named the photex. The reason why this of relevance to QM is because it appears to be a physically realistic candidate for the role of hidden variable and it is closely connected with the photon. One may be able to think of such an ever expanding toroidal electromagnetic field distribution as the physical model for the 'non-local particle' involved in the recent EPR experiments. This, then might realistically explain these experimental results, while simultaneously retaining both non-locality and locality depending on the rest frame, it is a candidate for the role of hidden variable.

II. RELATIVITY OF THE METRIC

We start with the relevant quote from the essay of Helmholtz on the relativity of space:[2]

"Let me first remind the reader that if all the linear dimensions of other bodies, and of our own, at the same time were diminished or increased, in like proportion, as for instance to half or double their size, we should with our means of space perception be utterly unaware of the change. This would also be the case if the expansion or contraction were different in different directions, provided that our own body changed in the same manner.

Think of the image of the world in a convex mirror. The common silvered globes set up in gardens give the essential features, only distorted by some optical irregularities. A well-made convex mirror of moderate aperture represents the objects in front of it as apparently solid and in fixed positions behind its surface.

But the images of the distant horizon and the sun in the sky lie behind the mirror at a limited distance, equal to its focal length. Between these and the surface of the mirror are found the images of all the other objects before it, but the images are diminished and flattened in proportion to the distance of their objects from the mirror. The flattening, or decrease in the third dimension, is relatively greater than the decrease of the surface dimensions. Yet every straight line or every plane in the outer world is represented by a [curved] straight line or plane in the image.

The image of a man measuring with a rule a straight line from the mirror will contract more and more the farther he went, but with his shrunken rule the man in the image will count out the same number of centimeters as the real man. And, in general, all geometric measurements of lines or angles made with regularly varying images of real instruments would yield exactly the same results as in the outer world, all congruent bodies would coincide on being applied to one another in the mirror as in the outer world, all lines of sight in the outer world will be represented by straight lines of sight in the mirror.

In short I do not see how men in the mirror are to discover that their bodies are not rigid solids and their experiences not good examples of the correctness of Euclid's axioms. But if they could look out upon our world as we look into theirs, without overstepping the boundary, they must declare it to be a picture in a spherical mirror, and would speak of us just as we speak of them; and if two inhabitants of the different worlds could communicate with one another, neither, as far as I can see, would be able to convince the other that he had the true, and the other the distorted, relations."

The above quotation can be illustrated also via the identical pair of Escher pictures which appear in Figure 1. In Figure 1a it is shown how Observer A, who is holding the reflecting sphere, sees his own surroundings and himself. His image in the mirror is labeled B. In Figure 1b it is shown how that image observer, Observer B, views the world of Observer A. Neither observer believes that his own surroundings are distorted but imputes such distortions to the other observer.

This relates also to a physical meaning for non-Euclidean metrics. Let us say that Observer A represents the ordinary case of a man in a flat Euclidean space. He is holding a silvered sphere in his hand while sitting in his study. He looks into the sphere and tries to describe the situation there. The volume of the sphere is thought by him to contain the volume within which Observer B exists. This is a well known example in optics which maps all points outside the sphere to virtual image points inside the sphere. The location of all image points inside the sphere can be derived from the sphere diameter and the location of the object points outside the sphere.

Observer A can thus, via the optical mapping equations set up both a local and global metric for all the images inside the sphere. The Euclidean geometry in his own world will map to a non-Euclidean geometry in the image world he sees in the sphere. Since the mappings are isomorphic, he can consider his own flat space transformed to that non-Euclidean space within the sphere.

On the other hand, because this situation is completely symmetric with respect to Observer B, B will go through the same procedure as Observer A did above. Thus A and B will each impute a rest frame distortion to their opposites, B and A, respectively. Since this can be put in the form of coordinate transformations which each applies to the other, the conclusion follows that when one labels a space, one is making a subjective judgment, and that non-Euclidean metrics are judgments which a 'non-inhabitant' of a particular space makes about that space. All Coordinate System Inhabitants (CSI), therefore, are observers who suffer distortions in themselves and their surroudings which they have no way of knowing. This implies that the presence of a non-Euclidean metric can be traced to a CSI external to a space who does not suffer the distortions of another CSI internal to that space. The metric which the external CSI writes would be non-Euclidean in the opinion of the internal CSI of that space.

A summary of the mathematical development of this is here given[1]:

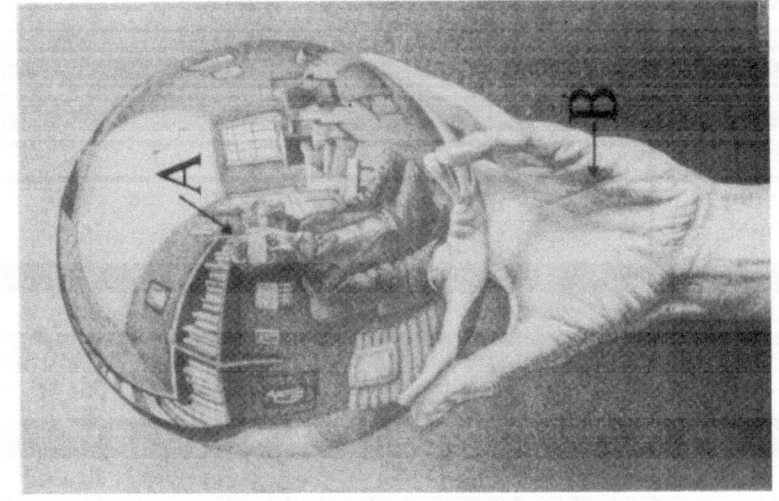

Figure 1b

Globe as seen by Observer B

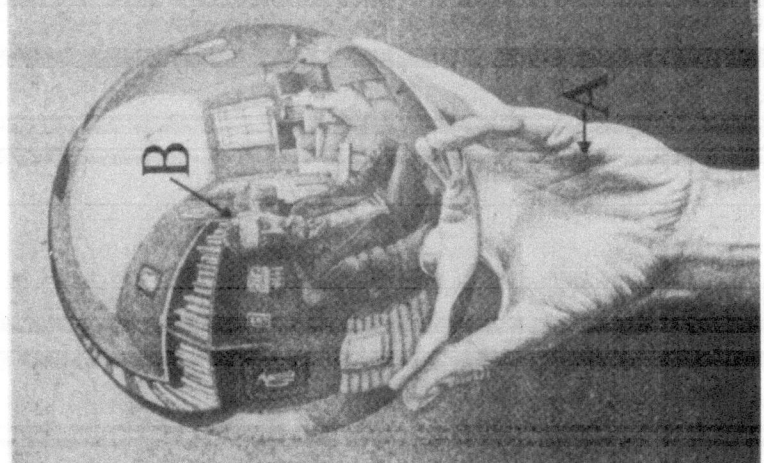

Figure 1a

Globe as seen by Observer A

(From "Self-Portrait" by M.C. Escher, lithograph, 1935).

A CSI is an observer to whom an arbitrary set of coordinates appear to have the characteristics usually associated with the Cartesian x, y, z set. The CSI will think the following about his 'native' set of coordinates:

1. The coordinate axes are orthogonal and extend to plus and minus infinity.

2. The one-dimensional variation of a point generates a straight line, the two-dimensional variation a plane, etc.

3. An increment of length is independent of position and orientation.

4. There is a point origin at which all coordinate values are zero, it is a singly connected space. There is also a point on each coordinate axis which has the value one, which defines the gauge.

5. $ds^2 = d\alpha^2 + d\beta^2 + d\gamma^2$, where α, β, γ are the coordinates of the arbitrary coordinate system., three dimensional in this case.

We assert that these 5 statements can be true about an arbitrary set of coordinates (in the opinion of its CSI). This would appear, however, to be generally untrue. For example for a three-dimensional cylindrical or spherical coordinate system, the angular dimensions are either discontinuous or periodic and the radial dimension has only positive values. Nevertheless, conformance with the above statements can be established by modifying via a series of linked sequential mappings the meanings for those coordinates that are initially irreconcilable with the above statements. This is here presented and applied.

Suppose the set of Cartesian coordinates (x, y, z) and another set of coordinates (α, β, γ) label a three-dimensional space. The relationship between these two coordinate systems giving x, y, and z explicitly are:

$$x = x(\alpha, \beta, \gamma), \qquad y = y(\alpha, \beta, \gamma), \qquad z = z(\alpha, \beta, \gamma) \qquad [1]$$

and the surfaces α, β, γ = a constant, form an orthogonal system. The differential element of length is ds, where:

$$ds^2 = dx^2 + dy^2 + dz^2 \qquad [2]$$

and the following functions may be constructed with the usual conditions on continuity and differentiability:

$$1/U = [(\partial x/\partial\alpha)^2 + (\partial y/\partial\alpha)^2 + (\partial z/\partial\alpha)^2]^{1/2}$$

$$1/V = [(\partial x/\partial\beta)^2 + (\partial y/\partial\beta)^2 + (\partial z/\partial\beta)^2]^{1/2} \qquad [3]$$

$$1/W = [(\partial x/\partial\gamma)^2 + (\partial y/\partial\gamma)^2 + (\partial z/\partial\gamma)^2]^{1/2}$$

so that ds^2 may be expressed as:

$$(ds)^2 = (d\,\alpha/\,U)^2 + (\,d\beta/\,V)^2 + (\,d\,\gamma/\,W)^2 \qquad [4]$$

On the basis of the previous discussion the above equations are merely those that would be written by an (x, y, z)-CSI (the inhabitant of an ordinary x, y, z Cartesian coordinate system). One may, however, write the following set of equations for an (α, β, γ)-CSI; the equation numbers will be same as above, but primed, in order to facilitate comparisions:

$$\alpha = \alpha(x,y,z), \qquad \beta = \beta(x,y,z), \qquad \gamma = \gamma(x,y,z) \qquad [1']$$

and the surfaces x, y, z = a constant, form an orthogonal system. The differential element of length will be called ds':

$$(ds')^2 = (d\,\alpha)^2 + (\,d\beta)^2 + (\,d\,\gamma)^2 \qquad [2']$$

and the following functions can be constructed:

$$1/U' = [(\partial\alpha/\partial x)^2 + (\partial\beta/\partial x)^2 + (\partial\gamma/\partial x)^2]^{1/2}$$

$$1/V' = [(\partial\alpha/\partial y)^2 + (\partial\beta/\partial y)^2 + (\partial\gamma/\partial y)^2]^{1/2} \qquad [3']$$

$$1/W' = [(\partial\alpha/\partial z)^2 + (\partial\beta/\partial z)^2 + (\partial\gamma/\partial z)^2]^{1/2}$$

so that ds'^2 may be expressed as:

$$(ds')^2 = (d\alpha/U')^2 + (d\beta/V')^2 + (d\gamma/W')^2 \qquad [4']$$

If the coordinates (α, β, γ) are merely another set of Cartesian coordinates which are displaced, shrunk, magnified, or rotated with respect to the (x, y, z) set, then the (α, β, γ) set will immediately satisfy the previous five CSI statements, and all of the primed equations above can be immediately applied. This corresponds to the case of a linear transformation linking the (x, y, z) and the (α, β, γ).

In the case where (α, β, γ) are angular or radial coordinates, then statement 1 of the five CSI statements obviously is not true. Whereas for (x, y, z):

one has $\qquad -\infty \leq (x,y,z) \leq +\infty \qquad [5]$

for the radial variable r and the angular variable θ one must write:

$$0 \leq r \leq +\infty \qquad \text{and} \qquad -\pi \leq \theta \leq +\pi \qquad [6]$$

where the principal values have been used for θ. However, if one performs the mappings:

$$r \rightarrow\rightarrow r' \qquad \text{and} \qquad \theta \rightarrow\rightarrow \theta' \qquad [7]$$

such that

59

$$- \infty \leq r' \leq + \infty \quad \text{and} \quad - \infty \leq \theta' \leq + \infty \qquad [8]$$

holds, then the r' and the θ' coordinates will satisfy the five CSI statements and the above transformation relations can be used. Such mappings, therefore, will be necessary for coordinate systems having radial, angular, or other non-Cartesian coordinates.

One must keep track of what is or is not apparent to any CSI. Thus the mappings above from r to r' and from θ to θ' are not apparent to the CSI inhabitant of these coordinates. This CSI will continue to label his coordinates as r and θ. In general, as will be shown below, when these mappings are performed, it will be found convenient to give a particular a numerical subscript to a coordinate, which changes value in each member of the sequence of mappings. This is used in order for the reader to clearly keep track of the mappings. The CSIs who are actually involved in the mapping will always use the same designation for the mapping. This will be clear in the following example.

Take the two dimensional space, see Figure 2a, where x, y and r, θ are as shown and are as seen by what we call an (x, y)-CSI. We show how this two-dimensional space can be transformed into a two-dimensional space where the r, θ axes have the same appearance, in the opinion of an (r, θ)-CSI as the x, y axes have to the (x, y)-CSI, and Figure 2b shows this space according to the (r, θ)-CSI.

It is important to note that this is not merely a switching of coordinate symbols, but rather if the (x, y)-CSI and (r, θ)-CSI were each asked to indicate those axes to which the five CSI statements would apply, each one would indicate the x, y and r, θ coordinate axes, respectively, shown in Figure 2a .

The mapping diagram, Figure 3, gives the overall features of the procedure. Starting at the upper left hand side, the (x, y)-CSI sees points in the plane of Figure 2a as, say, $x_0 + iy_0$, (in upper left hand corner) where the subscript merely permits, us, the observers of this procedure to keep track of the mappings. This same (x, y)-CSI can also use the r, θ characterization so that via a coordinate transformation to the lower left hand corner of Figure 3 one gets $r_0 e^{i\theta_0}$ as a point in the plane according to the (x, y)-CSI. The double headed arrow on the left connecting these designations denotes that this transformation is bi-lateral.

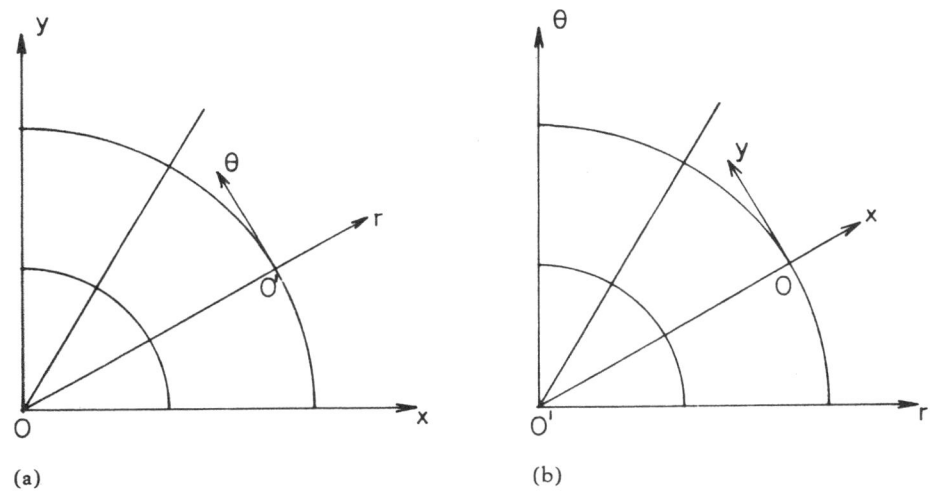

Figure 2. (x.y) and (r,θ) views of a plane. (a) (x.y)-CSI space A: (x_0,y_0) and (r_0,θ_0) spaces. (b) (r,θ)-space. D: (x_3,y_3).

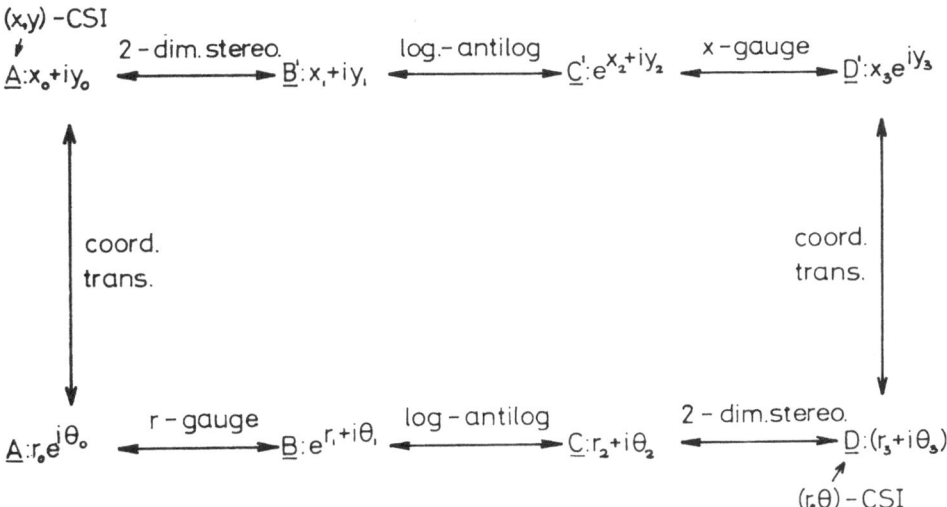

Figure 3. Mapping diagram

Now, we the observers of this procedure, leave the world of the (x,y)-CSI which consists of the perceptions given as A and B at the left hand side of the mapping diagram. We will be moving via mappings in a counterclockwise starting at the lower left corner. Here r_0 is transformed to $e^r 1$, which is therefore, an r-gauge mapping. This converts the range of r from zero to plus infinity into the range from minus infinity to plus infinity. This, according to the previous discussion, is in order to make the r variable conform to a Cartesian coordinate as per the five CSI statements. That point is shown on the mapping diagram as   B: $e^{r_1 + i\theta_1}$ for the convenience of we who are following these mappings.

A logarithmic transformation of the point B is now made so that the point now is called C or C: $r_2 + i\theta_2$.

At this point, if one does not use Principal values for the coordinate θ, then the range of θ can indeed be from $- \infty$ to $+ \infty$. This corresponds to the unfolding of the multileaved Reimannian surface which the original Figure 2a must have contained. The usual choice, however, is to use Principal values with the θ variation between, say, $\pm \pi$, see Figure 4. This range may be converted into variations between plus and minus infinity by performing a projective stereographic mapping, see Figure 5. Here the strip of Figure 4 has been taken into 3-dimensional space and rolled into a right circular cylinder, which is shown in cross section in Figure 5, such that the $+\pi$ and $-\pi$ edges touch each other. This cylinder has been made tangent to the plane D along the line where θ is zero. All of the points on the cylinder are then projected from the point a onto the plane D resulting in the change of range of θ from $\pm \pi$ to $\pm \infty$.

We have thus proceeded, in the Mapping Diagram, Figure 3, from the point A: $x_0 + iy_0$ in the upper left hand corner in a counterclockwise direction to the point D: $r_3 + i\theta_3$ in the lower right hand corner. The beginning and end points are for the route from the (x,y)-CSI to the (r,θ)-CSI point of view. At this point one notes that the forms $x_0 + iy_0$ and $r_3 + i\theta_3$ are identical. Each form is what each CSI will use to identify his native coordinates. The complete symmetry of this situation is apparent. This means that we have no right to consider the upper left hand corner designation of Figure 3 as any more valid than the lower right hand corner designation. We may, therefore, treat (r,θ) starting from D in an identical fashion as we treated (x, y) when we started from A. This will carry us in a counterclockwise direction back around the mapping diagram to our starting point. It thus shows that each CSI's viewpint is reciprocal.

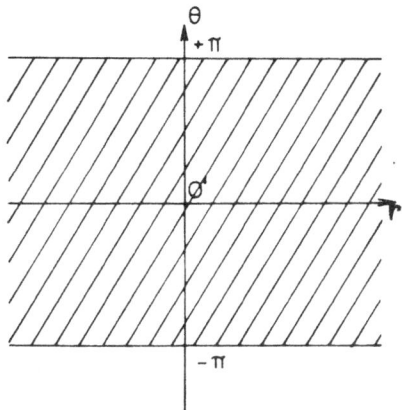

Figure 4. C space. C: (r_2, θ_2) space.

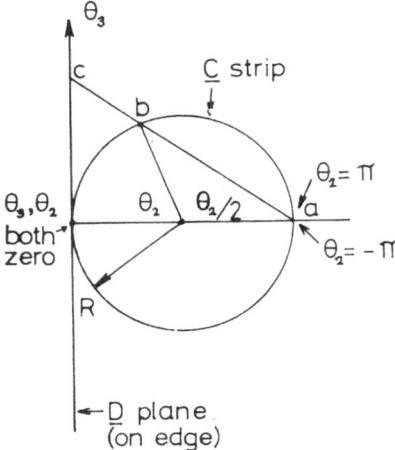

Figure 5. Two dimensional stereographic mapping to D.

The details of the algebraic relations which derive from these sequences of mappings are available[1]. It shows that the terms in the metric matrix are indeed different for different CSIs and that they consist of products of the mapping transformations such as the above, which were necessary to satisfy the 5 CSI statements. They will permit that the quantitative mathematical equations in one CSI viewpoint can be transformed to that of another CSI.

This technique can also be applied to three and four dimensional spaces. Many other considerations of these matters have resulted in a number of suggestions for other applications.[1]

III. TOROID TO SPHERE MAPPINGS

It is well known that mappings between a toroidal (doubly connected) space and a Cartesian (singly connected) space cannot be accomplished, in general, in an isomorphic manner or without singularities. Such mappings might be accomplished, however, on the basis of a more physical interpretation. Other discussions[4] of a physical meaning for different coordinate systems were based on having a space filled with a fluid whose flow corresponds to surfaces of constant coordinate value or to normals to these surfaces.

One thinks first of a physical smoke ring as consisting of fluid flow which gives it its nature. A set of coordinates can be set up which can match this flow. Now one may think of a (spinning) sphere and also set up a coordinate set which matches this flow. Can one be transformed into the other? Figure 6 shows a circular toroid. Figure 7 suggests how such a toroid might be mapped to a sphere.

First, one considers the torus as a smoke ring whose total energy is dependent on its volumetric energy density. The smoke ring can then be imagined as cut open, straightened, and stretched into a right circular cylinder whose cylindrical axis was the toroid's internal ring axis. This discontinuous mapping is followed by a continuous mapping which shrinks the cylindrical axis to a point subject to the condition that the distance from the shrinking axis to the surface of the changing cylinder remains remains constant. This will result in the final sphere at the bottom of Figure 7.

The singularities of the mapping from the toroid to the final sphere would exist first at the ends of the right circular cylinder and finally at the poles of

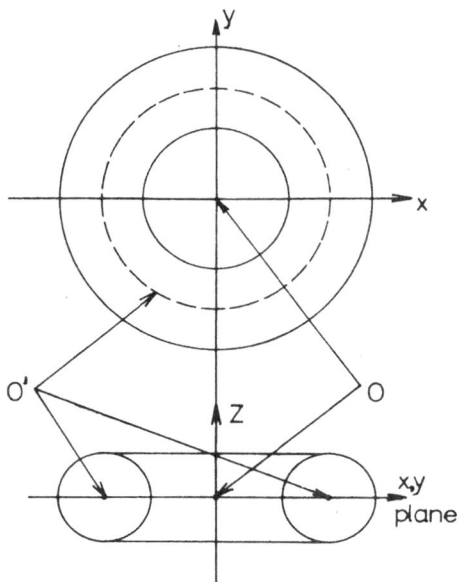

Figure 6. Top and side views of circular toroid.

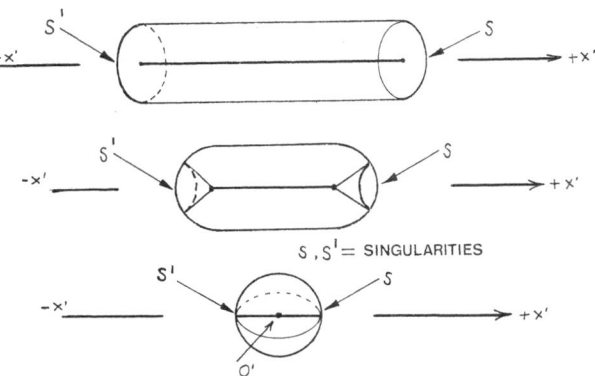

S ,S' = SINGULARITIES

Figure 7. Mapping a cut toroid to a sphere.

the final sphere. It is obvious that vanishingly small amounts of energy are involved at these singularities. In a physical sense, therefore, such a mapping can be consider to be isomorphic or non-singular in that energies and shapes would not change if one neglects the differential volume elements in which the singularities exist.

This mapping carries a ring (the ring axis of the toroid) into a point (the center of the sphere). These objects can, because of the physical isomorphism that is here suggested, serve as models for the 'non-local particle' and the 'canonical particle', respectively. Just as the canonical particle was described as the limit which a spherical object takes as its radius is reduced, so the 'non-local particle' can be the limit which a right circular toroid becomes as its toroidal thickness is reduced toward zero. This ring axis of a right circular toroid is the limiting form of a physical toroid. It is obviously non-local and it is completely within the toroid in a central kind of position, just as is the center of the sphere. The ring axis itself has no volume so that it might be represented by a delta function whose argument is a radial parameter (a - r), where a is the radius of the ring and r is measured from its center.

This suggestion argues that 'non-local particle' can given a meaning somewhat consistent with that of 'canonical particle' in that both are characterized by delta functions, but 2-dimensional and 3-dimensional ones, respectively. Thus, in the rest frames where the sphere and toroid appear, respectively, one has a local or non-local particle, respectively.

IV. NON-LOCAL AND LOCAL PARTICLES IN QM AND EPR

It has been suggested that dipole half wavelength field distributions are quantized and are the hidden variables.[1,4] This object has been named the photex. These are discrete ever expanding toroids of constant thickness like the continuously deforming kidney shaped objects which Hertz first sketched for his illustrations of dipole radiation.

According to this idea there is a sub-plenum or sea of these objects each with an energy deduced from Planck's constant of about 10^{-15} electron volts. These objects are emitted in all electron collisions and can be considered as spreading out in all directions from the point of generation. They serve to explain the electron double slit experiment and could play a role in EPR experiments.

Any EPR experiment must have a central particle generator from which the test particles start their trips to spatially separated detectors. At this point photexi are also generated whose parts will propagate at a radial velocity of c. Their ring axes can pass through two remote points in space at the same time. The energy of each photex is low but numbers of them could arrive simultaneously at remote spatial points. Since these objects are also the minimum energy quantum of interaction[1,4] each detector would need to have a high enough density of parts of these non-local particles landing on them in order to respond.

Their non-local character and the fact that they travel at the speed c obviates many conceptual difficulties which have hampered hidden variable explanations for EPR experiments. Just one conceptual difficulty, the uniqueness of photexi, does require consideration. As explained above, the remote detectors can only intercept a small portion of the circumferential extent of a photex. Since a minimum of energy is necessary to actuate a detector this actuation energy must come from the simultaneous reception and combining of a large number of small bits of these photexi, each with an energy of 10^{-15} electron volts. This means that although photex generation is unique, the local detection of radiation in small photex size chunks of energy is not unique. At one locality in space, one needs, therefore, a high enough density of bits of many photexi so that the local energy density is sufficient for the detection of a single photex. This means that the detected photex is composed of of of the many bits.

REFERENCES

1. Honig, W.M., Chapter VIII, in "The Quantum and Beyond", Philosophical Library, New York (1986), see also "Relativity of the Metric", F.O.P., 7:549-572,(1977).

2. Kohl, R., pp246-265, in "Selected Writings of Helmholtz, Wesleyan University Press, Ohio (1971).

3. Poincare, H., "Science and Hypothesis", Dover Press, New York (1952).

4. Honig, W.M., "A Minimum Photon 'Rest Mass'-Using Plank's Constant and Discontinuous Electromagnetic Waves", F.O.P., 4:367-380, (1974); see also Chapter IV in "The Quantum and Beyond" listed in reference 1 above.

A TWO-FLUID VACUUM, THE PHOTEX,

AND THE PHOTEX-PHOTON CONNECTION

William M. Honig

Western Australian Institute of Technology
Perth, Bentley, 6102,
Western Australia.

ABSTRACT

A general description is given of a fluidic model for physical reality. It conforms with relativistic invariance in an operational manner. Fluid models replace the canonical particle representations. A major feature of this work is a physical representation for the generation of electromagnetic waves which are quantized. This is closely related to the Hertzian dipole wave pictures. The energy of this discrete minimum electromagnetic dipole half wavelength field distribution is approximately 10^{-15} electron volts. Many quantum and other effects can be calculated via this approach but more important a number of experiments are suggested whose confirmation would be needed to bring such ideas into serious contention.

I. INTRODUCTION

A model for vacuum space is described in Section II and assumed to consist of the superposition of a negatively and a positively charged continuous fluid. This is the basic model for a fluid-filled universe. Relative differences in the density of these fluids and in their velocity fields are the components for the construction of physical fluid models for the fundamental particles.[1,2] A number of general considerations relevant to the use of such fluids are also discussed.

69

All this is based on the notion of Builder relativistic invariance which permits that canonical relativistic invariance be retained; in an Einsteinian sense in the cosmological rest frame but in an operational sense in all the other uniform relative motion rest frames.[3,4]

This neo-Lorentzian approach postulates an absolute rest frame (the rest frame of the universe) in which the velocity of light is c. In all other uniformly moving rest frames the time and space contractions are literal and not measurable by its inhabitants. The velocity of light in those frames depend on the frame's absolute velocity. The time and space contractions , however, are such that the measurement of the velocity of light in these frames will also always result in the value c.

By this means the 'non-invariant-appearing' fluid models for the fundamental particles acquire a theoretical and experimental validity consistent with the tenets of Special Relativity. The fluids can then be used to provide models for the fundamental particles including their fields. They should also be the physical cause for all canonical fields (there is no action-at-a-distance).

Section III describes how a spinning droplet electron can be constructed from these fluids and some of its interesting features. This model exhibits the Poincare stress, the anomolous magnetic moment, and results in the calculation of the complete self energy of the electron.[1]

The field of hydrodynamics provides some aid in visualizing how such spinning droplet electrons can generate (or shed) hydrodynamic-like toroidal vortices (similar to smoke rings) upon rectilinear acceleration or deceleration. After formation, these vortices would be similar in appearance to the initial toroidal half wavelength dipole field distributions first sketched by Heinrich Hertz. They would evolve then into the Hertzian continuously radially expanding kidney-shaped patterns which are likewise toroidal. This is described in Section IV together with many other features of these vortices which are introduced below.

The imputation is that electrons can emit such vortices in a discrete and nonlinear way and in a similar fashion to the vortex shedding behaviour which is seen in spinning spherical accelerated bodies immersed in hydrodynamic flow. Upon generation of these vortices from droplet electrons their subsequent time evolution is thought to follow that shown in the sketches of Hertz rather than that of fluid smoke rings, as, for instance, described by Lamb.[5]

This idea results in a number of experimental suggestions. Among these are a physical mechanism for the electron double slit experiment and a loss mechanism in electron "elastic" collisions, both of which can be quantitatively described and tested.

It also provides a physical meaning for h, Planck's constant, as the minimum energy quantum of interaction, independent of electromagnetic frequency and wavelength. It is also the minimum discrete energy which free electrons can emit or absorb. Such vortex generation by an electron leads to the physical derivation of the fine structure constant. This vortex entity has been named the <u>Photex</u> and is a component of or a decomposition of the photon.

The extremely small value of these vortex energies, as deduced from Planck's constant, is approximately 10^{-15} electron volts. They would be generated by electrons undergoing velocity reversals at potential walls or in collisions. The de Broglie relation can be used for the derivation of the thickness (half wavelength) of the generated Photex from the initial velocity of the generating electron.

This also appears to provide a physical interpretation for the commutation relations. Huge numbers of these photexi would provide a universal sea of subquantum energy. It has been shown how these ever expanding and discrete vortices can also be considered as non-local particles. These vortices (photexi) thus are good candidates for the role of hidden variables.

Section V describes how the photex can be related to the canonical photon.

II. FLUID MODEL FOR VACUUM SPACE

A dual (plus and minus) fluid filled universe gives to free space the qualities of a neutral plasma. It is important to note that the presence of these 2 fluids are axiomatic and they are invisible when undisturbed. We assume that neutral free space consists of such superposed quiescent continuous fluids with a charge density equal to that of, say, the classical electron, see Figure 1.

This will be used to as a microscopic model for space and the fundamental particles will be constructed out of the charge densities and velocity fields of parts of each of the vacuum fluids. All fields and interactions between particles will be via fluid interactions; thus none of the canonical "particles" should be necessary.

In terms of these fluids: droplets, bubbles, toroidal vortices, and a host of other discrete fluidic entities can be made to serve as spatially extended fluid models for the fundamental particles and their fields, i.e., electrons, nuclei, photons, electromagnetic field distributions, neutrinos, etc.

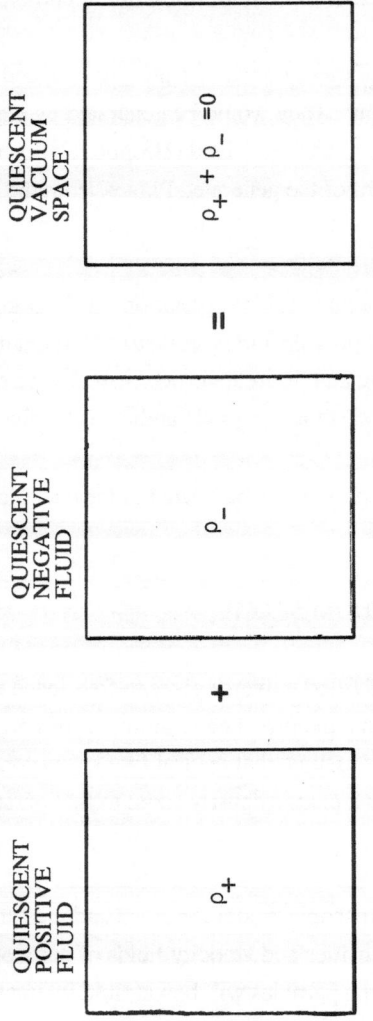

Figure 1. Fluids for Vacuum Space Model.

Such fluidic charge densities and their velocity fields can be described by a current 4-vector, J_4, and an associated potential 4-vector A_4. Both of these 4-vectors have a physical meaning; J_4 gives the charge density and the velocity fields of the fluid and A_4 gives the energy needed to set up these charge densities and their velocity fields. They are naturally expressed via the 4-vector or quaternion space function, A_4, which varies continuously throughout the region of a fluid model.

The local differences in the densities of the 2 fluids will cause a net spatial charge density from which the electric field, $\vec{E}$, can be derived. The velocity fields of these charge densities can provide a mechanism from which the magnetic field, H, can be derived. The electric and magnetic fields now become quantities that derive from the densities and velocities of the fluids.

These fluid models can be consistent with a Correspondance principle. The canonical fields which have been found for various particles should be the correct answer also when the the the appropriate portion of the fluid model of the particle is evaluated. The fluidic 4-vectors above, given for current and potential are not identical to their canonical counterparts. The canonical 4-vector of current will not be the same as the fluid current 4-vector since the former refers to microscopic charged <u>particles</u> from which macroscopic charge densities derive and the latter refers to continuous charge densities even at the microscopic level.

The 4-vectors of potential for the canonical and fluid case should be identical in those regions of space where there is a canonical answer. In regions such as the interior of fundamental particles, however, one is then free to continue the J_4 and A_4 functions in such a way as to explain or take into account the as yet unexplained features of the particles. The quantitative relationships between all these quantities and electric and magnetic fields have been given.[1,6] These fields are the usual electric and magnetic fields and their relationship with the 4-vector A_4 are also the usual ones in that the appropriate gradient and curl of A_4 will give these fields, respectively.

Each of the complete fluid models are constructible using no net charge and no net velocities or rotations. This is because both the internal parts of the particle and its external fields are part of its fluid model characterization. This means that the dual fluid vacuum space model is merely rearranged to make the fluid model for a particle: no net charge or net momentum exists for the complete model of the particle. Only energy is required to construct the model. One can also provide models for the anti-particles of any particle by reversing the unbalanced charge densities and the velocity fields.

According to this scheme, mass becomes a derived quantity, in that it is the energy necessary to construct the fluid model (its self energy) divided by c^2. Even in the case of the neutron, the neutrality of this particle is due to the net neutrality of its fluid core which can be composed of many pairs of equal and opposite charge and flow subvolumes. This will still require that energy be expended in the assembly of the model and it will, therefore, have a mass.

Since mass is a derived quality from the self energy of the fluid model, gravitation should have a mechanism in terms of these charged fluids. Bjerknes, Korn, and Prandtl have discussed similar fluidic mechanisms.[7,8,9,10] The discussion of the details of particle fluid models makes it plausible that the surface between oppositely charged regions can be the seat for a kind of plasma oscillation which could serve as the generator such a mechanism.[1]

III. FLUID ELECTRON MODEL

The electron consists of a negative charged spherical droplet of density equal to that of the original quiescent negative fluid. The positively charged fluid which was superposed on this in neutral vacuum space is distributed outside the droplet with a density that decreases in the radial direction. The droplet spins as in a rigid body (with curl) with an equatorial velocity of c, or rather a small amount less than this of c-(delta). This equatorial velocity will remain the same during the droplet's shape change with velocity which wiill be described. The positive charge flow outside the droplet occurs in a curl-free manner. See Figure 2, the upper right corner for the charge density profile and the lower left and lower right corner for the velocity profile.

Since these fluids replace electric and magnetic fields it is possible to deduce the necessary density and velocity variation needed in the fluids to give the correct value for the electrostatic field of the electron. A null value for the magnetic field external to the central droplet volume can also be derived in such a way that the magnetic field due to the droplet internal rotation provides stability for this model. It provides a radial force equal to that of the Poincare stress which can serve to keep the adjacent positive and negative charge densities separate, see Figure 2.

A detailed accounting of the total energy necessary to set up this model results in the full agreement of this fluidic self energy with the energy equivalent of the rest mass of the electron. At the same time it gives a value for the magnetic moment of the electron equal to its anomolous canonical value. The details of this have been given elsewhere.[1]

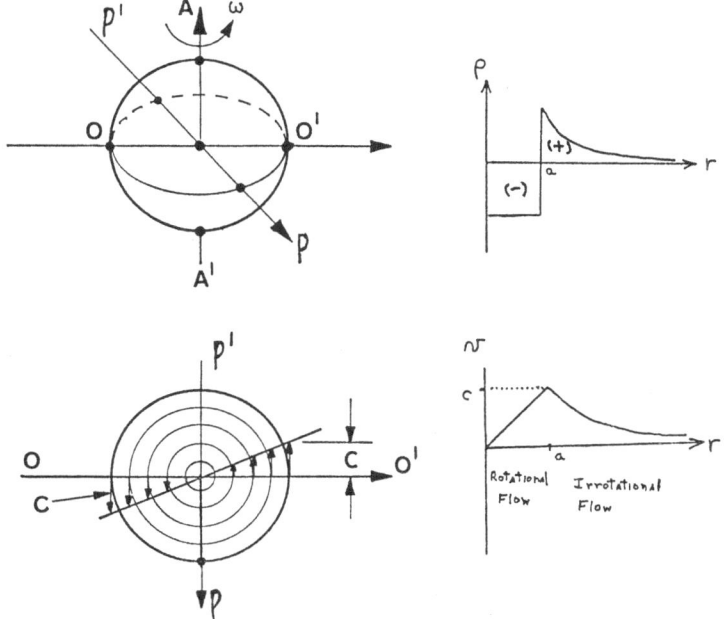

Figure 2. Flow Patterns in Electron Fluid Model.

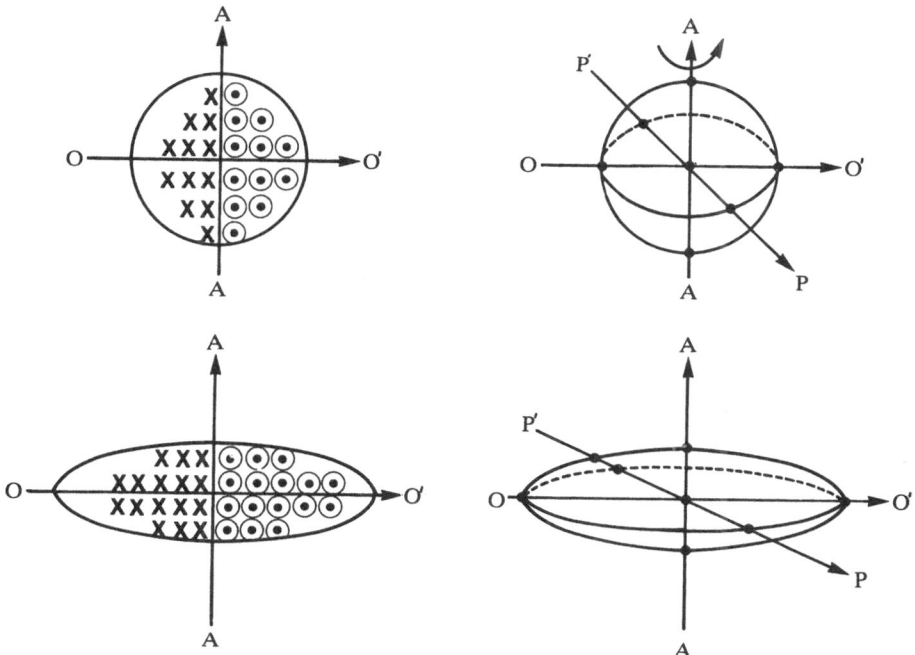

Figure 3. Electron Fluid Model Change with Velocity.

It should be emphasized that the great freedom which the use of fluids provides in the construction of arbitrary fluid densities and velocity fields is what makes it possible to do this. Until such time as these theoretical phenomena result in confirmed experiments which demonstrate their practical physical effects they must be viewed as conjectures.

One must remember that it is possible to speak of this model in this literal Newtonian fashion but still consider it as a relativistically invariant model. In any rest frame this electron droplet should be deforming with velocity. Such a constant volume droplet in motion would behave in similar way to that described first by Langevin.[11] The spin axis of the droplet will be parallel to the velocity vector. The spherical droplet will deform into a spheroid with increasing velocity so that it is shortened in the velocity direction and lengthened in all lateral directions.

The longitudinal shortening follows the Lorentz distance shortening formula so that because of the constant volume condition the electron droplet lateral diameter increases as the inverse square root of its shortening. Even though all of this is happening it would not be evident to an observer of this moving electron except for the increase in the measured mass of this electron. Hydrodynamically or rather fluidically, the reason for the mass increase is due to the greater oblateness of the spherical droplet; the increased kinetic energy of the electron is stored in the oblateness itself and the greater cross section of the droplet makes it harder to push through space, which we call its increased mass.

Figure 3 depicts this by showing cross sections and 3-D sketches of the spherical and oblate droplet which are in the upper and lower parts of the figure, respectively. The crosses and circles are merely to show the fluid flow inside the droplet: crosses, into the diagram and circles, out of the diagram.

If the electron is stationary in a rest frame, the motion of the rest frame will cause a deformation of the electron droplet (and everything in the rest frame). This, however, will not be measurable to the observers in that rest frame. In this way operational validity for invariance is retained while carrying on the analysis of this non-invariant appearing electron droplet model.

IV. THE SHEDDING OF A VORTEX BY THE ELECTRON DROPLET

ELECTROMAGNETIC DIPOLE WAVE GENERATION-THE PHOTEX

A sufficiently simple model for the generation of electromagnetic (EM) waves by charged particles (electrons) must first be given. For this reason, we restrict considerations here to electrons in rectilinear motion including both acceleration and deceleration, somewhat like the vibratory motion which is used as a condition for deriving dipole radiation.

The discussion up to this section has been a prelude to this discussion of how spinning droplets can generate dipole electromagnetic radiation upon acceleration and deceleration.

No simpler example exists for the consideration of electromagnetic wave generation than that of an electron in such rectilinear motion. The sketches first made by Heinrich Hertz of the Electric and Magnetic field distributions found about a charged particle in rectilinear vibratory motion are our starting point, see Figure 4a.

The great importance of this model must first be pointed out. Physics possesses no simpler example for the generation of EM fields with the advantage of a clear associated picture (Figure 4a) which illustrates the field details.

We assume an electron at O undergoing acceleration and deceleration in the z direction. The axially symmetric E fields are shown for the right hemisphere. The wavelength is shown larger than the classical electron radius. The fields sort of repeat for every half cycle but with a reversal in the direction of the E lines.

That is, each half cycle field distribution appears to be isomorphic to its position at subsequent times since each such distribution appears to evolve in time as a continuous 'flow' of the field lines. Initially, this appears to exist as a toroid of circular cross section which girdles the electron equator and then moves out in the radial direction, becoming deformed into the well known kidney shaped patterns shown in the figure, which Hertz first sketched.

If one compares this picture with the mathematical solution for dipole EM fields, as given in standard texts,[12,13] a number of striking contrasts
are evident. The pictured toroidal fields are not directly evident from the dipole field equations. The equations, which are complicated, include a number of terms with different names (i.e., radiation field, induction field, etc.,). The continuous deformation of the field distributions with time are not evident.

The longitudinal E fields, which are initally evident in the half cycle field distributions close to the origin decrease as the field distributions 'flow' and move out to greater radial distances. Stratton mentions his qualms about using the Poynting vector at infinite r and imputing those resulting solutions to dipole EM fields at finite r, which implies the 'flow' of field distribution lines.

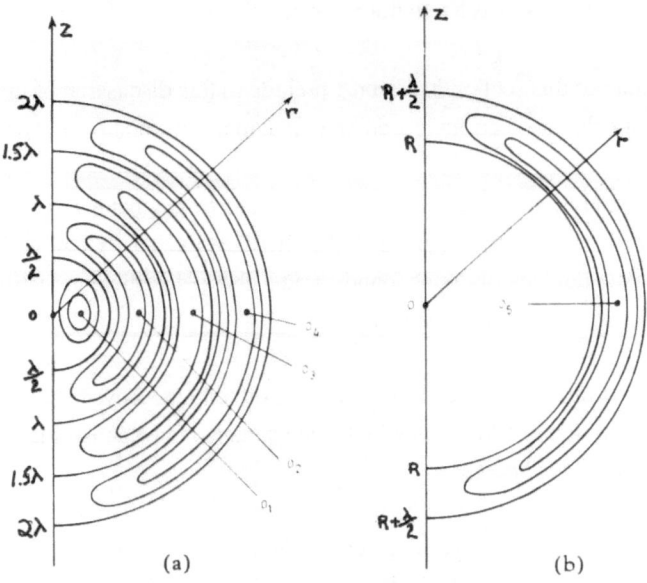

Figure 4. Hertzian Wave Pictures. (a) Hertzian Dipole Radiation.
(b) Single Photex Field.

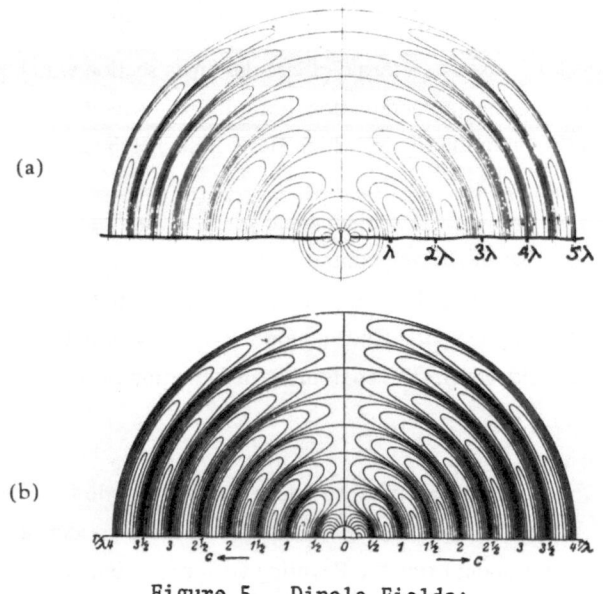

Figure 5. Dipole Fields:
(a) Near and Far,
(b) Far Only.

The diagrams in Figure 5a,b show the dipole field distributions with and without the near fields included. When the near fields are included it can be seen that the half wavelength of the fields is reached only after these fields have moved a few wavelengths from the origin. This also appears to show how physical and how fluidic is the time and space evolution of these fields. We continue the discussion in terms of the far field pictures of Fig. 5b but the near field pictures should be kept in mind.

Another advantage of considering these dipole field distributions is that it contains the "entire" field. Thus the usual consideration of electromagnetic energy in the "simple" form of plane waves is simply not a realistic model that comes from physical reality whereas the dipole field distributions are.

These remarks are part of the basis for our view that the Hertzian pictures are good guides for the construction of fluidic physical models for the generation and evolution of EM dipole fields. This dipole field case is also sufficiently general, because it has been shown that arbitrary EM fields can be built up from the superposition of dipole EM fields.

There are numerous examples from the field of hydrodynamics where rigid and non-rigid spherical bodies in a stationary or uniformly flowing fluid are given a swift 'kick'.[5] They respond by moving in the direction of the 'kick', but with the simultaneous non-linear generation of a hydrodynamic vortex (which is something like a 'smoke ring'). This usually is called the shedding of a vortex. It is well known in hydrodynamics that todoidal vortices have finite energies (which is not the case for even finite lengths of straight vortices).

Using the Hertzian pictures as a guide, it appears as if a charged droplet were being kicked back and forth in the z direction of Figures 4 and 5a and a contiguous number of vortices of alternately opposite rotation are being formed in the positive charge surrounding the electron droplet, which was described in the previous section.

Presumably, because of the nature of the dual fluid model, each vortex which is generated at the equator of the electron droplet moves out radially at the speed c becoming deformed into the kidney shaped field distributions. This is contrary to the behavior of hydrodynamic vortices which are either static or slowly increase in diameter (they also behave in an interesting manner when interacting with other vortices).[2,5]

We suggest that each such fluidic vortex correspond to a half wavelength dipole field distribution. Further, each such field distribution, should be considered as a separate, unique, discrete entity which may not be subdivided into anything smaller. Figure 4b shows this single vortex entity. A continous sequence of these is what the Hertzian pictures show. On the other hand, one can now, using this model, consider a finite wave train as consisting of an arbitrary number of contiguous vortices.

The process for the formation of these vortices is inherently non-linear and, once formed, each of them is no longer controlled by the generating electron droplet. Numerous physical examples illustrate this in hydrodynamics, i.e., eddies, whorls, smoke rings, etc.

This toroidal vortex is extended in space. It is thus obviously non-local, discrete, but within its own boundaries its charge densities and velocity fields are continuous. Thus the contiguous assembly of these vortices, however, would provide a continuous EM dipole field model as in Figures 4a, 5a,b.

Returning now to the single fluid hydrodyamic case, the energy of the toroidal vortex shed by a spherical body immersed in hydrodynamic flow, when it is 'kicked', is obviously finite. Referring now to the dual fluid electron droplet model, the EM vortex shed when the droplet is 'kicked' is likewise assumed to have a finite energy. In addition, however, this vortex energy is assumed to be a fixed amount. This condition is meant to hold as the duration of the 'kick' is varied. In this way all vortices that are generated will have the same total energy but with different cross section diameters or half wavelength dimensions. The volumetric energy density will, of course, decrease for increasing wavelength.

This means that if the electron droplet were 'kicked' more slowly the same total energy would be delivered to the vortex per 'kick' and its wavelength will increase (its frequency will decrease); a rapid 'kick' will do the reverse.

How are such 'kicks' to be administered? Consider the case of a tennis ball moving toward a wall where it decelerates to zero, deforms slightly at the wall when it reaches zero velocity, and then accelerates in the reverse direction to a velocity equal to its incident velocity (if there were no losses in this process).

It is here suggested that the electron droplet acts in a similar manner when confronted by a potential wall. Now, however, there will a vortex generated upon the droplet's deceleration to zero velocity at the wall and another vortex of opposite rotation, upon the droplet's acceleration in the opposite direction to almost its initial velocity. This is the model for what is meant by 'kick'. Figure 6 shows how would occur at a potential wall or boundary. The generation of these

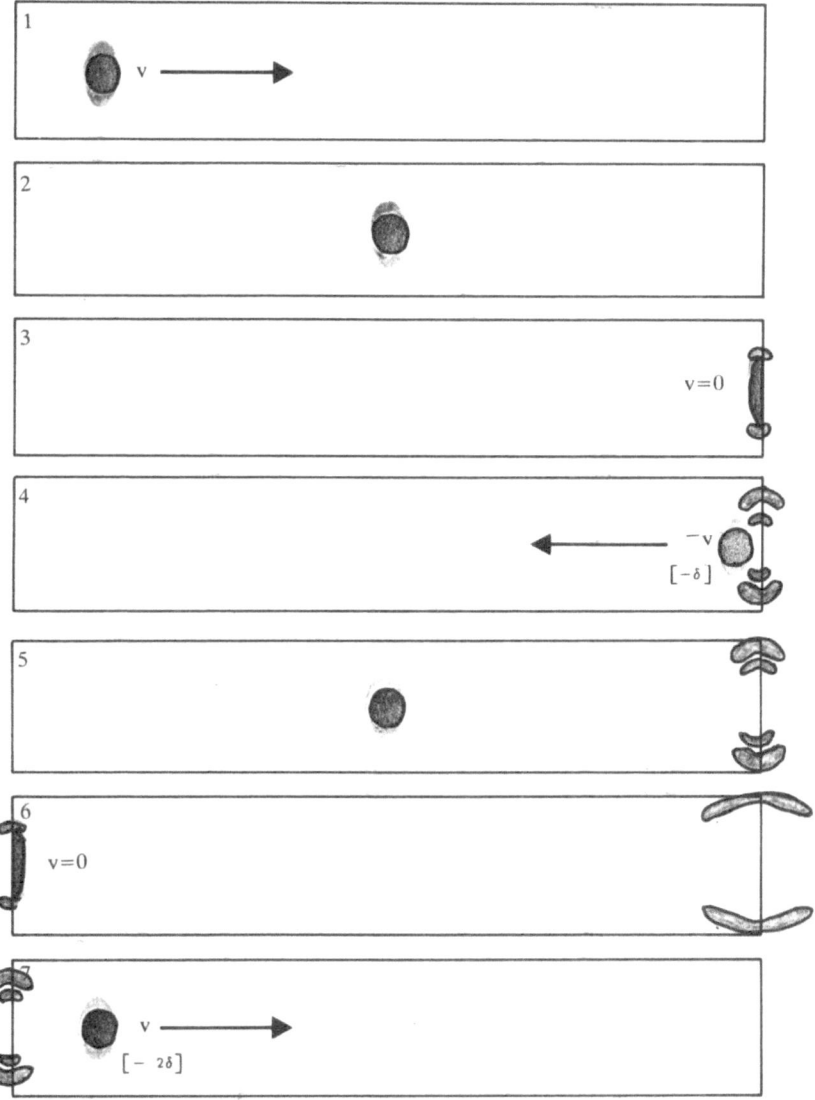

Figure 6. Electron Inastic Collisons, Photex Pair Generation.

81

discrete vortex pairs (each of which will be given the name PHOTEX) is, we believe, a physical phenomenon whose existence has not hitherto been suggested (see below).

If now one considers this happening at a set of different droplet velocities the meaning for the duration of that 'kick' is evident. It corresponds to the change in the droplet deformation time during which
the toroidal vortex is being generated. At the different droplet velocities vortices all with the same energy of formation but with different diameters are formed.

The usefulness for the de Broglie relations for this suggested phenomenon lies in the fact that the generating electron motion,
(its velocity) can be related to the diameter of the toroidal vortex and and hence to its half wavelength. Thus via:

$$p = h/\lambda \qquad \text{or} \quad p = \hbar k \qquad\qquad [\,1\,]$$

one connects p (mv), the momentum with the wavelength, λ.

Since a vortex corresponds to a half wavelength field distribution (and is also a half cycle phenomenon) one may relate the number of contiguous vortices passing a point in space in a second to the meaning of wavelength and frequency as follows:

lambda/2 = one vortex thickness or length,

wavelength = lambda = 2 vortex lengths = length of one vortex pair,

half cycle = one vortex, so that: one cycle = two vortices,

f = frequency = cycles per second,

2f = no. of half-cycles per second = no. of vortices per second.

The definition is made:

This vortex is to be called the Photex (plural is Photexi).

The quantum energy relation is

$$E = hf \qquad\qquad [2]$$

where E is in ergs, h in ergs per cycle per second, and f in cycles per second. Using the above ideas, f is not the frequency of a continuous wave train but represents a finite wavetrain consisting of a number of contiguous photexi. In this way we get:

$$E = (h/2)(2f) \qquad\qquad [3]$$

where E is in ergs, 2f is photex pairs per second, and h/2 is ergs per photex per second.

The major suggestion in this paper that h/2 be identified with the minimum and discrete energy that the electron droplet generates in each of its half wavelength dipole field vortices.

Both of the above equations would apply to an observer in an ordinary physical rest frame (PRF). If one defines a rest frame for the generated photex, this will be an electromagnetic rest frame (EMRF). Such a rest frame is only conceptual since it travels at the speed of light, that is the vortex ring is moving radially outward at c. One can, however, make the following interesting and dramatic correspondence between these two frames:

$$[\text{one photex in an EMRF}] \longleftrightarrow [\text{one photex/second in a PRF}] \qquad\qquad [4]$$

It then follows that:

$$[\text{h (ergs/cycle/second)}]_{PRF} \longleftrightarrow [\text{h (ergs/2 photexi/second)}]_{PRF}$$

$$[\quad " \qquad\quad " \qquad "]_{PRF} \longleftrightarrow [\text{h (ergs/2photexi)}]_{EMRF} \qquad\qquad [5]$$

This means that the meaning for h in the last expression above no longer contains time and its numerical value corresponds to 2 photexi directly. Details of this have been discussed[1]. We

83

note here, however, that the correct grammatical designation for electromagnetic waves must always refer to them as cycles or half cycles per second. This is because we are always in a PRF and can only experience electromagnetic waves as dynamic phenomema which travel at the velocity c. In the conceptual (but useful) EMRF, that which occurs in a PRF as a cycle per second must appear as a cycle to the inhabitant of the EMRF.

Converting h into two convenient energy measures for the photex we get:

Photex energy = 2 x 10^{-15} electron volts, approximately, [6]
            or .4 x 10^{-47} grams.

This is a minimum energy or mass based on the above discussion since the energy per photex is a minimum. Much additional discussion of this is contained in reference 1 and its references.

These ideas can be connected with the Proca relations. The relation:

$$(\omega/c)^2 = k^2$$ [7]

for EM waves was extended by Proca to include photons with a rest mass, m_x, with the modified relation:

$$(\omega/c)^2 = k^2 + m_x{}^2$$ [8]

The above equation is the dispersion relation that allows for the possibility that the photon may have a finite though small mass.

According to the ideas presented here the above dispersion equation can be written three ways:

$$(\omega/c)^2 = k^2$$ [9a]

$$= \sum_i^m m_i{}^2$$ [9b]

$$= k_j{}^2 + m_x{}^2$$ [9c]

This could now apply to finite dipole wave trains. In this case, k is the wave number for such a train (Eq. 9a above), m_i refers to the n photexi which are being summed, thus making up the full wave train (Eq. 9b above), and finally k_j is the original wave train minus one photex (or half cycle) and m_x is the energy (mass) of one photex (Eq. 9c above).

Many of the experimental methods which have been described for the measurement of photon mass are consistent with the ideas which have been described here[14]. This is because most of these methods depend upon a Proca-like equation such as Equation (9) above. These equations may equally well represent the photex which has been described with the feature that the photex always has the velocity c and need not travel at less than c as others have assumed for a photon with rest mass. In addition, the estimates of photon rest mass lie close to that of the photex rest mass of about 10^{-47} gms given above.

Figure 7 shows an experiment whereby if a moving 50 electron volt electron makes many velocity reversals it should lose about one electron volt of energy for every approximately 10^{15} collisions.[1] Solitary electron experiments as have been described by H. Dehmelt may soon permit such experiments to be done.

Figure 8 shows the electron double slit experiment and the presence of the photex pairs first seen in Figure 6. Their presence should be capable of deflecting other electrons passing through the slits so that the usual diffraction pattern forms. This would also be so even for solitary electrons because of the presence of the image electron and image photexi in the conducting face of the slit structures. Modifying the structures by varying the resistivity of the materials or even using insulators should make calculable changes in the final diffraction pattern.[1]

Figure 6 may also be used to provide a physical meaning for the commutation relation:

$$xp - px = ih \qquad [10\ ]$$

where some constants have been supressed. As shown in Figure 6, one can consider a electron incident upon a wall with say, a momentum p at a point x which can be the horizontal position in sketch 2. After it has rebounded from the wall and again reaches the same postion x in sketch 5, it will have generated a photex pair and thus will have lost an energy of about 2 times 10^{-15} electron volts. The term xp in the above equation can refer to sketch 2 whereas the term px can refer to sketch 5. Equation 6 will thus stand for the fact that the px product difference between the sketch 2 and sketch 5 situations corresponds to the loss by the electron of the

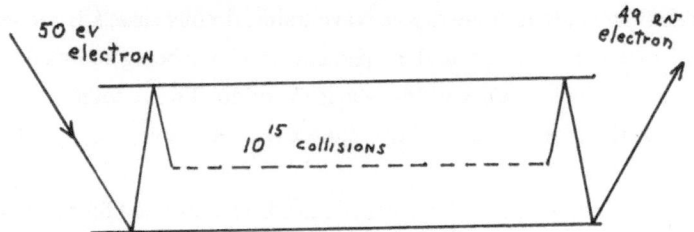

Figure 7.   Inelastic Electron Collision Experiment.

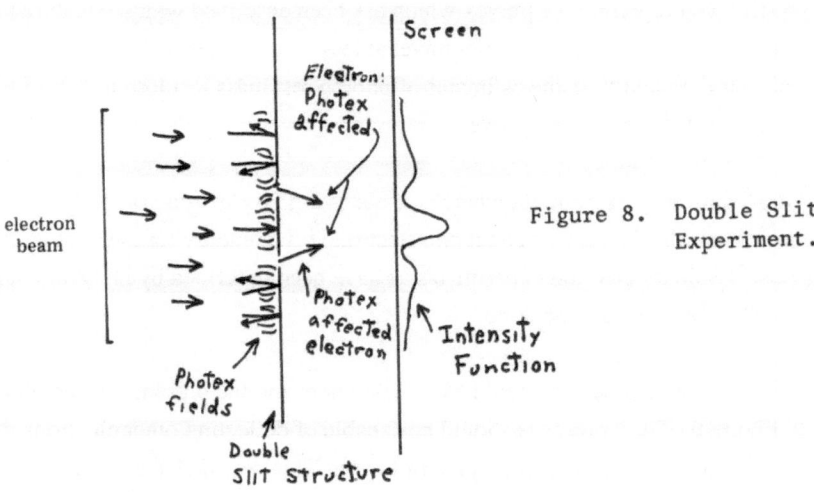

Figure 8.   Double Slit
Experiment.

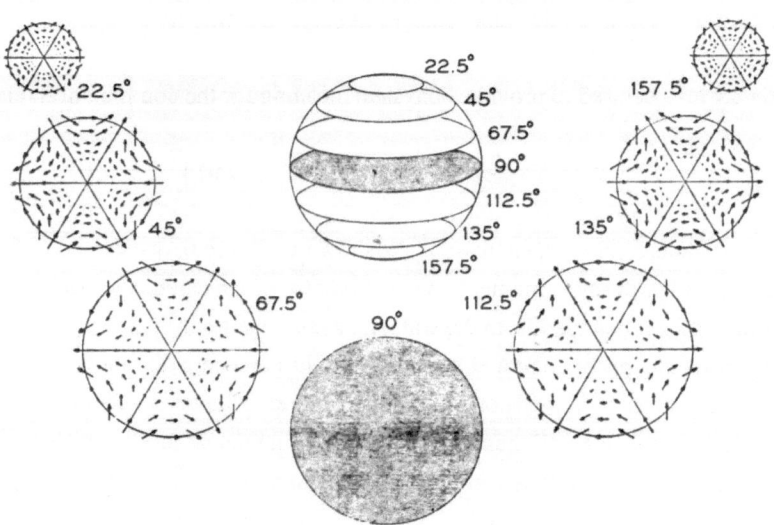

Figure 9.   Sample Spherical Modes.

above small amount of energy and although x is the same in both cases, p is only the same to within the small amount of 2 times 10^{-15} electron volts. Thus although each of the p terms in the above equation are sensibly equal their small difference is expressed on the right as h. The meaning for i is discussed elsewhere in this volume. See 'Logical Meanings in Quantum Mechanics for Imaginary and Transfinite Numbers and Exponentials'.

V. THE PHOTON-PHOTEX RELATIONSHIP

The canonical meaning for the photon is, of course, not the same as that of the photex which has been presented here. The photon energy corresponds to n photexi where n is the number of these photexi which can be put into a contiguous spatial relationship which can pass a point in space in one second. This, in fact, is the meaning of the canonical Planck equation, $E = hf$. The one half cycle decomposition of the photon energy which is the photex still does not explain how a photon can at one instant and at one point in space possess an energy hf. If one considers an electron with a kinetic energy E, it can be seen that it can emit a photex or photexi pairs which have the requisite wavelength and frequency behavior.

They can, moreover, generate huge numbers of these photexi which are almost identical in wavelength and frequency, since the small energy of 10^{-15} is very small compared to sensible electron energies. Since the time resolution of photon adsorption time is not less than approximately 10^{-8} seconds it may be possible for huge numbers of photexi to be generated within such times, but each photex would still have to contain many times the energy of the photex which has been postulated here.

On the other hand, in an atomic emission as distinct from the free electron fluid droplet behavior which has been described, single field distributions having energies of hf could be possible. The absorption of such a spatially extended field distribution by a much smaller electron droplet or atomic fluid model may be fluidically picturizable. This picture can be gained by considering the pictures in Figures 4 and 5 to be played in reverse to the time sequence which gave us the evolving patterns of those figures.

The photex exhibits a non-locality which would be necessary if it is to be considered for the role of hidden varaible. This is discussed elsewhere in this volume. See 'Relative Metrics and Physical Models for Non-Local Particles'.

We close with a sketch, Figure 9, of spherical charge and flow subvolumes taken from work in spin waves in ferrites. It is may prove useful to consider such fluid charge and flow regimes as the kind of fluid models which will have to be constructed for the heavier particles such as the proton, neutron, etc. In this case, higher order mode numbers could correspond to isotopic spin, strangeness, etc.

REFERENCES

1. Honig, W. M., Chapters 1, 2, and 3, in "The Quantum and Beyond", Pub. by Philosophical Library, New York, (1986) see also A Minimum 'Photon Rest Mass' - Using Planck's Constant and Discontinuous Electromagnetic Waves, Found. Physics, 4, 367-380, (1974).

2. Honig, W. M., in "Toroidal Electromagnetics", Ph.D. thesis, (1978), Catalog Number: 78-17,466, University Microfilms, Inc., Ann Arbor, Michigan, 48106, U.S.A.

3. Builder, G., Ether and Relativity, Aust. J. Phys., 11, 279-297 (1958) and The Constancy of the Velocity of Light, Aust. J. Phys., 11, 457-480 (1958).

4. Prokhovnik, S. J., An Introduction to the Neo-Lorentzian Relativity of Builder, Spec. Sci. Tech., 2, 225-229 (1979) and also in "The Logic of Special Relativity", Cambridge University Press (1967).

5. Lamb, H., in "Hydrodynamics", Dover, New York (1945).

6. Honig, W.M., Quaternionic Electromagnetic Wave Equation and a Dual Charge-Filled Space, Lett. Nuovo Cim., 19, 137-140 (1977).

7. Bjerknes, V. and C.A., in "Vorlesungen Uber Hydrodynamische Fernkrafte" and in "Die Kraftfelder" Braunschweig (1905) and (1909).

8. Korn, A., in "Eine Theorie der Gravitation und der Electrische Erscheinungen", E. Dummler, Berlin (1898).

9. Prandtl, E., pp.344, in "Essentials of Fluid Mechanics", Blackie & Son (1952).

10. Whittaker, E., see Index under Bjerknes and Korn in "History of the Theories of the Aether and Electricity", Nelson & Sons (1951).

11. Miller, A.I., A Study of Poincare's 'Sur la Dynamique de l'Electron', Arch. Hist. Exact Sciences, 10, 207-328 (1973); see also its extensive list of references.

12. Stratton, J.A., in "Electromagnetic Theory" McGraw-Hill, New York (1965).

13. Jackson, J.D., in Classical Electrodynamics, J. Wiley, York (1979).

14. Goldhaber, A. and Nieto, M., Rev. Mod. Phys., 43, 277 (1971) and its references.

DISCUSSION I.

Compiled and interpreted by D.W. Kraft and E. Panarella

Hunter. I wonder if Honig is aware that J.J. Thomson had a similar model of the photon from about the mid-1920s. We discovered some papers where he has an oscillating dipole and toroids very much like yours together with related material from Bjerknes and Korn.

Honig. Yes, and there was another Thomson about 1850 who had something similar. I reference it in my paper.

De Martini. Can you relate your paper to the kind of calculation that you have in systems like the electron-positron sea in which you can have effects of vortices and de Broglie waves?

Honig. I do not assume the existence of de Broglie waves; I'm assuming everything can be constructed from my dual fluids.

De Martini. Can fluid-dynamic model be understood in terms of a fluid like the vacuum field, like the Dirac ether, or something that is made up of electron-positron pairs?

Honig. It might be. My dual fluids are continuous at microscopic levels with respect to an electron itself. I'm making electrons out of these fluids and these fluids are continuous even down to that level. I'm hoping to explain everything with just these fluids.

De Martini. Can you explain photons in this way?

Honig. I did not reach the photons in this paper. I'll try to reach that on Wednesday, but there's obviously a connection. All I'm talking about here is energy per half cycle or per half wavelength which I call the photex. There is an intimate connection between them and I hope to explain at least a little of how a photon is related to the photon. In some cases I say that the photon is identical with the photex. What I mean is, when people talk about a photon mass of 10^{-47} grams, corresponding to 10^{-15} electron volts, that could be referring to my photex, but when one talks about photon excitations; photons of 3, 4, or 5 electron volts, they're not talking about a little 10^{-15} electron volt toroid. Remember I'm considering isolated electrons here; what you're talking about are atomic excitations. In that case a lot more happens because, as I have not said yet, my electron model is, a droplet when it is in isolation and is moving around. Such a fluid model electron to me appears to deform into some sort of open or closed bubble so that it encloses the nucleus, and as an atomic electron is some sort of spherical or deformed bubble which, when you strip it from the nucleus, coalesces then back into a droplet. Everything is fluid.

De Martini. I want to comment on Selleri's paper. I think that many of Selleri's proposals can be understood better from other pictures. For example, the spontaneous emission change in lifetime or intensity in front of a mirror, which is generally understood as an effect of reflection of vacuum fluctuations in front of the mirror. An alternative picture is the self-reaction of the atom. But this theory is very well understood and now everybody attributes this effect to the vacuum fluctuations. I worked out the theory and the experimental result becomes exactly like the quantum electrodynamics prediction. A second example is the Bohm-Aharonov effect. The best way to think of this effect is... just like... Loudon's people have seen this effect. It is a very interesting evidence of a Yang-Mills field of the photon. It is related to the very fundamental gauge theory for a particle of mass equal to zero. This is a common interpretation, so I don't know what the de Broglie waves can do. With respect to amplification of photons, a few authors that treat it, for instance it is treated in Loudon's book, do so very properly with the Glauber representation. If you use α-state representation, you don't work these problems with z and ϕ. You treat them with the complex representation and the experimental results agree perfectly with the theory, so the philosophy of what I'm saying is that there are many experiments that can be anticipated. But, one has to be very careful to work out the theory very well before starting them. For example, the experiment of Gozzini cannot be done because it is not ever possible with photons to single out a pure state. Usually, when one works with photons, one can never think to have only one photon. You have a chaotic or a coherent superposition of α-states or..., but never a pure state. One has to be very careful before starting with "gedanken" experiments that cannot be done.

Selleri. In answer to your first two points, I would stress that what I tried to do was to give a unified picture of different phenomena, instead of looking for a different picture for each phenomenon. As far as the third point goes, I disagree with you because not only I have followed what Gozzini has done and written, but the study has been repeated two other times; the first by Giovanelli of Milano and the other time by De Giorgio of Pavia. The three people who have made a study of the experiment think that the experiment can be done.

De Martini. But De Giorgio is a theoretician who assumes at the outset to have a single photon. This cannot be. One can never have a pure state. One cannot do the experiment.

Selleri. De Giorgio has done the calculations with a coherent pulse.

De Martini. I do not find in any experiment that has been done in optics, a system in which you can prepare only a single photon, and be sure of that. There is nothing against it, but, so far, it is an experiment beyond our capabilities.

Panarella. As an experimentalist, I have to agree with De Martini, that there are problems in creating single photons or single particles because we usually create many of them. But there are also problems in detecting them. We know very well that no known detector has 100% quantum efficiency. There's no way we can escape that reality. Maybe sometimes under special design specifications we can achieve up to 40% quantum efficiency, but certainly the experiments that have been done so far never dealt with such large quantum efficiency of detection. So then this brings about really two things simultaneously: we can't really make single photons or single particles and we can't really detect single particles; certainly we detect one particle, but there are many others of them around somewhere, and we pick up only one of them. This brings me to what I have been thinking for a number of years now is that my fellow experimentalists have deceived the theoreticians by injecting the notion that we are dealing with one particle at a time in our apparatus. They state that, with high probability, they are dealing with one photon or one particle. Well, I think that they have stretched their imagination. What they have in fact done is merely to hypothesize that they are dealing with a single particle at a time, and they should have said as much. So I have to agree from the reality point of view that it is very difficult to create single particles, without many others around, and it is even more difficult to detect single particles that are not isolated from the others.

Selleri. I would like to make another comment on this single photon business. First of all, what I've been discussing here was not under the assumption that the experiment had to be made with single photons, but I was talking about coherent pulses. However, I think it is also not true that it is impossible to approach a situation of single photons, which is, of course, a different question, but also an interesting question in itself. Because, you see, if we produce laser pulses, one after the other, and if we lower very much the average number of photons, then

there is that formula which I wrote for the probability $\langle n \rangle^n / n!$, which says that, if I take a very low value of $\langle n \rangle$, say 0.1, then I have a very high probability of having zero photons; of course that I would never detect. So the dominating contribution of what is left is from single photons. Of course I always have some impurity from 2, 3, 4 photons, but the impurity can be reduced as much as one wishes by taking a small enough value of $\langle n \rangle$. Since the disagreement between the two approaches to the stimulated emission coincidence is as large as 20%, one does not need to go to the limit of perfection where you have only one photon and nothing else; it is enough to go reasonably close to it, like having a detector registering 90% of the times single photons. So, De Martini, you require perfection - of course perfection's never possible - I think that is an ideological touch to say that we want perfection and then it is not possible, so we are wrong.

De Martini. If you want to start from laser pulses you have to be consistent with your picture and you have to use the correct formulations. This is not a correct formulation; it is not a complete formulation. You have to discuss not the P(n) but the $P(\alpha)$ distribution, where α is the Glauber state, the coherent state. If you use this formula, you come out with the right answer. As a matter of fact, if you go to 0.1 average number of photons in a coherent pulse, one loses coherence completely because one photon does not belong to a coherent or an incoherent state, it is just incoherent by itself. It is a coherent effect by itself. What I feel is that you are using an old-fashioned theory and not the good one to try to criticise. In order to see what Panarella was saying, I remind you that photons are Bosons, so they show a very distinct Bose condensation when they are emitted. So they tend to be emitted not only one, and another, and another, but rather in bunches. Certainly when you go to the limit of incoherency, the detection problem is very important, for you can never have 100% detection quantum efficiency. Now, you can keep speaking about the possibility of de Broglie waves, but I think you should look for something different. For example, one experiment that I am going to do is to look at de Broglie waves in a transversal way, rather than in a longitudinal way. This would be much more interesting, because it would eliminate all this problem of timing, the coherente time, etc. So this should be another possibility.

Aspden. I want to comment on Honig's paper. I've got considerable difficulty in visualizing a neutral mix of positive charges and negative charges in a vacuum which is wholly fluid and which doesn't contain any energy in its undisturbed state. Now if we're going to have a fluid model of this kind, and if we're also going to have positive and negative charges mixed in, I think the model could be simplified if we took the view that positive charge is fluid which is expanding and negative charge is fluid which is contracting; they bounce at any moment, and in fact, there is an oscillation but the oscillations that are in phase constitute charges of like polarity at any moment and oscillations which are in anti-phase would relate to unlike charges. Now, that seems to be a way of even explaining the nature of electric charge, which is an advantage.

The other factor about it all is this: We've heard today about empty waves - waves which somehow travel through the vacuum, but carry no energy. Well, we all know that there is a fundamental difference between electromagnetic waves and waves on water; that is, they're so different, they don't compare. But at the very deep level where we're talking about fluids in the vacuum, it might be possible to think more in terms of the type of wave you get on water. Now when you drop a drop of water onto a surface of water, it then generates a wave. Waves can travel and the energy in the system can be disturbed. I'm saying that maybe these ghost-like waves are actually oscillations of energy that is actually present in the vacuum in its undisturbed state. There's a kind of regular pattern here that is set up. So I think there might be something in developing models of this kind, providing we're looking to solve some of these problems that we are running into now.

Honig. I think that you're describing some of the ideas you'll be explaining. I should mention that that happens to correspond in a very analogous fashion to work of some of the early workers in pulsation systems and I'm sure you'll be explaining that in the future.

Vigier. I have three comments to make. First, the essential trouble with the photons is that they are Bosons and that you have that bunching effect. So you can never be sure you are really dealing with one, but of course this does not prevent you from doing experiments with amplifications, so I would not condemn experiments like the one proposed by Selleri because there is something very important and mysterious in that whole process. Are spontaneous emission and stimulated emission really different? Of course, I agree with De Martini that those experiments with the mirror and the lifetime of atoms are evidence of vacuum fluctuations. I think this is well-established. This strengthens Selleri's position that maybe the waves are responsible for the stimulated emission. That is at least a reasonable possibility. I don't know how far one can criticize his experiments, but I think his direction is worth exploring.

The second point is that with Bosons, you can never be sure that you are dealing with one particle at a time in the interferometer, but in the case of Fermions, and especially in the case of the neutrons which I mentioned this morning, of this you can be certainly sure. Rauch is sure that he is really dealing with neutrons one by one, and this type of experiment has gone way beyond everything that has been done until now. And I think this confirms very strongly Dirac's famous statement that particles interfere with themselves.

This now brings me to Honig's models. I like some of his ideas: that particles are extended structures and that there is no basic distinction between particles and fields, which is an old idea of Einstein who compared particles to singularities in the field. But, all

92

of what he said today resembled too much a return to the old classical ether models of the nineteenth century. Something fundamental has happened in between and this is the idea of the fact that isolated particles interfere with themselves. It's no use to say this is true for neutrons but not for other particles. There is no way in which you can interpret the double slit interference experiment by interactions between particles themselves. The isolated particle somehow interferes with itself. There's no way out of that, and no model can be produced which does not account for that fact. This is the boundary between classical and quantum mechanics and there's no way to bypass that in any simple way. So I would like to see how these photex models explain the interference. Beyond that, there's something even stranger in quantum mechanics and that is the existence of a 4-pi and not a 2-pi symmetry of spin-$\frac{1}{2}$ particles. Now, any model of an electron as an extended sphere is all right with me for it's probably true that particles are not dimensionless points. But you have to explain the 4-pi symmetry as correlated with spin. If you don't do that, your model is in trouble. Now Dirac has explained that 4-pi symmetry with an elastic, not a fluid, topology. So, if the particles are extended structures, and if you can compare the spin to an internal rotation, which I believe is true, then you have to have a model that is not purely fluidic.

I wish now to address Honig's model for the ether. With the zero point field having become an essential part of the quantum mechanical formalism, it has become the trend to reconsider the ether's existence. However, any model of the vacuum must not only be neutral, but must also be covariant with respect to all observers and should offer no resistance to motion. I know of only one model which possesses these features and that is the Dirac model of the ether, published in Nature in 1951. Dirac achieves covariance by dropping the static model of the ether; your static model cannot have a covariance structure built into it and therefore is not correct. But if through any small space time region you can pass a distribution of particles which have an equal surface density of four-momentum on the mass shell, then this is covariant because the Lorentz transformation transforms the light cone, the mass shell, and the stochastic distribution into themselves. The second condition requires that, since the vacuum offers no resistance to motion, you construct something like a superfluid model of the ether.

All of these models are very interesting but they must both reproduce the known results of quantum mechanics, and also go beyond it to predict new phenomena that are not within the quantum mechanical formalism and that are not contradicted by observation.

Honig. I would like to ask Dewdney about his film and computer simulations. Are you working on applications for these techniques?

Dewdney. Yes, whenever the film is shown, people suggest calculations which could be done.

Hunter. The problems you described were all scattering problems using wave packets. Can you apply it to bound states, say for the hydrogen atom?

Dewdney. Yes. We have done that. We found some strange features such as states with zero angular momentum. This was published by Bohm.

De Martini. Vigier stated that one-particle interference is a demonstration of de Broglie waves. But there is another picture which is described in a recent book by Nelson. He discusses a static and dynamic background field, perhaps a Dirac ether, which has nothing to do with the particle itself, but it feels all the boundary conditions. If you put a two-slit screen in this field, that prepares for the particle a set of trajectories, so when the particle arrives, it follows some of these trajectories.

Thus you have these two pictures. In one, the particle has a background associated with itself, in the other there is a static Dirac ether that feels all the boundary conditions of the experiment. If you close one slit, the vacuum field overrides itself to take out the possibility of any fringes.

<u>Vigier</u>. The Nelson formalism you refer to is the so-called stochastic re-interpretation of quantum mechanics. /It rests on the idea that such a medium can support organized collective motions which can be represented by the de Broglie waves. The stochastic analysis of such a vacuum gives you the wave equations of quantum mechanics, which, in this way, at least for the relativistic spin-zero case, can be deduced from the chaotic motion: exactly like you can deduce the equations of a sound wave in the chaos of the molecules in this room. In that sense, there's no difference between Nelson's point of view and ours. Now, de Broglie compares a particle to an oscillator surrounded by its own sound wave, much like an airplane flying at Mach one moves within its own sound wave. You need that wave if you consider the interference formed in a double slit experiment... the group velocity of the wave surrounding it. The wave goes to both slits and you pick up the interference pattern. You cannot build chaos and cancel out the collective motion on top of the chaos. In a paper we wrote collectively last September in Physical Review, we showed that the quantization laws for hydrogen can be deduced from such a picture by assuming that the electron has an integer number of oscillations around a Bohr orbit; from that model, you can quantize the action immediately. Thus the quantization is a straightforward consequence of the de Broglie wave picture. The particle must be accompanied by a wave, or else you cannot predict the interference pattern correctly. Here is where stochastic electrodynamics (SED) gets into difficulty, namely it has trouble in re-interpreting the hydrogen orbits as stable quantized orbits. This can be done only with the de Broglie wave, so you cannot eliminate the wave surrounding the particles.

<u>Wadlinger</u>. I believe one can gain much insight into the nature of radiation if the word 'wave' were included in the units. For example, the units of Planck's constant would then be joule-seconds per wave. This is discussed at length in my paper in the <u>Journal of Chemical Education</u>, November 1983.

<u>Kostro</u>. I have a question for Selleri concerning the empty waves. The distance covered by the de Broglie wave of a particle in a time equal to its period T is equal to its wavelength $\lambda_B = uT$, where u is the phase velocity of the de Broglie wave. If we assume that the particle has a trajectory then the distance covered by the particle in the time equal to T is $\ell = vT$. From the de Broglie relation we obtain $\lambda_B = h/mv$ where mv is the momentum of the particle. We can show that $\ell = \dfrac{h}{m u}$ where mu is not the particle momentum but is nevertheless a real physical quantity. Instead of writing mu we can write $\dfrac{h}{c^2} \nu u$. As one can see, the considered momentum depends upon the frequency and upon the phase velocity of the de Broglie wave and therefore it has to be considered as momentum of the de Broglie wave $\vec{p}_B = \dfrac{h}{c^2} \nu_B \vec{u}$. We can also show that the de Broglie wave

has energy $E_B = \frac{h}{c^2} \nu_B u^2$. If it is so, can we say that the de Broglie wave is empty, i.e. does not carry either momentum or energy? That's my question. For Einstein the real physical space constitutes a "total fluid", i.e. a field of all kinds of actions. The physical space, as such kind of field, has energy and mass as well. It is therefore "material", i.e., it constitutes an ether. de Broglie waves, as waves of Einstein relativistic ether, carry a special kind of momentum and energy.

Selleri. It is a matter of taste whether one wants to adopt this kind of energetic approach. In a sense it is unsatisfying to introduce empty waves - by the way, 'ghost-fields' is a joke of Einstein's - these fields are real.

Of course it is possible to attribute energy and momentum to the wave; however, as Vigier puts it, the wave has energy and momentum but they are not observable. Then we have a mystery: why does a wave have energy and momentum and why are they not observable? I prefer to think that the wave does not have energy or momentum, and that presents me with a philosophical problem. I am realistically minded so I have to ask if this contravenes realism. My feeling is that it does not because we talk about energy and momentum because we know they are conserved, and for no other reason. There could be a reality described by different physical entitites.

Vigier. Dewdney's films showed that some of the particles with energies lower than the barrier potential get by the barrier. This means that they absorbed energy from the wave. Unless you are going to say that this energy comes from nowhere, you have to say that there is a real distribution of energy and momentum in the wave which explains the quantum potential.

Now the argument that you only observe particles can be justified following Einstein's views, namely the singularities in the field represent the particles and you only observe the particles, and indirectly, the energy and momentum of the gravitational field. So that's a general property of the relation between fields and particles. If you accept that the particles move in space and time then the field gives energy to the particle and unless you are producing effects without causes, you are obliged to pick up something like what Kostro just said.

De Martini. You can have quantum tunneling.

Vigier. You can interpret it as a quantum tunneling according to the arguments of Heisenberg and Bohr, but then you have to accept Heisenberg's statement that energy-momentum is not conserved except statistically.

SUMMARY OF F. DE MARTINI'S PAPER, TESTING OF PARTICLE

DISTINGUISHABILITY WITH LASER INTERFERENCE EXPERIMENTS

William M. Honig

Western Australian Institute of Technology
Perth, Bentley, 6102
Western Australia

F. De Martini's contribution was a description of the first results
from his laboratory of the experiments in the title. He set up a laser
interferometer by splitting the laser beam into two arms which were then
recombined; a photon counter was placed so as to count photons in the recombined beam. One of the interferometer arms had a transparent, but translucent plastic disc placed in it, through which the light in that arm passed. Photon counting was then carried on when the original unsplit beam was
attenuated throughout a range of intensities. The final photon counter
readings varied from more than several thousand photons per second to the
lowest levels of less than ten photons per second. Each of these readings
was taken with the plastic disc stationary, and each such reading was considered as the standard or reference level for the particular setting of
attenuation of the unsplit beam.

The second part of the experiment consisted of repeating the above experimental procedure in identical fashion but with only one change. In this
part of the experiment, the plastic translucent disc was rotated at relatively low speeds, i.e., speeds that were low compared to the transit time
of the light through the apparatus. Even so, the measured photon count at
the higher attenuation settings was much lower than the previous photon
count when the plastic disc was stationary. A plot was then made of photons
per second on the horizontal axis (starting with the highest levels at the
origin and decreasing levels shown to the right) vs. the normalized photon
count. For the first test, this is a straight line at the normalized level
1.0, parallel to the horizontal axis. For the second test, however, the
plot starts at 1.0 at the high photon densities and decreases rapidly to
levels that approach zero at the lowest photon levels.

The interpretation of these results was the subject of some discussion.
One suggestion was that some sort of nonlinear modification of coherence
may be occurring at low photon flux. This interpretation was disputed and
it is apparent that additional experimental work and theoretical interpretation is necessary.

TOWARD A CAUSAL INTERPRETATION OF THE RELATIVISTIC QUANTUM MECHANICS OF

A SPINNING PARTICLE

Nicola Cufaro Petroni

Dipartimento di Fisica dell'Università e
Istituto Nazionale di Fisica Nucleare
Bari (Italy)

ABSTRACT

As a first step in the direction of a causal interpretation, we analyze the features of the second order wave equation for spin 12/ fields.
It is shown that this equation allows a coherent statistical interpretation by means of a conserved density which is positive definite. On this
basis we can also construct all the usual Hilbert space formalism of the
relativistic quantum mechanics. The relations with the fields ruled by
the first order Dirac equation are also discussed. Finally, the perspectives of the definition of a relativistic spin-dependent quantum potential,
of the connection with stochastic processes and the extension to the case
of two correlated particles are briefly discussed.

In this note we will briefly sketch an outline of a research[1], still
in progress, that will bring us to a complete deterministic interpretation
of the non local quantum interactions of the E.P.R. type [2], for the Bohm-
Aharonov [3] case of spinning correlated particles. We hope so also on the
basis of the fact that an analogous previous work on spinless relativistic
particles gives encouraging results[4].

Our interest in the second-order relativistic wave equation for spin
1/2 particles is based on the following considerations:

a) it is , in some sense, astonishing the fact that spin 1/2 particles
   should be described by means of a first order equation, whereas the
   integer spin fields are usually ruled by second order equations,
   which are the most natural quantum analog of the relativistic energy-momentum relations;

b) if we are looking for a causal interpretation of the relativistic
   quantum equations, a classical analogy is more easy found starting
   from second order wave equations.

Here we want to sketch the main lines followed and the first results
obtained in this research. By introducing the symbols

$$D_\mu = \frac{1}{mc}\,(i\hbar\partial_\mu - \frac{e}{c}\,A_\mu), \quad \not{D} = \gamma_\mu D^\mu$$

the second order equation we are talking about takes the from

$$(I - \not{p}^2) \; \psi(x) = 0,\tag{1}$$

ore the ore familiar one

$$[(i\hbar\partial_\mu - \frac{e}{c}A_\mu)(i\hbar\partial^\mu - \frac{e}{c}A^\mu) - \frac{e\hbar}{2c}F_{\mu\nu}\sigma^{\mu\nu} - m^2c^2]\;\psi(x) = 0.$$

Along with it we will consider the two first order equations

$$(I - \not{p}) \; \psi(x) = 0\tag{2}$$

$$(I + \not{p}) \; \psi(x) = 0\tag{3}$$

We will denote by means of F, D_-, D_+ the set of spinors solution respectively of (1), (2), (3). Now we can easy prove the following propositions:

1. the only spinor common to D_+ and D_- is the identically zero spinor;
2. D_+ and D_- are vector subspaces of the vector space F ;
3. there is a one-to-one correspondence between D_+ and D_- realized by means of the matrix γ_5;
4. F is the direct sum D_+ and D_-.

Among others, from the property 3. it follows that, while eq.(2) is the usual Dirac equation, eq. (3) is satisfied by $\psi_+ = \gamma_5\psi_-$ if ψ_- is solution of (2): it is the well-known correspondence between a positron wave function and an electron wave function moving backward in space-time with the sign of energy inverted.

All the theory can be derived from the Lagrangian density

$$L = \overline{\not{p}\psi}\; \not{p}\psi \; - \; \overline{\psi}\psi \; .\tag{4}$$

If we derive the usual current density from it

$$J_\mu(x) = 1/2 \; (\overline{\not{p}\psi}\;\gamma_\mu\psi + \overline{\psi}\gamma_\mu\;\not{p}\psi)\tag{5}$$

we find that its zero component

$$J_o(x) = \text{Re} \; (\psi^+\not{p}\psi)\tag{6}$$

is not positive definite. That is not an unexpected result: as everybody knows it is a feature common to all the second order wave equations like the Klein-Gordon equation. As a consequence, $J_o(x)$ can not be interpreted as a probability density and a coherent statistical interpretation of the formalism is forbidden on this basis. Moreover in a quantum theory we can always pose the following questions: Are the subspaces D_+ and D_- mutually orthogonal? Are the elements of F normalizable wave functions? How can we approximate the solution of a physical problem in F? To what extent we can reproduce in F the outcomes of the Dirac theory like, for example, the hydrogen atom spectrum? All these questions and others, along with that about the statistical interpretation, can find an answer only if we succeed in building a Hilbert space structure on F, based on a positive definite norm. Of course, non positive norms are conceivables, but how can we mantain the most usual results and what about their physical meaning? The fundamental achievement of Dirac can be read, from this standpoint, as a restriction to the subspace D_- as the set of the physically acceptable states, where $J_\mu(x) = \overline{\psi}\gamma_\mu\psi$ and hence $J_o(x) = \psi^+\psi > 0$.

Despite the impressive importance of this Dirac position, we want to stress here two remarks:

A) the Dirac restriction is too much severe: it eliminates, as unphysical a lot of states with positive density;

B) the Dirac restriction is unnecessary: a solution of all problems can be found in the whole space F .

In order to prove A) Let us take the free case ($A_\mu = 0$), namely

$$(\Box + \frac{m^2 c^2}{\hbar^2}) \; \psi(x) = 0 \tag{7}$$

whose plane wave solutions are

$$\psi_p(x) = N \, e^{ipx/\hbar} \; u$$

$$P_\mu P^\mu = m^2 c^2; \quad E = \pm mc^2 \sqrt{1 + (\frac{\vec{P}}{mc})^2} \tag{8}$$

where u is an arbitrary four-spinor, constant in the space-time. For the free Dirac solutions there is a further restriction on u:

$$(\not{p} \pm mc) \, u = 0 \tag{9}$$

with the sign dependent on the energy sign. We can now find plane waves (8), solutions of (7), with positive conserved density, without obeying the condition (9). In fact, if we label the u with a parameter $\varepsilon = \pm 1$, so that

$$u_\varepsilon = \begin{bmatrix} H_\varepsilon(\alpha)\xi \\ H_{-\varepsilon}(\alpha)\eta \end{bmatrix}$$

$$H_\varepsilon(\alpha) = \frac{e^\alpha + \varepsilon e^{-\alpha}}{2} \quad , \; \alpha \in [0,\infty]$$

with ξ, η two spinors such that $\xi^+\xi = \eta^+\eta = 1$, we will have $\bar{u}_\varepsilon u_\varepsilon = \varepsilon = \pm 1$. Now the current (5) becomes

$$J_\mu = Re(\bar{\psi}\gamma_\mu \frac{i\hbar}{mc} \not{\partial}\psi) = \varepsilon N^2 \frac{P_\mu}{mc}$$

and hence

$$J_o = \frac{N^2}{mc^2} \; \varepsilon \, E.$$

Consequently the most general positivity conditions is $\varepsilon = sgn(E)$ and the spinors

$$\psi_p(x) = Ne^{ipx/\hbar} u_{\varepsilon = sgn(E)}$$

always lead to positive J_o.
Hence, a solution at rest will be

$$\psi(x) = \begin{cases} N \, e^{-imc^2 t/\hbar} \begin{bmatrix} \cosh\alpha \; \xi \\ \sinh\alpha \; \eta \end{bmatrix} & ; \; E = mc^2 \\[3em] N \, e^{imc^2 t/\hbar} \begin{bmatrix} \sinh\alpha \; \xi \\ \cosh\alpha \; \eta \end{bmatrix} & ; \; E = - mc^2 \end{cases}$$

In other words, only $\alpha = 0$ is possible, under the Dirac condition (9), that, hence, must be considered too much restrictive.

To prove the statement B) it is enough to show, by direct calculations, that besides the current (5), the vector

$$j_\mu(x) = 1/2 \ (\bar{\psi} \ \gamma_\mu \psi + \overline{\slashed{D}\psi} \gamma_\mu \ \slashed{D}\psi) \tag{10}$$

is another conserved current density, whose zero component

$$j_o(x) = 1/2[\psi^+\psi + (\slashed{D}\psi)^+ \ \slashed{D} \ \psi \ ] > 0 \tag{11}$$

is always positive. We did not find it at first because it is not derived from our initial Lagrangian density. Anyway, $j_\mu(x)$ allows one to define a coherent scalar product, and all the machinery of the Hilbert space connected to it, and, of course, a statistical interpretation.

More light can be cast on this breakthrough by considering what the word "state" means here. If we look for the state of a system at a given time x^o, we can not consider it as completely determined by the knowledge of $\psi\big|_{x^o}$, because we are dealing with a second order equation, and hence $\psi\big|_{x^o}$ is not enough to determine $\psi(x)$ in all the space-time. In fact, we need here two initial conditions: $\psi\big|_{x^o}$ and $\partial_o\psi\big|_{x^o}$, or, in a covariant form

$$\psi\big|_{x^o} = \ \phi_1(\vec{r}) \ , \quad \slashed{D}\psi\big|_{x^o} = \phi_2(\vec{r}),$$

where ϕ_1, ϕ_2 are two arbitrary four-spinors, which, along with eq.(1), completely determine the spinor field $\psi(x)$ in all the space-time. Hence the vector space of the states at a given time x^o is the space of the eight-component double-spinors

$$\psi\big|_{x^o} = 1/\sqrt{2} \ \left\{ \begin{array}{c} \phi_1(\vec{r}) \\ \phi_2(\vec{r}) \end{array} \right\}$$

and the complete space-time dependence of such a double spinor is

$$\Psi(x) \ = \ 1/\sqrt{2} \ \left\{ \begin{array}{c} \psi(x) \\ \slashed{D}\psi(x) \end{array} \right\} \tag{12}$$

where $\psi(x)$ is are element of F . if now H is the vector space of these $\Psi(x)$, we can show that it coincides with the space of the doble spinors solutions of

$$(C_\mu \ D^\mu - C) \quad \Psi(x) = 0 \tag{13}$$

where

$$C_\mu = \begin{bmatrix} \gamma_\mu & 0 \\ 0 & \gamma_\mu \end{bmatrix} , \quad C = \begin{bmatrix} 0 & I \\ I & 0 \end{bmatrix}$$

and that it is a Hilbert space whose scalar product is defined by means of the positive density $j_o(x)$:

$$\langle\Psi|\Phi\rangle = 1/2\int d^3\vec{r} \ (\bar{\psi}\gamma_o\phi + \overline{\slashed{D}\psi} \ \gamma_o\slashed{D}\phi).$$

In other words (13) is perfectly equivalent to (1) and F is in a one to one correspondence with H . It is in H that we can build the operator algebra of the observables. As a consequence, in this framework, it can be shown that the energy spectrum of a quantum system obtained from (13) [or equivalently from (1)] is always the same as that derived from the Dirac equation (2); only the number of states for each eigenvalue is doubled, because now we must take solutions in D_- and D_+ .

As a conclusion we will sketch the further steps required for a deterministic interpretation of eq. (1). First of all let us point out that in
the non relativistic, spinless case, the position $\psi = R\,e^{is/\hbar}$ directly
leads to a splitting of the Schrödinger equation into a continuity and a
dynamical equation. The latter can be interpreted either as a generalized
Hamilton-Jacobi equation or as a velocity potential equation for a Made-
lung fluid. For spinning particles only the second interpretation can be
retained: in fact, for a classical spinning body, the Hamilton function S
should depend on space-time coordinates and Euler angles. But a spinor
$\psi(x)$ shows no dependence on angular variables at all. On the other hand,
in fluidodynamics, a spinning fluid can be described by means of a number of fields, all of them functions of the space-time only, that can be
accomodated into the spinor components. In this case a velocity field
$V_\mu(x)$ will be described by means of a velocity potential s(x) plus a couple of Clebsch parameters $\chi(x)$, $\omega(x)$, in the following way

$$v_\mu(x) = \partial_\mu s(x) + \omega(x)\,\partial_\mu \chi(x)\ . \tag{14}$$

Hence the forthcoming step will be the decomposition of the four-spinors
$\psi(x)$ in terms of velocity potentials and Clebsch parameters, all connected with complex Euler angles. In fact, the complexification of these parameters rests on the basis of the well-known isomorfism between the group
of the Lorentz transformation and that of the rotations in a there dimensional complex Euclidean space. In the representation where

$$\gamma_0 = \begin{bmatrix} 0 & I \\ I & 0 \end{bmatrix}, \qquad \gamma_k = \begin{bmatrix} 0 & -\sigma_k \\ \sigma_k & 0 \end{bmatrix}, \qquad \gamma_5 = \begin{bmatrix} I & 0 \\ 0 & -I \end{bmatrix}$$

a tentative, but not yet final, decomposition is

$$\psi(x) = R(x)\,e^{is(x)/\hbar}\,\phi(x)$$

where R is the 4x4 matrix

$$R(x) = p(x) + iq(x)\,\gamma_5$$

S is a real velocity potential and

$$\phi(x) = 1/\sqrt{2} \begin{bmatrix} \cos u\ e^{iw} \\ i\sin u\ e^{-iw} \\ \cos u^*\ e^{iw*} \\ i\sin u^*\ e^{-iw*} \end{bmatrix}$$

is a spinor satisfying the relations

$$\bar\phi \phi = 1, \qquad \bar\phi \gamma_5\,\phi = 0.$$

That leads to forms like (14) for the velocity field. The final step is,
of course, the derivation, from (1), of the equations for all the parameters involved, that would lead to the definition of a relativistic, spin-
dependent quantum potential.
Furthermore, the generalization to the case of two particles of the eq. (1),
would enable one to decide if the resulting non local quantum potential can
satisfy the compatibility conditions of the relativistic predictive mechanics[6]. If it will be so, we could realistically hope to build up a coherent, deterministic, relativistic, non local theory of the quantum phenomena. Finally we will remark that it seems likely that our starting point
of a second order wave equation will allow a coherent connection with sub-
quantum stochastic processes, as previous works have already shown [7].

REFERENCES

1. N.Cufaro Petroni, Ph. Gueret, J.P. Vigier: N.Cim. $\underline{81B}$ (1984) 243
   N.Cufaro Petroni, Ph. Gueret, J.P. Vigier: Phys. Rev. $\underline{D30}$ (1984) 495
   N.Cufaro Petroni, Ph. Gueret, A.Kyprianidis, J.P.Vigier: Lett.N.Cim. $\underline{42}$ (1985) 362
   N.Cufaro Petroni, Ph.Gueret, A.Kyprianidis, J.P. Vigier: Phys. Rev. D $\underline{31}$ (1985) 3157
   N.Cufaro Petroni, Ph.Gueret, A.Kyprianidis, J.P. Vigier: Phys. Rev. D $\underline{33}$ (1986) 1674
2. A.Einstein, B.Podolsky, N. Rosen: Phys. Rev. $\underline{47}$ (1935) 777
3. D.Bohm, Y.Aharonov: Phys. Rev. $\underline{108}$ (1957) 1070
4. N.Cufaro Petroni, Ph. Droz-Vincent, J.P. Vigier: Lett.N.Cim. $\underline{31}$ (1981) 415
   N.Cufaro Petroni, J.P. Vigier: Phys. Lett. $\underline{93A}$ (1983) 383
5. H.Lamb, "Hydrodynamics", (Cambridge Univ. Press, New York, 1953)
6. Ph. Droz-Vincent: Phys. Scrip. $\underline{2}$ (1970) 129
   Ph. Droz-Vincent: Ann. I.H. Poincaré $\underline{27}$(1977) 407
7. L. De La Peña-Auerbach: J. Math. Phys. $\underline{12}$ (1971) 453

NONLINEAR BEHAVIOUR OF LIGHT AT VERY LOW INTENSITIES: THE "PHOTON

CLUMP" MODEL*

E. Panarella

National Research Council
Ottawa, Canada K1A 0R6

ABSTRACT

   This is essentially a review paper that deals initially with the
analysis of the wave-particle duality notion for photons, as accepted
today, namely as that of a particle and/or an associated wave. This
model will be shown to be lacking at present adequate, direct and
conclusive experimental proof, which is so crucial for the very existence
of that model, despite the numerous experiments that have been done in
the past and some novel experiments that are presented here. Starting
from this situation of deficiency, a model of interacting photons is
introduced from plausible arguments and an interaction law is derived.
It will be shown that the main consequence of this model is that the
single photon concept, as understood today, has to be abandoned in favour
of a model of collection of photons, a "clump", in which the individual
photons are arranged in a geometrical wave pattern, much like a wave
distribution. In this sense, therefore, the photon clump is both a
particle and a wave and the model thus reconciles two concepts which are
normally a source of some debate, namely the Dirac's notion that a photon
interferes only with itself, and the pilot wave concept of the Broglie.
In fact, as far as the interference of a single photon with itself is
concerned, such statement can now be reinterpreted as meaning that a
photon clump contains all the elements of interference in itself, namely
maxima and minima, because in the clump there are maxima and minima of
photon number density distribution. As far as the pilot wave of de
Broglie is concerned, its reality and its real meaning are readily
retrieved from this model. It is, in fact, the interaction among photons
responsible for guiding them on a wave distribution within a clump.
Rather than being separate entities, the particles and their wave
geometrical distribution are coexisting and inseparable properties of
what should now be said an element or an "atom" of light.

---

*Brief accounts of parts of this work have appeared in the literature and
 are abstracted here: Ann. Fond. Louis de Broglie, 6, 197 (1981); ibid,
 10, 1 (1985); Bull. Amer. Phys. Soc. 23, 858 (1978), ibid, 26, 279
 (1981); Spec. Sc. Tech. 5, 501 (1982); ibid 5, 509 (1982); ibid 6, 383
 (1983); ibid 8, 35 (1985).

TABLE OF CONTENTS

I. INTRODUCTION

The aim of this review article is to shed new light on the wave-particle duality. We shall do so by presenting new experimental and theoretical results on this subject, rather than discussing past achievements, because we believe that all arguments, both in favour and against the wave-particle duality, have been already and abundantly expressed in the literature and a generally accepted opinion has been formed which considers the wave aspect of particles as real as the particles themselves.

We shall analyze the wave-particle duality by examining first its idealized content. Then, we shall contrast its ideal formulation with the requirement of experimental verification. We shall then provide proof that such experimental verification is difficult, if not impossible, to obtain. We shall report the results from three new experiments of diffraction of statistically independent laser photons at different light intensities which invariably do not provide the experimental confirmation of the wave-particle duality for single photons. Moreover, through an historical analysis of all the experiments done in the past (Sec. II.4.1), which seemed to provide a confirmation of the wave-particle hypothesis, we shall prove that such confirmation was never really obtained.

Our experimental results will be initially interpreted by assuming that the wave-particle hypothesis is correct and finding reasons for the departure of the results from the expected ones. In fact, in the case of the first experiment, in which we used the photographic technique to reveal the photons, we shall borrow from the accepted photographic theory all the tools for justifying the results. However, in the case of our second and third experiment, in which the photons were revealed with the photoelectric technique, we shall not do so because, by assuming again that the wave-particle hypothesis is correct and justifying the deviation from the expected results with some further assumption, it is realized that a proof of the correctness of the original assumption cannot be achieved in this way.

The lack of a direct experimental demonstration of the wave-particle duality for single photons leads us to consider the hypothesis that perhaps isolated photons do not exist. We shall then put forward a model of light in which a photon is invariably accompanied by other photons, all clumped together. If the individual photons in a clump are arranged on a distribution with maxima and minima of number density (i.e., a wave distribution), one is able to retrieve from this model not only an explanation for our experimental results, but also for those of Hanbury-Brown and Twiss[1], of Pfleegor and Mandel[2], of Clauser[3], and of Grangier, Roger and Aspect[4]. Moreover, in the light of this model, Dirac's dictum that a photon interferes only with itself[5] must be reinterpreted as meaning that a clump or cluster of photons has already imprinted in it all the characteristics of interference or diffraction. Consequently, an interferometer must be viewed now as an instrument that does not do anything to the photons to let them interfere (because they have already interfered and positioned themselves on a wave geometrical arrangement with maxima and minima of distribution, even before entering the interferometer) but, by changing slightly the direction of motion of two outgoing clumps or conglomerates of photons originating from a single clump, makes them change the initial geometrical arrangement into an arrangement which can be clearly seen as a wave pattern. In short, an interferometer acts as an amplifier of the fringe separation or as a microscope to see more easily the interference or diffraction pattern already existing in the clumps of photons.

As to the existence and origin of such clumps of photons, this matter has been already dealt with classically to some extent by Dicke[6], who pointed out that individual atoms in a source of thermal light cannot emit photons independently of each other, because they are constantly interacting with a common radiation field. Therefore, incoherent photons are not emitted as random isolated particles, but have certain characteristic bunching properties[7].

We will take up this matter, however, from a novel point of view[8], namely from an analysis of the Heisenberg Uncertainty Principle for photons. It will be shown that an interaction law for photons can be derived from this analysis which leads naturally to the bunching or clumping effect.

The Heisenberg Uncertainty Principle is a fundamental building block in modern physics. Since it plays a basic role in our derivation of the interaction law for photons we would like to stress what in essence the principle is, namely a statement of the unavoidable interaction between measuring apparatus and measured system, which limits the precision with which certain pairs of physical variables can be simultaneously measured. Although the interaction is essential for the observation and measurement, its central role in the Heisenberg principle has not been emphasized because the principle can be incorporated "in toto", i.e. without explicit relevance to the underlying cause of it, in the definition of a particle as a wave-packet. We shall prove that this definition is tantamount to a postulation of the Heisenberg principle. Although such postulation is theoretically legitimate, when the role of the interaction is re-examined, a derivation (rather than a postulation) of the principle is obtained.

The interaction between system and apparatus, which makes the measurement possible, is not an instantaneous process. As we shall prove, the interaction takes place continuously, thus continuously affecting the system, be it an atom, an electron, a neutron or a photon, and such influence goes on irrespective of whether some knowledge of the system is required or not. If a photon, for instance, is brought close to an electron for the purpose of observation, the two particles do not experience a collision type of interaction, but rather they influence one another continuously, and such continuous interaction takes place anyway, whether or not the observation or measurement of some parameters regarding the electron (position, for instance) is required at all.

The present study will therefore attempt to determine the role of this interaction and to establish the interaction law. One finds that such "Heisenberg interaction" is different from, and in addition to, any other known particle-particle interaction. As to the law of interaction, it is obtained from an analysis of the experiments upon which the derivation of the Heisenberg principle is based. It will be shown that, as a first approximation, the law can be expressed as

$$p(r) = h/r$$

where r is the separation between the interacting particles (i.e., the target or observed particle and the observing particle) and $p(r)$ is the amount of momentum transferred from one particle to the other along the r-direction (h = constant). It is then shown that, when this law is applied to a collection of photons, for instance, it leads to the general form of Kirchhoff's equation (Sec. IV.5), of fundamental importance in optics for understanding diffraction phenomena of light, or to Helmholtz' formula, in case of monochromatic photons (Sec. IV.9). Subsequent

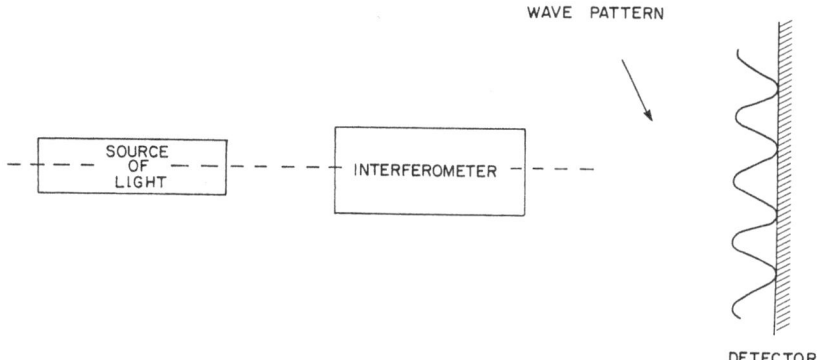

WAVE PATTERN

SOURCE OF LIGHT

INTERFEROMETER

DETECTOR

Fig. 1   The experimental arrangement required to demonstrate
the wave nature of photons consists of a source of
light, an interferometer and a detector. The
interferometer selects and displays the wave aspect of
the photons. The detector reveals such wave aspect as
an interference pattern.

discussion shows that from this result one may infer that photons, as
particles, arrange themselves on a diffraction pattern, with maxima and
minima, not because they are guided by waves, but because they are
constantly under the influence of such mutual "Heisenberg interaction".

In summary, the purpose of this review paper is two-fold: on the
one hand it will re-examine the accepted wave-particle duality concept,
including its historical rising, within the context of the numerous
experiments that have been done in the past and some novel experiments
which are reported here, all aimed at proving the validity of such
concept. Then, starting from the conclusion that such concept is not
warranted, we shall provide a new physical interpretation of Heisenberg's
uncertainty principle which will lead us to an interaction law for
photons, which in turn will lead us to the wave properties of light being
derived without need to resort to the wave-particle duality of photons.

II. EXPERIMENTAL

II.1. Wave-Particle Duality. General Considerations

II.1.1. Ideal situation. The ideal situation that the
wave-particle concept implicitly conveys is illustrated in Fig. 1.
Confining our discussion, without loss of generality, to the case of
photons, consider a source of light that continuously emits photons.
These photons then cross an instrument, an "interferometer", that is
capable of selecting and displaying the wave aspect of the photons.
Finally, a detector reveals such wave aspect as an interference pattern.

More in detail, in the ideal situation that we are considering, each
individual photon, when crossing the interferometer, seems to experience
a sudden change of state, from a characteristic individual entity or
unity to a diffuse and wide configuration resembling a wave pattern. Out
of the interferometer, a means is provided for revealing such a pattern
by a detector capable of revealing parts of a photon. Therefore, if we
define as "signal" the recording of each part of a photon and by "signal

amplitude" the total number of signals revealed by the detector, a strict proportionality of signal amplitude with number of photons emitted by the light source exists. In other words, the signal amplitude grows <u>linearly</u> with light intensity.

II.1.2. <u>Departure from the ideal situation.</u> A departure from the ideal situation is realized when one considers that a detector cannot reveal parts of a photon, but that it needs a whole photon in order to generate a signal. The ideal situation, however, can be modified to take this fact into account. What is needed is to consider the interferometer as an ideal device capable of rearranging the direction of motion of each individual photon according to a prescribed probability law so that more photons will travel along particular directions than in other directions. With this modification, each photon conserves its entity, only its direction of motion is modified.

If the source of light emits now N photons, these are all carefully recorded by the detector, which essentially acts as a counter. Again, in this case too, the detector will provide a signal of amplitude growing <u>linearly</u> with the light intensity.

II.1.3. <u>Real situation.</u> It is unfortunate that the wave-particle duality hypothesis does not go beyond the foregoing considerations. It believes that any detector (including the photographic and the photoelectric detectors) is linear in the sense of providing a signal for each photon arriving on it or that a one to one relation exists between signal amplitude and light intensity.

We would like to examine here if problems arise when a real detector, such as a photographic or photoelectric detector, is used. As we shall soon prove, these detectors in practice need more than one photon to yield a signal and the wave-particle duality hypothesis for single photons cannot receive experimental demonstration with these detectors.

II.2. <u>The Case of a Detector Requiring More than One Photon to Yield a Signal</u>

This is certainly the case of a photographic detector and normally the case of a photoelectric detector.

Let us deal first with the photographic detector. Modern photographic theory clearly points out[9-11] that a photographic grain cannot become developable unless it absorbs at least 3 or 4 photons within the finite memory time of the grain. Hence, single photons cannot be recorded by a photographic detector. Only packets or clumps of 3 or 4 photons at least are recorded. The consequence is that, if an interference pattern appears on a photographic detector, it cannot be ascribed to individual photons. Rather, it has to be ascribed to packets of photons. We will show in Sec. II.3.1.1 that a strong nonlinearity exists of photographic grain activation with light intensity or photon flux. In other words, the build-up of an interference or diffraction pattern on a photographic plate does not proceed linearly with the number of photons reaching the plate. Consequently, one of the fundamental assumptions upon which the wave-particle concept rests, namely linearity of signal amplitude with light intensity, cannot receive experimental confirmation with the photographic technique and the wave nature of a single photon cannot thus be demonstrated.

In the case of the photoelectric detection, we are facing more or less the same problem. This is because a photosensitive surface with 100

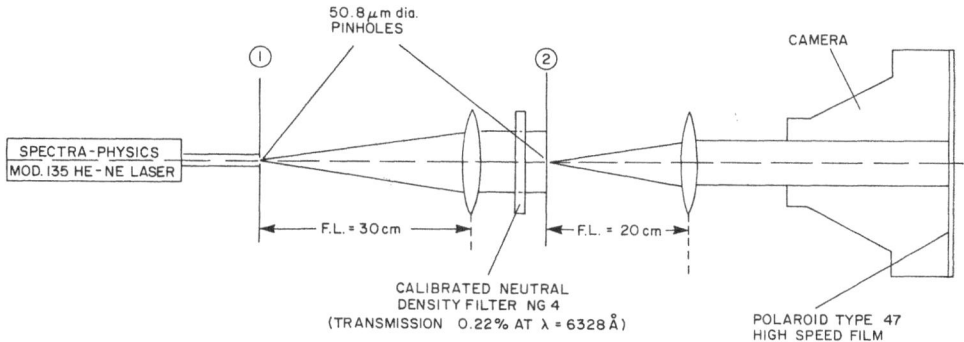

Fig. 2 Experimental apparatus used to reveal the effect of the degree of statistical independence on the photon distribution on a photographic film. Without the neutral density filter along the light path, a clear diffraction pattern can be recorded on the film. With the neutral filter inserted, the diffraction pattern does not appear as clearly as before, even when the total number of photons inpinging on the film is more than two orders of magnitude larger than before.

percent quantum efficiency just does not exist. This means that, if 100 photons impinge on a photoemissive surface, less than 100 electrons are emitted. A typical quantum efficiency[12] is of the order of 10^{-3} electrons per incident photon*. A typical photodiode, such as RCA 926, has a radiant sensitivity (at 4200 Å) of 1.9×10^{-3} amp. watt^{-1}, which means that 5.62×10^{-3} electrons are released per incident photon, or that 177 photons are required before an electron is released[13]. So, if 177 photons cross the interferometer of Fig. 1 before being capable of liberating an electron in a photodetector, how do we know that interference is produced by one photon or by two, three photons, etc. clumped together? We would be able to affirm for sure that a single photon carries its own wave-packet only if the photodetector, in this case, would yield a signal proportional to the light intensity, as explained in Sec. II.1.2. Unfortunately, as we shall soon prove in the following sections where the experiments are reported, photons crossing the interferometer in apparent isolation do not provide signals proportional to the light intensity and therefore we cannot affirm for sure that a single photon has wave property.

II.3. Experiments of Diffraction of Statistically Independent Laser Photons

II.3.1. Photographic detection of the diffraction pattern.
Consider the experimental apparatus of Fig. 2 which has been used to produce statistically independent photons. A 5 mW cw TEM_{oo} mode Spectra-Physics Mod. 135 He-Ne laser was the source of light. The laser emitted a Gaussian beam of radius a = 0.35 mm at $1/e^2$ points. The peak light intensity in the central part of the beam was:

$$I_p = \frac{2P_o}{\pi a^2} = 2.59 \text{ W.cm}^{-2} \ (P_o = 5 \times 10^{-3} \text{ W})$$

*Although we use here typical values for the purpose of exemplification, the conclusions reached do not change if one uses higher values of quantum efficiencies.

The light intensity profile was smoothed out by means of a pinhole of diameter $d = 5.08 \times 10^{-3}$ cm positioned at the center of the beam, at the point of maximum light intensity. The resultant emerging bright central disc of the Airy pattern was collimated by means of a simple double-convex lens located at a distance from the pinhole equal to the lens focal length $f = 30$ cm. The intensity of light at the center of the Airy pattern resulted in[14]:

$$I_o = \frac{AP_1}{\lambda^2 f^2} = 2.95 \times 10^{-4} \text{ W.cm}^{-2}$$

where $A = \frac{\pi d^2}{4}$ is the pinhole area and $P_1 = I_p A$. A second pinhole of diameter $d = 5.08 \times 10^{-3}$ cm drilled in aluminum foil 1.27×10^{-3} cm thick was positioned at the center of the Airy disc. Since the light intensity across this pinhole was essentially constant, the photon flux entering the pinhole was 1.90×10^{10} photons.sec^{-1}.

The diffracted light out of this second pinhole was then recollimated by means of a simple double-convex lens located at a distance from the pinhole equal to the lens focal length $f = 20$ cm and the diffraction pattern was recorded by means of a camera equipped with Polaroid type 47 high speed film. The resulting intensity of light at the center of the second Airy disc was 7.57×10^{-8} W.cm^{-2}. The reasonable assumption was then made that such intensity was constant over a small circular area of radius equal to the diameter of the pinhole. Hence, the photon flux resulted in being 1.95×10^7 photons.sec^{-1}.

The objective of the experiment was the following. Irrespective of the photon flux reaching a detector, the diffraction pattern is considered to be the result of the superposition of the patterns created by each individual photon, which diffracts only with itself. On the other hand, if one photon were sufficient to activate a photographic grain, an identical number of photons reaching the film should provide identical diffraction patterns. Fig. 3 shows the experimental results. Figure 3a was obtained with the apparatus just described. The photograph was exposed for 20 sec and 3.91×10^8 photons produced the clearly defined diffraction pattern shown in the figure. We then reduced the intensity of light by inserting a calibrated neutral density filter (type NG4-homogeneous filter-transmission 0.22% at $\lambda = 6328$ Å) along the light path (see Fig. 2). The intensity of light crossing the second pinhole was reduced in this way by a factor of 454 and only 4.29×10^4 photons reached the film per second. In order to have the same diffraction pattern as in Fig. 3a, it was calculated that an exposure time of 2h32m was required. The first experiment performed with such exposure time failed to provide the expected result in that the film did not record any light at all. Only when the exposure was increased to 17h36m, or when 2.72×10^9 photons reached the plate (i.e. a number of photons almost an order of magnitude larger than before) were we able to obtain a meaningful photograph (Fig. 3b). Finally, when the exposure time was pushed up to over 2 weeks (more exactly, 336h20m, or 5.19×10^{10} photons on the plate) the resultant photograph was better defined, although the expected diffraction pattern did not appear, as Fig. 3c shows.

These experimental results bring therefore new evidence that a diffraction pattern on a photographic plate is not preserved when the intensity of light is extremely low, even when the total number of photons reaching the film is larger than that which is capable of producing a clear diffraction pattern. In other words, a diffraction pattern does not build up linearly with light intensity, as the wave-particle duality requires. In the next section, we will start from the

a   b   c

Fig. 3   a) Regular diffraction pattern obtained with a total of 3.91×10^8 statistically independent photons reaching the photographic film (20 sec exposure time); b) Picture obtained when a total of 2.72×10^9 photons reach the photographic film (17h36m exposure time); c) Picture obtained when a total of 5.19×10^{10} photons reach the photographic film (336h20m exposure time). The (b) and (c) pictures show that the diffraction pattern is missing, although the number of photons inpinging on the film is ~1 order of magnitude, or even 2 orders of magnitude, respectively, larger than that which was capable of producing a clear diffraction pattern in (a).

assumption that the wave-particle duality hypothesis is correct and explain the experimental results with the assumption that the photographic detector acts as a coincidence counter of the laser light.

II.3.1.1 _Explanation of the photographic results._ The foregoing experimental results can be explained if one refers to the theory of photographic grain developability, as put forward by Rosenblum[15] and experimentally verified by Polovtseva et al.[16]. The theory is centered around a model of photographic detector as a coincidence counter, so that when 4 or more photons arrive near the grain within a time τ, the grain is rendered developable. Rosenblum[15] identifies such interval τ as the coherence time of the light, rather than a characteristic time of the emulsion material. The reason is, we believe, that, if the latter were the correct parameter, no difference should exist in the response of photographic films irradiated with lights of equal intensity and duration but different coherent times. This is not so, as Polovtseva et al.[16] have shown. For our purposes, however, we need not commit ourselves to a particular interpretation, but simply try to verify if Rosenblum's definition leads to an explanation of the experimental results.

Let a flux of λ laser photons per unit time arrive at a photographic plate. If the laser photon statistics are treated as Poisson[17], the probability that a given grain receives at least four photons during time τ is:

$$p_4 = \sum_{n=4}^{\infty} P(n,\bar{n}) = 1 - \sum_{n=0}^{3} P(n,\bar{n}) \qquad (1)$$

The Poisson distribution is:

$$P(n,\bar{n}) = \frac{\bar{n}^n}{n!} e^{-\bar{n}} \qquad (2)$$

where $\bar{n} = \lambda\tau$ and n are the average number and the actual number of photons reaching the grain in time τ, respectively.

The probability of 4 or more photons arriving at a grain during time τ is given by (1). In total exposure time T there are T/τ such time intervals. If we make the simplifying assumption that all photons contained in each interval τ travel in a specific direction towards a new grain on the photographic plate*, the expected number of grain activations during the entire exposure T is:

$$E = \frac{T}{\tau} p_4 = \frac{T}{\tau} \left[1 - \sum_{n=0}^{3} P(n,\bar{n})\right]$$

$$= \frac{T}{\tau} \left[1 - e^{\bar{n}}\left(1 + \bar{n} + \frac{1}{2}\bar{n}^2 + \frac{1}{6}\bar{n}^3\right)\right]$$

$$= \frac{T}{\tau} \left\{1 - e^{-\lambda\tau}\left[1 + \lambda\tau + \frac{1}{2}(\lambda\tau)^2 + \frac{1}{6}(\lambda\tau)^3\right]\right\} \qquad (3)$$

A useful approximation of (3), in case $\lambda\tau \ll 1$, is obtained as follows:

$$E = \frac{T}{\tau} p_4 = \frac{T}{\tau}\sum_{n=4}^{\infty} P(n,\bar{n}) = \frac{T}{\tau} e^{-\lambda\tau}\sum_{n=4}^{\infty}\frac{(\lambda\tau)^n}{n!} \approx \frac{T}{\tau}\frac{(\lambda\tau)^4}{4!}\left(1 + \frac{\lambda\tau}{5}\right) \qquad (4)$$

The short term coherence time τ of the commercial He-Ne laser used in the experiment is the inverse of its linewidth, which is of the order of 1 MHz. Hence $\tau \simeq 10^{-6}$ sec. Inserting now in (3) the following figures:

$$T = 20 \text{ sec}; \qquad \lambda = 1.95 \times 10^7 \text{ sec}^{-1}$$

one finds that the number of grain activations is:

$$E = 1.99 \times 10^7$$

Assuming that this number is sufficient to reveal a clear diffraction pattern on a photographic plate, as Fig. 3a shows, we are interested in knowing what would be the E-value when the light flux is reduced by a factor of ~500 and becomes $\lambda = 4.29 \times 10^4$ sec^{-1}. Since $\lambda\tau = 4.29 \times 10^{-2} \ll 1$, we insert this figure in Eq. (4) and get:

$$E = 2.84$$

In other words, if the light flux is reduced ~500-fold, this model predicts very few grain activations during the same exposure time T. To rebuild the pattern to the same level as before, one would require not 500-fold but 7.02×10^6-fold exposure increase. In fact, if one deduces T from Eq. (4):

---

*This simplifying assumption is removed in the joint paper with T.E. Phipps, Jr. (Spec. Sc. Tech. 5, 509, 1982) where three models of grain activation are considered.

$$T = \frac{24E\tau}{(\lambda\tau)^4 \left(1 + \frac{\lambda\tau}{5}\right)} \tag{5}$$

and inserts the following figures:

$$E = 1.99 \times 10^7; \qquad \tau = 10^{-6} \text{ sec}; \qquad \lambda = 4.29 \times 10^4 \text{ sec}^{-1}$$

one gets:

$$T = 1.40 \times 10^8 \text{ sec} = 3.89 \times 10^4 \text{ hrs} = 54.01 \text{ months}$$

This proves that there is a strong nonlinearity of statistical expectation of grain activation with light intensity, thus explaining why the same diffraction pattern as in Fig. 3a did not appear in Fig. 3c, which was irradiated for only 336h20m.

In conclusion, both the experiment and the theory point out that packets of at least four photons are required for diffraction effects to be revealed by a photographic plate. Single photons are not recorded and their dual nature cannot be demonstrated with the photographic technique.

II.3.2 <u>Photoelectric detection of the diffraction pattern.</u> There is apparently a way, however, of proving that a single photon has wave nature. What is required is to repeat the experiment previously described, using this time a detector capable of recording single photons, such as a photoemissive surface, and observe if identical diffraction patterns appear in situations of different light intensities. If the phenomenon is linear with light intensity, the wave-particle hypothesis is proven; if not, the duality concept remains an unproven hypothesis. Clearly, since we are dealing with a particle phenomenon, we would like to have a criterion to decide whether or not we have, with high probability, only one photon within the interferometer, or several. We shall assume, in fact, that diffraction or interference effects can take place with any number of photons, including one, and in the next section we shall provide the details of the analysis used to derive such criterion.

II.3.2.1 <u>Probability analysis.</u> Our main concern is to calculate the probability of interference effects to occur with two or more photons and then to compare this probability with that of occurrence of the same phenomenon with one photon at a time within an interferometer.

Consider a flux of λ photons per unit time crossing an interferometer. If $\bar{n} = \lambda\tau$ is the average number of photons present within the interferometer during time τ, defined as that time interval during which the presence of at least R photons is both necessary and sufficient for interference effects to take place (we take as τ the time of flight of the photons within the interferometer), and if the photon statistics are treated as Poisson, then the probability that the interferometer be occupied by at least R photons during time τ is:

$$P_R = \sum_{n=R}^{\infty} P(n,\bar{n}) = 1 - \sum_{n=0}^{R-1} P(n,\bar{n}) \tag{6}$$

where $P(n,\bar{n})$ is the Poisson distribution given by:

$$P(n,\bar{n}) = \frac{\bar{n}^n}{n!} e^{-\bar{n}} \tag{7}$$

Interference effects appear as particular geometrical distributions of photon detection events. Let us treat these collections of events as Poisson (i.e. independent of photon statistics). Since interference can take place only when certain conditions are satisfied, namely R photons to be present within the interferometer during time τ, the probability of interference during total time $T \gg \tau$ is then $P(T) = 1 - e^{-N}$, where N is the expected number of times the event in question occurs during time T^{18}. Consider the event of the presence of a photon at the entrance of the interferometer at an arbitrary instant t in the interval (0,T). The chance that during the interval (t, t + τ) at least R-1 additional photons appear within the interferometer is p_{R-1}. Together with the first, these constitute R photons during interval τ, so interference can take place. In time T, the expected number of photons arriving at the entrance of the interferometer, each capable of thus initiating the interference process, is λT. Hence

$$N = \lambda T p_1 \quad \text{for} \quad R = 2$$

$$P(T) = 1 - e^{-\lambda T p_1} \tag{8}$$

which is the probability of interference during T, assuming that at least 2 photons are required. p_1 is given by Eq. (6):

$$p_1 = 1 - e^{-\lambda \tau} \tag{9}$$

Hence

$$P(T) = 1 - e^{-\lambda \tau (1 - e^{-\lambda \tau})} \tag{10}$$

If interference is initiated by one photon alone, the analysis proceeds as before. We consider the event of the presence of a photon at the entrance of the interferometer at an arbitrary instant t in the interval (0,T). The chance that during the interval (t, t + τ) no other photon enters the interferometer is $1 - p_1 = e^{-\lambda \tau}$. In time T, the expected number of photons arriving at the entrance of the interferometer, each capable alone of thus initiating the interference process, is λT. Hence

$$N = \lambda T (1 - p_1) = \lambda T e^{-\lambda \tau}$$

$$P(T) = 1 - e^{-\lambda T e^{-\lambda \tau}} \tag{11}$$

We are now in a position to compare the two probabilities given by Eqs. (10) and (11). For the probability of interference with one photon alone within the interferometer to be greater than the probability of interference with two or more photons, the following inequality must be satisfied:

$$\lambda T e^{-\lambda \tau} > \lambda T (1 - e^{-\lambda \tau})$$

which reduces to

$$2 e^{-\lambda \tau} > 1$$

$$\ln 2 > \lambda \tau$$

or

$$\lambda \tau < 0.69 \tag{12}$$

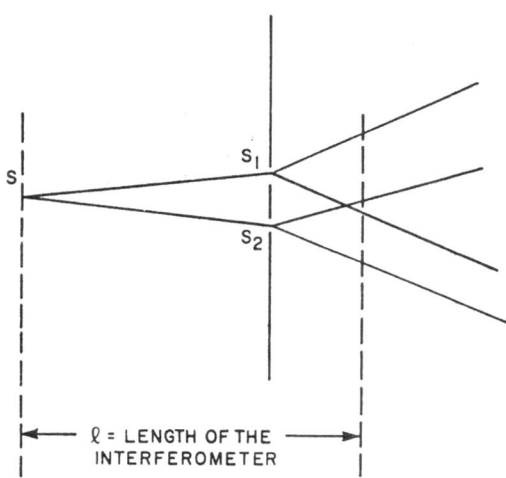

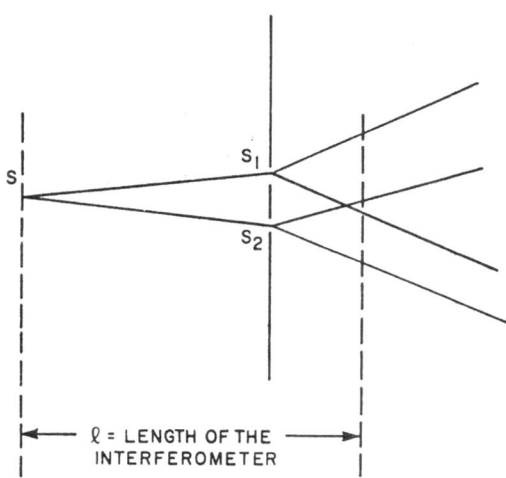

ℓ = LENGTH OF THE
INTERFEROMETER

Fig. 4   Two-pinhole interferometry with a Young apparatus. The
interferometer length ℓ is the distance from the light
source S to the area, past the pinhole, where the
fringes appear.

Inequality (12) tells us that, if the flux λ times the transit time
within the interferometer is less than 0.69, the probability of
interference with one photon exceeds the probability that interference
was caused by two or more photons.

In our case λ is known. In order to apply Ineq. (12) to our
experimental results, we need to know the transit time τ through the
interferometer. In other words, we need to know what is in our case the
interferometer and how much its length is. Consider first Fig. 4 which
shows a Young apparatus for interferometry with two pinholes. Light
coming from a source S goes through two separate pinholes S_1 and
S_2, beyond which interference fringes appear in the area where the two
emerging beams overlap. On the other hand, by definition, an
interferometer is any suitable apparatus that divides a beam into two
beams, which are then superposed. Therefore, for a Young apparatus the
length ℓ of the interferometer is the distance from the source of light S
to the area, past the pinholes, where the fringes appear. Our apparatus,
as Fig. 2 shows, is essentially a Young interferometer: the first
pinhole acts as a light source, and the second pinhole can be considered
as two overlapping pinholes in a Young interferometer. Since the fringes
are formed immediately after pinhole No. 2, the length of our
interferometer is the distance between pinhole No. 1 and No. 2, i.e.
ℓ = 42 cm. The transit time across the interferometer is therefore:

$$\tau = \frac{\ell}{c} = \frac{42 \text{ cm}}{3 \times 10^{10} \text{ cm.sec}^{-1}} = 1.4 \times 10^{-9} \text{ sec}$$

We thus know both λ and τ and these figures will be used in the next
section, when we shall analyze the experimental results.

II.3.2.2 <u>Photoelectric detection and oscilloscope recording of the
diffraction pattern</u>. The experimental apparatus used for the
photoelectric detection of the photons is essentially the one previously
described. Only the camera has been replaced by a high gain
photomultiplier mounted on a motor-driven translation unit (Fig. 5). For

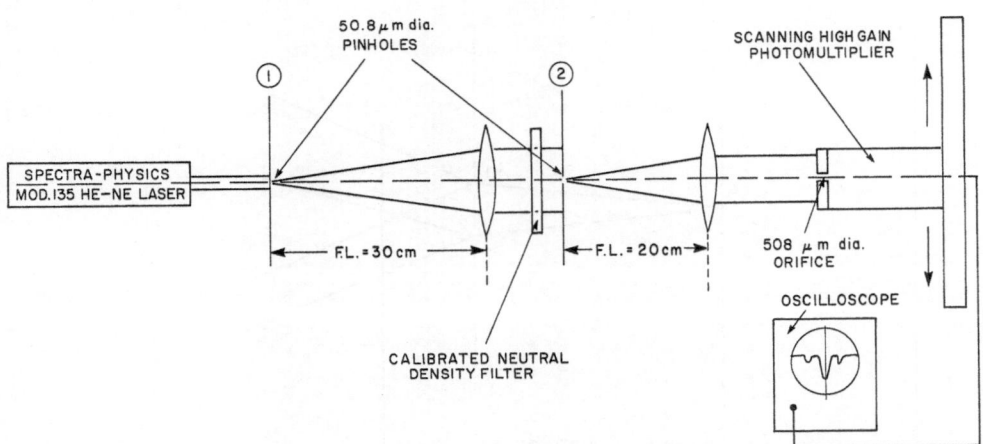

Fig. 5 Experimental apparatus used to reveal the effect of the
degree of statistical independence on the photon distribution on a diffraction pattern. With a photon flux
$\lambda = 1.90 \times 10^{10}$ photons/sec within the interferometer
(i.e. between pinholes 1 and 2), a clear diffraction
pattern is recorded on the oscilloscope. When a neutral density filter of transmission T = 2.09% is
inserted in the light path, so that the light flux is
reduced by a factor of 48 to $\lambda = 4 \times 10^8$ photons/sec,
the same clear diffraction pattern as before does not
appear, despite an overall increase of amplification of
the detection system by a factor of 441.

good fringe resolution, the photomultiplier is provided with a small
orifice of 5.08×10^{-2} cm diameter drilled on its front cover. The
fringe pattern is then vertically scanned and recorded on an
oscilloscope.

More in detail, the detection system consisted of a fourteen-stage,
flat-faceplate RCA photomultiplier type 7265 having a multialkali
photodiode ([Cs]Na$_2$KSb) with S-20 response. The photomultiplier current
amplification was 2×10^7. The tube was normally operated at 2000 V,
that is below the maximum permissible voltage of 2400 V, in order to
reduce the dark current from thermionic emission and to increase the
signal-to-noise ratio[19]. However, when maximum amplification was
required, the tube was operated at 2400 V. In order to further reduce
the dark current, the photomultiplier was cooled with a blanket of dry
ice to -15°C. Light uniformity over the photocathode area was achieved
by inserting a diffuser within the photomultiplier case, right behind the
entrance orifice. Finally, the signal from the photomultiplier was sent
to a Tektronik type 555 oscilloscope where it was recorded by means of a
type D high-gain preamplifier unit.

Fig. 6 reports the experimental results. As in the case when the
diffraction pattern was recorded with photographic film, Fig. 6a (left)
shows that, with the beam unimpeded by any filter and photon flux
$\lambda = 1.90 \times 10^{10}$ photons.sec^{-1} within the interferometer, the diffraction
pattern (Airy pattern) is clearly defined and is composed of a central
peak surrounded by two subsidiary maxima. The latters can be seen more
clearly in Fig. 6a (right), where the central peak has been amplified by

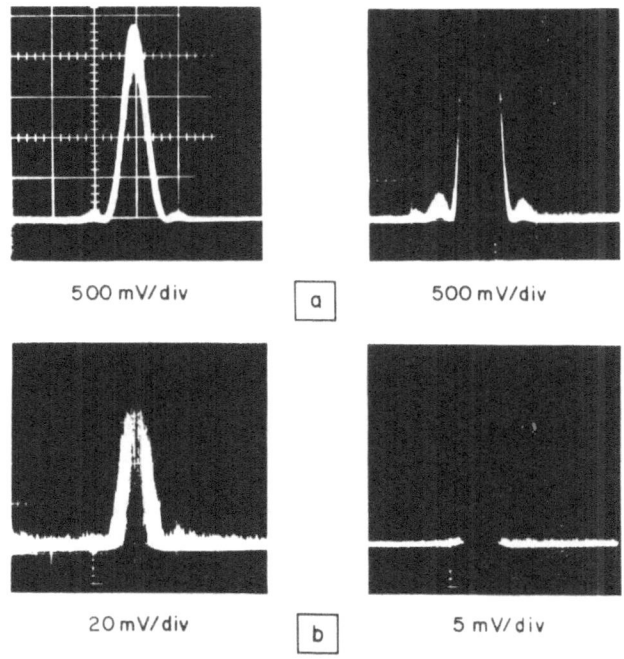

500 mV/div $\boxed{a}$ 500 mV/div

20 mV/div $\boxed{b}$ 5 mV/div

TIME SCALE : 0.5 sec/div

Fig. 6   a) Regular diffraction pattern obtained with a photon flux $\lambda = 1.90 \times 10^{10}$ photons/sec within the interferometer. b) The diffraction pattern is affected and the lateral fringes do not appear when the light flux is reduced to $\lambda = 4 \times 10^{8}$ photons/sec, despite the fact that the amplification of the detection system has been increased 441-fold while the light flux went down only 48-fold from (a) and (b).

a factor 2.55 to ~10.2 divisions (by 'division' we mean, of course, the separation between two consecutive solid horizontal lines on the photographic grid) from the original ~4 divisions, by increasing the photomultiplier voltage from 2000 V to 2200 V. According to classical optics[20], the first subsidiary maximum on the diffraction pattern should have an amplitude equal to 0.0175 times the central peak amplitude, that is

$0.0175 \times 10.2 = 0.178$ division

As Fig. 6a (right) shows, this is indeed so and the first subsidiary maximum is clearly seen. Actually, even the second subsidiary maximum is revealed, whose amplitude is[20]:

$0.0042 \times 10.2 = 0.042$ division

The fringes in this case are justified by considering that, with high probability, more than one photon are present within the interferometer at any one time. In fact, the probability analysis carried out in Sec. II.3.2.1 shows that, for the case of Fig. 6a:

$$\lambda\tau = 1.90 \times 10^{10} \times 1.4 \times 10^{-9} = 26.6 \gg 0.69$$

This means that the probability of interference with two or more photons is by far greater than the probability of interference with one photon and the fringes might then be created by packets of photons rather than single photons.

Now, if the same experiment is repeated with reduced light intensity, the fringes (or subsidiary maxima) should be seen again provided the amplification is sufficiently high to yield a fringe amplitude of 0.178 division or higher. We inserted therefore in the light path (Fig. 5) a calibrated neutral density filter of transmission T = 2.09% at the laser wavelength, thus reducing the light intensity within the interferometer by a factor of ~48 to ~4 × 10^8 photons.sec^{-1} ($\lambda\tau$ of the probability analysis is now 0.56 < 0.69) and obtained the picture of Fig. 6b (left) which shows only the central peak of amplitude ~2.5 divisions. No sign of fringes or subsidiary maxima is present in this picture. In an attempt to retrieve the fringes, the amplification of the photomultiplier was increased by a factor of 4.41 to its maximum value, by allowing the maximum permissible photomultiplier voltage (2400 V). Also, the oscilloscope amplification was increased by a factor of 4 from 20 mV/division to 5 mV/division in going from Fig. 6b (left) to 6b (right) (this means that the oscilloscope amplification was increased by 100 from the initial 500 mV/division – Fig. 6a – to 5 mV/division – Fig. 6b right). Despite an overall amplification of 441 in going from Fig. 6a (left) to Fig. 6b (right), the subsidiary maxima did not appear. This is surprising because the amplitude of the first subsidiary maximum on Fig. 6b (right) should have been:

0.0175 × 2.5 × 4.41 × 4 divisions = 0.77 divisions

i.e. larger than in Fig. 6a (right), where it was detected. Consequently it seems that the expected fringes did not exist.

In order to analyze more in depth these unexpected results and to discover if a valid reason exists for the absence of the fringes, we have reproduced in Fig. 7 the photographs of Fig. 6a and 6b (right) and added , another oscilloscope record (Fig. 7c) obtained with a much lower light flux of 7 × 10^7 photons.sec^{-1} within the interferometer, i.e. a photon flux lower by a factor of ~273 than the initial one. Moreover, beside each oscilloscope record, we show the same diffraction pattern drawn with a thin line passing in the middle of the baseline or in the middle of the broadened trace.

Now, if one looks at Fig. 7a, and more specifically at the figure on the right, one observes that the first subsidiary maximum has an amplitude of ~250 mV. If one reduces the light intensity by a factor of ~48, the amplitude of such subsidiary maximum should be reduced accordingly to ~5 mV. This signal is of sufficient amplitude and a clear upward displacement of the baseline in Fig. 7b at the position of the fringes should have occurred.*

---

*The absence of the fringes cannot be justified on the ground that, since the photomultiplier gives very short pulses, they do not overlap at the position of the fringes when the light intensity is weak, thus precluding the trace to elevate from the baseline. In fact, at the position B on the central peak at the same height as point A of the fringe (see Figs. 7a and 7b) the light intensity is just as weak. Although in B the trace does not elevate from the baseline by as much as it should, namely ~5 mV, still an upward displacement takes place by ~1 mV, whereas nothing of this happens on the fringes (A. Gozzini, C.W. McCutchen, E.S. Hanff, private communication).

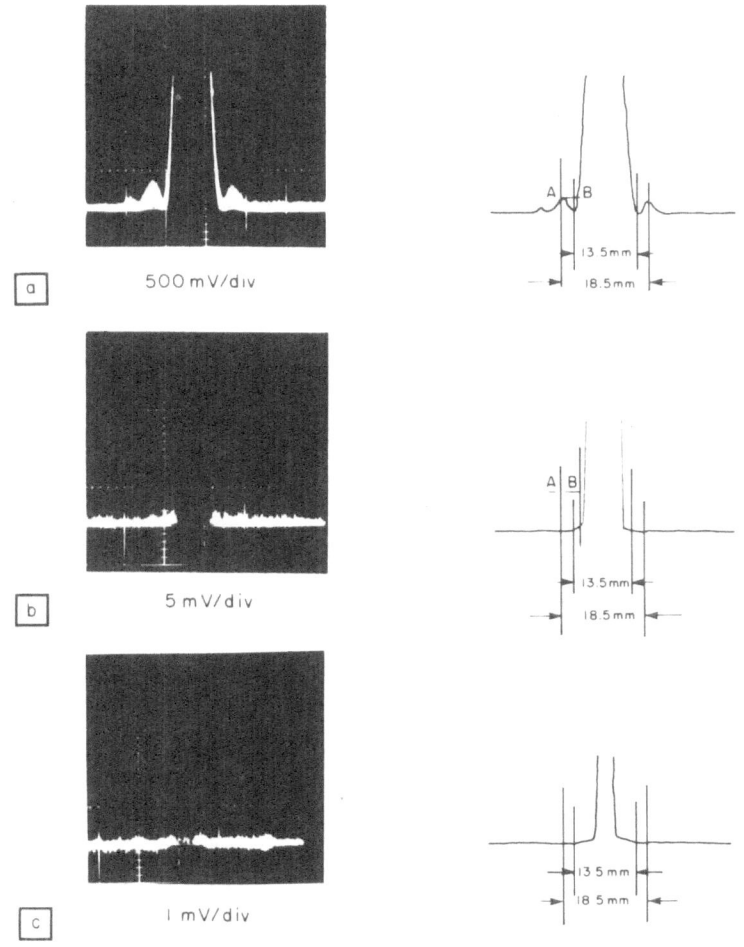

500 mV/div

5 mV/div

I mV/div

TIME SCALE : 0.5 sec/div

Fig. 7   The regular diffraction pattern obtained in (a) with a
photon flux $\lambda = 1.90 \times 10^{10}$ photons/sec within the
interferometer is not preserved and the lateral fringes
do not appear when the light flux is reduced to
$\lambda = 4 \times 10^8$ photons/sec (b) and to $\lambda = 7 \times 10^7$
photons/sec (c), despite the fact that the
amplification of the detection system has been
increased 441-fold from (a) to (b), and 2200-fold from
(a) to (c), while the light intensity went down only
48-fold and 273-fold, respectively.

On the other hand, the experimental results reported in Fig. 7 seem
to indicate a departure from the predictions of wave optics and an
apparent approach to the predictions of geometrical optics. Such
indication is provided in particular by Figs. 7b to 7c which show a
sudden discontinuity of light intensity at points which are closer to the
optical axis of the system as the light intensity goes down. In other
words, although the central fringe seems to be maintaining always, at the
light intensities we have investigated, a width of 13.5 mm, the light

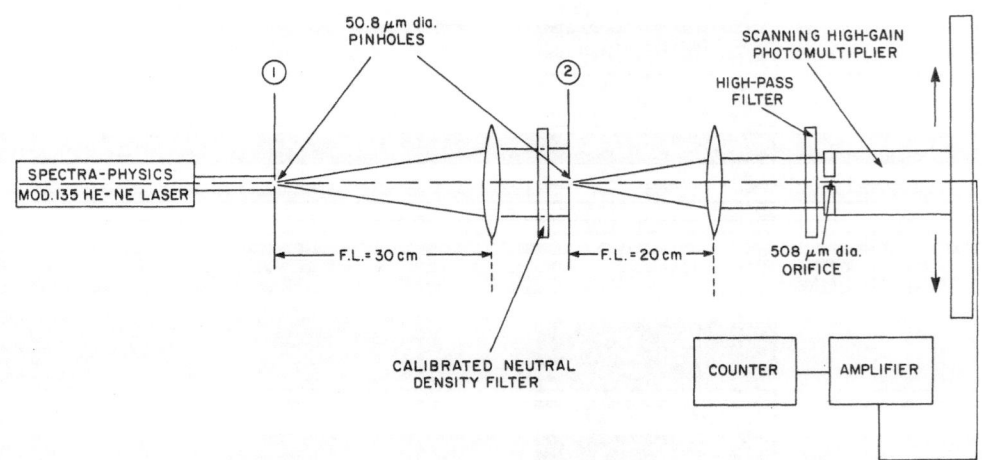

Fig. 8   Experimental apparatus used to reveal the effect of the
         degree of statistical independence on the photon
         distribution on a diffraction pattern. With a photon
         flux $\lambda = 1.90 \times 10^{10}$ photons/sec within the
         interferometer (i.e. between pinholes 1 and 2), a
         clear diffraction pattern is revealed with a counting
         time of 2×10^{-3} sec. After reducing the light
         intensity 769-fold with a neutral density filter
         inserted in the light path, the same diffraction
         pattern does not appear, despite having increased the
         counting time 1000-fold to 2 sec.

intensity distribution presents a sudden discontinuity (very similar to a
shadow effect) which is closer to the geometrical axis of the system as
the light intensity goes down.

To conclude this section, it seems that the absence of fringes is
due to nonlinearity of detection at very low light intensity, and this
assumption will receive a confirmation from the experiment to be reported
in the next section.

II.3.2.3 <u>Photoelectric detection and photon counting along a
diameter of the diffraction pattern</u>. Our latest experiment, in which we
counted the photons along a diameter of the diffraction pattern, was done
with basically the same experimental apparatus as previously described
(see Fig. 8). The photomultiplier was now operated at a constant voltage
of 2050 V. In order to eliminate any stray light entering the photomultiplier, the entire apparatus containing the laser and related optics was
enclosed within a black box, so that only a small opening was available
for the laser beam to get out of pinhole No. 2. As to the residual light
from the laser discharge tube going through the pinhole, it was cut
almost completely out by placing in front of the photomultiplier a high-
pass filter having transmission 84% at the laser wavelength $\lambda = 6328$ Å
and rapidly falling down to 0.03% at $\lambda = 5540$ Å . Finally, the entire
experiment was carried out in a small windowless dark room completely
shielded from any external light.

The experiment consisted in moving the photomultiplier by equi-
distant steps of 5/1000 of an inch (= 1.27×10^{-2} cm) and arresting it at
each step just for the time required for pulse counting. The counting
was done with a Tennelec 546P Scaler and 541A Timer, the signal from the

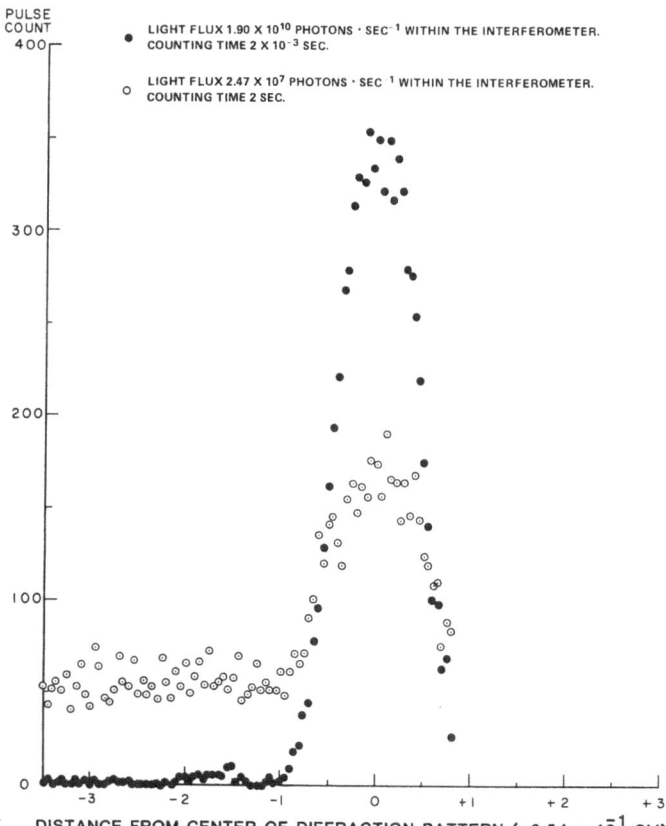

PULSE
COUNT

● LIGHT FLUX 1.90 X 10¹⁰ PHOTONS · SEC⁻¹ WITHIN THE INTERFEROMETER.
COUNTING TIME 2 X 10⁻³ SEC.

○ LIGHT FLUX 2.47 X 10⁷ PHOTONS · SEC⁻¹ WITHIN THE INTERFEROMETER.
COUNTING TIME 2 SEC.

DISTANCE FROM CENTER OF DIFFRACTION PATTERN (x2.54 x 10⁻¹ CM)

Fig. 9 <u>Solid circles</u>: regular diffraction pattern obtained
with a photon flux $\lambda = 1.90 \times 10^{10}$ photons/sec within
the interferometer and counting time 2×10^{-3} sec;
<u>Open circles</u>: the diffraction pattern does not have
the same amplitude as before when the light flux is
reduced 769-fold, despite having increased the counting
time 1000-fold to 2 sec.

photomultiplier having been amplified by a factor of 10 through an
amplifier having input resistance 1000 Ω.

The counting time was chosen rather short, 2×10^{-3} sec and 2 sec
for the two experiments that we ran, respectively, because this offered
some distinct advantages over long counting times. For one thing, one
avoids in this way problems of photomultiplier fatigue and decrease of
sensitivity[19]. For another, the dark count can be greatly reduced with
an appropriate choice of short counting time.

The experimental results are reported on Fig. 9. The solid circles
represent the counts obtained when the photon flux within the
interferometer (i.e. between pinholes 1 and 2) was 1.90×10^{10}
photons.sec⁻¹ (the average photon separation is 1.57 cm, much less than
the length of the interferometer 42 cm) and the counting time 2×10^{-3}
sec. The open circles are the counts obtained when the photon flux was
decreased 769-fold to 2.47×10^7 photons.sec⁻¹ by the insertion of a
calibrated neutral density filter along the light path (the average

photon separation is now 1214 cm, much greater than the interferometer length) and the counting time increased 1000-fold to 2 sec. One can see that the two diffraction patterns do not overlap (actually, the second pattern should be 30% greater than the first because of the factor $1000/769 = 1.3$). On the other hand, a well defined diffraction pattern appears in the first instance - the high light intensity case - with a clear fringe or subsidiary maximum on the left side of the central peak (the other on the right is absent because we did not scan the full diffraction pattern). Also, the fringe amplitude is what one would expect[20], namely 0.0175 times the central peak amplitude:

$$0.0175 \times 335 = 6 \text{ counts}$$

In the second instance, - the low light intensity case -, the diffraction pattern, besides lacking the lateral fringe, which can be justified because its amplitude is below the noise level, does not have the expected central peak amplitude of

$$\frac{1000}{769} \times 335 + 52 \text{ (average noise)} = 487 \text{ counts}$$

but only an amplitude of 163 counts.

In order to have a measure of the detection nonlinearity, we subtracted the noise-free signal amplitude of the low light intensity case from the expected noise-free signal, and divided the difference by the former amplitude:

$$\frac{435 \text{ (expected)} - 111 \text{ (found)}}{111 \text{ (found)}} = 2.91 = 291\%$$

a quite large nonlinearity.

In conclusion, these experimental results confirm the nonlinearity of photoelectric detection of the previous Sec. II.3.2.2. Moreover, they indicate that such nonlinearity, at very low light intensities, is no different, as far as the effects are concerned, from the nonlinearity of the photographic detection and that both constitute an obstacle for proving that we are dealing with a single particle phenomenon.

II.4. Discussion

In Sec. II.1 we stated that the wave-particle duality for single photons can be demonstrated only if a wave-phenomenon, such as an interference or diffraction pattern, is unequivocally associated with a single particle phenomenon, for which linearity of photon detection with light intensity is required. All three experiments reported above have shown that, at very low light intensities, the phenomenon is nonlinear. Moreover, they indicate that the flux for which the nonlinearities start to appear is of the order of 10^4 photons.sec^{-1} at the detector. It is interesting to find that, apparently, never before the linearity of photomultipliers response at such low light fluxes has been carefully investigated[19]. Fig. 10 reports the linearity characteristics of typical RCA photomultipliers[19]. It is to be noticed that these instruments are linear within a large range of photon fluxes (10^5–10^{13} sec^{-1}), but their linearity characteristics have not been tested right where they should be for our purposes of verifying the wave-particle duality hypothesis, namely below 10^4 photons/sec. In summary, because of the nonlinearities found, the wave-particle concept for single photons remains at the "status quo ante", namely as that of a theoretical hypothesis or postulate.

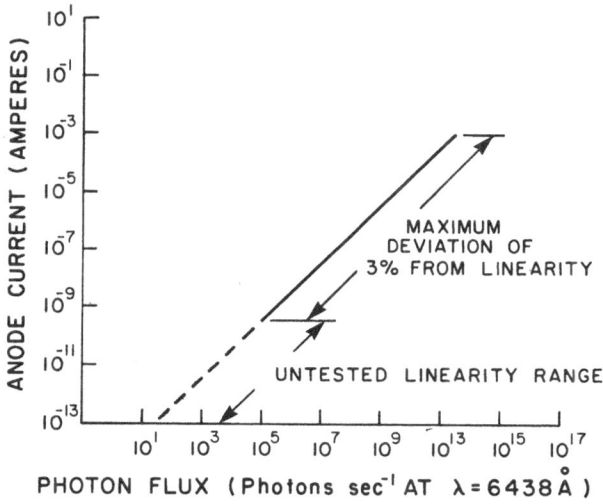

Fig. 10 Linearity characteristics of RCA photomultipliers[19].

One could explain, of course, the photoelectric results reported here in the same manner as it was done with the photographic results, in terms of some possible cause for the detection nonlinearity. One of these possible causes, for instance, is that the higher the light flux, the higher the noise generated within the photomultiplier. The problem with this approach is that it does not serve its purpose. In fact, the justification of the nonlinearity in this way will require the assumption that the wave-particle duality hypothesis is correct and that linearity of photoelectric detection with light intensity is to be expected. But then, any justification of the departure from such linearity cannot be used to prove the original hypothesis. To put it more clearly, a hypothesis (the wave-particle duality) cannot be proven by starting with the assumption that the wave-particle duality hypothesis is correct. What is required, in other words, is a direct and clear demonstration of linearity of photon detection with light intensity (at very low light intensities) in order to prove that we are dealing with a single particle phenomenon.

In the case of several photons within the interferometer, or in what we would call the regular intensity case, the wave-particle duality is proven: clear interference fringes appear and the phenomenon is linear. It is unfortunate, however, that we cannot unequivocally ascribe the wave phenomenon to single particles because there are many of them within the interferometer which could collectively act to create the fringes.

To summarize our results, Fig. 11 reports in graphical form, for comparative purposes, the two diffraction patterns obtained in the latest of our experiment with the photon counting technique. We observe that, at regular photon flux (= 1.95×10^7 sec^{-1}), such that the total number of photons reaching the detector is 39000, we obtain diffraction pattern B which peaks at ~350 counts. When we lower the photon flux to 2.53×10^4 sec^{-1} and let a larger number of photons (= 50600) reach the detector, we obtain diffraction pattern A of smaller amplitude (= 130 counts). The nonlinearity is clearly present.

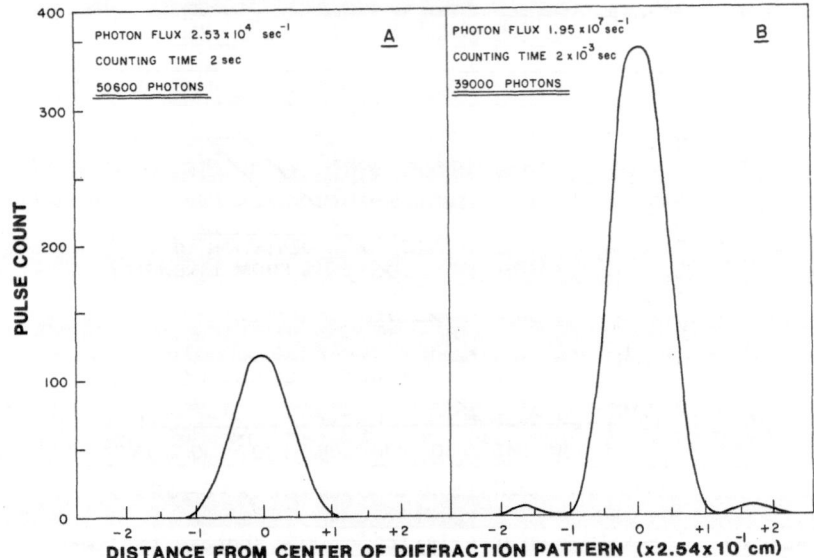

Fig. 11 At regular photon flux (= 1.95×10^7 photons/sec),
such that the total number of photons reaching the
detector is 39000, we obtain diffraction pattern B
which peaks at $\simeq 350$ counts. When we lower the
photon-flux to 2.53×10^4 photons/sec and let a larger
number of photons (= 50600) reach the detector, we
obtain diffraction pattern A of smaller amplitude
($\simeq 130$ counts). The nonlinearity is clearly present.

Such nonlinearity is not unique to our experiments. In a paper by
Reynolds, Spartalian and Scarl[21] (Fig. 12) diffraction patterns A and B,
were obtained with two photon fluxes of 200 sec^{-1} and 30 sec^{-1} (see the
Table below the diffraction patterns) such that the total number of
photons was 72000 and 14400, respectively, the ratio between these two
numbers being 5. We have measured the ratio of the densities of the two
photographs and found it to be 8.125. Thus, the nonlinearity present is
62.5%, a quite noticeable nonlinearity.

In conclusion, the series of experiments reported here on the
detection of diffraction patterns from a laser source at different low
light intensities confirms the wave nature of collections of photons but
tends to dispute it, or not to provide a clear proof of it, for single
photons.

It remains now to be seen why, in the past, several other
investigators obtained results which were apparently at variance from
ours. The purpose of the following section is to make a historical
analysis of those experiments in the light of what they intended to
prove.

II.4.1 <u>Historical analysis of the experiments and results upon
which the wave-particle idea was born</u>. The aim of this section is to
show that, historically, the experimental demonstration of the
association of a wave to a single particle was believed to be obtained
from the analysis of experiments that in no way proved that single
particles had wave nature.

A   B

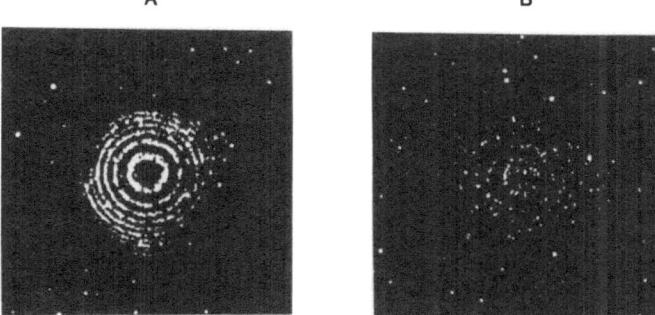

FIGURE	PHOTONS/SEC	EXPOSURE TIME	TOTAL NUMBER OF PHOTONS	RATIO
A	200	6m = 360 sec	72000	5
B	30	8m = 480 sec	14400	

Fig. 12   Diffraction patterns obtained by Reynolds, Spartalian
and Scarl[21] with a photon flux of 200 photons/sec (A)
and 30 photons/sec (B), such that the total number of
photons was 72000 and 14000, respectively, the ratio
between these two numbers being 5. By a density
measurement of the two diffraction patterns, it is
found that the ratio is 8.125. The nonlinearity is
therefore 62.5%, a quite noticeable nonlinearity.

When these experiments were performed (1909-1927) the quantum or
corpuscular nature of light had been well established through the
fundamental work of Einstein[22] which associated a frequency to the energy
of a photon in order to explain photoelectric effect phenomena. In 1923
de Broglie[23], recognizing that for light there exists a corpuscular
aspect and a wave aspect united by the relationship: energy = h times
frequency (h = Planck's constant), came naturally to suppose that, for
matter as well, such corpuscular and wave aspects should be
simultaneously present and postulated the relation $p = h/\lambda$ to account for
the wavelength of a particle having momentum p. It is to be noted that
in the above expressions for energy and momentum of a single particle the
frequency and the wavelength, respectively, are essentially
multiplication factors which can be defined, and measured, only if we
have a large number of particles disposed on a diffraction or
interference pattern. In other words, for a single particle, the concept
of frequency and wavelength would lose meaning in the absence of other
particles.

When these concepts were emerging and being developed, the idea that
a single particle might have wave nature even in the absence of other
particles came quite naturally from the evidence offered by some
experiments, initially done by Taylor[24], and later by Gans and Miguez[25],
Zeeman[26] and Dempster and Batho[27] whereby interference or diffraction
patterns had been obtained at intensities of light so low that it was

127

reasonable to assume that each and every photon recorded on a photographic plate had crossed the interferometer in isolation. Since it was believed that the interference pattern had been constructed by one photon at a time, it seemed that each photon was endowed with some sort of wave nature, in the sense of being able to choose, or being guided in a particular direction and of landing on a prescribed point of the interference pattern. The detector, namely the photographic detector, was believed at that time to be a linear device in the sense of being able of recording every single photon, or, more precisely, a number of photons proportional to the light intensity. In other words, the appearance of a wave phenomenon like the interference fringes on a photographic plate, together with the assumption that the detector was linear, were believed to be sufficient and conclusive arguments in favour of the idea that each and every photon was endowed with a wave nature.

Of the above mentioned experiments those by Taylor[24], and by Dempster and Batho[27] are the most relevant, and have been widely quoted, especially the latter, as proof that single photons have wave properties. The experiment of Taylor was very simple. He used the shadow of a needle as diffraction pattern and a narrow slit placed in front of a gas flame as a source of light. The intensity of light was reduced by means of smoked glass screens, and the detection of the diffraction pattern was done with photographic plates. In order to know the intensity of light for each screen used, these were calibrated with the same apparatus by assuming that the photographic detector was linear. In other words, by changing the screens inserted between the source of light and the photographic plate, the times of exposure necessary to have the same blackening of the plate were taken as inversely proportional to the light intensities. However, as Taylor himself points out[24], this assumption "...showed...to be true (only) if the light was not very feeble". Therefore, at the feeble light intensities that he used in the course of his experiments, the linearity of the detector, which was taken for granted, was not really true and the appearance of the diffraction patterns was not a proof that it was produced by single photons. And that indeed Taylor was using very low light intensities is confirmed by the fact that the longest exposure time for his plates was 2000 hours or about 3 months. More specifically, he quotes using light of flux 5×10^{-6} ergs/sec, which corresponds roughly to 1.26×10^6 photons/sec at $\lambda = 5000$ Å, with an average photon separation of 2.38×10^4 cm, and such average separation was then assumed to be the true separation.

Much more sophisticated experiments were those of Dempster and Batho[27]. In their first experiment, these authors used an echelon grating and a prism in order to obtain interference patterns of the light coming from a narrow slit placed in front of a discharge tube containing helium at low pressure. By carefully controlling the current through the tube they were able to vary the intensity of the helium line emitted by the discharge, the calibration of such source being done by comparison with a black-body radiator of known temperature. Photographs of the interference patterns were taken at various intensities and exposure times. The longest time was 24 hrs during which the source emitted an average of 95 photons/sec, with an average photon separation of 3.15×10^8 cm. This average separation was then assumed to be the true photon separation, thus leading the authors to conclude that the interference pattern observed was due to the fact that "...a single light quantum can produce effects that are due to its passing through several steps of the echelon simultaneously ...". In other words, although these authors had no proof that the initial assumption, namely that the photons were separated, was correct, they used it to ascribe the phenomenon they were observing, namely the interference pattern, to separated or isolated photons.

The second experiment by Dempster and Batho[27] put more stringent conditions on the light source, by further reducing the current flowing through it to values such that only one atom at a time was believed to emit radiation. They came to this conclusion from the consideration that, "...if the time of each radiation emission process (which they knew to be $\le 5 \times 10^{-8}$ sec) multiplied by their number was much less than one second, the individual quantum of radiation from any atom must be separated in time from the radiation emitted by other atoms". In other words, they considered as true time separation of the emission of the photons what was instead just an average separation. And with this assumption they proceeded with the experiment in which they now used a Fabry-Perot etalon as interferometer, and found that the interference fringes on a photographic plate appeared even with a light intensity so low that 26 hours exposure time was needed.

The experiments by Taylor[24] and by Dempster and Batho[27], briefly summarized here, as well as those by Gans and Miguez[25], and by Zeeman[26], made all use of photographic plates as light detectors. What they observed therefore was never interference or diffraction patterns created by single photons, but by groups or clumps of photons. This is because, as we have shown in Sec. II.3.1 and subsequent theoretical analysis, a photographic film, being a nonlinear detector, is not capable of revealing single or isolated photons. In Sec. II.2 we said that a photographic grain [9-11], when exposed to light, does not become developable unless it absorbs at least three or four photons within a time which can be assumed to be of the order of the coherence time of the light[15]. This means that a photographic grain acts as an R-fold coincidence counter, where R is of the order of 3, 4 or more[28]. The coherence time τ, which is the inverse of the bandwidth, in the typical case of thermal light (as used by the above authors) of peak wavelength $\lambda_p = 5000$ Å and bandwidth $\Delta\nu = 10^{12}$ sec^{-1} (corresponding to $\Delta\lambda = 10$ Å) is of the order of 1 psec. Photons, therefore, in order to be detected, have to be confined within a distance of the order of the coherence length $c.\tau = 3 \times 10^{-2}$ cm from the target grain. This coincidence counter model of photographic detection has been successful in explaining the effect known as "reciprocity failure" at both high and low intensity of light,* whereby a photographic detector does not respond to just the total number of photons fT (f = photon flux; T = exposure time) incident during time T but, due to the short time of persistence of the counting mechanism, can record a count only when three or more photons impinge on the grain area within that time of persistence. With extremely low light flux very few events are recorded and the detector shows clearly its nonlinearity, because the probability of having at least three photons near the grain area within the short counting time is negligible.

Such detection nonlinearity has profound consequences. Had they known, at the time the wave-particle duality hypothesis was being put forward, that the experiments which seemed to prove directly and conclusively such hypothesis for single particles had never recorded single photons but bunches or packets of at least three photons (and certainly more), the wave particle concept would have been more logically ascribed to collections of particles rather than single particles. In short, the widely referred to experiments by Taylor[24] and by Dempster and Batho[27] in no way provided a proof that single photons have wave nature.

---

*Actually we believe that such an experimental nonlinear effect triggered the research that ultimately led to the photon counter model.

In order to justify, however, why a photographic film was believed to be a linear detector, we think that such assumption was the most logical one at that time because previous experiments on the photoelectric effect[29] had proven the linearity of the latter phenomenon. Actually, the very notion of a photon as a single entity with well defined particle properties was a natural consequence of the linearity of the photoelectric effect. And the photographic effect was believed to be identical to the photoelectric effect.

Unfortunately, it was not realized that the photoelectric effect investigated at that time with such light sources as arcs, sparks and mercury vapour lamps[30], from which intensities of light greater than 0.007 foot.candle = 3.06×10^{10} photon.cm^{-2}.sec (at $\lambda = 5500$ Å) were used, was so far away from the extremely low light intensities of the experiments of Taylor[24] and of Dempster and Batho[27] that nothing could be said about the linearity of the phenomenon in this latter range. In other words, besides considering the photographic effect as linear, which we now know is not, they assumed that the linearity stemmed from the photoelectric effect as linear, which they had not investigated at low light intensities.

Proceeding further with this historical review, it should be said that, even if the photographic detection, as we now know, had not really revealed single photons, still several experiments of photoelectric detection done in subsequent times by Janossy and Naray[31], and Pfleegor and Mandel[2], to quote the most relevant of them, experiments done with extremely low light intensities, seemed to have been able to prove that single photons have wave nature, because invariably an interference pattern appeared even when the photons were supposed to cross in isolation the interferometer. Therefore, the matter seemed to be settled anyway.

Again, no one of these experiments proved their thesis because invariably they were done at a single light intensity, no matter how low, and the linearity of the phenomenon vs. light intensity, which is so important in order to know whether or not we are dealing with a single particle phenomenon, was not verified. Moreover, even the recent coincidence experiments by Clauser[3], and by Grangier, Roger and Aspect[4] are not conclusive in this regard, as we shall soon prove.

Let us describe briefly the experiments. The one by Janossy and Naray[31] made use of a Michelson interferometer and an interference pattern was obtained with a light source consisting of a selected line ($\lambda = 5461$ Å) of a mercury discharge tube. A photomultiplier coupled to a photon counting system was used to plot the interference pattern. The photon flux from the source was 2×10^{10} photons/sec and this flux was reduced to 2×10^{6} photons/sec with a calibrated grey filter. Such filter was placed alternately before and after the interferometer, along the optical path (see Fig. 1 for a better understanding of the experiment), and in both cases the same interference pattern was recorded by the photodetector. This led the authors to conclude that, since the interference pattern was not affected by the intensity of light within the interferometer, isolated photons within the interferometer can be simultaneously present in both arms of the interferometer and can interfere.

Clearly, this conclusion would be right only if the authors had provided us with a proof that the photons were isolated within the interferometer. In order to prove that they were dealing with single, or isolated photons, these authors should have illuminated the detector with different intensities of light I and exposure times Δt (or counting times), such that the product I Δt remained constant and should have observed whether or not the interference pattern remained the same at the very low light intensities they were using. If it remained the same, the phenomenon was linear with light intensity, and therefore it was a single photon phenomenon, and the association of a wave (the interference pattern) to single or isolated particle would have been justified. Lacking such an experimental demonstration, the authors could only claim that they had observed a wave phenomenon at very low light intensity, but had no way to ascribe it to single or isolated particles.

The experiment of Pfleegor and Mandel[2] made use of two independent light sources, namely two single-mode lasers (Fig. 13), and interference effects were detected with a photon counting technique, the photon counting beginning only when the frequencies of the two lasers happened to be almost the same, the difference being less than 50 kHz. The counting was done for 20 μsec, which was the coherence time of the lasers used. These authors found that interference effects appeared even under conditions where the light intensity was so low that the mean interval between photons was great compared with their transit time through the apparatus. This led them to conclude that "the effect cannot be readily described in terms of one photon from one source interfering with one from the other", and that the famous statement by Dirac that "each photon interferes only with itself... appears to be as appropriate in the context of this experiment as under the more usual conditions of interferometry".

Clearly, in this case too, these authors drew their conclusion from an assumption, namely that the mean separation between photons was the true separation, which was unwarranted. Therefore their conclusion that a photon interferes only with itself remains another unproven assumption, not much different from the original assumption, namely that the photons were isolated within the interferometer.

Moreover, if one looks carefully at the Pfleegor and Mandel experimental apparatus, as shown in Fig. 13, one finds the claim by these authors that the photons reached the interference detector in isolation, to be unwarranted. In fact, the beat frequency detector and the 50 KHz filter used to open the gate of the two photomultipliers that revealed the interference fringes could be activated only if a large number of photons of frequency difference less than 50 KHz where picked up simultaneously by the two beamsplitters S. These photons crossed simultaneously the two strong attenuators and a few of these photons reached simultaneously the interference detector.

Let us now examine some very recent experiments done by Clauser[3] and by Grangier, Roger and Aspect[4]. These fall into the category of the so-called "coincidence experiments", whose motivation is the following. In order to prove that in an interferometer one is really dealing with one photon at a time, it is sufficient to reveal what happens to a photon right at the time when it enters the interferometer through a beam splitter. Clearly, this isolated photon has to choose one or the other path, namely the transmitted or the reflected path. It cannot, of course, choose both because a photon is an indivisible entity. Consequently, if one positions behind the beam splitter two photomultipliers, one to detect the transmitted photons and the other to reveal the reflected photons, their signals will never be coincident, if the photons are isolated. And this is what these authors claim to have found.

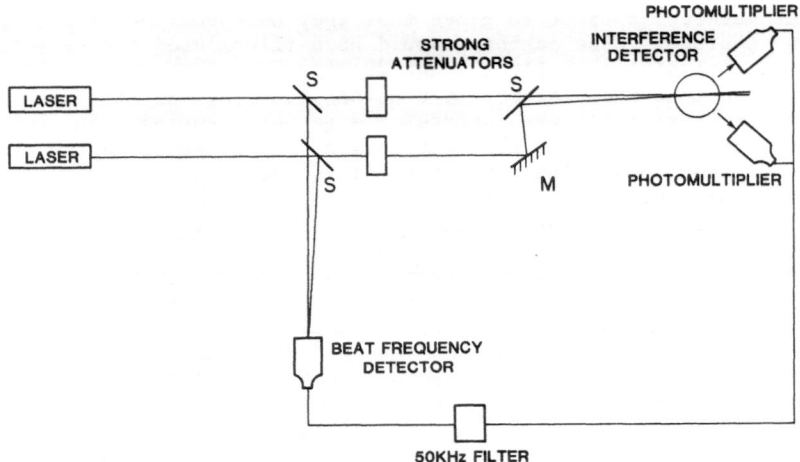

Fig. 13 Pfleegor and Mandel[2] experiment with independent light sources.

A few words of caution are in order, however, in these experiments. As we shall prove in the following, the limited quantum efficiency of the detectors precludes the knowledge that one is dealing with single photons, and actually the experimental results seem to indicate that the photons are not isolated at all. In other words, although these authors can rightly claim that they have ruled out any "classical" splitting of the photons in two by the beam splitter, they have not proven that they were dealing with single or isolated photons within the interferometer, i.e. between the beam splitter and the photodetectors.

Let us examine carefully the experiment by Clauser[3], as shown in Fig. 14. This author had 4 photomultipliers PM_1a, PM_1b, PM_2a and PM_2b viewing cascade photons emitted from a source S and travelling on opposite directions towards beam splitters BS_1 and BS_2. Coincidences are found by both pairs of photomultipliers PM_1a, PM_1b, and PM_2a, PM_2b. These are termed "accidental" coincidences, induced by emission from two different excited source atoms. "Real" coincidences are instead defined as those found by photomultiplier pairs PM_1a, PM_2b, and PM_1b, PM_2a because they view, through interference filters, different photons of wavelength λ_1 and λ_2 emitted by the same cascading atoms.

Keeping these definitions in mind, we would like to verify whether or not this author has been able to prove that one photon goes one path, or the other. Clearly, the experiment did not bring any light on this question because the photomultipliers used had a limited quantum efficiency. Considering the case of photomultipliers PM_1a and PM_1b, which have 15% quantum efficiency, this means that they need, an average, between 6 and 7 photons before releasing an electron. This has the consequence that one cannot have any knowledge about the arrangement of these 6 or 7 photons, namely, if they are isolated one from the other, or if they travel in group, etc. Actually, the "accidental" coincidences found can be really due to a splitting of the groups in two components at the beamsplitter.

This analysis leads to further conclusions. For instance, if the two photomultipliers had higher quantum efficiency than the above, say 30% as is the case of photomultipliers PM_2a and PM_2b, since they will require now only between 3 and 4 photons in order to reveal one of them,

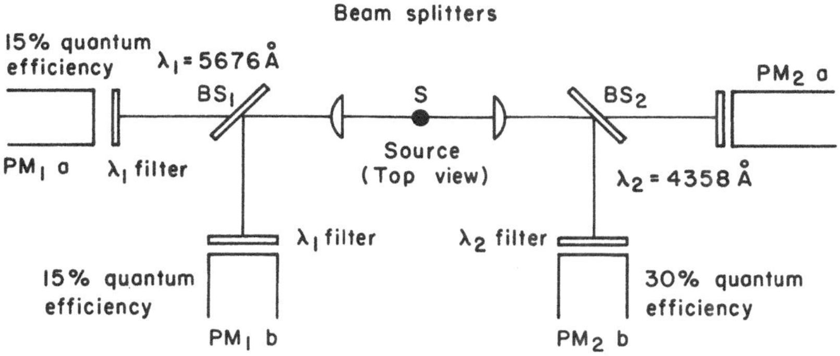

Fig. 14 Clauser's[3] experiment with cascading photons.

the number of coincidences will necessarily increase. This is indeed
what is found with photomultipliers pair PM_2a, PM_2b. Here again,
however, we can say nothing about the arrangement of those 3 or 4
photons, i.e. if they are isolated one from the other of if they go
through the beamsplitter as a group.

If we now introduce a time delay τ in the measurement of the
coincidences, the number of coincidences will be the same as for zero
time delay because the emission from the source goes on continuously and
the two photomultipliers will consider as coincident those two photons in
the group of N required photons that have a time separation equat to τ.
This is what the experiment reports.

If the coincidences are detected with photomultipliers PM_1a and
PM_2b, the first having 15% quantum efficiency and the second having 30%
efficiency, the number of coincidences, for the same time delay τ will be
larger than that of the pair PM_1a, PM_1b, but smaller than that of the
pair PM_2a, PM_2b. This is what the experiment reports.

Finally, if the coincidences between PM_1a and PM_2b are revealed at
zero time delay, since now for sure the group of 3 or 4 photons of
wavelength λ_1 and λ_2 are emitted simultaneously by the same atoms, the
number of coincidences detected will be determined by the photomultiplier
with the highest efficiency, and this is again what has been found in the
experiment.

In conclusion, it must be stressed that the experiment by Clauser[3]
has great importance, in the context, however, of the limitation of the
efficiencies of the detectors, which prelude any detailed knowledge about
whether or not one is dealing with one photon at a time within the
interferometer.

The recent experiment by Grangier, Roger and Aspect[4], as depicted in
Fig. 15, leads to the same conclusion. In this experiment, a source S
emits pairs of photons of different frequencies ν_1 and ν_2. The detection
of ν_1 by photomultiplier PM_1 acts as a trigger for a gate generator,
which enables two photomultipliers PM_t and PM_r located along the
transmitted and the reflected path from beamsplitter BS, respectively, to
view ν_2 for a duration 2ω, where ω is the cascade lifetime. These
authors aimed at comparing the "classical" prediction of photon splitting
at beamsplitter BS with the quantum mechanical prediction that a photon
cannot be split at a beamsplitter. The results clearly favoured the
latter predictions, which therefore ruled out the splitting of a photon.

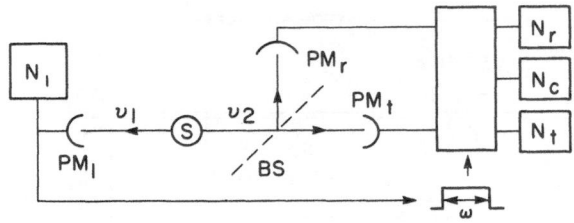

Fig. 15 "Single-photon" coincidence experiment by Grangier et al.[4]

From this conclusion these authors however arrived by inference that the photon goes "alone", i.e. isolated from all the others, in the path between beamsplitter and photomultipliers. This conclusion is unwarranted. The reason is the following. The photodetectors used, for which the quantum efficiency was not stated, have limited quantum efficiency. Therefore they need N photons before releasing an electron. Assuming, for instance, a 10% quantum efficiency, one electron is relased on average for every 10 photons impinging on the photocathode. This therefore precludes any knowledge about the arrangement of these 10 photons, i.e. whether they are isolated one from the other or grouped together. On the other hand, the very low intensity source of light used in this type of experiment (= 4×10^7 photons/sec into the 4π steradiant) combined with the very short gate time (= 4×10^{-9} sec) made it very difficult for the two photomultipliers to detect a large number of coincidences, but these were detected anyway. Hence, it is difficult to understand how these authors can claim that they were dealing with single photons. Moreover, it is difficult to accept the claim, following the results from another experiment as depicted in Fig. 16, in which a Mech-Zehnder interferometer was built around beamsplitter BS1 and the fringes were detected with photomultipliers MZ1 and MZ2, that the fringes were created by "single photons", or one photon at a time within the interferometer, when the proof that they were dealing with single photons was never provided.

The detection of coincidence in the experiments by Clauser[3], and by Grangier, Roger and Aspect[4], is in line with the finding of the famous experiment by Hanbury-Brown and Twiss[1] (Fig. 17). In this experiment the 4358 Å line from the mercury lamp was isolated by a system of filters, and the beam was divided by a beam splitter to illuminate the cathodes of two photomultipliers, one fixed and the other movable on a horizontal slide. Two amplifiers and a correlator completed the apparatus. The results showed "...beyond question that the photons in two coherent beams of light are correlated, and that this correlation is preserved in the process of photoelectric emission."

To summarize, in order to prove conclusively that single photons have wave nature (and this applies just as well to other particles, such as electrons and neutrons), one needs to have the following two conditions satisfied:

1. An interference pattern should appear at a light intensity so low that the assumption that the photons are isolated within the interferometer is a plausible one. The presence of the interference fringes in this case would prove that we are dealing with a wave

134

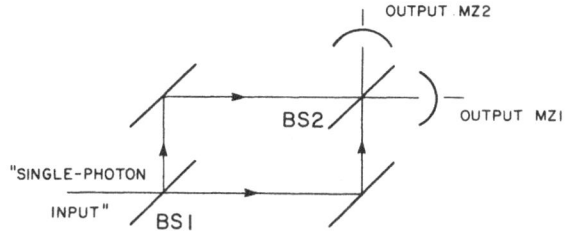

Fig. 16 Interference effects with "single photons"[4].

phenomenon, but it would not prove that it is a single-particle phenomenon;

2. If the experiment is repeated now at lower (higher) light intensity and longer (shorter) recording times, such that the total number of photons is the same as before, the same interference pattern should appear. This would prove that the phenomenon is linear, or that one is dealing with a single-particle phenomenon, thus granting the association of a wave to a single particle.

In conclusion, the reference that is always made to the experiments done in the past as proof that single photons have wave nature is not entirely appropriate, because all those experiments satisfied only the first of the above two conditions and therefore the wave pattern obtained could not be unequivocally ascribed to single photons. In Sec. III below, we shall take up this matter again and put forward some arguments, based on the results of our experiments, of those of the classical Hanbury-Brown and Twiss[1] experiment on photon correlations, and of those examined in the previous historical section II.4.1, in favour of an explanation that interference and diffraction are collective, rather than single-particle, phenomena. In other words, while it will be confirmed that collections of particles have wave nature, as proven by all experiments carried out either in the past or in the present, it will be argued that single particles either don't have wave nature or, if they have it, it has not been demonstrated as yet.

III. THE "PHOTON CLUMP" MODEL

Several facts are now available from experiments done with light that, when put together, provide an interpretation according to a model of light that we shall soon advance. But, let us first summarize the experimental facts:

1. When one deals with regular light intensities, clear interference takes place and the phenomenon is linear with light intensity.

2. At low light intensities, the wave phenomenon is still clear, but the phenomenon is nonlinear with light intensity.

3. The experiments of Hanbury-Brown and Twiss[1], who detected a correlation between light beams of narrow spectral width, show that the photons are clumped, or that they travel in packets, rather than in a uniform stream.

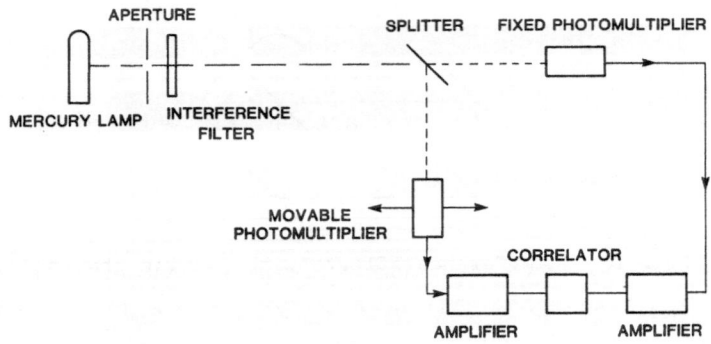

Fig. 17   Hanbury-Brown and Twiss experiment[1].

4. The experiment of Pfleegor and Mandel[2] shows that one gets interference even with independent light sources.

5. The experiments by Clauser[3] and by Grangier, Roger and Aspect[4] show that one gets photon "accidental" coincidences at a beam splitter, thus proving that "accidental" correlations between light beams exist beyond a beam splitter.

All the foregoing facts, as we shall soon prove, indicate that the single-particle concept of the photon should be modified into a model in which the photon is invariably accompanied by a collection of other photons, all clumped together, and that the individual photons in the clump are arranged in a geometrical wave pattern, with maxima of distribution in one area and minima in another. As said in the Introduction, an interferometer is then an instrument that does not do anything to the photons to let them interfere (because they have already imprinted in the clump that carries them all the characteristics of interference) but, by changing slightly the direction of motion of two outgoing clumps of photons originating from a single clump, makes them change the initial geometrical arrangement into an arrangement which can be clearly seen as a wave pattern. In the light of the above, the famous statement by Dirac[5] that a photon interferes only with itself can be interpreted as meaning that a cluster of photons carries with it a wave pattern of photon distribution.

In the following the photon clump model will be initially outlined mainly in pictorial form, considering the various situations in which the clump is split at a beam splitter, or when it enters an interferometer, or when it crosses a pinhole etc. Once this model has been qualitatively described, it will be confronted with the set of experiments that we have analyzed in Sec. II.4.1 in order to see how far it can go in their interpretation. Finally, in Part IV of this review paper, it will be shown that the theoretical analysis leads straightforwardly to a justification of the "photon clump" model.

III.1. Model for a Clump of Photons

Fig. 18 reports in pictorial form a clump of photons out of a light source. It is made up of strings of photons separated by a constant distance equal to λ in the direction of motion. Transversally, i.e. perpendicularly to the direction of motion, the strings of photons have distances increasing as one moves away from the center string. The photon clump has all the characteristics of a wave distribution of photons within it, with maxima and minima. Clearly, one cannot see this

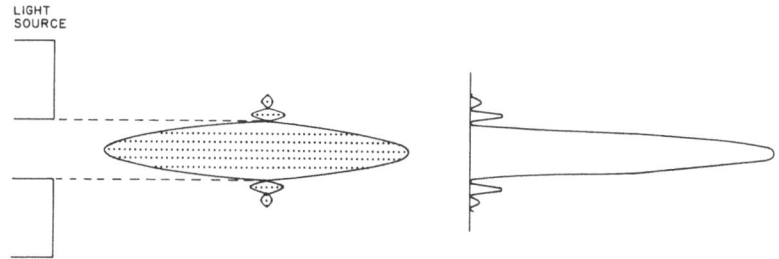

LIGHT
SOURCE

Fig. 18   Model for a clump of photons.

distribution in a normal clump, because the maxima and minima are
narrowly separated. Because of the structure of a clump, its overall
photon distribution look like a regular diffraction pattern at normal
light intensities.

III.1.1. <u>Transmission of a clump through a thin filter.</u> Fig. 19
shows that photon clumps crossing a low density thin filter, say having
50% transmission, keep intact their structure, only the number of photons
decreases in each clump as a function of the filter transmission. In
other words, with a thin filters the transmission is linear with light
intensity.

III.1.2. <u>Transmission of a clump through a thick filter.</u> Different
is the situation when clumps of photons tend to cross a thick filter, of
very high density and low transmission (Fig. 20). The clumps, although
they try to preserve their structure, tend to loose the weakest of all
photons attached to them, namely the side photons. Moreover, not all of
them emerge from the strong filter with a number of photons linearly
related to the filter transmission. These clumps now contain less
photons than expected from the filter transmission. In other words, with
a thick filter the transmission is nonlinear with light intensity.

III.1.3. <u>Splitting of a clump into two clumps at a beamsplitter.</u> A
beamsplitter breaks the original clump into two smaller clumps (Fig. 21),
the reflected clump and the transmitted one. Each clump has the same
number of photons. Only when the original clump is made up of a very
small number of photons, the transmitted and the reflected clumps can
have an unequal number of photons.

III.1.4. <u>Interference of two clumps originating from a single
clump.</u> The original clump is split at the entrance beam splitter (Fig.
22) into two clumps, which recombine out of the interferometer. Since
the two clumps, on recombination, intersect at a very small angle, a
rearrangement takes place in the strings of photons and these now are
more widely separated than in the original clump. The strings of
photons, being now widely separated, can be clearly observed as fringes.

III.1.5. <u>Diffusion of a clump of photons passing through a pinhole.</u>
When entering a pinhole, a clump of photons is left with much less
photons than in the original clump. The clump now diffuses in space and
expands (Fig. 23).

III.1.6. <u>Formation of clumps from photons randomly emitted by a
light source.</u> Fig. 24 shows how the tendency of photons to bunch
together after they have been emitted from a light source leads to the
formation of clumps.

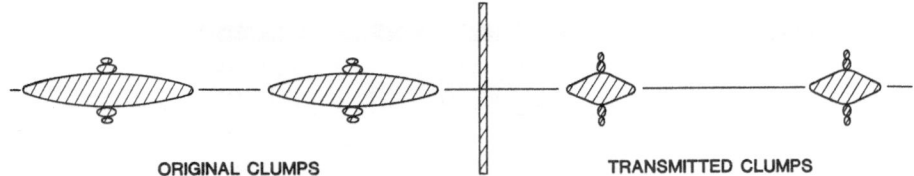

FILTER – TRANSMISSION 50%

ORIGINAL CLUMPS   TRANSMITTED CLUMPS

Fig. 19   Transmission of clumps through a thin filter.

From the above qualitative model of a clump of photons and its behaviour in various experimental situations we shall proceed and employ it in the interpretation of the previoulsy mentioned optical experiments.

III.2. Explanation of the Experimental Results with the Assistance of the Photon Clump Model

As far as the experimental nonlinearities found in our experiments as well as in the experiment by Reynolds et al[21], these are explained by the fact that, at extremely low light intensities, the clumps of photons tend to loose their structure and to break up in smaller clumps containing less photons than expected.

III.2.1. <u>The Hanbury-Brown and Twiss experiment</u>. Fig. 25 shows how the Hanbury-Brown and Twiss experiment[1] is explained with the photon clump model. The original clump is split into two smaller clumps at the beam splitter. These proceed towards the photomulipliers which will detect coincidences when the photomultipliers' openings, as seen from the source of light, are superimposed. When the movable photomultiplier is displaced and the photomultipliers' openings are no longer superimposed, the coincidences deteriorate, as expected.

III.2.2. <u>The Pfleegor and Mandel experiment</u>[2]. The beam splitters S along the optical paths of the two lasers (Fig. 26) detect simultaneously two clumps of photons and open the gate that activates the two photomultipliers used to reveal the fringes. The two clumps of photons, after passing through the strong attenuators, will emerge as two smaller clumps which will reach the interference area simultaneously.

III.2.3. <u>The Grangier, Roger and Aspect experiments</u>[4]. The "single-photon" coincidence experiment by Grangier et al[4], as well as Clauser's experiment[3], are easily explained with the clump model of light (Fig. 27). The clump ν_1, which is made up of several photons, triggers photomultiplier PM_1 which opens the gate of the counters N_r, N_t, and N_c. The coincidences found are real coincidences of the two clumps ν_2 emerging from beam splitter BS. Many more coincidences would have been detected if photomultipliers PM_r and PM_t had 100% quantum efficiency, or if the light source used by the investigators were of intensity higher than 4×10^7 photons/sec. Moreover, the alignment of photomultipliers PM_r and PM_t is critical for the detection of coincidences, as the experiment of Hanbury-Brown and Twiss[1] clearly points out.

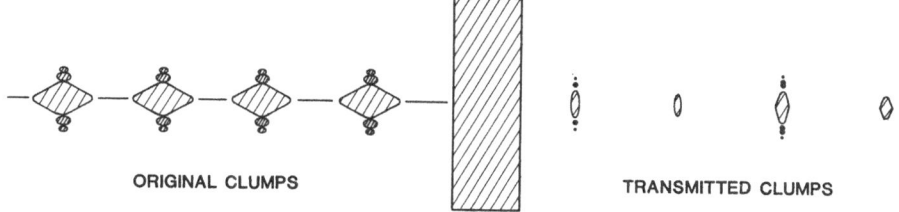

FILTER – TRANSMISSION 0.02%

ORIGINAL CLUMPS

TRANSMITTED CLUMPS

Fig. 20 Transmission of clumps through a thick filter.

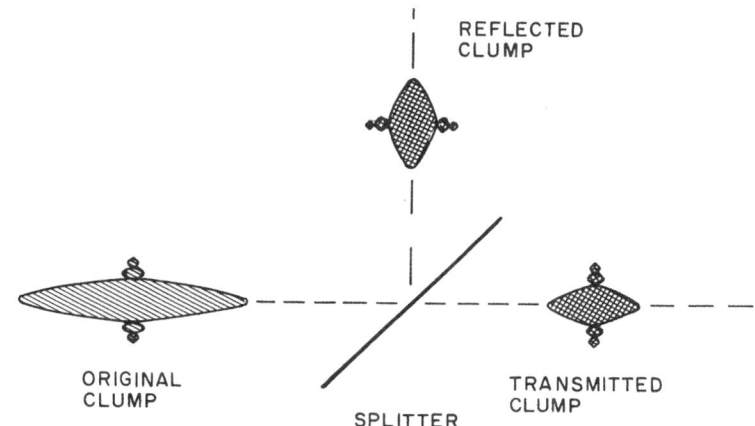

REFLECTED CLUMP

ORIGINAL CLUMP

SPLITTER

TRANSMITTED CLUMP

Fig. 21 Splitting of a clump into two clumps.

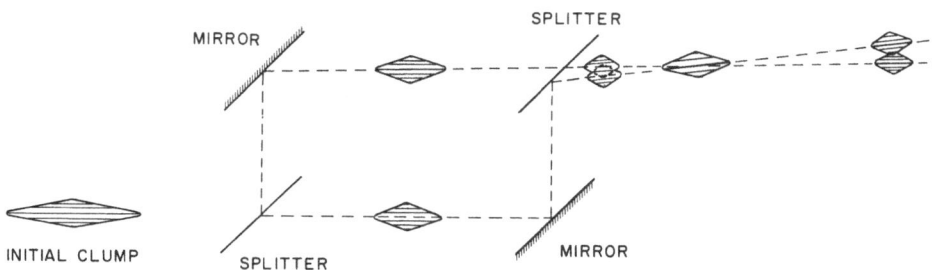

MIRROR

SPLITTER

INITIAL CLUMP

SPLITTER

MIRROR

Fig. 22 Interference of two clumps originating from a single clump.

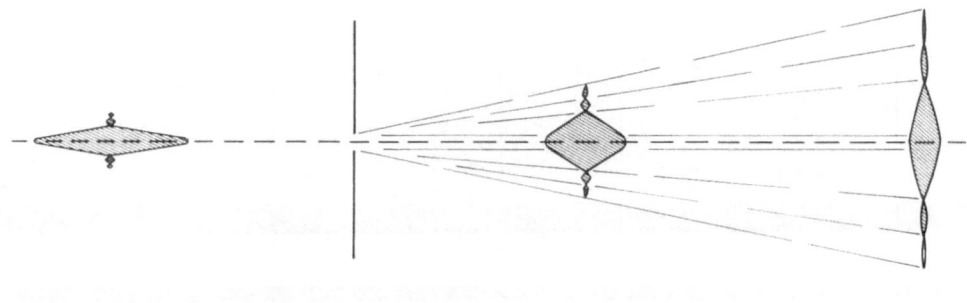

PINHOLE   EXPANSION OF A CLUMP

Fig. 23 Diffusion of a clump of photons through a pinhole.

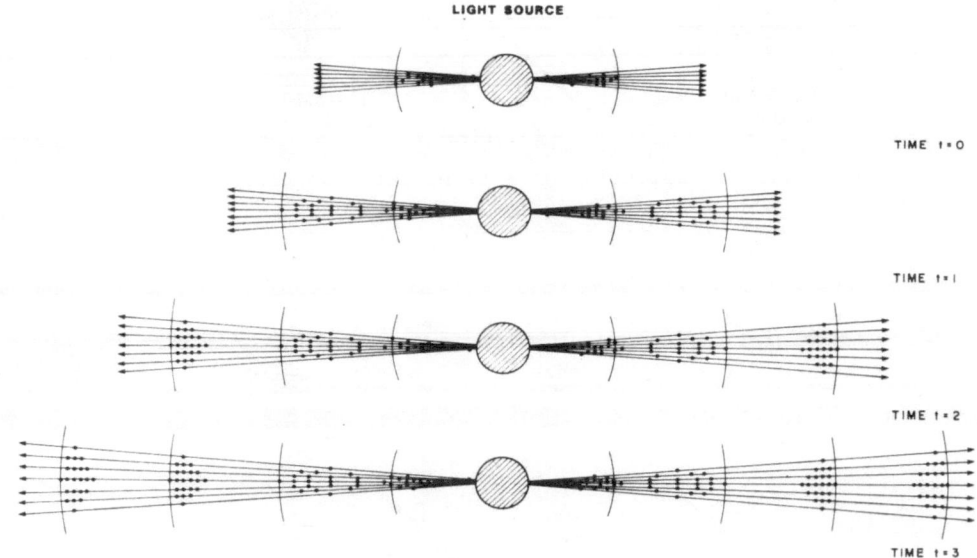

Fig. 24 Formation of clumps from photons randomly emitted by a
         light source.

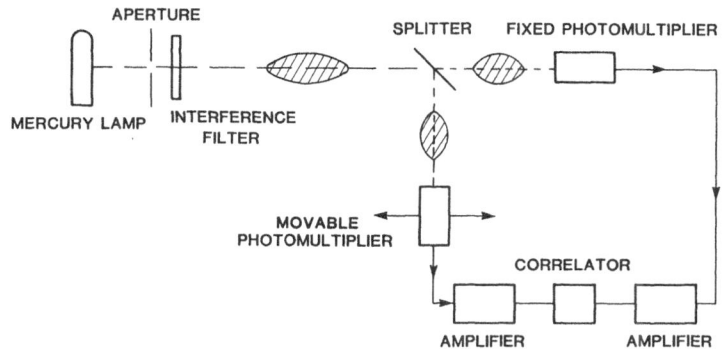

Fig. 25 Explanation of the Hanbury-Brown and Twiss experiment[1]
with the photon clump model.

As far as the interference fringes detected in the same experiments
are concerned (Fig. 28), they are the result of the superposition of two
clumps at a time at the exit of the Mach-Zehnder interferometer, the two
clumps originating from a single clump entering the interferometer.

IV. THEORETICAL

The photon clump model can only be justified if one is able to
derive a photon-photon interaction force leading to the following
effects:

1. The photons in a clump must stick together.

3. The transversal separation (i.e. perpendicular to the direction of
motion) is variable.

4. The sticking separation of two colliding clumps of photons is a
function of the collison angle, or relative velocity $\frac{dr}{dt}$.

It is towards the derivation of an interaction law having the above
properties that the following sections are dedicated.

IV.1. Analysis of the Heisenberg Principle

The interaction law will be obtained from a study of the Heisenberg
principle and the way it was derived. One of the derivations, and indeed
the most famous one, was given by Heisenberg himself[32]. He considered an
electron moving along an axis x. In order to observe the electron and to
determine its position, a microscope is used. Any photon used to observe
the electron transmits a momentum to the electron which is uncertain by

$$\Delta p_x = \frac{h}{\lambda} \sin \theta$$

where θ is the microscope angular aperture. On the other hand, since the
resolving power of the microscope is

$$\Delta x \cong \frac{\lambda}{\sin \theta}$$

one gets

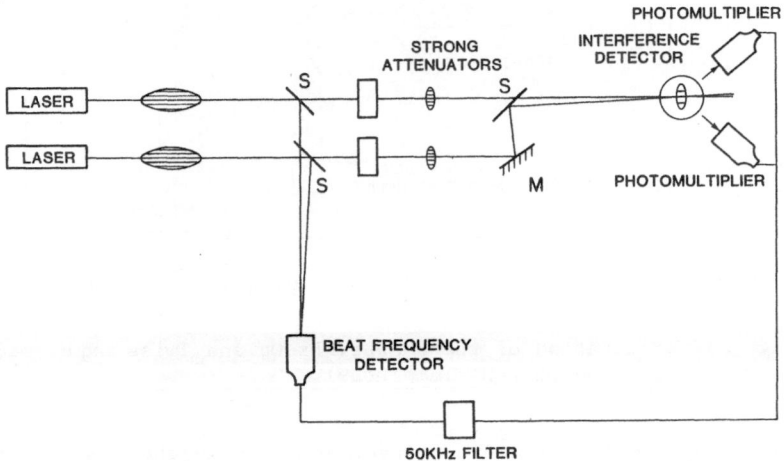

Fig. 26   Explanation of the Pfleegor and Mandel experiment[2]
with the photon clump model.

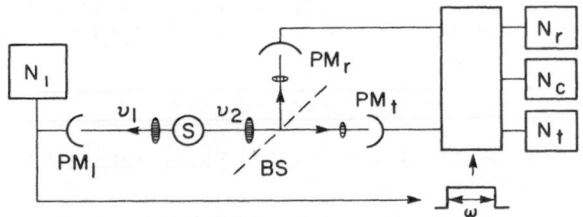

Fig. 27   Explanation of the coincidence experiment by Grangier
et al.[3] with the photon clump model.

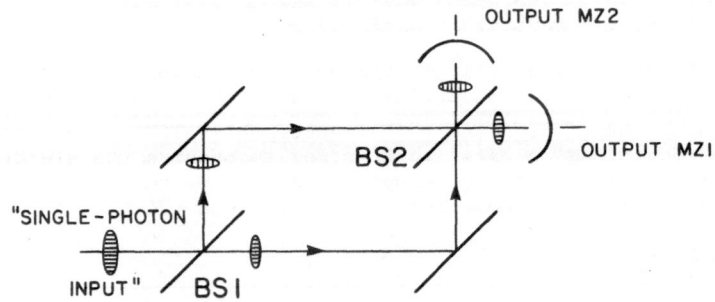

Fig. 28   Explanation of the "single-photon" interference
experiment by Grangier et al.[3] with the photon clump
model.

$$\Delta p_x \cdot \Delta x \simeq h \tag{13}$$

It is clear therefore that Heisenberg's principle can be formulated only because, in this case, a photon interacts with an electron and momentum is transferred from the former to the latter. If the two particles ignored one another, the photon would have proceeded undisturbed by the electron and the indeterminacy principle could not have been established. That an observer, in turn, picks up the information carried by the scattered photon and makes what is called a "measurement" does not change the physical reality that the photon interacted and transferred momentum to the electron. The question of observability vs. physical reality is therefore irrelevant here.

The Heisenberg principle does not need to be derived from experiments. It can be postulated as true, by hypothesizing that it is intrinsically associated with any particle, which must then have an inherent undefined position and an undefined momentum. This postulate alone, however, is not sufficient. Additional assumptions are required in order to:

1. make the particle position and momentum uncertainties Δx and Δp, respectively, satisfy relation (13);

2. make the particle uncertainties conceptually acceptable (a single particle cannot, intuitively, have a spread Δx of position and Δp of momentum — a collection of particles can).

One finds that, by ascribing to the particle another nature — a wave nature — and prescribing a functional relation

$$p = h/\lambda \tag{14}$$

between particle's momentum p and wavelength λ, the foregoing requirements are satisfied. In fact, the spread of particle momentum Δp becomes now a spread of particle wavelength:

$$\Delta p = h \Delta(1/\lambda) \tag{15}$$

which is conceptually acceptable because waves can have a distribution of frequencies around a central frequency. Moreover, since the waves associated with the particle, in order to be able to represent the position of the particle, must be localized within an interval Δx, they must interfere destructively outside this interval. The interval Δx, therefore, must contain at least one wavelength more of the wave λ_1 than of λ_2, where λ_1 and λ_2 are the shortest and the longest of all possible waves associated with the particle, respectively[33]. In other words:

$$\left(\frac{\Delta x}{\lambda_1} - \frac{\Delta x}{\lambda_2}\right) \geqslant 1 \tag{16}$$

Multiplying both sides of inequality (16) by h, one gets:

$$\Delta x \cdot h \left(\frac{1}{\lambda_1} - \frac{1}{\lambda_2}\right) \geqslant h$$

which, on account of assumptions (14) and (15), leads to

$$\Delta x \cdot \Delta p \geqslant h$$

i.e., to the Heisenberg principle. This shows that the association of a group of waves (a wave-packet) with a particle, together with assumption

(14), leads to, or is consistent with, the Heisenberg principle. Conversely, the postulation of the latter leads to the wave-particle duality.

A principle is believed to be true as long as its validity is verified by experiments. No doubt can be cast upon the validity of the Heisenberg principle because of the demonstrated agreement of quantum electrodynamics with experimentation. And the authority of the proponents of the intimate connection between the wave and the particle nature of matter, namely Bohr, Born and Heisenberg, reinforced the argument in favour of the postulation of the Heisenberg principle, despite the conceptual difficulties and objections raised by the wave-particle duality even in minds of no less authority, namely Einstein, De Broglie, and Schrödinger.

We believe that the most successful of the postulates, capable of predicting and verifying a wealth of experimental results, is not satisfactory as long as it is not assisted by intuition. To exemplify, the principle of conservation of energy is an intuitively acceptable principle. Likewise, the principle of conservation of momentum, that of mass, etc. are principles justified by intuition.

The Heisenberg uncertainty principle, by hinging so directly upon the wave-particle duality, lacks that degree of persuasiveness to make it acceptable 'toto corde' or without reservation. We believe that the association of a wave packet with a particle represents a useful mathematical model that lacks physical reality. The purpose of our study is to attempt to find the physical reality behind that mathematical model.

IV.2. Derivation of the Heisenberg Principle. The Case of Photons

The derivation of the Heisenberg principle, even from a simple experiment like the "γ-ray microscope" experiment described earlier, conveys two important concepts:

a) the principle does not need to be postulated, but it can be experimentally derived, and

b) there is physical reality in the interaction of particles necessary for their observation.

Therefore, if we are aiming at an understanding of the Heisenberg principle in terms of physical reality, and not at mathematical modeling, we have to base our reasoning on experiments. In doing so we should stay away from "gedanken" experiments, which are fraught with danger about their ability to represent a possible, real experiment.

The requirement that the principle be derived from real experiments is satisfied by all known particles. These, in fact, are able to interact with other particles, thus leading to the experimental derivation of the Heisenberg principle. Photons are the only exception. When one deals with photons, for them no real experiment seems to exist from which the principle can be derived. The reason is that, in order to establish an uncertainty principle for photons from the generalization of the outcome of real experiments, one should try to proceed in the same way as we previously did for electrons and have the photon acting as a target for other particles. If one considers a possible experiment in which the photon acts as a target for a beam of electrons, for instance, this would imply the detection of an "inverse-Compton effect", involving the scattering of electrons by photons, and such an effect has not been

144

experimentally verified as yet*. Moreover, even if such "inverse-Compton effect" were found, the determination of the photon position from the scattering of the electrons, which is necessary for the formulation of the uncertainty principle, would just be impossible, because no experiment can ever provide an image of a photon. It seems, therefore, that the Heisenberg principle for photons cannot be derived but only postulated.

We would like to offer here the following derivation of the Heisenberg principle for photons, based on the concepts previously outlined and on an experiment routinely performed. We refer to Fig. 29 which shows the classical experiment of a beam of photons crossing a narrow slit. After crossing the slit, the x-momentum of each and every photon is changed from zero to anywhere between $-p_x$ and $+p_x$. If we disregard, as proposed, the wave-particle model of interpretation of this phenomenon and if we reinstate the role of the interaction, the change of photon momentum can occur only if the photons interact, some of their momentum being transferred to the surrounding photons. Since the x-component of the momentum acquired by each and every photon, after crossing the slit, ranges between zero and $\pm p_x$, this means that such momentum is not precisely known and therefore is uncertain by:

$$\Delta p_x = p \sin \theta \qquad (17)$$

where p is the original photon momentum and θ is the deflection angle. On the other hand, in order to establish another relation between the various parameters, one notices that, by gradually decreasing the slit width, the following relation holds:

$$\Delta x \cdot \sin \theta = k \qquad (18)$$

where k is known because Δx and θ are measurable quantities. In other words, by reducing the slit width Δx, we experimentally find that the maximum deflection angle θ increases. But Δx is the uncertainty of the photon position. Hence, inserting sin θ from (18) into (17), we find

$$\Delta p_x = p \cdot \frac{k}{\Delta x} \qquad (19)$$

and

$$\Delta p_x \cdot \Delta x = p \cdot k = const \qquad (20)$$

We find experimentally therefore that the product of the uncertainties of position and momentum is a constant, as previously found. In other words, without postulating the wave-particle duality for photons and analyzing the single-slit experiment in terms of interacting photons, one derives the same uncertainty principle as previously found when the momentum of a photon was postulated to be h/λ in terms of its wave nature. This interpretation of interacting photons, therefore, provides

---

*Since we have excluded "gedanken" experiments from our considerations, we need real experiments in order to derive the Heisenberg principle for photons. Unfortunately the "inverse-Compton effect" is still awaiting an experimental verification, although such an effect certainly exists. The effect might be observed using an experimental set up proposed by Kapitza and Dirac[34] with standing light waves. Despite several attempts to observe the Kapitza-Dirac effect, none of them has provided convincing evidence of its detection (see the review paper by Eberly[35]).

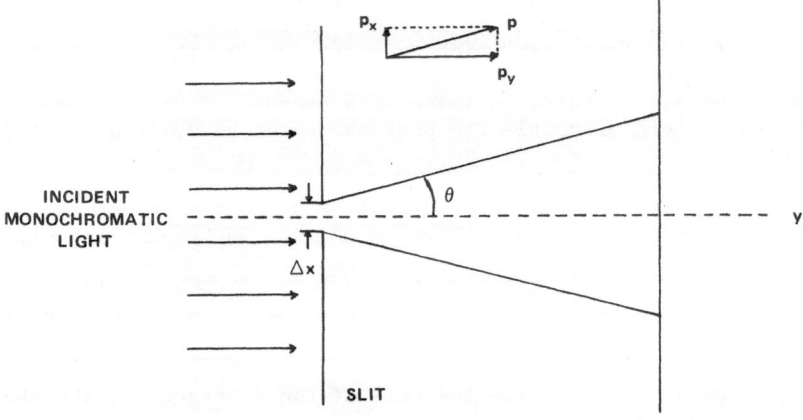

Fig. 29 Diffraction of light through a slit. The photons
crossing the slit are deflected and the maximum
angular deflection θ is related to the photon momentum
p according to: $\Delta p_x = p \sin\theta$. Another relation that
applies here is: $\Delta x \sin\theta = k$, where Δx is the slit
width and k is a constant.

the required experimental proof that photons obey the uncertainty
principle.

IV.3. Analysis of the Single-Slit Experiment When the Photons Cross the Slit One at a Time

Apparently, the hypothesis of interacting photons cannot be accepted
because it does not explain diffraction effects obtained when isolated
photons cross the slit one at a time. In this case, in fact, the photons
cannot interact with other photons and the diffraction pattern can be
explained only with the wave-particle duality of the photons. Likewise,
the formulation of the uncertainty principle in terms of interacting
photons, based as it is on diffraction effects caused by interaction, is
impossible.

If one considers carefully this objection, however, one finds that
it would have validity only if the existence of isolated photons was a
proven fact, and not just on assumption. In other words, only if a clear
demonstration of the existence of isolated photons were provided by
experiments, which is not the case thus far, then the objection would be
valid. Even when one analyzes the experiments of detection of single
photons, in this instance too one cannot be sure that these single
photons are isolated (by isolation we mean, of course, at a distance from
one another that they cannot reasonably interact). This is because the
limited quantum efficiency of any detector precludes the detection of all
photons. In addition, the limited time resolution of even the fastest of
all possible detectors is another factor that precludes the possibility
of verifying whether the photons are isolated or not. On the other hand,
if one examines the available literature, one finds that photons cannot
be radiated as isolated particles from any source. Rather, one always
deals with sources of light whose individual atoms cannot emit photons

146

independently of each other, because they are constantly interacting with a common radiation field[6]. This implies that photons are not emitted at random but have certain characteristic bunching properties[36]. This "clumping" effect for photons, on the other hand, has been successfully demonstrated by Hanbury-Brown and Twiss[1] who detected a correlation between light beams of narrow spectral width. In summary, the evidence in favour of collection of photons, as opposed to isolated photons, is by far preponderant. Moreover, once an interaction law for photons will be derived in the next section, it will be found that it leads naturally to the clumping effect for photons.

IV. 4. Derivation of the Interaction Law

We have proven thus far that the Heisenberg principle refers to interacting particles. The principle has never been contradicted. It has the status of a physical law experimentally found. The interpretation, however, of the terms Δp and Δx has always been given in terms of uncertainties of the outcome of the measurement or observation. Since we have already recognized that the observation is not necessary and have admitted that the Heisenberg's relation is a consequence of an interaction between particles, the interpretation of the principle must of necessity change. Let us refer again to Fig. 29. Assume that the intensity of light is such that only two photons, at a particular instant of time, cross the slit. Because of the mutual interaction that tends to push the two photons away one from the other, and because of the absence of other photons from the immediate surrounding, the two photons position themselves at a distance equal to the slit width Δx just before emerging from the slit. Immediately after, when they are unrestricted by any physical bound, the component of each photon's momentum in the x-direction changes by an amount $\Delta p_x = p \sin \theta$. We have to assume the maximum amount of momentum change in order not to violate the Heisenberg principle. In fact, assuming the deflection angle θ' to be less than θ, we would have:

$$\Delta p_x' = p \sin \theta' \tag{21}$$

On the other hand, experimentally one finds:

$$\Delta x = \frac{k}{\sin \theta}, \tag{22}$$

where Δx is the slit width, and this relation is independent of the intensity of light, i.e. it is valid even with two photons crossing the slit. Multiplying (21) by (22), one gets:

$$\Delta p_x' \cdot \Delta x = pk \frac{\sin \theta'}{\sin \theta} < pk \text{ because } \theta' < \theta$$

Since this would violate the Heisenberg principle, θ' must be equal to θ.

From the foregoing, it can be said that each photon receives an amount of momentum from the other in the x-direction equal to Δp_x and the product of $\Delta x = x$ (slit width or photon separation) and Δp_x brings back the known formula $\Delta x \cdot \Delta p_x \simeq$ const. Based on the new interpretation of its constituent terms, this formula can be written in a more concise way:

$$x \cdot p_x \simeq \text{const.}$$

or

$$p_x \simeq \frac{const}{x} \qquad (23)$$

where p_x is now the momentum in the x-direction (measured in the laboratory frame of reference, see Sec. IV.7) transferred from one photon to
another photon located at a distance x from the first.

Equation (23) is then the interaction law between these two particular photons just emerging from the slit, x being the photon separation.
We might generalize Eq. (23) by writing it in the following form:

$$p_r = \frac{const}{r} \qquad (23a)$$

This means that, rather than considering a slit of width extending only
in the x-direction (Fig. 29), we consider a circular opening of diameter
r and find the photon momentum transfer along the r-direction. We
further generalize Eq. (23a) by writing it in vectorial form, thus
pointing out that the momentum transfer occurs only in the r-direction,
and not in any other direction. A vectorial form of the equation is the
following:

$$\vec{p}_r = \frac{const}{r} \vec{r}_u \qquad (24)$$

where $\vec{r}_u$ is a unit vector in the r-direction emanating from the position
P of one of the two photons. Finally, we postulate that Eq. (24) is the
interaction law between any two photons separated by a distance r*. In
other words, we assume that Eq. (24) applies to any two photons, no matter where they are. Equation (24) is therefore a formula of general
validity representing a law of interaction or of momentum transfer
between any two photons separated by a distrance r. Since the motion of
a photon is perturbed in this way by the presence of all surrounding
photons, we are interested in determining the amount of the perturbation,
i.e. the displacement of the position of that photon from the one it
would have if all other photons were absent.

IV.5. Derivation of Kirchhoff's Formula

We would like to inquire whether Eq. (24) is capable of explaining
diffraction phenomena, i.e. the arrangement of photons on a wave pattern,
successfully explained so far only in terms of the wave nature of light.
Because each photon is now subject to a field (24) of interaction or
field of momentum transfer generated by each and every surrounding
photon, the perturbation of its motion can be derived by making use of
standard potential theory[37] and applying Green's theorem to a region R
containing the position P(x,y,z) of a test photon:

$$\int_R (U\nabla^2 V - V\nabla^2 U)\, dv = \int_S \left(U\, \frac{\partial V}{\partial n} - V\, \frac{\partial U}{\partial n}\right) ds \qquad (25)$$

---

*Equation (24) is a universal law applicable to all particles, and not
just photons, because no reference exists to mass, size, density,
initial momentum, etc. of the interacting particles. Clearly, this law
of interaction is in addition to any other known particle-particle
interaction which may also lead to momentum-energy transfer and may in
fact be predominant. In Appendix A we shall compare the interaction
force derived from Eq. (24) with the electrostatic and gravitational
forces.

In this identity we take as a function V one of the three cartesian components of the momentum vector $\vec{p}_r$ (Eq. 24):

$$p_x = \frac{const}{r} \sin \theta \cos \phi; \quad p_y = \frac{const}{r} \sin \theta \sin \phi; \quad p_z = \frac{const}{r} \cos \theta \quad (24a)$$

where r, θ and ϕ are spherical polar coordinates with origin at the test photon. Hence

$$V = \frac{K}{r} \qquad (24b)$$

where the value of the constant K depends on the particular cartesian component (24a) we have chosen. The analysis that follows therefore refers to a cartesian component of the momentum vector $\vec{p}_r$. Since V has a singularity for r = 0, the identity (25) cannot be applied to the whole region R. Therefore, we surround P with a small sphere σ with P as a centre and remove from R the interior of the sphere. For the resulting region R' we have, since V is harmonic in R':

$$-\int_{R'} \frac{1}{r} \nabla^2 U dv = \int_S \left(U \frac{\partial}{\partial n} \frac{1}{r} - \frac{1}{r} \frac{\partial U}{\partial n} \right) ds + \int_\sigma \left(U \frac{\partial}{\partial n} \frac{1}{r} - \frac{1}{r} \frac{\partial U}{\partial n} \right) ds \qquad (26)$$

where n denotes the normal to the boundary of R, pointing outward from R', so that on σ it has the direction opposite to the radius r. Hence, the last integral may be written

$$\int_\Omega \left(U \frac{1}{r^2} + \frac{1}{r} \frac{\partial U}{\partial r} \right) r^2 d\Omega = \bar{U} \cdot 4\pi + \int_\Omega r \frac{\partial U}{\partial r} d\Omega \qquad (27)$$

Here $\bar{U}$ is a value of U at some point of σ, and the integration is with respect to the solid angle subtended at P by the element of σ. As the radius of σ approaches zero, the limit of the integral over σ in (26) is $4\pi U(P)$ and the volume integral on the left converges to the integral over R. We thus arrive at:

$$U(P) = -\frac{1}{4\pi} \int_R \frac{\nabla^2 U}{r} dv + \frac{1}{4\pi} \int_S \frac{\partial U}{\partial n} \frac{1}{r} ds - \frac{1}{4\pi} \int_S U \frac{\partial}{\partial n} \frac{1}{r} ds \qquad (28)$$

This is the expression to be satisfied by the second function U in Green's theorem if the first function V is equal to $\frac{K}{r}$.

We shall take the function U to represent the total momentum transferred to the test photon by all other photons. We are interested in knowing the form to be taken by the function U in order to satisfy our physical conditions and equation (28). Since we are dealing with a stream of photons moving with velocity c, whose position is changing with time t, the total momentum U applied to the test photon at point P has to depend on time t also:

$$U = U(P,t) \qquad (29)$$

Moreover, since simultaneity of cause and effect (action at a distance) is excluded here, any signal emitted by the moving photons will be transmitted with finite velocity c and will be received at the point P after a time r/c.* Hence, the integration in (28) must be performed not

---

*Here, we are treating this matter in the same way as the field of a moving source[38].

at time t, but at the retarded time t - r/c. The function U to be inserted in (28) is then:

$$U = U\left(x,y,z,t - \frac{r}{c}\right) = [U] \tag{30}$$

where the square brackets indicate the retarded value of the function.

The perturbation or "optical disturbance" produced at a point P by a collection of streaming photons is given by $U(x,y,z,t)$. In order to find this function, we consider initially only one photon travelling with velocity $\vec{c}$. In polar coordinates, the field of interaction (or field of momentum transfer) of such a moving source is expressed as[38]:

$$p(\vec{r},t) = \frac{const}{\rho - \frac{\vec{c} \cdot \vec{\rho}}{c}} \tag{31}$$

where $\rho = |\vec{r} - \vec{c}t_0| = c(t-t_0)$ and t_0 represents a time such that a signal emitted by the photon at t_0 will arrive at $\vec{r}$ at time t.

In a cartesian coordinate system, in which the x-axis coincides with the velocity vector $\vec{c}$, expression (31) becomes:

$$p(x,y,z,t) = \frac{const}{c(t-t_0) - \frac{c(x-ct_0)}{c}} = \frac{const}{ct-x} \tag{32}$$

Equation (32) for the field of interaction (or momentum transfer) of the moving photon is a function that satisfies the wave equation:

$$\nabla^2 p = \frac{1}{c^2} \frac{\partial^2 p}{\partial t^2} \tag{33}$$

The field of a collection of photons, because of the linearity of the wave equation (33), is the sum of the fields of each photon. Hence the function U also satisfies the wave equation:

$$\nabla^2 U = \frac{1}{c^2} \frac{\partial^2 U}{\partial t^2} \tag{34}$$

In summary, it has been verified that the application of Green's theorem to a region R containing the position of the test photon leads to Eq. (28), when the function V has been chosen to be the one expressed by (24b), a cartesian component of the momentum vector $\vec{p}_r$. For the physical conditions provided by a stream of photons moving with velocity $\vec{c}$, the second function to be inserted into Green's formula is $U(x,y,z,t)$, which represents the momentum transferred from all the streaming photons to the test photon, and which must be calculated at the retarded time t - r/c in the integration of (28). It was also proven that the field of interaction (or momentum transfer) (32) for a moving photon satisfies the wave equation (33). Therefore, the function $U(x,y,z,t)$ also satisfies the wave equation (34). This analysis is sufficient to lead, by straightforward but tedious calculation,[39] to:

$$U(x,y,z,t) = - \frac{1}{4\pi} \int_S \left\{ [U] \frac{\partial}{\partial n} \frac{1}{r} - \frac{1}{r} \left[\frac{\partial U}{\partial n}\right] - \frac{1}{cr} \frac{\partial r}{\partial n} \left[\frac{\partial U}{\partial t}\right] \right\} ds \tag{35}$$

This is the well-known general form of Kirchhoff's theorem, as found in any textbook on classical optics[40]. At variance from the classical case, however, it expresses now a type of "optical disturbance" which is not related to light intensity or amplitude but to the "total interaction or total momentum transfer" produced by a collection of photons, randomly distributed in space and time, to a test photon positioned at a point (P,t). Such momentum transfer displaces the test photon by an amount given by Eq. (35). We shall be seeing more clearly in Sec. IV.9, when we will be dealing with the derivation of the Helmholtz' formula, that such displacement makes the particles reposition themselves on a distribution of maxima and minima of number density, much like a wave distribution. Moreover, since Eq. (35) does not depend on such physical properties of the photons as energy and momentum, and therefore it is valid for any photon, we shall see that, in the particular case of monoenergetic (or monochromatic) photons, a characteristic length appears in the formula, which is a parameter in all respects equivalent to the wavelength λ in the classical wave theory of light. The next sections will deal at length with this particular case, and it will be proven that diffraction phenomena can be explained as a geometrical arrangement of interacting photons.

IV. 6. Derivation of the Interaction Force

We previously stated that the interaction law between two photons is given by Eq. (24):

$$\vec{p}_r = \frac{\text{const}}{r} \vec{r}_u \qquad (24)$$

where r is the separation between the two photons and $\vec{r}_u$ is a unit vector in the r-direction emanating from the position of one of the two photons.

We shall now derive the interaction force F. From

$$\vec{F} = \frac{d\vec{p}_r}{dt} \qquad (36)$$

one gets:

$$\vec{F} = \frac{d\vec{p}_r}{dt} = \frac{d}{dt}\left(\frac{h}{r}\right)\vec{r}_u + \frac{h}{r}\frac{d\vec{r}_u}{dt}$$

$$= -\frac{h}{r^2}\frac{dr}{dt}\vec{r}_u + \frac{h}{r}\frac{d\vec{r}_u}{dt} \qquad (37)$$

where the constant appearing in (24) has now been designated as h. The physical meaning of expression (37) can be illustrated from the examination of two simple cases, namely the case of two photons converging towards a focal point (Fig. 30a) and the case of two counterpropagating photons (Fig. 30). In both cases $\vec{r}_u$ = const and $\frac{d\vec{r}_u}{dt} = 0$. Hence:

$$\vec{F} = -\frac{h}{r^2}\frac{dr}{dt}\vec{r}_u \qquad (38)$$

Since $\vec{p}_r$ in (24) is measured in the fixed laboratory frame of reference, $\frac{dr}{dt}$ in (38) is also measured in such frame. It is then advantageous to consider one photon fixed and the other moving against it in the r-direction in these two cases. Expression (38) states that the force is

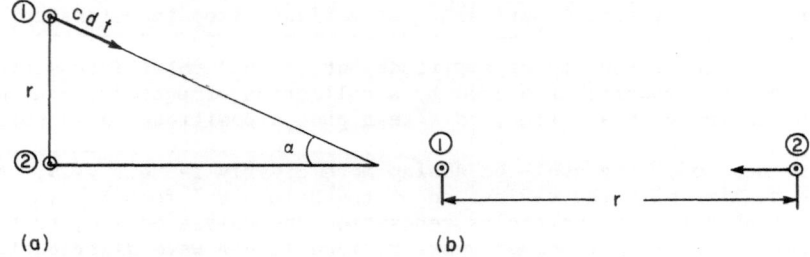

(a)   (b)

Fig. 30   Derivation of the relative velocity of photons
converging towards a common focal area (case a), or
belonging to counterpropagating beams (case b).

inversely proportional to the square of the distance r between the interacting photons and directly proportional to their relative velocity $\frac{dr}{dt}$. In the case of two photons directed towards a focal point (Fig. 30a) the relative velocity is

$$\frac{dr}{dt} = -c \, \sin \frac{\alpha}{2} = \text{const} \qquad (39)$$

and the interaction force increases as the photons converge towards the focal area, because $F \propto 1/r^2$. The force is obviously repulsive because it is impossible to focus to a point a beam of light. If the same two photons belong to counterpropagating beams (Fig. 30b), their relative velocity is

$$\frac{dr}{dt} = -c \qquad (40)$$

The force is still repulsive. However, immediately after the two photons cross each other, the force becomes attractive because $\frac{dr}{dt}$ changes sign and becomes positive.

In summary, the force acting between two photons can be either attractive or repulsive, depending on whether the photons move away one from the other or approach one another, respectively. To obtain a simple mechanical representation of such an interaction force, it is as if the photons were connected by ideal tiny springs (Fig. 31a), the compression of the spring, which takes place when the photons approach one another, yielding a repulsive force, and the stretching of the spring, which takes place when the photons move away one from the other, yielding an attractive force. This is a very crude model, and perhaps a better one would be given by an ideal damper element (Fig. 31b), in which the force is proportional to the relative velocity of the damper's terminals. At any rate, we shall not attempt here to conjecture on the nature of such a force, but rather assume its existence and verify its ability to explain well known experimental results.

IV. 7.   The Equilibrium Case (F=0) in a Collection of Streaming Monochromatic Photons.

Let us now consider the case of a source of light which continously emits collinear photons. Assume that the light is quasi-monochromatic, i.e., the photons initially have a slight spread of

(a)   (b)

Fig. 31   a) Model of photons connected by a tiny spring. The
stretching of the spring yields an attractive force
between the two photons, whereas the compression of
the spring yields a repulsive force.
b) Model of photons connected by a damper element. In
this case the force acting between the two photons is
proportional to their relative velocity.

momentum around a central momentum p_o. Because no two photons have
exactly the same velocity, they either get closer during the motion, or
they separate further. In the former case a repulsive force will act on
them; in the latter, the force will be attractive. We are interested in
knowing if they attain and, if so, what is the separation between the
photons at the time of equilibrium when $F = 0$. To this end, let us
consider the simple case of two photons of momentum p_o and $p_o + \Delta p$,
respectively, emitted by a common source (Fig. 32). At a particular
instant of time, the photons are separated by a distance equal to r. The
interaction law states that the amount of momentum transferred from one
photon to the other is:

$$p_r = m\, v_r = \frac{h}{r} \tag{24a}$$

where v_r is the velocity, measured in the laboratory frame of reference
in the r-direction, acquired by one photon because of the presence of the
other photon (it might be useful to re-examine Fig. 29 in order to be
convinced that v_r is indeed the velocity in the laboratory frame of
reference). If the velocity v_p of the latter photon in the r-direction
were zero, v_r would be given by:

$$v_r = \frac{dr}{dt}$$

When $v_p \neq 0$:

$$v_r = \frac{dr}{dt} + v_p$$

where $\frac{dr}{dt}$ is the velocity of one photon relative to the other. Hence,
eq. (24a) can be written in this way:

$$m\left(\frac{dr}{dt} + v_p\right) = \frac{h}{r}$$

or

$$\frac{dr}{dt} + v_p = \frac{h}{m}\frac{1}{r} \tag{41}$$

Fig. 32   The case of two photons having a small difference of
          momentum Δp.

Let us assume that $\dfrac{dr}{dt}$ is the velocity of the forward photon relative to
the rear one and v_p is the velocity of the latter. Equation (41) can be
solved for t as follows:

$$dt = \frac{mr \cdot dr}{h - mv_p \cdot r} \qquad (42)$$

$$t = \int \frac{mr \cdot dr}{h - mv_p \cdot r} \qquad (43)$$

The integration of (43) can be performed after transforming the integrand
as follows:

$$\frac{r}{\frac{h}{m} - v_p \cdot r} = \frac{A}{\frac{h}{m} - v_p \cdot r} + B = \frac{A + B\frac{h}{m} - Bv_p \cdot r}{\frac{h}{m} - v_p \cdot r} \qquad (44)$$

Equation (44) is satisfied if

$$-Bv_p = 1 \text{ and } A = -B\frac{h}{m} = \frac{1}{v_p}\frac{h}{m} \qquad (45)$$

The integration of (43) now yields:

$$t = \int \left(\frac{\frac{h}{mv_p}}{\frac{h}{m} - v_p \cdot r} - \frac{1}{v_p} \right) dr$$

$$= \frac{1}{v_p} \int \left(\frac{h}{h - mv_p \cdot r} - 1 \right) dr$$

$$= \frac{1}{v_p} \left[-\frac{h}{mv_p} \ell n(h - mv_p \cdot r) - r \right] + t_o \qquad (46)$$

Equation (46) shows that, as $t - t_o \to \infty$ (t_o is an arbitrary initial
time):

$$-\ell n(h - mv_p \cdot r) = \ell n \frac{1}{(h - mv_p \cdot r)} \to \infty \qquad (47)$$

and therefore

154

$$(h - mv_p \cdot r) \rightarrow 0 \tag{48}$$

If Eq. (48) is solved for r:

$$r_0 \rightarrow \frac{h}{mv_p} \quad \text{when } t \rightarrow \infty \tag{49}$$

where r_0 is the minimum (or equilibrium) distance between the two photons. But mv_p is the momentum p_0 of the rear photon. Hence

$$r_0 \rightarrow \frac{h}{p_0} = \text{const} \tag{50}$$

i.e., the two photons position themselves at the equilibrium distance $r_0 = \lambda = \text{const}$, having designated as λ such distance. Equation (50) tells us that the two particles, in the elementary one-dimensional analysis provided here, tend to be locked together at an equilibrium distance equal to λ. Clearly, in a collection of particles, this would induce a conglomeration or clustering of the particles, i.e. a lattice structure. In other words, isolated photons do not exist in this model. It is interesting to remark that this finding is consistent with the arguments of Sec. IV.3 and with the results of the experiment on nonlinearities reported in Part II of this paper, which are strongly indicative of collective, rather than single particle effects.

In conclusion, the first two properties required by the photon clump model, namely that the photons stick together and that the longitudinal separation between two contiguous photons be a constant are brought about by the interaction relation (24a) in case of photons collinearly emitted by a light source.

IV.8. The Case of Photons not Collinearly Emitted by a Light Source or the Case of two Clumps of Photons Crossing at an Angle α

Fig. 33 shows the case of two photons of converging paths. Equation (24a) must be written in this case as:

$$p_r = mv_r = m \left(\frac{dt}{dt} + v_1 \right) = \frac{h}{r}$$

where v_1 is the velocity along the r-direction of the frame of reference attached to photon (1) when photon (2) is at infinite distance from (1). The final separation of the self-collimated photons is:

$$a = \frac{h}{mv_1}$$

since $\frac{dr}{dt} = 0$ now. The photons therefore stick together at a distance a, which is a function of v_1, where a is always larger than λ.

The case of two clumps of photons intersecting at an angle α is no different from that of two isolated photons. Although it is not possible in the present elementary analysis to provide a detailed account of how the photons arrange themselves in the combined clump, it is not difficult to imagine that the resultant strings of photons will have a wider separation than in the original clump.

IV.9. Derivation of the Helmholtz Equation

The general wave type rearrangement of photons in a clump, which began to appear from the derivation of the Kirchhoff's formula, will

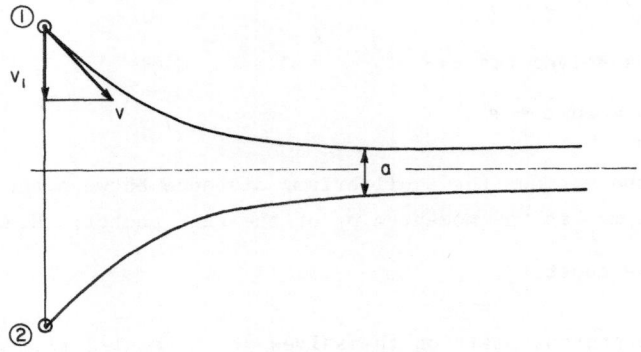

Fig. 33 The case of two photons of converging paths.

receive a more appropriate confirmation from the derivation of the
Helmholtz equation from the same basic principles of interacting
photons.

Let us then proceed with the derivation of the Helmholtz equation.
We know that each photon is the carrier of a field of interaction (or
momentum transfer) as expressed by Eq. (32):

$$p(x,y,z,t) = \frac{h}{ct - x} \tag{32}$$

The analysis of this equation (and of Eq. (31) from which it
originates) shows that the denominator cannot become negative because a
signal emitted by the moving photon cannot travel any faster than the
photon itself. In other words

$$ct \geqslant x$$

Consider now a photon emitted, at time $t = 0$, by a source located at
$x = 0$ (Fig. 34). Because $x \leqslant ct$, the field cannot be felt at points
outside the sphere having radius ct. There is an additional physical
limitation to the allowed values of x due to the fact that the maximum
momentum the moving photon is allowed to transfer is its own momentum
$p_0 = h/\lambda$. Therefore, an upper bound exists on the allowed values of
$ct - x$, and such bound is λ:

$$x \leqslant ct - \lambda$$

This means that another photon cannot penetrate into the spherical
segment of height λ shown in the figure. Therefore the field can be felt
only within the dashed area, and the solid line of the graph in the upper
part of Fig. 34 is a plot of such field. In short, no interaction
propagates ahead of the moving photon, whereas an increasing amount of
momentum can be transferred to an oncoming photon from behind, up to the
maximum amount h/λ.

Let us now consider the case of a test photon subjected to a
periodic collision with a train of monochromatic photons emitted by a
single source (Fig. 35). The test photon can be provided by an external
light beam interacting with the train of photons. It is therefore that
photon which occupies the chosen position in space at the time under
consideration (in other words, we are dealing with a succession of
photons at a fixed place in the laboratory). We are interested in

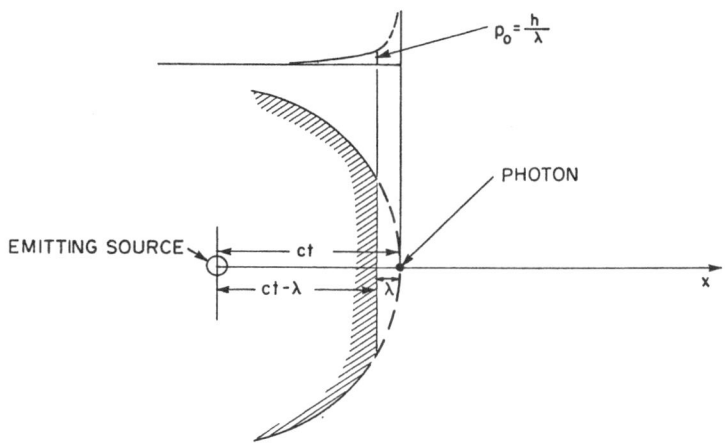

Fig. 34   The field of interaction of a moving photon can be
          felt by another photon only within the dashed volume.
          A plot of the field is shown as a solid line in the
          upper graph.

knowing the time dependence of the field experienced by the test photon
as it collides with each and every photon of the train. Arbitrarily, we
shall take 100λ as the distance after which the influence of the field
of the test photon on the oncoming photons is negligible,* so that the
scattering angle θ is also negligible, being only $0.57°$:

$$\tan^{-1}\frac{\lambda}{100\,\lambda} = 0.57 \simeq \frac{1}{2}\text{ degree}$$

The field experienced by a test photon positioned at (x,y,z) can now be
derived. As shown before, such field is created only by those photons
that are to the right of the test photon. If we assume again that 100λ
is the distance after which the influence of the field of these photons
on the test photon is negligible**, the latter is subjected to the
following field:

$$P(x,y,z,t) = \frac{h}{ct-x+\lambda} + \frac{h}{ct-x+2\lambda} + \ldots + \frac{h}{ct-x+100\lambda} = \sum_{n=1}^{100}\frac{h}{ct-x+n\lambda}\, , \qquad (51)$$

Equation (51) therefore expresses the perturbation experienced by a test
photon as a function of its position x and time t. Eq. (51) can be
transformed into:

$$P(x,t) = \sum_{n=1}^{100}\frac{h}{\lambda\frac{t}{T} - x + n\lambda} = \frac{h}{\lambda}\sum_{n=1}^{100}\frac{1}{\frac{t}{T} - \frac{x}{\lambda} + n} \qquad (52)$$

---

*Remember that an oncoming photon cannot penetrate into a volume of
 radius smaller than λ containing the test photon.

**We could have equally taken 1000λ, or $10^{4}\lambda$ etc. as cut-off, on the
 ground that the summation cannot go to infinity, because the streaming
 photons will be absorbed somewhere in space. The series therefore
 cannot diverge.

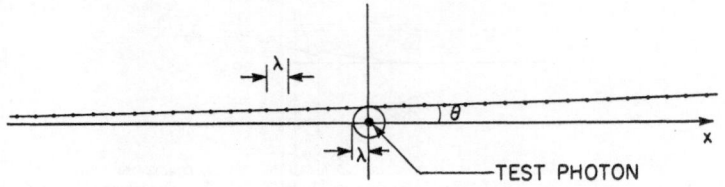

Fig. 35 Scattering of monochromatic photons by a succession or
test photons at a fixed place in the laboratory.

because $c = \frac{\lambda}{T}$. This is the value of the function for one fundamental
time period T and space period λ. The function repeats itself for equal
periods. Therefore, it is a periodic function, whose time and space
Fourier series representations are:

$$P(t) = \frac{1}{2} a_o^t + \sum_{m=1}^{\infty} \left(a_m^t \cos \frac{2\pi}{T} mt + b_m^t \sin \frac{2\pi}{T} mt \right) \tag{53}$$

and

$$P(x) = \frac{1}{2} a_0^x + \sum_{m=1}^{\infty} \left(a_m^x \cos \frac{2\pi}{\lambda} mx + b_m^x \sin \frac{2\pi}{\lambda} mx \right) \tag{54}$$

respectively.

If we take x = 0, the Fourier coefficients of (53) are:

$$a_m^t = \frac{2h}{\lambda T} \sum_{n=1}^{100} \int_o^T \frac{\cos \frac{2\pi}{T} mt}{\frac{t}{T} + n} \, dt \tag{55}$$

$$b_m^t = \frac{2h}{\lambda T} \sum_{n=1}^{100} \int_o^T \frac{\sin \frac{2\pi}{T} mt}{\frac{t}{T} + n} \, dt \tag{56}$$

If we take t = 0 and consider only x ⩽ 0 (because ct ⩾ x), the Fourier
coefficients of (54) are:

$$a_m^x = \frac{2h}{\lambda T} \sum_{n=1}^{100} \int_o^\lambda \frac{\cos \frac{2\pi}{\lambda} mx}{\frac{x}{\lambda} + n} \, dx \tag{57}$$

$$b_m^x = \frac{2h}{\lambda T} \sum_{n=1}^{100} \int_o^\lambda \frac{\sin \frac{2\pi}{\lambda} mx}{\frac{x}{\lambda} + n} \, dx \tag{58}$$

We can make the calculation of the coefficients for the two Fourier
representations (53) and (54) identical by a change of variables. In the
case of (55) and (56) we write

$$\frac{t}{T} = y \; ; \quad dt = Tdy$$

158

and therefore

$$a_m^t = \frac{2h}{\lambda} \sum_{n=1}^{100} \int_0^1 \frac{\cos 2\pi\, my}{y + n}\, dy = \frac{2h}{\lambda} A_m \qquad (55')$$

$$b_m^t = \frac{2h}{\lambda} \sum_{n=1}^{100} \int_0^1 \frac{\sin 2\pi\, my}{y + n}\, dy = \frac{2h}{\lambda} B_m \qquad (56')$$

In the case of (57) and (58) we write

$$\frac{x}{\lambda} = y$$

$$dx = \lambda dy$$

and therefore

$$a_m^x = \frac{2h}{T} \sum_{n=1}^{100} \int_0^1 \frac{\cos 2\pi\, my}{y + n}\, dy = \frac{2h}{T} A_m \qquad (57')$$

$$b_m^x = \frac{2h}{T} \sum_{n=1}^{100} \int_0^1 \frac{\sin 2\pi\, my}{y + n}\, dy = \frac{2h}{T} B_m \qquad (58')$$

The coefficients A_m and B_m, for $m = 1$ to 5, have been calculated by numerical integration. They are:

$$A_1 = 2.25 \times 10^{-2} \qquad\qquad B_1 = 1.51 \times 10^{-1}$$

$$A_2 = 6.11 \times 10^{-3} \qquad\qquad B_2 = 7.78 \times 10^{-2}$$

$$A_3 = 2.76 \times 10^{-3} \qquad\qquad B_3 = 5.22 \times 10^{-2}$$

$$A_4 = 1.56 \times 10^{-3} \qquad\qquad B_4 = 3.92 \times 10^{-2}$$

$$A_5 = 1.00 \times 10^{-3} \qquad\qquad B_5 = 3.14 \times 10^{-2}$$

The perturbation experienced by the test photons as a function of x (at a fixed time t) can be obtained by a plot of Eq. (54). Such plot, correct up to the 5th order in the Fourier representation of (52), is shown in Fig. 36. One can see that the perturbation or displacement experienced by the test photons located at x, x+λ, x+2λ,.... etc. repeats itself and is therefore a periodic signal (period λ) with a fast rise and a slow decay. It would be interesting to see this signal corresponding to what is obtained from actual experiments with light beams. We know already that light beams interfere and maxima and minima of photon distribution appear, as shown in Fig. 36. In order to have a qualitative comparison between the experimental and the theoretical signals, we have analyzed a set of interference fringes created by a pulsed ruby laser source, as shown in the inset of Fig. 37. The duration of the light pulse is only 30 nsec = 3×10^{-8} sec in this case, and therefore it is negligible, so that the fringes provide a space representation of the density of photons at a fixed time t. Each density maximum clearly represents the position where most of the photons fall, or the position of the photons which have been subjected to maximum momentum perturbation. A density scan of these interference fringes shows that indeed the experimental signal (Fig. 37) qualitatively agrees with that theoretically predicted (Fig. 36). In fact, it has similar fast rise and slow decay. In other words, it seems that the space distribution of the interference pattern from a quasi-monochromatic source of light is well represented by the Fourier

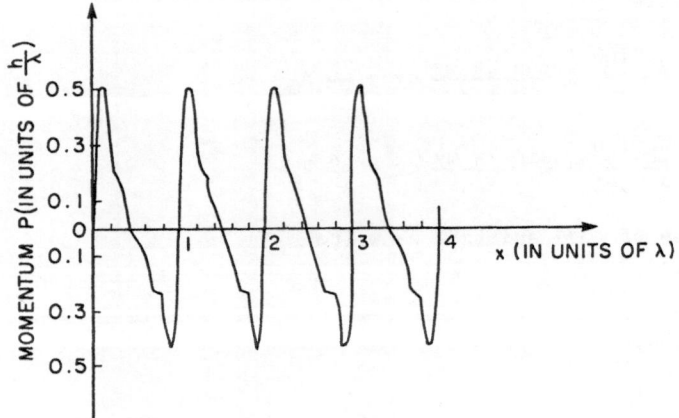

Fig. 36 Perturbation of the position of a test photon as a function of the position x. The perturbation plots as a periodic signal (period λ) having a fast rise time and slow decay.

series (54) of Eq. (52), correct up to the 5th harmonic, as predicted by the present interaction theory of light.

If we assume now that the fundamental function is preponderant,* i.e., if we neglect all harmonics and the constant term $1/2 \, a_o^t$, the time representation of the signal (53) is

$$P(t) = a_1^t \cos \frac{2\pi}{T} t + b_1^t \sin \frac{2\pi}{T} t \qquad (59)$$

In complex representation, (59) becomes:

$$P(t) = c_1^t \exp \left(i \frac{2\pi}{T} t \right) = U(t) \qquad (60)$$

where

$$c_1^t = \overline{\sqrt{(a_1^t)^2 + (b_1^t)^2}}$$

Kirchhoff's general formula (35):

$$U(z,y,z,t) = - \frac{1}{4\pi} \int_S \left\{ [U] \frac{\partial}{\partial n} \frac{1}{r} - \frac{1}{r} \left[\frac{\partial U}{\partial n} \right] - \frac{1}{cr} \frac{\partial r}{\partial n} \left[\frac{\partial U}{\partial t} \right] \right\} ds, \qquad (35)$$

transforms into:

$$U(x,y,z,t) = - \frac{1}{4\pi} \int_S \left\{ [U] \frac{\partial r}{\partial n} \frac{d}{dr} \left(\frac{1}{r} \right) - \frac{1}{r} \left[\frac{\partial U}{\partial n} \right] - \frac{1}{cr} \frac{\partial r}{\partial n} ikc [U] \right\} ds$$

$$= - \frac{1}{4\pi} \int_S \left\{ [U] \frac{\partial r}{\partial n} \left(\frac{d}{dr} (\frac{1}{r}) - \frac{ik}{r} \right) - \frac{1}{r} \left[\frac{\partial U}{\partial n} \right] \right\} ds$$

$$= - \frac{1}{4\pi} \int_S \left\{ [U] \frac{\partial r}{\partial n} (\frac{-1 - ikr}{r^2}) - \frac{1}{r} \left[\frac{\partial U}{\partial n} \right] \right\} ds$$

---

*This is a reasonable assumption, in that the first harmonic is at least twice as small as the fundamental function, and all the other harmonics rapidly decrease in amplitude.

$$= -\frac{1}{4\pi} \int_S \left\{ [U] \frac{\partial r}{\partial n} \left(\frac{-e^{-ikr} - ikr e^{-ikr}}{r^2 e^{-ikr}} \right) - \frac{1}{r} \left[\frac{\partial U}{\partial n} \right] \right\} ds$$

$$= -\frac{1}{4\pi} \int_S \left\{ [U] e^{ikr} \frac{\partial r}{\partial n} \frac{d}{dr} \left(\frac{e^{-ikr}}{r} \right) - \frac{1}{r} \left[\frac{\partial U}{\partial n} \right] \right\} ds$$

$$= -\frac{1}{4\pi} \int_S \left\{ [U] e^{ikr} \frac{\partial}{\partial n} \left(\frac{e^{-ikr}}{r} \right) - \frac{1}{r} \left[\frac{\partial U}{\partial n} \right] \right\} ds$$

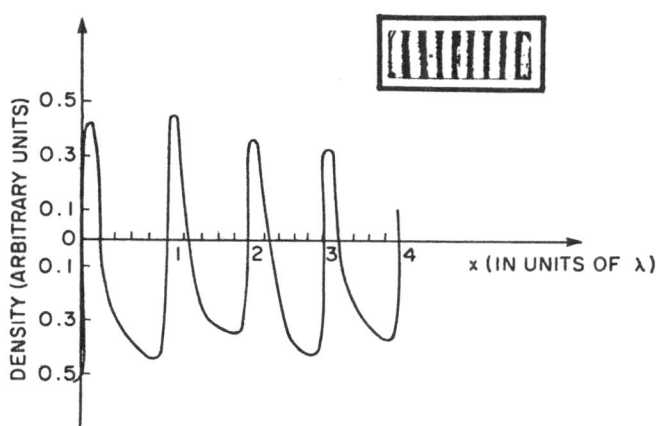

Fig. 37 Density of fringes created with a monochromatic pulse
of light (a ruby laser beam of 30 nsec duration).
Since the density of the fringes represents the space
distribution of the photons, a strong similarity
exists betweent his periodic signal and the previous
of Fig. 36 obtained from the hypothesis of interacting
photons.

because:

$$\frac{\partial U}{\partial t} = i \frac{2\pi}{T} U$$

and

$$\left[\frac{\partial U}{\partial t} \right] = i \frac{2\pi}{T} [U] = ikc [U]$$

where $k = 2\pi/\lambda$.

Since, from (60):

$$[U] = c_1^t e^{ikc \left(t - \frac{r}{c} \right)} = c_1^t e^{ikct} e^{-ikr} \tag{61}$$

then:

$$U(x,y,z,t) = -\frac{1}{4\pi} \int_S \{c_1^t e^{ikct} e^{-ikr} e^{ikr} \frac{\partial}{\partial n} \left(\frac{e^{-ikr}}{r}\right) - \frac{1}{r} \left[\frac{\partial U}{\partial n}\right]\} \, ds$$

$$= -\frac{1}{4\pi} \int_S \{c_1^t \frac{\partial}{\partial n} \left(\frac{e^{-ik(r-ct)}}{r}\right) - \frac{e^{ikc\left(t - \frac{r}{c}\right)}}{r} \frac{\partial c_1^t}{\partial n}\} \, ds$$

$$= -\frac{1}{4\pi} \int_S \{c_1^t \frac{\partial}{\partial n} \left(\frac{e^{-ik(r-ct)}}{r}\right) - \frac{e^{-ik(r-ct)}}{r} \frac{\partial c_1^t}{\partial n}\} \, ds \qquad (62)$$

But this is Helmholtz formula[41], extensively used in Optics to explain diffraction phenomena with monochromatic light. Therefore, it seems that, in this case too, the hypothesis of interacting photons leads to the same equation as the wave theory of light.

V. CONCLUSIONS

The present review paper has analyzed the wave-particle duality concept in its historical context and its factual supporting context. It has been proven that, in the historical context, the wave-particle concept has never really gone beyond the status of a theoretical hypothesis, albeit rich in predictive ability. In its factual supporting context, it has been demonstrated that the wave-particle hypothesis can receive experimental confirmation only if a preliminary assumption is adopted, namely that the photons are widely separated in an interferometer. Clearly, this assumption invalidates the direct verification of the original fundamental hypothesis.

Such a lack of conclusive experimental proof of the wave-particle duality has led us to carrying out three direct experiments, whose results have been reported in this review paper.

Despite the use of advanced instrumentation, or probably because the hypothesis does not lend itself to a direct experimental proof, all three experiments have left the duality concept at the "status quo ante", namely as that of a theoretical hypothesis.

Taking then a positive approach, it has been advanced the notion that wave phenomena, such as interference and diffraction, are perhaps collective phenomena, due to many particles. A model has been proposed, the "clump" model, whereby a photon is invariably accompanied by a collection of other photons, all clumped together, with the individual photons in the clump arranged in a geometrical wave pattern, with maxima and minima of photon distribution.

This model has been shown to explain, at least qualitatively, the available optical experimental results. Moreover, when photons are considered as interacting particles, and on interaction law is derived from plausible arguments as a dispersion-free version of the Heisenberg principle in which equality replaces inequality and Δ-quantities become sharp, one is able to find from this law that indeed photons tend to form clumps and that the distribution of photons in the clumps obeys the Kirchhoff and Helmholtz equations, as obeyed when one postulates wave properties to photons.

Although in its infancy, this "clump" model is rich in predictions, at least one of which has been already verified. In particular, some of the predictions are:

1. The tendency to form clumps, with maxima and minima of photon distribution, suggests tht the only available means for retrieving the particle properties of photons, without their associated wave aspect, is to break up a clump both transversally, i.e. perpendicularly to the direction of its motion, and longitudinally. In other words, it is not enough to use a very small pinhole to reduce the number of photons in a clump. The string of photons going through the pinhole needs to be broken down with a transverse chopper in order to separate one photon from the following one;

2. If one does not proceed in the foregoing way, and simply reduces to very low values the light intensity with a strong filter, for instance, although the individual clumps will now have a very small number of photons, as shown in Fig. 20, their intrinsic wave distribution, when a sufficient number of clumps is collected (on a screen, for instance) will tend to reappear;

3. The interaction relation (24a), which is at the basis of the photon clump model, leads to eq. (50), whereby the separation between successive photons is equal to λ. This implies that a photon cannot occupy a volume smaller than $\sim\lambda^3$. When a photon, because of the presence of other photons surrounding it, is forced to occupy a volume smaller than $\sim\lambda^3$, the photon wavelength must change or the frequency must increase. In Appendix B, as well as in Ref. 43, it is shown that there is already experimental evidence for this effect.

In conclusion, we believe that our study can be considered a "nonlocal hidden-variable theory"[42]. One must add, however, as Röhrlich[44] has recently pointed out, that the only way for any "nonlocal hidden-variable theory" to become a viable scientific alternative to quantum mechanics is to provide predictions which are not borne out by the latter theory. The predictions listed above are of this kind. The challenge now rests on the test of experimentation.

ACKNOWLEDGEMENT

We are indebted to many people who have provided, in the course of the past several years, valuable feedback in the development of the ideas expressed in this paper. In particular, we would like to thank: Drs. M. Vijay, E.S. Hanff, P. Savic, J. Lau, K. Sala, and T.E. Phipps, Jr.

The experiments reported in the paper were made possible because of the continuous and highly competent technical assistance by V. Guty. Without his help, fundamental questions addressed to nature would have had a much delayed answer.

APPENDIX A - Comparison of the Proposed Force of Interaction with the Electrostatic and Gravitational Forces

Let us consider two electrons separated by a distance r and moving with relative velocity $\frac{dr}{dt}$:

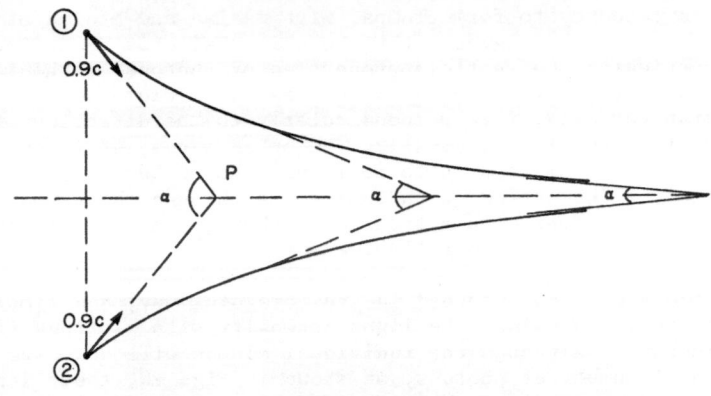

Fig. 38   The case of two electrons initially moving towards a
focal point P.

Electrostatic force $|F_C| = \dfrac{1}{4\pi\varepsilon}\dfrac{e^2}{r^2} = 2.30 \times 10^{-28}\ r^{-2}$ newton

Gravitational force $|F_G| = \gamma\dfrac{m_e^2}{r^2} = 5.40 \times 10^{-71}\ r^{-2}$ newton

Interaction force $|F_I| = \dfrac{h}{r^2}\dfrac{dr}{dt} = 6.62 \times 10^{-34}\ r^{-2}\dfrac{dr}{dt}$ newton

We are interested in knowing the magnitude of the interaction force relative to the electrostatic and gravitational forces, respectively:

a) $F_I/F_C = 2.87 \times 10^{-6}\ \left(\dfrac{dr}{dt}\right)$ (1A)

b) $F_I/F_G = 1.22 \times 10^{37}\ \left(\dfrac{dr}{dt}\right)$ (2A)

Relations (1A) and (2A) show that, when $\dfrac{dr}{dt}$ is large, the ratios F_I/F_C and F_I/F_G can become quite large.   Consider the case of two electrons initially far apart and moving towards a focal point P with velocity 0.9c (Fig. 38).   Their relative velocity is given by formula (39):

$$\dfrac{dr}{dt} = -\ \times 0.9c \times \sin\dfrac{\alpha}{2}\ .$$ (3A)

α in this case is quite large and $\left|\dfrac{dr}{dt}\right| \approx 0.9c.$ Hence:

$F_I/F_C \approx 7.74 \times 10^2$ (4A)

$F_I/F_G \approx 3.29 \times 10^{45}$ (5A)

Despite the fact that F_I is 774 times greater than F_C, and 3.29×10^{45} times F_G in this case, all these forces are quite weak because of the r^{-2} dependence.   On the other hand, when the electrons are at their closest distance, $\alpha \sim 0$ and F_I is negligible in comparison to F and F_G.

In conclusion, the interaction force is predominant over the electrostatic and gravitational forces when the two electrons are far apart, but it is negligible when the electrons are at their closest distance. In general, Eqs. (1A) and (2A) tell us that the relative velocity of the two particles is the important factor. For $\frac{dr}{dt} < 3 \times 10^6$ m/sec $= \frac{1}{100}$ c the electrostatic force dominate over the interaction force. For $\frac{dr}{dt} > \frac{1}{100}$ c the opposite begins to occur.

APPENDIX B - Threshold Value of Light Intensity for Photon Energy Variation

In the elementary analysis provided in this paper it has been shown that a photon cannot approach another one any closer than a characteristic distance λ, which can be assumed to be the equivalent of the wavelength λ in the classical wave theory of light. This implies that a photon occupies a volume of space equal to or greater than $\sim\lambda^3$. In terms of photon number density N, for $\lambda = 5000\text{Å}$, the maximum allowed value is therefore:

$$N = \frac{1}{\frac{4\pi}{3}\left(\frac{\lambda}{2}\right)^3} = 1.52 \times 10^{13} \text{ cm}^{-3}$$

or, in terms of photon flux F:

$$F = N \cdot c = 4.56 \times 10^{23} \text{ cm}^{-2} \cdot \text{sec}^{-1}$$

or light intensity I:

$$I = F \cdot h\nu = F \cdot h\frac{c}{\lambda} = 1.81 \times 10^5 \text{ W} \cdot \text{cm}^{-2}$$

However, it is well known that, at the focus of high intensity laser beams, these values are exceeded. If we take the fundamental view that two photons cannot come any closer than λ unless a specific mechanism allows this to occur (it might perhaps be photon-photon inelastic scattering), this implies that some photons, in the course of focussing, have their wavelength reduced or frequency raised, thus gaining energy at the expense, of course, of energy from surrounding photons. This hypothesis is verifiable. Experiments exist in connection with ionization of gases by focussed laser beams which seem to point out to a photon energy increase at this particular threshold of $\sim 10^5$ W·cm^{-2}. In fact, at this particular light intensity[45] and never less than this[46], the gases begin to be ionized, although their ionization potential is well above the original energy of the photons, when emitted by the laser source. Hence, some photons seem to be having gained energy in the course of focussing.

REFERENCES

1. R. Hanbury-Brown and R.Q. Twiss, Proc. Roy. Soc. A242, 300 (1957); A243, 291 (1958).

2. R.L. Pfleegor and L. Mandel, Phys. Rev. 159, 1084 (1967).
3. J.F. Clauser, Phys. Rev. 119, 853 (1974).
4. P. Grangier, G. Roger, and A. Aspect, Europhys. Lett 1, 173 (1986).
5. P.A.M. Dirac, "The Principles of Quantum Mechanics" (Clarendon Press, Oxford 1958), p. 9.
6. R.H. Dicke, Phys. Rev. 93, 99 (1954).
7. L. Mandel and E. Wolf, Rev. Mod. Phys. 37, 231 (1965).
8. E. Panarella, Ann. Fond. Louis de Broglie, 10, 1 (1985).
9. J.C. Dainty and R. Shaw, "Image Science" (Academic Press, London), 1974, p. 34.
10. P. Kowaliski, "Applied Photographic Theory" (John Wiley & Sons), 1972, p. 320.
11. C.E.K. Mees and T.H. James, "The Theory of the Photographic Process", 3rd ed. (The Macmillan Company, New York), 1966, p. 76.
12. "RCA Phototubes and Photocells - Technical Manual PT-60", 1963, Radio Corporation of America, p. 5.
13. Ref. 12, p. 108.
14. L. Levi, "Applied Optics - A Guide to Optical System Design", Vol. 1 (John Wiley & Sons, New York), 1968, p. 87.
15. W.M. Rosenblum, Jour. Opt. Soc. Amer. 58, 60 (1968).
16. G.L. Polovtseva, A.A. Dybine, and V.V. Lipatov, Opt. and Spectr. 33, 183 (1972).
17. F.T. Arecchi, Phys. Rev. Lett. 15, 912 (1965).
18. T.E. Phipps, Jr. (private communication); T.C. Fry "Probability and its Engineering Uses" (D. Van Nostrand Co., Princeton), 2nd Edit. 1965, Ch. 9.
19. R.W. Engstrom, Jour. Opt. Soc. Am. 37, 420 (1947).
20. M. Born and E. Wolf, "Principle of Optics", 3rd ed. (Pergamon Press, Oxford) 1965, p. 397.
21. G.T. Reynolds, K. Spartalian, and D.B. Scarl, Nuovo Cim. 61B, 335 (1969)
22. A. Einstein, Annln. Phys. 17, 132 (1905).
23. L. de Broglie, Compt. Ren. Hebd. Séance Acad. Sci. Paris, 177, 506, 548, 630 (1930); Thèse de Doctorat (Masson, Paris, 1924).
24. G.I. Taylor, Proc. Camb. Phyl. Soc. Math. Phys. 15, 114 (1909).
25. R. Gans and A.P. Miguez, Ann. Phys. 52, 291 (1917).
26. P. Zeeman, Physica Eindhoven 325 (November 1925).
27. A.J. Dempster and H.F. Batho, Phys. Rev. 30, 644 (1927).
28. H.J. Zweig and D.P. Gaver, IBM Jour. Res. Dev. 9, 100 (1965).
29. H.S. Allen, "Photoelectricity", (Longmans, Green) 1925, p. 126.
30. H.S. Allen, op. cit., p. 119.
31. L. Janossy and Zs. Naray, Acta Phys. Hung. 7, 403 (1957).
32. W. Heisenberg, "The Physical Principles of the Quantum Theory" (The University of Chicago Press, Chicago), 1930, p. 31.
33. R.B. Lindsay and H. Margenau, "Foundations of Physics" (John Wiley & Sons, New York) 1949, p. 399.
34. P.L. Kapitza and P.A.M. Dirac, Proc. Phil. Soc. (Cambridge), 29, 297 (1933).
35. J.H. Eberly, Progress in Optics, Vol. 7, p. 361, Ed. E. Wolf (North-Holland Publishing Co., Amsterdam) 1969.
36. L. Mandel and E. Wolf, Rev. Mod. Phys. 37, 231 (1965)
37. O.D. Kellog, "Foundations of Potential Theory" (Frederick Ungar Publishing Co., New York), p. 218.
38. P.M. Morse and H. Feshbach, "Methods of Theoretical Physics", (McGraw-Hill, New York), 1953, part I, p. 841.
39. B.B. Baker and E.T. Copson, "The Mathematical Theory of Huygens' Principle", (Oxford University Press), 1939, p. 38.
40. Ref. 20. p. 378.
41. Ref. 39, p. 36.
42. F.J. Belifante, "A Survey of Hidden-Variables Theories" (Pergamon Press), 1973.

43. E. Panarella, other paper in these Proceedings.
44. F. Röhrlich, Science 221, 1251 (1983).
45. P. Agostini and P. Bensoussar, Appl. Phys. Lett. 24, 216 (1974).
46. B. Held, G. Mainfray, C. Manus, and J. Morellec, Phys. Rev. Lett. 28, 130 (1972); P. Agostini, G. Barjot, G. Mainfray, C. Manus and J. Thebault, IEEE Jour. Quantum Electr. QE-6, 782 (1970); G. Petite, J. Morellec, and D. Normand, Jour. de Phys. 40, 115 (1979); M. Miyazaki, and H. Kashiwagi, Phys. Rev. A18, 635 (1978).

GEOMETRY DEPENDENT PREDICTIONS OF THE QUANTUM
POTENTIAL MODEL FOR THE ANOMALOUS PHOTOELECTRIC EFFECT

A. Kyprianidis

Institut Henri Poincaré
Laboratoire de Physique Théorique
11, rue P. et M. Curie, 75005 Paris

INTRODUCTION

In a set of older and recent experiments[1-4] extremely high intensity
laser beams are focused onto metals or gases and anomalous photoelectric
emission and gas photo-ionization is observed. The anomalous character of
the effect consists of the fact that outcoming photoelectrons are observed
although the single photon energy is lower than the work function of the
material. The first attempt to interpret this effect was based on a light
intensity dependent approach, the multiphoton theory[5]. This theory makes
a definite prediction for the photoelectron current i as a function of the
light intensity I, namely $i \propto I^n$ where n is the integer part of $W/h\nu + 1$, W
being the work function of the material. This was however disprooved by
experiment[6] since the latter showed a linear relation between the current
and the light intensity thus discarding the multiphoton hypothesis.

EFFECTIVE PHOTON AND QUANTUM POTENTIAL MODEL

Since a multiphoton explanation seemed to be inadequate for this
phenomenon, a single photon model was advanced by Panarella[7], the effecti-
ve-photon-model (EPM), in which an energy enhancement is postulated for
the photons in a high intensity beam, i.e.

$$\varepsilon = h\nu \cdot \exp \beta_\nu f(I) \approx \frac{h\nu}{1-\beta_\nu f(I)} \qquad (1)$$

where $h\nu$ is the normal photon energy, β_ν is a constant and $f(I)$ is a function of the light intensity. Whenever now $\beta_{\nu} \ell (I)$ significantly differs from zero, which can occur at the focal point of focused laser beams, enhanced energy photons are born according to eq.(1). Furthermore an intensity threshold exists for this effect[8] : anomalous effects arise when more than one photon occupy a volume of diameter equal to λ , or at photon number densities much greater than $6\,\pi^{-1}\,\lambda^{-3}$. Note that this model (EPM) is intensity dependent and geometry independent.

On the other hand it was shown by Allen[9] that this effect can be deduced from first quantum mechanical principles, namely the uncertainty relations. Roughly speaking we can see that by focussing we decrease the position uncertainty of the photons, thereby increasing their momentum uncertainty and through the energy-momentum relation their energy as well. This geometry dependent approach predicts the existence of enhanced energy photons in focused beams for a radius of the focal point of

$$ r < \frac{\lambda}{2\sin\theta} \qquad (2) $$

where θ is the lens half-cone angle. Of course, this is a necessary condition for the existence of the effect and no probability estimation for the photoelectric current is given, the latter being naturally intensity dependent.

Based on this simple quantum mechanical prediction, we recently proposed[10,11] a theoretical account for the anomalous photoelectric effect in the frame of the causal interpretation of quantum mechanics using an extension of the quantum potential model (QPM) of Bohm[12]. The idea behind this approach is simple and can be easily illustrated in the case of a potential barrier : the quantum potential of the wave packet approaching the barrier is modified by the presence of the barrier and modifies the energy of the particle so that it can tunnel through it, although its energy is inferior to the barrier height. The energy or potential changes are not "visible" (measurable) on their own but their consequences are. The analogy with the present configuration of the anomalous photoelectric effect is striking ; one can think of the quantum potential of a focussed laser beam to provide photons with enough energy or correspondingly to reduce the work function in the metal in such a way, so as to permit the photoelectrons to escape out of the metal although $h\nu < W$. This seems at first sight to go along the lines of Panarella's EPM but there are some major differences : The EPM is a geometry independent, intensity dependent theory, while the QPM is an intensity

170

independent, geometry dependent model, exactly as is the explanation of the effect by Allen. The quantum potential is not intensity dependent but only geometry dependent since $Q \propto \Delta\sqrt{\rho}/\sqrt{\rho}$ ($\rho = \psi^*\psi$). Nevertheless the number of emitted photoelectrons is intensity dependent as one naturally expects. In the following we will shortly expose the main theoretical aspects of the model in the frame of the Einstein-de Broglie theory of light[13,14].

EINSTEIN-DE BROGLIE THEORY OF LIGHT

In this theory the photon is considered as an oscillating localized particle with a non-zero mass $m_\gamma \neq 0$ ($m_\gamma \ll 10^{-48}$ g) which moves on the average along the lines of flow of a continuous wave field described by the complex four-vector wave field $A_\mu = R \cdot \exp[i\frac{S}{\hbar}] \cdot a_\mu$, where R and S are real functions of the coordinates x_μ , and a_μ is a real four-vector with $a_\mu a^\mu = 1$. Then one can define for the system a Lagrangian which reads :

$$\mathcal{L} = -\frac{1}{4} F_{\mu\nu}^* F^{\mu\nu} - \frac{1}{2} \mu_\gamma^2 A_\mu^* A^\mu \qquad (3)$$

where the field term is added to the usual Maxwell-term

$$\mathcal{L}_M = -\frac{1}{4} F_{\mu\nu}^* F^{\mu\nu} \quad , \quad F_{\mu\nu} = \partial_\mu A_\nu - \partial_\nu A_\mu \qquad (4)$$

and $\mu_\gamma = m_\gamma \frac{c}{\hbar}$. This Lagrangian leads to the field equations

$$\partial_\mu F^{\mu\nu} = \mu_\gamma^2 A^\nu \qquad (5)$$

which due to the mass-term imply a transverse gauge $\partial_\mu A^\mu = 0$. This wave equation when contracted by A_ν^* yields the relativistic Hamilton-Jacobi equation

$$\partial_\mu S \, \partial^\mu S + m_\gamma^2 c^2 + \hbar^2 \partial_\mu a_\nu \partial^\mu a^\nu - \hbar^2 \frac{\Box R}{R} = 0 \qquad (6)$$

which with $p^\mu = \partial^\mu S$ can be put into the following form

$$P_\mu p^\mu + m_\gamma^2 c^2 + T + Q = 0 \qquad (7)$$

Here $Q = -\hbar^2 \Box R/R$ is the relativistic form of Bohm's quantum potential and $T = \partial_\mu a_\nu \partial^\mu a^\nu$ is a spin-dependent quantum potential for the spin 1 case, which vanishes in any linearly polarized beam, i.e. when a_ν = const. One can now immediately deduce from this equation that the energy of a photon $E = h\nu = -\partial S/\partial t$ is not, in general, a constant of the motion along a photon path, except when Q and T vanish identically, as in the case e.g. of linearly polarized plane waves where Q = T = 0.

In order to use this general theoretical pattern for simplified calculations we can now proceed as follows. We can rewrite the eq. (6) in the following form ($\partial_\mu a_\nu$ = 0):

$$E = \sqrt{(\nabla S)^2 c^2 + m_r^2 c^4} = c\sqrt{\left(\frac{\partial S}{c\partial t}\right)^2 + \hbar^2 \frac{\Box R}{R}} = \sqrt{(h\nu)^2 + \hbar^2 c^2 \frac{\Box R}{R}} \qquad (8)$$

As we readily deduce, the photon energy seems to be augmented by the amount of the quantum potential. Furthermore, whenever we apply this formalism in a stationary context, i.e. calculation on the focal point of a focused laser beam or stationary interference patterns, we can drop the time dependence of the quantum potential. Furthermore we will postulate that this energy enhancement will account for the overcoming of the work function by the electrons, therefore setting $E \geqslant W$.

In order to give a simplified treatment of the known experiments and since we dispose of a variety of Gaussian shaped solutions of the Schrödinger equation, which we can apply directly to the cases under consideration, we perform a linearization of eq.(6) (with $\partial_\mu a_\nu$ = 0) and put it in a Schrödinger form. The resulting equation can be written as follows

$$\frac{\partial S}{\partial t} + \frac{(\nabla S)^2}{2\bar{m}} - \frac{\hbar^2}{2\bar{m}} \frac{\Delta R}{R} = 0 \qquad (9)$$

where we have set $2\bar{m} = h\nu/c^2$ and $-\partial S/\partial t = h\nu$ and furthermore neglected the photon rest mass term. Our aim is now to find the critical points beyond which the quantum potential attains a certain value which is high enough to provide the photon with sufficient energy so as to overcome the work function. These critical points can be determined by the following equation

$$\frac{\hbar^2 c^2}{h\nu} \cdot \frac{\Delta R}{R} = W - h\nu \qquad (10)$$

Having this in mind we can proceed to examine a series of proposed or performed experiments.

The most interesting one is a recent experiment performed by Panarella et al.[15] and deals with a focused low-intensity He-Ne laser beam. The interesting feature of this experiment is that it provides us with a lower limit for the probability to obtain enhanced energy photons, i.e. 10^{-9}, and that hereon a negative result prediction exists for the EPM-theory. The interesting data are the following : A laser of 3mW power and $h\nu = 1,96$ eV, has a 7.7×10^{-2} cm diameter. After passing through a lens it attains a focal radius of 1.7×10^{-4} cm and an intensity of $I = 3.3 \times 10^{4} \frac{W}{cm^2}$, a value which lies beyond the threshold intensity of $I_{th} = 7.07 \times 10^{4} \frac{W}{cm^2}$ of EPM. It is therefore interesting to calculate the QPM prediction for this specific experiment. Assuming that the laser beam has a Gaussian shape $I = Io \exp(-2x^2/\sigma^2)$ and a variance $\sigma \sim 10^{-4}$ cm, i.e. in the order of magnitude of the focal radius, we can use the eq. (10) to determine the critical point beyond which the enhanced energy photons appear. Knowing that the Quantum Potential of a Gaussian is parabola shaped

$$\frac{1}{R} \frac{\partial^2 R}{\partial x^2} = \left(\frac{2x}{\sigma^2}\right)^2 - \frac{2}{\sigma^2} \tag{11}$$

we get for the critical value Xo

$$\frac{x_o}{\sigma} = \pm \sqrt{2 + \sigma^2 \frac{h\nu(W-h\nu)}{\hbar^2 c^2}} \tag{12}$$

which for the specific data given above yields a value $\frac{x_o}{\sigma} \sim 6$.

The probability of observing the anomalous photoelectric effect is now given by

$$p = \frac{\Delta n}{n} = \frac{2 \int_{x_o}^{\infty} I_o \exp\left(-\frac{2x^2}{\sigma^2}\right) dx}{\int_{-\infty}^{+\infty} I_o \exp\left(-\frac{2x^2}{\sigma^2}\right) dx} = \frac{2}{\sqrt{\pi}} erfc\left[\sqrt{2} \frac{x_o}{\sigma}\right] \tag{13}$$

which for $\frac{x_o}{\sigma} \sim 6$ gives p in the order of magnitude $p \sim 10^{-30}$.

Consider now an increase of the laser power density by a factor of 10, which can be achieved either by augmenting the power P, i.e. $P' = 10P$, or by reducing σ, i.e. $\sigma' = \sigma/\sqrt{10}$. The first case produces no change in p but a change in the number of the active photons, i.e. $n' = 10n$, which is of course insignificant since $p \sim 10^{-30}$. The second possibility produces a considerable effect, because now $\frac{x_0}{\sigma} \sim 1,9$ and $p' \sim 10^{-3}$ and the number of active photons increases in addition to the enhanced probability by another factor of 10. Notice that for this second possibility the way σ is reduced, i.e. focal or afocal system is indifferent. Only the final geometry counts. The original $p \sim 10^{-30}$ is of course an almost vanishing probability which cannot produce any observable effect. But we can draw some interesting features from the above calculation :

(1) The probability depends on the energy difference (W-hv) and on the geometry of the focused beam. In fact the effect is very sensitive to the spatial intensity distribution , especially here on σ since we have a Gaussian profile distribution.

(2) The probability does not depend on the power of the laser if all other characteristics of the beam remain invariant.

(3) Of course the number of emitted photons n is intensity dependent, i.e. $n \propto pI$ and hence the intensity variation has an effect on the photo-electric current.

The preceeding calculation and the points we have made enable us to state that it is not indifferent how we achieve the high power density of the laser. Focussing of the beam increases the probability of emission non-linearly while a simple power increase causes a linear increase of the photoelectron number, because the geometry of the beam remains unaffected. It is even more striking that this effect is not tied to focussing but simply to a geometry change : the same effect will emerge if σ is re-duced by means of an afocal system, in perfect agreement with Held et al.'s[16] observations on Cs atoms. This indicates that the concept of light power density is not appropriate as a parameter of the theoretical prediction since the effects of power magnitude and geometry variations have quantitatively and qualitatively different influences on the result.

If we keep this scheme as a valid approximation to the real situa-tion then we can estimate the threshold intensity at which a detectable emission should occur. Take e.g. the detection limit for a reasonable experiment to be $p \sim 10^{-8}$. This probability is obtained for $\frac{x_0}{\sigma} \sim 3$ or correspondingly for $\sigma' = \sigma/2$ and a power density I' = 4I, which yields 1,5 times the Panarella threshold. Since only order of magnitude predictions can be established in the frame of this approximation we consider this result quite satisfactory, also in view of the fact that no adjustable parameters exist in our model.

It is very interesting to note that this kind of anomalous photoelectric effect is not restricted to laser beams of Gaussian profile but applies as well for interfering beams. This can be demonstrated easily by taking a double slit configuration and calculating the intensity interference pattern in the standard way. Assume again two Gaussian packets emerging from the double slit and consider the component of their motion parallel to the x coordinate of the screen. The two partial beams can be written as follows, if one neglects common phase factors and normalization coefficients :

$$\psi_1 = \exp\left[i K(x+a)\right] \quad \cdot \quad \exp\left[-\frac{(x+a)^2}{2\sigma^2}\right]$$

$$\psi_2 = \exp\left[-i K(x+a)\right] \quad \cdot \quad \exp\left[-\frac{(x-a)^2}{2\sigma^2}\right] \tag{14}$$

where 2a is the separation of the centers of the packets on the screen. One can now calculate the intensity pattern on the screen which reads (irrelevant factors are ommited):

$$I \propto \exp\left[-\frac{x^2}{\sigma^2}\right] \cdot \left(\cosh\frac{2xa}{\sigma^2} + \cos 2Kx\right) \tag{15}$$

The quantum potential has been calculated for this pattern by Philippidis et al.[17] and shows the following significant feature. Sharp minima exist at the interference minima $2kx = (2n+1)\pi$ and broad maxima are situated on the interference maxima $2kx = 2n\pi$. At the first few interference minima the quantum potential has a singular behaviour and a very low minimum, while the depth of the minimum decreases with increasing x. Around these minima the Q.P. can supply the photon with enough additional energy to overcome the work function. As a simple illustration consider the singular case a = 0, where the centers of the two packets coincide at the screen and the Quantum Potential takes the form

$$Q = \left(\frac{x}{\sigma^2}\right)^2 - \frac{1}{\sigma^2} - K^2 + 2\frac{K}{\sigma^2} x \cdot \tan Kx \tag{16}$$

This singular case yields infinities for Q at $2kx = (2n+1)\pi$ and can be used as a basis for probability considerations concerning the anomalous photoelectric emission of interfering laser beams. The whole procedure can be put in a scheme analogous to the probability calculations of the preceeding section, and will be subject of future publications.

It is worth mentioning in this context of anomalous photoelectric emission that a significant difference arises between coherent and incoherent light. This is not based on their power density difference but on geometrical and statistical considerations. We have namely shown[18] that incoherent light should obey Bose-Einstein statistics, instead of the Maxwell distribution present in the coherent light, a fact that yields a reduced visibility of the fringes for the interference pattern, namely (for a = 0)

$$I \propto \exp\left[-\frac{x^2}{\sigma^2}\right] \cdot \left(1 + \frac{\pi}{4}\cos 2Kx\right) \tag{17}$$

As can be readily seen from this formula the Quantum Potential has now a non-singular behaviour : The appearing denominators of the form $(1 + \frac{\pi}{4}\cos 2kx)^n$, $n = 1,2$, do not vanish at the intensity minima because of the $\pi/4$ factor in front of the cosine term. This is in clear contrast to the case described by eq. (16). Since now the QP values are well bounded, no anomalous photoelectric effect can be produced using an incoherent light source in interference experiments.

RESPONSE CURRENT OF A PULSED LASER

Let us finally treat in the frame of this model the response current of anomalous photoelectric emission due to a laser pulse. The pulse has a gaussian profile in the following sense : The light intensity has a gaussian distribution in any cross section of the beam $I(r) = I_1 \exp\left(-\frac{2r^2}{\sigma^2}\right)$ and an arbitrary distribution in the longitudinal direction $I_1 = I_1(z-ct)$. If we calculate the quantum potential for this pulse we find :

$$\frac{\Box R}{R} = \left(\frac{2r}{\sigma^2}\right)^2 - \frac{4}{\sigma^2} \tag{18}$$

i.e. independent of the longitudinal coordinate. The condition for photoelectric emission is now equivalent to the one established in eq. (12), namely :

$$r > r_0 \quad \text{where} \quad \frac{r_0}{\sigma} = \frac{1}{2}\sqrt{4 + \sigma^2 \frac{h\nu(W-h\nu)}{\hbar^2 c^2}} \tag{19}$$

If we wish now to calculate the photoelectric response current, namely $i \propto \frac{dn}{dt}$, then we have to determine the variation of the number of active photons and hence emitted photoelectrons with time for a fixed z value. This can be now put as follows

$$\frac{dn}{dt} = \frac{d}{dt} \int_{-\infty}^{t} d\tau \int_{r=r_0}^{\infty} dr \int_{0}^{2\pi} d\phi \, r \, I_1(z-c\tau) \exp\left[-\frac{2r^2}{\sigma^2}\right] =$$

$$= I_1(z-ct) \frac{\pi\sigma^2}{2} \exp\left[-\frac{2r_0^2}{\sigma^2}\right]$$

(20)

where r_0 is given by eq.(19). This result invalidates Panarella's former criticisms on our model[19] and establishes the fact that the response current is a true replica of the longitudinal pulse profile in perfect agreement with Panarella's experimental results[6]. Furthermore it shows that the current itself is sensitive to the geometry of the cross section of the beam, and is consequently subjected to all restrictions elaborated in the previous considerations on focused beams.

CONCLUSIONS

We wish to conclude with three remarks.
Let us first of all stress what we consider as a big advantage of the Quantum Potential Model. It is a model derived from first quantum mechanical principles and contains no ad hoc postulates or adjustable parameters. It can therefore be subjected to experimental tests and its eventual agreement or disagreement with them can reveal its advantages or shortcomings. But let us also equivalently stress that the approximate versions presented here do not necessarily reproduce the whole range of possibilities of the model. They simply confirm that even in a first approximation the QPM is useful in giving quantitative predictions. Furthermore, the characteristic geometry dependence it predicts clearly distinguishes the QPM from Panarella's EPM, a fact that can and should be tested by experiment.
Our second remark concerns the physical nature of the effect. It is still open in the frame of this formalism whether the action of the Quantum Potential consists in an increase of the photon energy, i.e. a frequency shift, or if it alternatively reduces the work function to an effective value below the invariant photon energy. For the time being, a choice between the two versions stated above, seems to be a matter of taste. However, a non approximate version of the model could indeed oblige us to distinguish between these two aspects.
This point leads us to our last remark, namely that non-approximate calculations are needed in order to arrive at accurate quantitative pre-

dictions on the basis of this model. This implies first of all the explicit solution of the Proca equation for different states of polarization of laser light, in order to account for the spin-dependent contribution to anomalous photoelectric emission. This effect which was neglected in the present considerations seems to open a vast area of problems which will be examined in future publications.

ACKNOWLEDGEMENTS   The author would like to thank Prof. Vigier, and Drs. C. Dewdney, P.R. Holland and S. Roy for many helpful discussions and suggestions, the French Government for a research grant that made this work possible and the Institute H. Poincaré for its hospitality.

REFERENCES

1. Gy. Farkas, I. Kertesz, Zs. Naray and P. Varga, Phys.Lett.A 21 : 475 (1967).
2. A.J. Alcock and M.C. Richardson, Phys.Rev.Lett. 21 : 667 (1968).
3. E.M. Logothetis and P.L. Hartman, Phys.Rev. 187 : 460 (1969).
4. E. Panarella, Found.Phys., 7 : 405 (1977).
5. H.B. Bebb and A. Gold, Phys.Rev., 143 : 1 (1966).
6. E. Panarella, Lett.Nuov.Cim., 3 : 417 (1972).
7. E. Panarella, Found.Phys., 4 : 227 (1974).
8. E. Panarella, Ann.Found.L. de Broglie, 10 : 1 (1985).
9. A.D. Allen, Found.Phys., 7 : 609 (1977).
10. C. Dewdney, A. Garuccio, A. Kyprianidis and J.P. Vigier, Phys.Lett.A., 105 : 15 (1984).
11. C. Dewdney, A. Kyprianidis, J.P. Vigier, and M.A. Dubois, Lett.Nuov. Cim. 41 : 177 (1984).
12. D. Bohm, Phys.Rev., 85 : 166 (1952).
13. L. de Broglie, "La mécanique ondulatoire du photon", Hermann, Paris (1940).
14. A. Einstein, Annalen der Physik (Leipzig), 17 : 132 (1905) ; ibid. 18 : 639 (1905).

    A. Einstein, Zeitschr.Phys. 18 : 121 (1917).
15. A. Panarella, P. Saric, V. Guty and J.N. Lancaster, Experimental Discrimination between the effective photon model of intensity dependent photon energy and the intensity independent models, preprint, subm. to Phys.Rev.Lett. (1986).
16. B. Held, G. Mairfray, G. Manus and J. Morellec, Proc.10th Intern. Conf.Phen.Ionized Gases, Oxford, U.K. Sept. 1971, p. 45.
17. C. Philippidis, C. Dewdney and B.J. Hiley, Nuov.Cim. B, 52 : 15 (1979).

18. F. De Martini, A. Kyprianidis, D. Sardelis and J.P. Vigier, Quantum mechanical causal action-at-a-distance correlations in optical beam splitting devices and interference experiments : The Bohr-Einstein controversy, preprint, subm. to Nuov.Cim.B.

19. E. Panarella, Effective photon and quantum potential interpretations of the anomalous photoelectric effect, NRC preprint (1985).

18.

19.

20.

E.P.R.-TYPE EXPERIMENTS AND ENHANCED PHOTON DETECTION

A. Garuccio

Dipartimento di Fisica - Universita di Bari
INFN - Sezione di Bari
70126 Bari, Italy

ABSTRACT

Local realistic models exist which reproduce the quantum mechanical predictions for single photons and approximate closely the quantum mechanical predictions for pairs of correlated photons; in these models the quantum efficiency is interpreted as a true average of detection probabilities different from photon to photon. A large set of local realistic models with enhanced photon detection is discussed. In these models the apparent violations of locality arise from polarization correlations establishied in the detectors. This means that the introduction of a third polarizer gives rise to different results from ordinary quantum theory. The numerical values of such effects in the case of real experimental apparatuses (Strirling, Catania) is establishied.

INTRODUCTION

Bell's inequality [1], in 1965, settled a border between the so-called local realist theories and Quantum Mechanics (QM): if additional (hidden) parameters are used in order to get a more detailed description of the physical world than provided by QM, then the value of a certain physical quantity is constrained to obey an inequality which is, in turn, violated by QM. Later on Clauser, Horne, Shimony and Holt (CHSH) [2] and then Clauser and Horne (CH) [3] made the original Bell inequality liable to experimental investigation with atomic cascade photons at the price of an extra assumption: For every atomic emmission, the probability of a count with a polarizer in place is less than or equal to the probability with the polarizer removed.

This assumption, due to CH and somehow weaker than the original CHSH one , is known as no-enhancement hypothesis (NEH).

In the last yars this hypothesis has been critized on many gronds. The first attempt in this direction has been made in 1983 [4] by showing that it is possible to fit the experimental data for atomic cascade experiments [5] with a local realistic model involving an enhancement angular variable. Later on Marshall showed that enhancement detection models may be extremely difficult to discriminate from Q.M.[6].

At present the research on enhacement is developing in different directions: Marshall ad Santos [7] and Ferrero and Santos [8] put forward enhanced detection models fitting the experimental data for atomic cascade tests [5]. These models give rise to a slight modification of Malus' law for EPR correlated photons. Other Authors[9,10], on the other hand, studied a broad class of enhanced photon detection (EPD) models in which the validity of the single photon physics is not questioned at all. Aim of this paper is to discuss the investigation in this direction and to show how the idea of enhanced photon detection can be put to a stringent test.

In order to clarify the main idea as well as possible, we discuss before a simple example.

A SIMPLE EPD-MODEL

In this example every "photon" is assumed to have two physical attributes: a polarization vector l and a detection vector λ . The probability of the photon crossing a polarizer with axis a is assumed as λ -independent and given by

$$C = \cos^2(l-a) \tag{1}$$

where l-a is the angle formed by l and a (Malus law). The polarization vector l is naturally assumed to become a, after crossing the polarizer.

The probability of photon detection by a photomultiplier with quantum efficiency η is assumed dependent both on λ and on the photon polarization and is assumed given by

$$D = 8\eta\cos^2(\lambda-a)/3 \tag{2}$$

for a photon which has crossed a polarizer with axis a . Equation (2) can be readily written:

$$D = \eta[1 + 4\cos 2(\lambda-a)/3 + \cos 4(\lambda-a)/3] \tag{3}$$

whereby it is evident that the λ -averaged value of D is just η , i.e. the quantum efficiency. Assuming that a statistical ensemble of photons (polarized or unpolarized) has λ distributed uniformly from 0 to π , it can be seen immediately that for single photons D can <u>always</u> be substituted by η : the whole single-photon physics is therefore reproduced by our model.

In the 0-1-0 cascade of calcium used for experimental studies of EPR-type situations <u>pairs</u> of photons emitted by the same atom, travel in opposite directions, impinge over polarizers with axes a and b and eventually enter the photomultipliers with quantum efficiencies η_2 and η_1 . For such pairs considered within <u>this</u> EPD-model, we assume that polarization vectors l and l` rotate in opposite directions around the common propagation axis and are uncorrelated, so that factors such as (1) should be averaged independently over l and l`.
Furthermore we assume that :
    i) the vector λ is the same for the two photons of a given pair;

ii) it is distributed uniformly from 0 to π over the statistical ensemble (as before);

iii) it is left unchanged by the photon crossing of a polarizer.

The probability of a double count is thus given by

$$\omega(a-b) = \langle\cos^2(1-a)\rangle\langle\cos^2(1'-b)\rangle\int_0^\pi d\lambda\, [\eta_1 8\cos^4(\lambda-a)/3][\eta_2 8\cos^4(\lambda-b)/3]/\pi$$

Using (3) we obtain :

$$\omega(a-b) = [1 + 8\cos2(a-b)/9 + \cos4(a-b)/18]\eta_1\eta_2/4$$

This probability agrees well with the result of the first Orsay experiment[5e]

$$\omega(a-b) = [0.249 + 0.218\cos2(a-b)]\eta_1\eta_2$$

The presence of a small extra term proportional to $\cos4(a-b)$ is a necessary consequence of our model and can lead to experimental controls of this proposal.

A LARGE CLASS OF EPD-MODELS

The model given in previous section is only one of the many EPD-models leading to apparent violations of Bell`s inequality because of the correlations of λ-enhanced photon detection processes, which can be conceived. A minimal set of assumptions capable of characterizing EPD-models is the following:

(1) Every photon is endowed with a linear polarization vector 1 fixing the photon transmission probability C(1-a) through a polarizer with axis a.

(2) Every photon has also a detection variable which fixes the detection probability $D(\lambda-a)$ after the photon has acquired the linear polarization a.

(3) The variable λ is distributed uniformly from 0 to π ,in all the statistical ensembles used in experiments.

(4) The variable λ has no interaction with the crystal when the photon crosses a polarizer, consequently the photon emerges with exactly the same λ it had before hitting the polarizer. Moreover we assume that the transmission probability C(1-a) does not depend on λ.

(5) In the case of atomic cascades (like the 0-1-0 calcium cascades used in recent experiments) we assume that the two photons emitted by the same atom have the same λ and an arbitrary correlation between the polarization vectors 1 and 1'.

Of course the distribution $\rho(1,1')$ of the polarization of two photons is always non-negative and satisfies the normalization condition

$$\int_0^\pi d1 \int_0^\pi d1'\, \rho(1,1') = 1$$

In this way λ looks like a "particle" property being transmitted unchanged with the energy itself, while 1 resembles much a "wave" property such as the linear polarization of a classical field. The latter conclusion is supported also from the evidence produced by recent experiments performed in Pisa by A. Gozzini[11] and collaborators. For these reasons the EPD-models seem nearer to the original dualistic picture of the electro-magnetic field than ordinary quantum theory which is based uniquely on polarization.

We will first analyze the simple case of ideal polarizers; an attempt will be made thereafter, to tackle the general case with real polarizers. The probability of a double count with ideal polarizers set at a and b respectively is

$$\omega(a-b) = T(a-b)R(a-b) \tag{4}$$

where

$$T(a-b) = \int_0^\pi d1 \int_0^\pi d1' \, \rho(1,1') C_1(1-a) C_2(1-b) \tag{5}$$

and

$$R(a-b) = \int_0^\pi d\lambda \, D_1(\lambda-a) D_2(\lambda-b)$$

are the double transmission and double detection probability respectively (the labes 1 and 2 refer to the first and second photon of the atomic cascade respectively). It is a well-known fact that all the correlation functions having the general structure (5) with the said conditions for ρ and with C_1, C_2 positive definite and smaller than one obey inequalities of Bell's type. In particular T satisfies the Freedman and Clauser inequality[5a]:

$$3T(a-b) \geq T(3(a-b)). \tag{6}$$

If we want our EPD-model to approach the QM predictions within a few percent, then

$$T(a-b)R(a-b) \simeq \eta_1 \eta_2 (1+\cos 2(a-b))/4 \tag{7}$$

η_i being the quantum efficiency of the i-th photomultiplier (typical values for η_i are 15-20%). From (7) we get:

$$R(a-b) \simeq \eta_1 \eta_2 (1+\cos 2(a-b))/(4T(a-b)) \tag{8}$$

If now we insert a third polarizer, set at an angle a', between the second one (set at a) and the second photomultiplier the probability of a double detection with 3 polarizers in then:

$$\omega_3^{EPD}(a',a,b) = T(a-b)\cos^2(a'-a)R(a'-b) \tag{9}$$

while QM predicts:

$$\omega_3^{QM}(a',a,b) = \eta_1 \eta_2 (1+\cos 2(a-b))\cos^2(a'-a)/4 \tag{10}$$

The ratio between the two prediction is:

$$\gamma(a',a,b) = \omega_3^{QM}(a',a,b)/\omega_3^{EPD}(a',a,b) = \tag{11}$$

$$= (1+\cos 2(a-b))/(4T(a-b)R(a'-b))$$

which, due to (8), becomes

$$\gamma(a',a,b) \simeq [(1+\cos 2(a-b))/(3(1+\cos 2(a'-b)))][3T(a'-b)/T(a-b)] \tag{12}$$

If the values $(a'-b) = \theta$ and $(a-b) = 3\theta$ are chosen, then, from (6), we get:

$$\gamma(a',a,b) \geq (1+\cos6\theta)/(3(1+\cos2\theta)) \qquad (13)$$

this curve is plotted in fig.1.

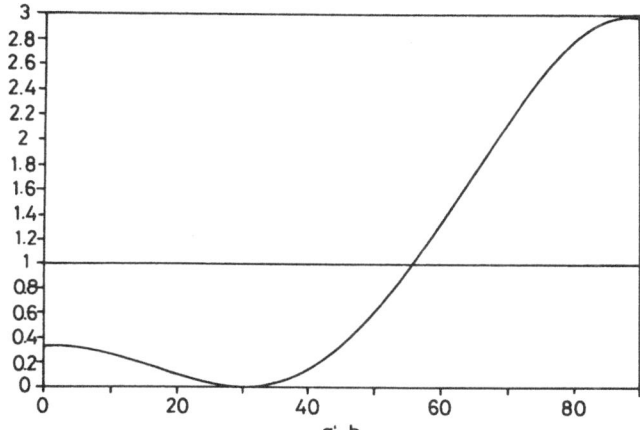

Fig. 1. Lower limit for the ratio of EPD-model to quantum mechanics as given from equation (13) in the case of ideal polarizers.In the region 70-80⁰ an experimental choice between the two theoretical models seems easily possible.

Its meaning is the following: if the quantity ω_3 is experimentally measured,and found in good agreement with ω_3^{QM} in the θ-range in which $\gamma > 1$ then the class of EPD models here considered is ruled out. Indeed (13) gives a lower limit for EPD-models compatibly with Einstein locaity and realism (which are the only two hypotheses necessary to derive the inequality (6)).

A TEST OF EPD-MODELS FOR REAL POLARIZERS

The sharp conclusion of previous section holds for ideal polarizers. Unfortunately, when polarizers' imperfections are taken into account, the mathematical expression for γ becomes very entangled and it is not easy to obtain an equivalent expression for (13).

What we can do is to consider a particular EPD-model of the class considered and study its behaviour when polarizers' imperfections are taken into account. In this way we obtain an upper bound for the lower limit sought.

Inequality (13) was obtained by applying formula (6). Moreover fig.1 shows that the most interesting region (where the gap between QM and EPD-models is larger) is $60^0 < (a'-b) < 90^0$. In this range formula (6) holds as an equality for the function

$$T^B(a-b) = (1-2|b-a|/\pi)/2 \qquad (14)$$

which is the Bell linear correlation function for photons. Therefore, for $60^0 < (a'-b) < 90^0$, (13) holds as an equality if the transmission probability is given by (14).

When the polarizers are real inequality (6) must be replaced by [5a]

$$3T(a-b) - T(3(a-b)) \geq \delta \qquad (15)$$

with $\delta = (\varepsilon_+^1 + \varepsilon_+^2)/2 - 1$, where $\varepsilon_\pm^i = \varepsilon_M^i \pm \varepsilon_m^i$ are polarizers' efficiency parameters (i labels the polarizer), and T^B is not a solution of the equality in (15). Therefore, we have to try to obtain a new function satisfying (15) as an equality with arbitrary polarizers' efficiencies. This can be accomplished modifying the function (14) into the symmetric fuction $T'(a-b)$ shown in fig.2.

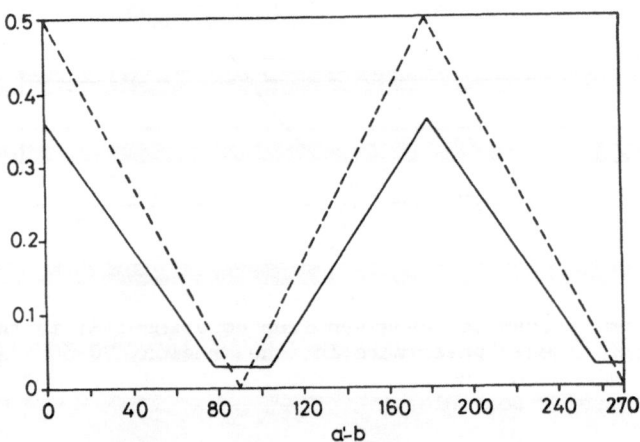

Fig. 2. Comparison between the new function $T'(a-b)$ and original Bell's function $T(a-b)$ (hatchen).

The function $T'(a-b)$ satisfies the (15) as equality in the the ranges $60° < a-b < 90° - \sigma$, $180° < a-b < 180°-3\sigma$ where

$$\sigma = (\varepsilon_+^1 \varepsilon_+^2 / \varepsilon_-^1 \varepsilon_-^2 - 1 - 2\delta/\varepsilon_-^1 \varepsilon_-^2)\pi/4 \qquad (16)$$

therefore the analysis we will carry out will be limited to these ranges.

It is important to stress that no physical meaning is attached to the highly artificial correlation function $T'(a-b)$, which has, among others, the very unpleasant feature that it implies a loss of photons somewhere in the polarizers. One could devise a very unnatural photon model leading to T' as correlation function, and could even "lift" the plateau between $90-\sigma$ and $90+\sigma$ by means of some other strange mechanism, in order to "recover" the photons lost, but this would be worthless. T' is rather to be considered as the natural extension of the "limit" correlation function T^B of formula (14) when polarizers' imperfections are considered. It satisfies (15) as an equality and if the angle σ of formula (16) is not too big, one can hope that the "sensible range" $60° < a-b < 90° - \sigma$ be wide enough to include some angles for which a discrimination between QM and EPD-models is still possible. As we will see, this will be the case for the experimental parameters of the groups in ref. 12.

The prediction for the coincidence probability with 3 polarizers for the particular EPD-model just considered is:

186

$$\bar{\omega}_3^{EPD} = \{(\varepsilon_m^3 - \varepsilon_+^3\cos^2(a'-a))\varepsilon_M^2(2\varepsilon_M^1 - \varepsilon_-^1 4(a+b-\sigma)/\pi))/4 +$$

$$+(\varepsilon_M^3 - \varepsilon_-^3\cos^2(a'-a))\varepsilon_m^2(2\varepsilon_m^1 + \varepsilon_-^1 4(a-b+\sigma))/4\}R(a'-b) \qquad (17)$$

In deriving (18) it has been assumed that Malus law holds, that $\varepsilon_M^2(\varepsilon_m^2)$ are the ratio of photons that have crossed the second polarizer with the right (wrong) polarization and that $R(\theta) = R(\theta+\pi/2)$: This last assumption was deduced by general arguments in reference 13.

The QM prediction for the coincidence probability with 3 polarizers is

$$\omega_3^{QM} = n_1 n_2 [(\varepsilon_+^1\varepsilon_+^2 + \varepsilon_-^1\varepsilon_-^2\cos2(a-b))\varepsilon_+^3 +$$

$$+ (\varepsilon_+^1\varepsilon_-^2 + \varepsilon_-^1\varepsilon_+^2\cos2(a-b))\varepsilon_-^3\cos2(a'-a)]/8 \qquad (18)$$

and can be easily derived by applying the same method as in (17). The ratio

$$\gamma'(a',a,b) = \omega_3^{QM}(a',a,b)/\bar{\omega}_3^{EPD}(a',a,b) \qquad (19)$$

is drawn in fig.3 for (a-b) = 3(a'-b) and for the polarizers' efficiencies of the Stirling and the Catania experiments. For the Catania experiment $90^0 - \sigma \simeq 80^0$ and the maximum value for γ' is 1.254 at $(a'-b) \simeq 69^0$, while for the Stirling experiment $90^0 - \sigma \simeq 76^0$ and $\gamma' = 1.079$ at $(a'-b) \simeq 68^0$.

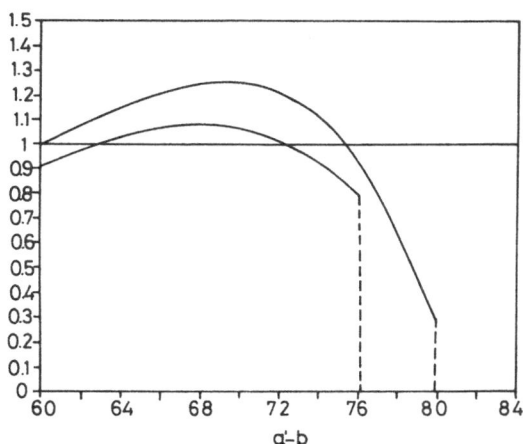

Fig. 3. Ratio of QM to the particular EPD-model considered for Catania(upper line) and Stirling(lower line) experiments.

Note that only the "sensible" interval $60^0 < (a'-b) < 90^0 - \sigma$ has been considered: the whole reasoning leading to formula (17), (18) and (19) is meaningless out of that range. Some more comments about fig. 3. The curve γ' has been obtained by considering a particular (even though "suitable") EPD -model. The limit for γ for all of the EPD model of the class considered in this paper (and in ref.[9]) is therefore upper-bounded by γ'. Moreover, an experimental choice between QM and the EPD-models considered seems still possible, so that the main idea of reference 9 may not be given up. Unluckily, this choice is much harder to carry out, experimentally, than suggested in ref. 9: indeed the curve γ , in fig.1, gets much lowered when real polarizers are considered, and the angular range in which a discrimination is possible (namely where $\gamma > 1$) gets shrunk. It is remarkable that the upper bound γ' to γ takes values less than 1 for $(a'-b) = 60^0$ and $(a'-b) = 90^0 - \sigma$ in both the Catania and Stirling case. This suggest that the angles for which $\gamma > 1$ are unlikely to be sought out of the range here considered.

CONCLUSIONS

In conclusion we wish to stess the physical idea upon which the EPD proposal rests. If one accepts the "no-enhancement hypothesis"(NEH) normally made in connection with EPR-type experiments, the problem can be considered satisfactorily solved by the available experimental evidence, except for the photon rescattering problem[14,15] which is however more of an experimental than of a conceptual nature. The only conceptually rich way-out of the conclusion that no separable reality can be attributed to the quantum systems,is therefore to be found in the refusal of NEH.

The EPD models assume therefore explicitly the negation of NEH: for some pairs of emissions (i.e., for some values of l, λ) the probability of a count with a polarizer in place is larger than the corresponding probability with the polarizer removed. It can be proved that if in eq.(5) the detection probabilities D_1 and D_2 do not depend on a and b, then Bell's type inequalities are satisfied by $\omega(a-b)$. It follows that only polarization correlations arising from enhanced detection probabilities can simulate the quantum mechanical perdictions for $\omega(a-b)$. In the set of models discussed above this is evident: only the presence of $R(a-b)$ allows one to approach closely the quantum predictions, while $T(a-b)$ satisfies explicitly the inequalities usually deduced from the assumption of separable reality.

We have therefore discussed the hard-core of the possible EPD-models: Barring the possibility that correlations between the detection parameters λ ,λ' of the two photons are extablished in the polarizers (this would amount to a nonlocal effect of the type which we are trying to avoid), we had to assume that these correlations existed when the pair of photons was created and that they were not destroyed by the photon crossing of the polarizers. For this reason the hypothesis of no interaction between λ and the polarizer was made.

Of course models in which not all the assumptions (1)-(5) above are made and which approach closely the quantum predictions should be possible, for instance by allowing for a mild decorrelation of the detection parameters in the polarizers. However, in all such models it remains necessarily true that the apparent violations of local realism arise from polarization correlation established in the detectors. But this simply means that the introduction of a third polarizer (which alters the photon polarization correlation just before detection) should give rise to non quantum mechanical effects in all cases.

References

1. J.S. Bell, Physics (N.Y.) 1 (1964) 195

2. J.F.Clauser, M.A.Horme, A.Shimony and R.A. Holt, Phys.Rev. Lett. $\underline{23}$ (1969) 880

3. J.F. Clauser and M.A. Horne, Phys. Rev. $\underline{D10}$ (1974) 526

4. T.W. Marshall, E.Santos and F.Selleri, Phys. Lett. $\underline{98A}$ (1983)5

5. S.J.Freedman ad J.F. Clauser, Phys.Rev.Lett. $\underline{28}$(1972) 938
    R.A. Holt and F.M.Pikin, Harvard University preprint (1974);
    J.F. Clauser, Phys. Rev. Lett 37 (1976)1223; Nuovo Cimento 33B (1976) 740
    E.s. Fry and R.C. Thompson, Phys. Rev. Lett. 37 (1976) 465;
    A.Aspect, P.Grangier and G.Roger, Phys. Rev. Lett. 47(1981) 460; 49 (1982) 91;
    A.Aspect, J.Dalibard and G. Roger. Phys. Rev. Lett 49(1982) 1804.
    W.Perrie, A.J. Duncan, H.J.Beyer and H.Kleixpoppen, Phys. Rev.Lett. $\underline{54}$ (1985) 1790.

6. T.W.Marshall, Phys. Lett. $\underline{99A}$ (1983) 163; $\underline{100A}$ (1984) 225

7. T.W.Marshall and E.Santos, Phys. Lett. $\underline{107A}$ (1985) 164

8. M.Ferrero and E.Santos, Phys.Lett. $\underline{108A}$ (1985) 373

9. A.Garuccio and F.Seleri, Phys. Lett. $\underline{103A}$ (1984) 99

10. F.Selleri, Phys. Lett. $\underline{108A}$ (1985) 197

11. A.Gozzini, Article in: "The Wave-Particle Dualism", S.Diner et al. editors, reidel, Dordrecht (1984)

12. Striling and Catania groups, private communications.

13. F.Selleri, in the proceedings of the conference in Joensen-Finland, World Publ. Singapore.

14. T.W. Marshall, E. Santos and F. Selleri; Lett. Nuovo Cimento $\underline{38}$, 417 (1983).
    F. Selleri; Lett. Nuovo Cimento, $\underline{39}$, 258 (1984)

15. S.Pascazio, Nuovo Cim. $\underline{5D}$, 23 (1985)

DISCUSSION II

Compiled and interpreted by D.W. Kraft and E. Panarella

Vigier. I wish to make some comments on De Martini's experimental results. First, this is an experiment which is not against quantum mechanics but which goes beyond quantum mechanics and which proves that the causal interpretation of quantum mechanics gives new predictions which can be tested by experiment and that's why I attach a great importance to these results; they should be tested and retested. This is the first time that the new interpretation provides a new prediction which can be tested by experiment, so the Einstein-de Broglie model has, in my opinion, received its first direct experimental test.

The nice thing about the De Martini experiments is that, as he said this morning, when he destroys the coherence of the laser beam, it is equivalent, if you count just the individual modes, to observing an incoherent source at the fantastic temperature of a few million degrees. The experiment shows that this incoherent light at H-bomb temperatures still statisfies the Bose-Einstein statistics and therefore that the Planck distribution is valid at fantastic temperatures, and this had never been tested before. This, in my opinion, is an argument against Panarella's model because if you modify Planck's Law by introducing the intensity of the photon in Planck's law, then you are in conflict with Bose-Einstein statistics, and therefore the only hope to interpret the Panarella effect is along the line which has been presented by Kyprianidis because the validity of the Bose-Einstein statistics at fantastic temperatures is proof that this is the only path open.

Third, what is very strange is that the distribution between the horizontal line of the Maxwell-Boltzmann statistics and the curves of the Bose-Einstein statistics prediction is evidence of the reality of the quantum potential. This is not new in the literature; there is a famous statement by Einstein who, answering the criticism of Ehrenfest, said, "the idea of indistinguishability can be considered as a trick to pick up the correct statistics. It just means that there must be between the elements of the statistics some forces of a mysterious nature." So, the experiments that you have seen this morning, if they are true, can be considered as direct evidence in favor of the real existence of the quantum potential.

Marshall. Vigier began by saying that "...if you think that light is made of particles...". De Martini, when he addressed his questions to me this morning, said that he thought that I ought to try to explain his results, and I think that you will see from the title of my talk, "Wave-Particle Duality Without Particles," that I don't think that light is made up of particles. I won't go into all the particle-like behavior that one has to explain if one thinks that; there is a problem about explaining certain particle-like behavior. But, from the point of view of someone who thinks, as I do, that light is more or less what Faraday and Maxwell thought it was, the problem of interpreting De Martini's results is really not very severe. My first reaction is that I can't quite see why anyone should think that there is any problem about explaining those results. In terms of the wave picture of light, one obtains the Maxwell-Boltzmann horizontal line result because one thinks that the laser produces a perfect sine wave. Actually it is not quite perfect; in fact if one introduces the idea of Stochastic Electrodynamics, it is modeled by an oscillating wave packet which is represented by a Gaussian-Wigner distribution in the appropriate variables. But to a very good approximation you've got a perfect sine wave with a constant intensity. In that case, all you have to do is to make the normal semi-classical assumption that the probability of activation of the photodetector in a given time interval is the product of the intensity and the time interval and then you get that which De Martini referred to as the Maxwell-Boltzmann distribution. But I would refer to it as a Poisson distribution, not of photons, but of photoelectrons. Now this is the most organized form of light, i.e., a more or less perfect sine wave. As De Martini has pointed out, from a thermal light source you get something with a rather short coherence time which has been modeled by a Gaussian autocorrelation. It is this which gives, in a very natural way, the Bose-Einstein distribution of photoelectrons. We ought to be reminded of Heitler's statement in his Quantum Theory of Radiation where he echoes the Einstein remark quoted by Vigier, that it is necessary to have the Bose-Einstein distribution for photons in order to recapture the normal field properties of the Maxwell field. So for me this all is quite straightforward. The Bose-Einstein distribution of photons is a fiction which has been put in by those people who have deluded themselves into thinking that light is made up of photons. It isn't made up of photons; it is a continuous electromagnetic wave.

Now, I don't quite understand what this condenser is that De Martini referred to, but clearly it is a kind of chaotizing device; we are basically taking this highly organized laser light and we're chopping it up and putting it together again. It's just like shuffling a pack of cards. For when you shuffle a pack of cards, you produce some sort of Gaussian distribution. Then, what's so surprising? You are going from a pure sine wave to Gaussian light, and so you are going from what those people who have deluded themselves that light is made up of photons, think is a transition from Maxwell-Boltzmann to Bose-Einstein. That's how I see this process.

De Martini. Marshall should not deny certain experimental evidence for photons. If you have a few photons impinging on a photographic plate, you obtain a little dot; why is this dot here, rather than at some other point? Second, with respect to the visibility of this dot - there is something that must carry the energy, $h\nu$. Other evidence is that of the bubble chambers, photographic emulsions, etc. where, if you see a track of a gamma ray, this track means something.

Marshall. I've never seen a track ascribable to a gamma ray.

De Martini. In a scintillator, you can have a gamma ray. With respect to the experiment I described this morning, I agree with you that it is not clear, because this experiment is made to verify the wave properties of light. But what is striking is that in order to explain the results, you have to take into account the statistics and these are related to the particle properties of light. If you have only wave properties, how can you think in terms of Bose-Einstein and Maxwell-Boltzmann statistics?

Marshall. Bose-Einstein statistics is a Gaussian distribution of the way...

Vigier. It doesn't reduce to a Gaussian distribution. You can get a perfectly good Gaussian distribution from classical Maxwell-Boltzmann statistics but there are properties of the Bose-Einstein statistics that you cannot get from the Maxwell-Boltzmann assumptions; if that were not so, why then would Bose and Einstein ever have come into the picture?

De Martini. There is a theory by Glauber in which he shows that, for this kind of experiment, you can chop this kind of thing and you get actually the Gaussian distribution. But let me ask Marshall: Are you able to explain these results just on the basis of stochastic electrodynamics?

Marshall. I don't have the advantage of having designed the experiment, so I shall need a while to think about it. However, my initial reaction is that there is not a great deal of problem about this. The qualitative answer I've given you seems to satisfy me as far as explaining your experiment. Now, tomorrow morning I'm going to take some of the most difficult experiments to explain, using a wave picture, and I'm going to explain them using a wave picture. In fact I'm going to take some of the most widely discussed pieces of evidence for the particle nature of light and I will show you how you can explain them by the wave picture. I hope that you will discuss with me the evidence that I present for those particular phenomena, but I wouldn't be surprised if this audience will point out that I have overlooked certain pieces of evidence from the last fifty years. Well, I'm sorry if that's the case - you'll have to wait another fifty years so that the "wave party" can grow to the same dimensions that the establishment school of physics has grown to, and then you will get the explanation for all these other phenomena. But, I will give you wave explanations of the most difficult phenomena.

Selleri. In this debate I agree with Vigier and disagree with Marshall. I think there is plenty of evidence for wave-particle dualism and I don't believe that it is possible to give a totally stochastic or wave-like explanation of these phenomena.

I want to comment on the talk by Garuccio and to stress that the Einstein-Podolsky-Rosen paradox is an extremely interesting field to enter at this time, full of exciting ideas. The first most important point is that we have never before seen a theory which is so strongly committed to a philosophy that even its empirical predictions have a philosophical content. This is because quantum mechanics has a mathematical structure and from this structure you get predictions, and the predictions are violations of Bell's inequality, so they are violations of any philosophy which can justify Bell's inequality. Now the philosophy happens to be one which I find to be extremely reasonable, which is realism and separability. Quantum mechanics, as it appears in textbooks today, is so philosophically committed that it is incompatible even with local realism. Now, we all know that a lot of experiments have been done to clarify this problem and the large majority of them have been in agreement with quantum theory. However, it may not be clear to everyone that the interpretation of these experiments always required additional assumptions. These additional assumptions were of a very arbitrary nature and had no empirical support. So you see, Bell's inequality for real experiments is not violated by quantum mechanics; in order to obtain a violation, one needs a different inequality; such a different inequality was deduced from local realism plus additional assumptions. Now the experimental situation seems to be that the new inequality is violated. On this we can probably agree. But if that is

so, then we deduce that either local realism or the additional
assumptions are wrong. I prefer to say that the additional assumptions
are wrong; they have no empirical basis whatsoever. So we, in Bari, have
started to play a different game with the additional assumptions, and we
made a different one, assumption number four in Garuccio's paper. In
that assumption, we say that nothing happens in the polarizers and that
the new things eventually take place in the detectors. Now if one does
that, Garuccio has shown that we get complete agreement with all
experiments performed up to now, but disagreement with a new class of
experiments, as yet unpublished, which has three polarizers. The first
work on this was performed by Santos, Marshall, Garuccio, myself and
others. So there is a new game to be played here, a very physical game,
to find out whether Einstein locality, i.e. local realism, or quantum
theory is correct; both cannot be correct, and the problem is completely
open now. I think that perhaps we are starting to see here a little
piece of future physics when we shall overcome quantum theory to make it
local and realistic again. Of course the game we are playing in Bari is
not the only one possible; we think it is reasonable but perhaps there
are others. For instance, there is a very interesting line of research
in Spain by Santos and Ferrero; they make different assumptions, that
something happens in the polarizers, and they deduce that for real
detectors, there can be complete compatibility between quantum theory and
local realism for all types of possible experime..ts. But then one can
conceive different processes like the pi-meson going to $K°$, $\bar{K}°$; perhaps
the problem of detectors is not so serious there, 'thus, the field is
completely open and it has not yet been decided whether local realism or
quantum theory is correct. It is a rigorously established theorem that
both cannot be correct. So it is very exciting and please work on it.

Surdin. I have two comments for Panarella. The first one is that the
detectors for neutrons and electrons are quite different from those for
photons; the efficiencies of these detectors are much higher than for
photons so I think probably that one may consider that there is one
neutron or one electron isolated from the other in their paths. The
second observation is the following: you said that the coherence time
for the first experiment was one microsecond, and that your photon flux
was 10^8 per second. That means that the photons are not independent of
each other, because they are correlated. When you reduce your flux by a
factor of 500, then you are at the verge of practically having no
correlation anymore. I believe that this plays some role in the fact
that you observe in the first case the interference fringes, or that you
do not observe the fringes in the second case.

Panarella. Your remarks are two-pronged. The first is related to
neutron detectors. I don't believe claims of 99.9% detection
efficiencies. This is the flipping efficiency and has nothing to do with
the quantum efficiency of detection of neutrons. While 100% neutron
detection quantum efficiency is a necessary condition, it is by no means
sufficient because one must also be able to time resolve with an
oscilloscope, or other instruments, these neutrons in order to identify
individual events and verify that they are really separated in time.
Therefore I believe that the evidence presented for single neutron
detection is not sufficient to prove the point.

    As for the correlations, I have simply presented experimental
results which theoreticians can now study and for which they can provide
explanations.

Datzeff. I would like to ask Panarella about the origin of the
nonlinearities, which are very important. I am interested in this area

because I have been working with electrons and electron interference is very similar to photon interference. I have seen nonlinearities of electron interference of the same type that you mention.

Panarella. Whatever their origin, be it the limitation of the detector, the filter, or the photomultiplier which has a nonlinear quantum efficiency, we are not able to prove directly and conclusively the association of a wave to a particle. I do not dispute the wave-particle duality; I simply say that we must reserve judgement until such time as better experiments can prove the point.

Vigier. It is a crucial question whether you can associate one particle with one wave in an interference phenomenon. Now I deny your statement that Rauch has never written that, because you will find in my paper a quotation taken from Rauch's latest work about to come out in Physical Review, A. I quote:

> "The beam intensity of the neutrons has been reduced by chopping to 2-3 neutrons/second with a velocity of 2073 m·sec^{-1} (with an error by ±1.4%). Since the Compton wavelength is of the order of 10^{-13} cm, this cleary establishes that only one wavepacket can be present at a time in the device (silicon monocrystal) ~10 cm long since its passage takes ~30 μsec. Any detected neutron (with an efficiency >99%) is thus observed long before the next has left its uranium atom within the pile. Observed interference in such a situation clearly confirm that quantum particles (here neutrons) only interfere with themselves (since the neutrons are detected individually)."

There you are. Rauch has proved that in the interference pattern he measures in the device one particle at a time, associated with one wave. This can be checked and rechecked and he has done experiments which show that the spin flipping device is efficient to 96%, and that two detectors of individual neutrons are efficient to 99%. Of course I am not discussing photons. I agree with De Martini that nobody has measured an individual photon in an interference device, but as for the neutron, it has been done and so the wave-particle duality is a fact.

Panarella. Let me deal with Rauch's statement. What he really said was that he has an absorber, which is a detector, and he picks up all the neutrons which are present, and that's the meaning of quantum efficiency for him. That is not the meaning that we attach to any other device. So he has simply said that all the neutrons are absorbed, and that the neutrons are separated, on average, by so many meters and therefore there are, on average, or with high probability, one neutron at a time within his interferometer. That's all that he is saying. And, as I said, that's a necessary but not sufficient condition, because he should have been able to time resolve these neutrons to see if they were separated in time.

Vigier. But that's exactly what he's done. He sees two neutrons per second. The average number of neutrons which come in are one per second. Now a probability calculation can never tell you that there is a 100% probability that at one time there will not be two neutrons at a time in the interferometer. But the fact that you have no more than two neutrons per second, and that the Compton wavelength is 10^{-13} cm, then it is fantastic to say that he has, with 99% probability, an average of more than one neutron per second. You will not interpret the interference pattern by interactions between the particles in an interferometer. Dirac was right.

Panarella. As I said, 100% quantum efficiency is only a necessary (but not sufficient) condition. One needs also to time resolve these neutrons and then, one has to observe the interference patterns build up linearly with neutron flux, at very low fluxes.

De Martini. I think what Cufaro-Petroni has done is very interesting. I want to ask him if he has checked the physical consistency of the other solution, beyond the Dirac equation. Does this anticipate some new kind of particle, or is it a repetition of the Dirac solution? I also want to ask how you now define the wave function, because now it is no longer the probability density, because of the second order derivative, so you can no longer use the Born interpretation of the wave function.

Cufaro-Petroni. It is not so easy to say. First of all, I tried to test if there are some different predictions. Of course I did make all possible quantum mechanical calculations, and maybe there are testable discrepancies with Dirac's theory.

De Martini. What do the hyperbolic sines and cosines mean with respect to Dirac's theory? Do they correspond to some kind of shape of the wave packet?

Cufaro-Petroni. No. I made two tests. The first was to calculate the Mott electromagnetic cross-section for the Dirac case, which is in standard textbooks, and to calculate in the framework of this second order equation; I find that the cross-sections are the same. The calculations are different, but the cross-sections are the same. The other test I made was the spectrum of hydrogen-like atoms, and the spectrum is exactly the same. I didn't calculate the transition probabilities so I don't know if they are different. It is possible to make these and other calculations, and maybe discrepancies will be discovered. But as for now, in my opinion, these calculations lead to the same results as for the Dirac equation.

Vigier. I refer to the experiments of Panarella. He does not discuss the effect of the filter in his measurements. Suppose that somehow the filter destroys the coherence of the laser beam; then he might have something of what De Martini has observed.

Panarella. Whatever part of the apparatus behaves nonlinearly, this does not change the fact that we have a problem. We are limited by the instrumentation itself in proving the wave-particle duality.

Vigier. This is a problem with light, because the technology is difficult and one cannot isolate one photon from the other.

De Martini. I want to make a comment on nonlinearities. There is nothing mysterious in the nonlinearities that Panarella has found. If one is looking at intensity, i.e. the charges liberated by a number of photons, one has to be very careful, because this is one kind of measurement, whereas another one is photon counting. In one set of oscilloscope records of Panarella, the central peak of his interferogram measures intensity, whereas on the lateral fringes he was really making a photon counting measurement. So, he was really doing two completely different measurements at the same time. But, let us consider again the case of measurement of intensity. Of course, if one goes down in intensity, one ends up with nonlinearities, because the detection of a single microscopic event is absolutely nonlinear: it is either on or off. It is a digital experiment. Because of the quantum nature of the phenomenon, one has a pulse or not. On the other hand, the detection apparatus is doing a measurement, by which one gains information.

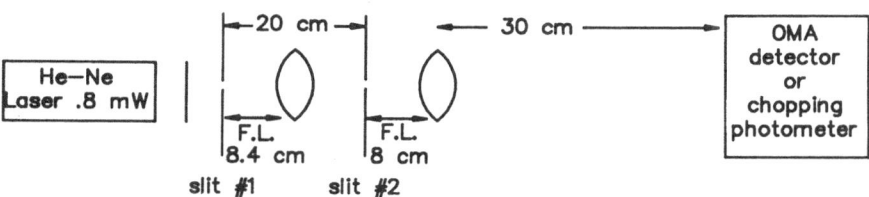

Fig. 1 Experimental apparatus used to investigate single slit
diffraction at low light levels.

Everytime one gains information, something is lost. If one wants to gain
information from a detection apparatus, one has to have a quantum
efficiency <1. So, I agree with Panarella that, if one wants to gain
information from a detecting apparatus, be it a photomultiplier or a
neutron detector, one has to have a quantum efficiency <1. Therefore one
has to be very careful about what Rauch said. In essence, he said that
all the neutrons are absorbed, but at the point where they are absorbed,
energy is lost and one loses time information. In the relation $\Delta\varepsilon \cdot \Delta t$,
the quantity Δt becomes enormous when $\Delta\varepsilon$ becomes very small.

Vigier. Rauch was operating with slow neutrons and the depth of his
detector is such that all neutrons are going to be stopped and detected.
He has lost very little information about time because he operates with
very slow neutrons. In any case, this should be discussed with Rauch
himself.

Jeffers. I would like to thank the organizers of the Symposium for the
opportunity to present now some experimental results I wasn't officially
scheduled to present on the formal program. I like to report on some
experiments that have been specifically designed to test two of
Dr. Panarella's claims and in this work I had some help from my
colleagues Dr. Wadlinger and Dr. Hunter. The two claims that I want to
investigate were described in Dr. Panarella's paper. Specifically, the
claim that, when one goes down in light flux, the side lobes of the
diffraction patterns disappear and the flux at which this occurs is
reported to be of the order of 4×10^8 photons/sec. The second claim is
that, when one goes down from 10^{10} photons$\cdot$sec^{-1} to 10^7 photons$\cdot$sec^{-1} one
sees strong nonlinearity.

The apparatus used in the first series of experiments is shown in
Fig. 1. It is similar to Panarella's with the exception that single
slits (as opposed to apertures) are used and different detectors are
employed. The slits employed here are ~100 μm wide. Optical fluxes were
measured at both slits and in the plane of the detector using a United
Technologies Model 40A Opto-meter. For our arrangement, the critical
"photon at a time" flux is C/L i.e. ~$7.9.10^8$ photons/sec.

Fig. 2 shows the classical single slit diffraction pattern computed
for our experimental parameters. For the initial experiments, the
detector employed was an intensified silicon vidicon. Images were
recorded for a variety of flux levels from 1.3×10^7 to 1.3×10^6

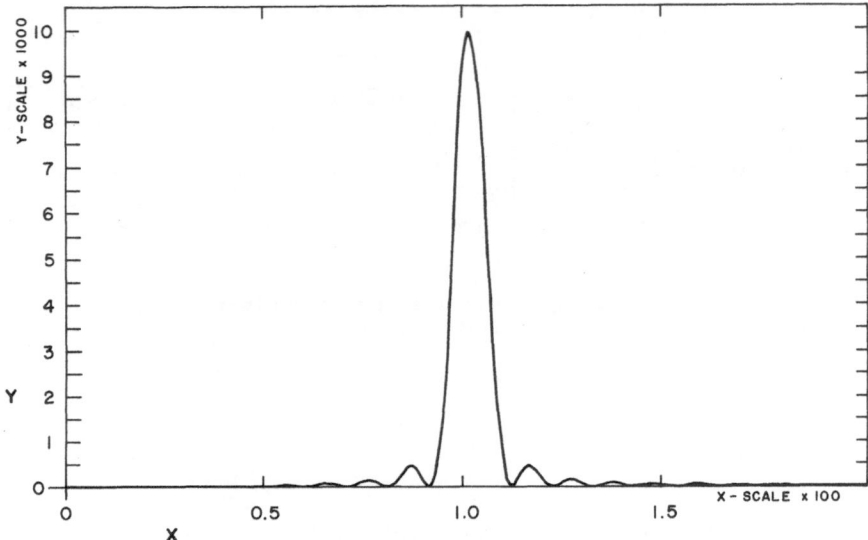

Fig. 2   Classical single slit diffraction pattern.

photons/sec as shown in Fig. 3.   There is some asymmetry in the pattern
possibly due to non-uniform illumination of the slit but the important
result is that the side-lobes are clearly visible in all records as shown
in the montage of Fig. 3(f).   As a test for linearity, the total number
of counts recorded in each pattern was determined and a plot (Fig. 4)
made of photon flux vs. counts/sec.   Over this range of flux, a nonlinear
result is obtained.   However, it has been previously established that at
low (and high) flux levels this detector responds nonlinearly.   At low
flux levels, this is due to changes in the efficiency with which the
electron beam discharges the target.   These results do not bear on the
linearity between the radiation and the photoemissive cathode.   This is
discussed in more detail below.

    A further experiment was performed with respect to the diffraction
pattern side lobes using a detector originally designed for astronomical
photometry.   This is a chopping photometer which comprises a single-stage
image intensifier (ITT Model F.4708).   An exit slit is contacted to the
output face of the image tube.   A photomultiplier tube (EMI 6094 S) is
mounted immediately behind the exit slit.   A pair of deflection coils is
mounted around the image intensifier and a square wave voltage of
variable frequency and amplitude is applied to the coils.   The width of
the exit slit was chosen to be exactly equal to the width of the
diffraction central maximum at its base.   The amplitude of the square
wave voltage was adjusted to give a deflection of the electron image in
the image tube exactly equal to the slit width.   In this way, the
detector alternately measures the flux in the central maximum and the
first two side lobes of the pattern.   The frequency of deflection was set
at 0.25 Hz.   One hundred readings of the signal were recorded followed
by one hundred of the instrumental background.   Table 1 shows the results
for a range of fluxes from $6.3.10^9$ to $6.3.10^6$ photons/sec. The results
are presented as the ratio of the mean signal (background subtracted) of
the central maximum to the side lobes and as the mean of the ratios of

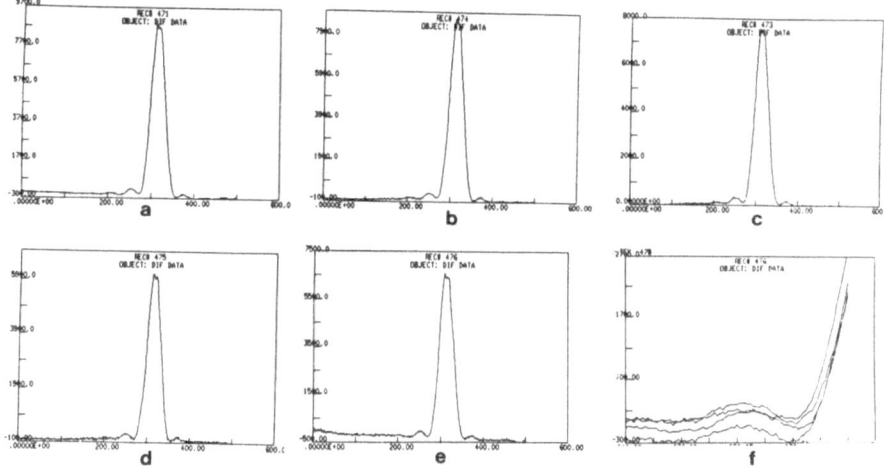

Fig. 3 Single slit diffraction pattern recorded using an
intensified, silicon vidicon at the following flux
levels (a) 1.34×10^7 photons/sec (b) 1.13×10^7
photons/sec (c) 9.2×10^6 photons/sec (d) 2.1×10^6
photons/sec (e) 1.34×10^6 photons/sec (f) Montage of
region of first side lobes.

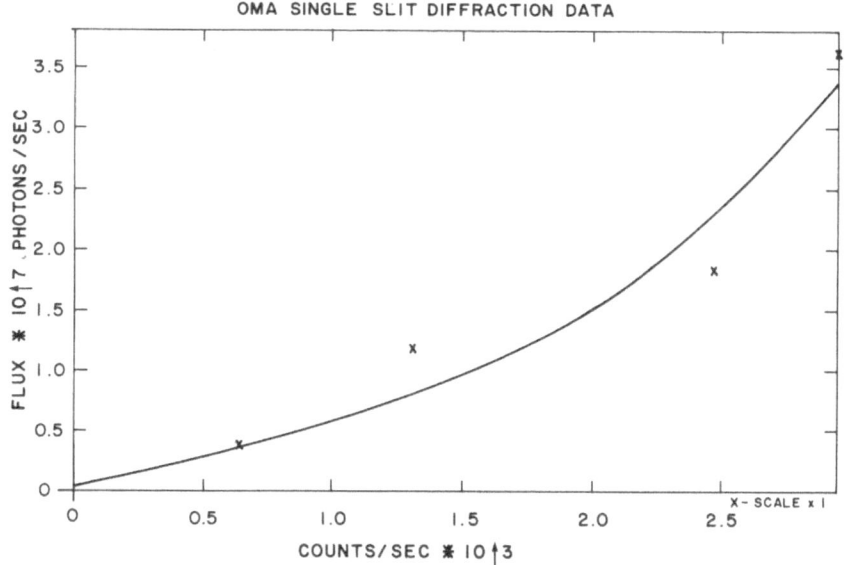

Fig. 4 Vidicon output as function of photon flux.

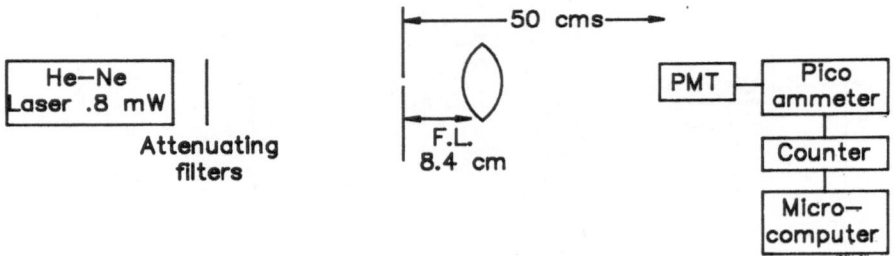

Fig. 5   Experimental apparatus used to investigate single slit
diffraction at low light levels.

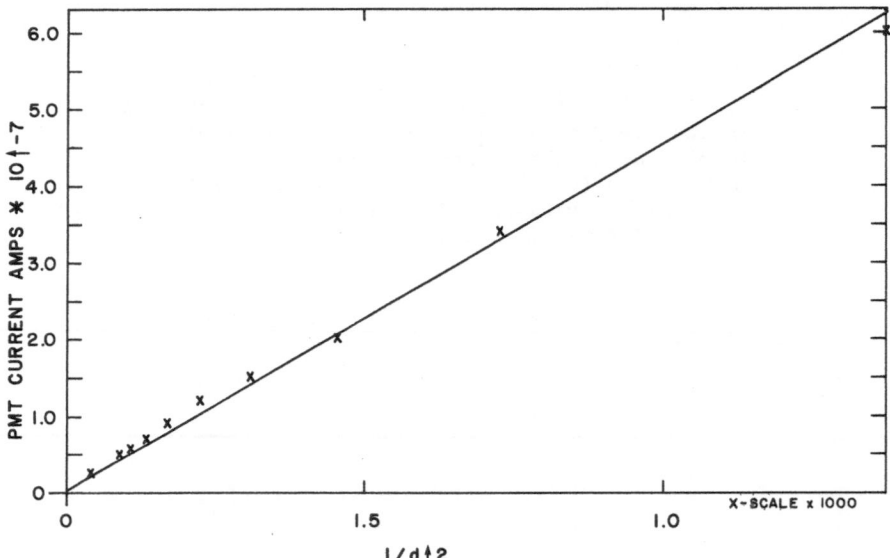

Fig. 6   Check of photomultiplier tube response using the
inverse square law.

TABLE I

FLUX AT SECOND SLIT μ W	FLUX AT SECOND SLIT PHOTONS/SEC	RATIO OF MEANS, CENTRAL MAX/SIDE LOBES	MEAN OF RATIOS CENTRAL MAX/SIDE LOBES
2.10^{-3}	6.3×10^9	$4.96 \pm .11$	6.18 ± 2.8
2.10^{-4}	6.3×10^8	$3.9 \pm .04$	4.86 ± 2.2
2.10^{-5}	6.3×10^7	$4.3 \pm .05$	10.4 ± 4.7
2.10^{-6}	6.3×10^6	3.5 ± 1.02	3.9 ± 1.1

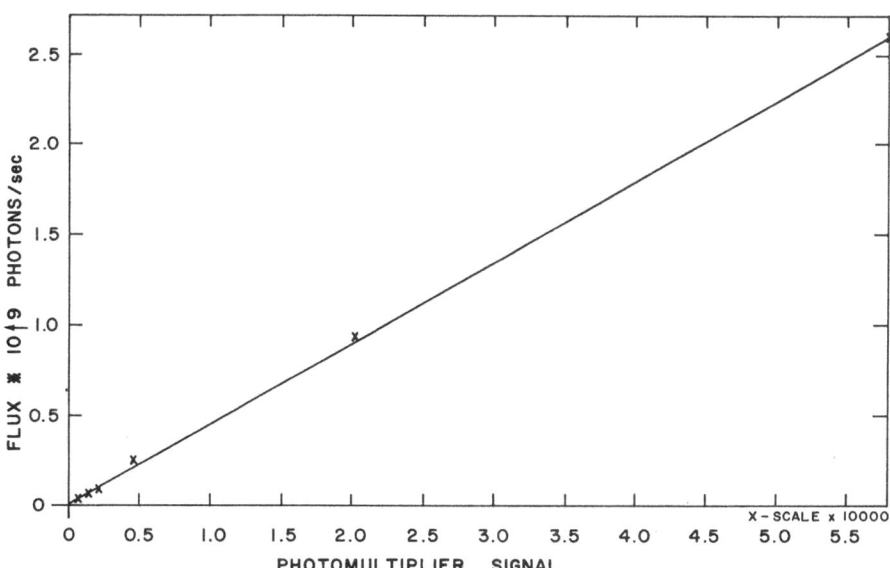

Fig. 7   Response of photomultiplier tube to varying flux of
         diffracted light.

the central maximum to the side lobes. Within the standard deviations,
there is no significant dependance on the optical flux level.

To test for linearity, the experimental arrangement shown in Fig. 5
was used. Here the central maximum of the diffraction pattern and three
side lobes were imaged on to an EMI 6094 photomultiplier tube (uncooled,
operated at 2000 volts). The response of the photomultiplier tube to
undiffracted light was tested using an incandescent bulb placed at
varying distances from the photomultiplier tube. Fig. 6 shows the
photomultiplier current as a function of $1/D^2$. The relationship is quite
linear as expected.

Fig. 7 shows the response of the photomultiplier to varying fluxes
of diffracted light and again the relationship is found to be quite
linear.

In conclusion, three separate single slit diffraction experiments
have been performed at low flux levels to test for changes in the
intensity distribution in the diffraction pattern and for nonlinearities.
No such effects have been observed.

<u>Panarella.</u> Since my work has been criticized by Jeffers, I would like to
comment on his work as follows:

1. Jeffers finds a nonlinear behavior of his first detector
    (Fig. 4) but, rather than mentioning that such nonlinearity is
    of fundamental importance because it precludes the experimental
    demonstration of the wave-particle duality for photons, he
    explains it as a "...change in the efficiency with which the
    electron beam discharges the target". Jeffers does not realize
    that any justification of the nonlinearities is irrelevant

because, in order to unequivocally prove the wave-particle duality hypothesis for single photons, the nonlinearities have to be absent at very low light levels, otherwise the hypothesis remains at the "status quo ante" namely as a hypothesis not directly and conclusively verifiable through an experiment.

2. To operate a photomultiplier without cooling is an invitation to stay in its linear regime, because the noise is so high that necessarily one has to use a high intensity light source in order to see a signal. Clearly, cooling the photomultiplier involves some extra work and patience, which evidently was not available to Jeffers.

3. Jeffers presents results that have been available for almost 40 years. The linearity test of a photomultiplier in the range that he explored (see Fig. 6) can be found in any photomultipliers handbook (see "RCA Phototubes and Photocells - Technical Manual PT-60", 3rd ed. - The Macmillian Company, New York - 1966, p. 64), which are in turn extracted from a famous paper by R.W. Engstron, J. Opt. Soc. Am. $\underline{37}$, 420 (1947). He did not go beyond what was available forty years ago. Actually, his investigation of photomultiplier linearity stopped short by an order of magnitude from what was available forty years ago. In fact, he checked the linearity of his photomultiplier down to 2×10^{-8} A, whereas such linearity has been verified down to $\sim10^{-9}$ A (see Fig. 10 of my paper). I believe that my main contribution in this field has been to extend the range of photomultiplier analysis to extremely low light intensities and to rediscover and to assess the importance for the consequences in quantum mechanics of what was already known, namely that photomultipliers behave nonlinearly at these intensities of light.

4. Finally, it is misleading on the part of Jeffers to claim that he has worked in the same range of light intensities as myself, when he uses as lowest photon flux 10^6 photons/sec into the detector, whereas I use 10^3 photons/sec, three orders of magnitude lower than his photon flux. It is the photodetector that presents problems of nonlinearity, not the interferometer.

De Martini. I think the methods of measurements of Jeffers are completely different from those of Panarella. Panarella is using a photomultiplier and you are using a different apparatus.

Jeffers. In the last experiment I used a photomultiplier.

De Martini. But you must have nonlinearities as soon as you arrive to the photon counting regime. You are going in fact to detect single, or very few photons.

Jeffers. This is true, if you are to detect a photon or not. Either the electron is released or not in this case. But if you have a flux of photons that changes, the average current generated on the photocathode will be linear with the flux.

De Martini. But there is certainly one region in which you approach the photon counting regime from the very large number of electrons regime. This is Panarella's region. I believe in what Panarella said because it should be un-physical if it were not like that. One is passing through a regime of intersection between photon counting and intensity measurements.

Panarella. Clearly, these are two different regimes.

STOCHASTIC OPTICS: WAVE-PARTICLE DUALITY WITHOUT PARTICLES

Trevor W. Marshall*

36, Victoria Avenue
Didsbury
Manchester M20 8RA

1.INTRODUCTION

There is an irony about the article of Einstein, Podolsky and Rosen[1]. It contains the most fundamental criticism ever made of quantum *particle* mechanics: that it is non local. It says nothing about quantum *field* theory, yet almost all the experimental attempts to demonstrate quantum non locality are based on pairs of electromagnetic signals[2-6]. In the widely accepted "analysis" of these experiments the signals are treated as equivalent to the particles of the E.P.R. article. For the best known of the E.P.R. authors this would have been a natural enough assumption. He, after all, invented the "light quantum", later to become the "photon", and apparently never doubted the validity of that concept. But we have learned that semiclassical radiation theories of various kinds can explain not only the photoelectric effect[7], for which Einstein invented the quantum theory, but also such diverse phenomena as the Lamb shift[8], the Casimir effect[9] and super-Poisson distributions of photoelectrons ("photon bunching")[10]. We do not have to assume that quantum electrodynamics has the last word on the nature of light. We do not have to assume that the "photon" is an E.P.R. particle.

Among the classic articles on the Bell inequalities, I believe that Clauser and Horne's[11] is the clearest. If you examine it carefully you will even find in it the point (footnote 23) at which semiclassical radiation theories intervene.† The biggest contribution of these authors, however, was to give us the concept we must have if we are to make a local realist analysis of the cascade coincidence experiments: the concept of *enhancement*.

Although this is not a philosophical article, I should make clear my philosophical position. I believe, along with Einstein[12], that an "analysis" which is nonlocal cannot be scientific and therefore is not really an analysis at all. That is why I used inverted commas in referring above to the widely accepted "analysis". If I were to be

---

* On study leave from Dept. of Mathematics, University, Manchester M13 9PL
† Note, however, that Clauser and Horne were not correct in stating that semiclassical radiation theories all satisfy the hypothesis of no enhancement. This is, in fact true only of those semiclassical theories which exclude the possibility of a real zero-point field.

consistent I would henceforth not put in the words "local" and "realist" when referring to my analysis, because local realist analysis is the *only* analysis. However, the debased state of the scientific culture after sixty years of Copenhagen dominance is such that it will be necessary to keep these two words in as a constant reminder.

The contribution of semiclassical radiation theory to the discussion of E.P.R. will be to show how the, at first sight, implausible idea of enhancement may be given a natural explanation in relation to the light signals of individual atoms. We shall see that the particular type of semiclassical theory which achieves this connection is stochastic electrodynamics. It is the modern version of Planck's second theory and is what Planck[13] and Nernst[14] used in their energetic attempts to oppose, in its infancy, the light quantum hypothesis of Einstein. This is the irony of E.P.R; in order to carry through the criticisms made by the mature Einstein against quantum theory, we have to reject the most "revolutionary" idea of the young Einstein, and concede that the "conservatives" Planck and Nernst were correct.

But now that all the positions of power in physics are occupied by supporters of the Copenhagen school, such labels are misleading. What we now find is that the insistence on local realist analysis, initiated by Einstein, Podolsky and Rosen, results in a thorough reinvestigation not only of the cascade coincidence experiments, but eventually of all the evidence, from 1905 to the present day, purporting to establish the reality of photons. Viewed in this way, the ideas of the mature Einstein, as reflected in the E.P.R. article, are far more revolutionary than those of the young man who wrote the article on the photoelectric effect.

2. STOCHASTIC LOCAL REALISM AND THE NEED FOR ENHANCEMENT

Let us study the diagram of the E.P.R.B. experiment (Fig. 1). We write "E.P.R.B.", because it is based on an improved and simplified form, devised by David Bohm, of the E.P.R. experiment.

The source emits a succession of pairs of signals, one red and one green (see footnote overleaf)*. Each pair is normally assumed to be independent of its predecessors, and any given pair is "assumed" to be specifiable by a set of random variables λ. This latter "assumption" is that of realism and, within the local realist analysis, is no assumption

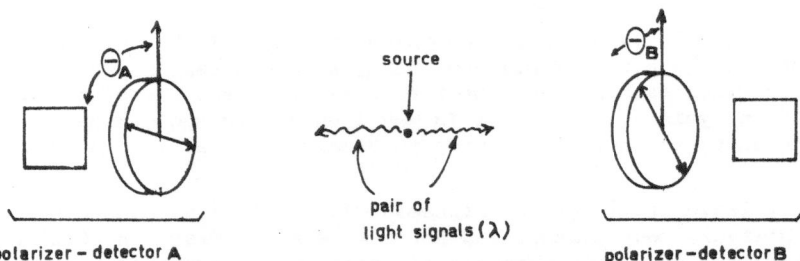

polarizer – detector A   source   polarizer – detector B
                          pair of
                          light signals (λ)

FIG.1. Diagram of E.P.R.B. experiment.

204

at all. The polarizer-detector A is fitted with a polarizer of variable orientation Θ_A, and B has a similar one of orientation Θ_B. It is "assumed" that A registers the arrival of the red signal with probability $P_1(\Theta_A,\lambda)$, and that B registers, independently, the arrival of the green signal with probability $P_2(\Theta_B,\lambda)$. These "assumptions" are all consequences of locality and, within the local realist analysis, are not assumptions at all.

It now follows that the singles rate, R_1, of registrations at A is given by

$$R_1/R_0 = p_1 = \int \rho(\lambda)\ P_1(\Theta_A,\lambda)d\lambda, \tag{2.1}$$

where R_0 is the rate at which pairs are emitted, and $\rho(\lambda)$ is the probability density of the signal variables. The singles rate, R_2, at B is given by a similar expression, while the coincidence rate, R_{12}, is given by

$$R_{12}/R_0 = p_{12}(\Theta_A-\Theta_B) = \int \rho(\lambda)\ P_1(\Theta_A,\lambda)\ P_2(\Theta_B,\lambda)d\lambda. \tag{2.2}$$

The assumption that p_1 and p_2 are independent of Θ_A and Θ_B, while p_{12} is a function of the difference between these angles, is a natural consequence of rotational invariance, and is well verified experimentally.

Now according to quantum theory[2,3] the probability p_1 may be factorized into the probability of the red "photon" passing the polarizer times the probability of its activating the detector. If the polarizer is ideal and the detector efficiency is η_1, this gives

$$p_1 = \frac{1}{2}\ \eta_1 \tag{2.3}$$

with a similar expression for p_2. Also the probability p_{12} factorizes

$$p_{12}(\Theta_A-\Theta_B) = \eta_1\eta_2 q_{12}(\Theta_A-\Theta_B)\ ,\qquad \text{where} \tag{2.4}$$

$$q_{12}(\Theta_A-\Theta_B) = \frac{1}{2}\ \cos^2(\Theta_A-\Theta_B). \tag{2.5}$$

Equation (2.4) is interpreted as follows: p_{12} is the joint probability of a pair of "photons" both passing their respective polarizers and both being detected; *since it is considered that the processes of passing the polarizers and of being detected are independent*, we may regard q_{12} as the joint probability that a pair of photons both pass through their respective polarizers.

If we accepted the italicized clause we would say that there existed probabilities $Q_1(\Theta_A,\lambda)$ and $Q_2(\Theta_B,\lambda)$ such that

$$\int Q_1(\Theta_A,\lambda)\rho(\lambda)d\lambda = \int Q_2(\Theta_B,\lambda)\rho(\lambda)d\lambda = \tfrac{1}{2},$$

$$q_{12}(\Theta_A,\Theta_B) = \int \rho(\lambda)Q_1(\Theta_A,\ \lambda)Q_2(\Theta_B,\lambda)d\lambda. \tag{2.6}$$

---

* Not all versions of the E.P.R.B. experiment use visible light signals. The Stirling[6] uses signals in the near ultraviolet and the versions based on signals from positronium annihilation[2] use gamma rays. There is no problem about extending our analysis to the first of these, but the second type cannot[15] be made a basis for a test of the Bell inequalities because of the need for supplementary assumptions even more ad hoc than that of no enhancement.

Clauser and Horne[11] showed that, with q_{12} given by (2.5), no such probabilities can exist for any $\rho(\lambda)$ and this is the modern form of the E.P.R. "paradox".

Within the local realist analysis there is no paradox; we simply infer that the right hand side of (2.5) cannot be a joint probability and therefore that *the above italicized clause is incorrect*.

Now for the experimental evidence. It is well verified that, after making some fairly small corrections for the imperfection of real polarizers, p_1, p_2 and p_{12} do indeed satisfy (2.3), (2.4) and (2.5). However, this does not invalidate the conclusion we have just reached, because in contrast to (2.6), it is not at all difficult to find probabilities P_1 and P_2 and a probability density ρ such that (2.1) and (2.2) are satisfied.

Here is the simplest such model I know of[16]:

$$\rho(\lambda) = \frac{1}{\pi} \ (0 < \lambda < \pi) \tag{2.7}$$

$$P_1(\theta_A, \lambda) = \begin{cases} \eta_1 & (|\lambda - \theta_A| < \pi/4) \\ 0 & (\pi/4 < |\lambda - \theta_A| < \pi/2) \end{cases} \tag{2.8}$$

$$\text{and} \quad P_1(\theta_A + \pi, \lambda) = P_1(\theta_A, \lambda), \tag{2.9}$$

$$P_2(\theta_B, \lambda) = \begin{cases} \tfrac{1}{2}\pi\eta_2 \cos(2\theta_B - 2\lambda) & (|\lambda - \theta_B| < \pi/4) \\ 0 & (\pi/4 < |\lambda - \theta_B| < \pi/2) \end{cases} \tag{2.10}$$

$$\text{and} \quad P_2(\theta_B + \pi, \lambda) = P_2(\theta_B, \lambda). \tag{2.11}$$

This example shows clearly, the difference between (2.2) and (2.6). The quantities Q_1 and Q_2 could be defined as

$$Q_1 = P_1/\eta_1, \ Q_2 = P_2/\eta_2, \tag{2.12}$$

but then, although (2.6) would be satisfied, we could not interpret Q_1 as a probability, because

$$Q_2(\theta_B, \theta_B) = \pi/2 > 1. \tag{2.13}$$

There is, however, no difficulty about interpreting P_1 as a probabiltiy because $\tfrac{1}{2}\pi\eta$ is appreciably less than one for all existing optical detectors.

The property demonstrated by (2.13) is a general one which, following Clauser and Horne, we shall call *enhancement*. These authors showed that any local realist model satisfying an additional condition called "no enhancement":

$$P_1(\theta_A, \lambda) < P_1(\infty, \lambda), \ p_2(\theta_B, \lambda) < P_2(\infty, \lambda), \tag{2.14}$$

where "∞" indicates that the polarizer part of the polarizer—detector has been removed, must satisfy a Bell—type inequality which is both violated by quantum theory and refuted by the experimental evidence. The above model has

$$P_2(\infty, \lambda) = \eta_2, \ P_2(\lambda, \lambda) = \tfrac{1}{2}\pi\eta_2 \tag{2.15}$$

so the second of (2.14) is not satisfied. *There are some signals for which the detection probability increases when a polarizer is placed in front of the detector.* This is enhancement. In order to make further (local realist) progress, we must say that the experimental evidence establishes enhancement as a definite physical phenomenon. The task which now has to be undertaken is to find a satisfactory explanation of this phenomenon and to relate it to other phenomena.

A general point should be borne in mind concerning the stochastic theory of the following sections, and this is that we should not (within the local realist programme) be seeking to reproduce all the predictions of the quantum theory, but *only those predictions which have been experimentally verified, and even then agreement is only to be expected within the experimental errors.* It is often overlooked that these errors are in some important cases quite substantial, and also I believe[17] that insufficient attention has been paid to the statistical theory required to evaluate the experimental data correctly. These remarks are especially pertinent when we are dealing with orientations for which quantum theory predicts zero or near-zero values, for example $(\theta_A - \theta_B) = \pi/2$ in (2.5).

3. THE REAL ZERO-POINT FIELD − STOCHASTIC ELECTRODYNAMICS[18]

The real zero-point field was proposed by Planck in 1911[13] in opposition to Einstein's light quantum hypothesis. The latter dates from 1905, but it did not become dominant until after Einstein's paper on spontaneous and stimulated emission in 1917[19], where he argued that the light quanta have definite values of momentum as well as energy. The term which Einstein used at that time to describe the light quanta was "needle radiation" (Nadelstrahlung). I shall try to show that the modern version of Planck's theory has now evolved to the point where it can assimilate this idea of Planck's young opponent. Such a development offers the possibility of incorporating quantum optics[20], with its whole aparatus of "photon-counting" distributions, into a stochastic classical theory based entirely on Maxwell's equations, thereby avoiding completely the need for field quantization.

Note that although this implies a kind of reconciliation between Planck's classical field and Einstein's light quanta, it does *not* admit the idea of the photon as any kind of a particle. The needle may look like a particle when viewed along its direction of travel, but looked at from a lateral direction it is certainly not a particle. It has a length of about 30 nanoseconds (the lifetime of a typical atomic excited state) or 10 metres. This means that, with modern photodetectors having resolving times of less than one nanosecond, we should be observing fractions of a photon. According to the analysis of stochastic electroydnamics and its offshoot, stochastic optics[21], the "split photon" is precisely what we are seeing when we observe the phenomenon currently classified as "photon antibunching"[22]. Indeed, this is the type of phenomenon to which we must transfer our attention after we have explained enhancement. However, in the present article we shall limit ourselves to understanding enhancement and relating it to one other time-independent phenomenon. Readers interested in the time-dependent analysis must consult reference[21].

The real zero-point field[23-27] has a vector potential represented by

$$A(\underset{\sim}{r},t) = V^{-\tfrac{1}{2}} \sum_{\underset{\sim}{k}} \sum_{\lambda} c(\hbar/2\omega)^{\tfrac{1}{2}} [a,\underset{\sim}{k},\lambda)\underset{\sim}{\epsilon}(\underset{\sim}{k},\lambda)e^{i\underset{\sim}{k}\cdot\underset{\sim}{r}-i\omega t} +$$

$$+ a^{\star}(\underset{\sim}{k},\lambda)\,\underset{\sim}{\epsilon}\,(\underset{\sim}{k},\lambda)e^{-i\underset{\sim}{k}\cdot\underset{\sim}{r}+i\omega t}\;]\quad,$$

(3.1)

where V is an (arbitrary) normalization volume, c the speed of light, $\hbar$ a constant, $\underset{\sim}{\epsilon}(\underset{\sim}{k},\lambda)$ a (polarization) unit vector, $\lambda = 1,2$, and $a(\underset{\sim}{k},\lambda)$

complex, Gaussian, independent random variables, each with zero mean and random phase, the variance being

$$\langle a(\underset{\sim}{k},\lambda)a^{*}(\underset{\sim}{k},\lambda)\rangle = \tfrac{1}{2}. \qquad (3.2)$$

Such a field has an energy density per unit frequency interval

$$\rho(\omega) = \hbar\omega^3/(2\pi^2 c^3). \qquad (3.3)$$

This field has been shown to be (a) Lorentz invariant (b) spatially isotropic in all Lorentz frames (c) invariant under adiabatic compression and (d) invariant under scattering by a dipole oscillator moving with arbitrary constant velocity. It is the only field possessing all of these properties, except that the constant $\hbar$ is arbitrary. Several quantum properties manifest themselves when certain material objects are immersed in this field, provided $\hbar$ is given its normal quantum theoretic value (Planck's constant divided by 2π).

The objects most extensively studied in stochastic electrodynamics have been the charged harmonic oscillator, the free particle, and macroscopic dielectrics. The first of them is by far the simplest, because it interacts with a narrow band of frequencies of the zero point field. One finds that an equilibrium is established, in which the radiative loss due to the oscillating particle's acceleration is balanced by the energy it picks up from the zero-point field. The joint position- momentum probability density is precisely the Wigner distribution for a quantum harmonic oscillator in its ground state. Furthermore, if we suppose that the zero-point field in a cavity at temperature T has a spectrum which is the sum of (3.3) plus the Planck spectrum, we obtain the Wigner distribution for the appropriate mixture of excited states.

Although both the position and momentum distributions for the above systems are therefore apparently "correct" according to quantum mechanics, the same cannot be said for any of the other common dynamical variables. For example, while the energy has its "correct" mean value, at zero temperature, of $\tfrac{1}{2}\hbar\omega$, this quantity has a nonzero dispersion. Similarly, while all the compnents of angular momentum have zero mean, their mean squares, and hence the expectation value of the quantity L^2, are not zero. There is therefore no possibility of establishing complete equivalence with the quantum formalism. Nevertheless, such "incorrect" results may help us to improve our understanding of the system. For example, if we add a uniform magnetic field, we find that the contribution to the magnetic moment arising from the nonzero value of L^2 gives precisely the diamagnetic susceptibility of a harmonic oscillator. This solves a rather old problem in classical statistical mechanics, known as the Bohr–van Leeuwen theorem, according to which diamagnetism is impossible. Through the study of the free particle in stochastic electrodynamics it has also been possible to obtain the correct expression for the Landau diamagnetism of free particles.

The other body of results in stochastic electrodynamics has resulted from a study of macroscopic dielectrics. The simplest system of this type is a pair of parallel conducting plates. Because of the new boundary conditions on the zero-point field imposed by the presence of the plates, there is a change in the total field energy, resulting in an attraction between the plates. This is known as the Casimir effect. It may be regarded as a particular example of a long-range van der Waals force, and a similar treatment may be carried out using plates of other shapes and dielectric properties. Although the calculations resemble closely the corresponding calculations based on quantum electrodynamics, it is fair to say that many workers in this area recognize the extra depth of

208

understanding provided by the stochastic interpretation of the formalism.

We can point then to some insights provided by the assumption of a real zero-point field. However, it must be admitted that, compared with some expectations among the originators of stochastic electrodynamics, including myself, the theory has failed. It gives a qualitative explanation for the stability of atoms, and indeed it is possible to show that the hydrogen atom has more or less the correct size in the theory. But it gives no explanation for the sharp line spectra. Similarly while there exist several arguments for the Planck-plus-zero-point spectrum in a cavity at nonzero temperature, the problem of the equilibrium between matter and radiation still retains some anomalous features known from the earliest days of quantum theory.

The success of stochastic electrodynamics in the interpretation of the Casimir effect, and its failure in the hydrogen atom, suggest that the theory of emission and absorption of radiation needs some modification, but that the interaction of radiation with macroscopic bodies is the same as predicted in classical electrodynamics, except for the presence of the real zero-point field. This has led us to develop stochastic optics: the theory of phenomena involving electromagnetic radiation as it interacts with macroscopic bodies, without attempting to give a microscopic theory of these interactions, but including the zero-point field. We shall not include phenomena involving radiation of long wavelength simply because these can be studied by purely classical theory (The zero-point field is negligible according to (3.3) in this case).

We also do not include radiation with short wavelength because it cannot be considered to interact collectively with macroscopic bodies. Then we are left with visible, near infrared and near ultraviolet light, which is just the domain of optics.

4. THE ATOM AS BLACK BOX - STOCHASTIC OPTICS[28]

Although we do not pretend to know exactly how the atom interacts with electromagnetic radiation, we can say that, in stochastic optics, no emission of radiation is really spontaneous; rather we may regard a "spontaneous" emission as one which is stimulated solely by the zero-point field. With only this knowledge, however, we obtain a qualitative understanding of the needle radiation. If we assume that the amplitude of the outgoing spherical wave from the atom is of the same order of magnitude as that of the (plane) components of the zero-point field stimulating it (see Fig. 2), then the two waves will interfere constructively over only a small part of their common wave front. The range of angles in which the signal has an intensity appreciably greater than that of the zero-point field is given by

$$R(1-\cos\theta) < \tfrac{1}{2}\lambda = \pi/k. \qquad (4.1)$$

For $R = 1$ m and $\lambda = 500$ nm, θ is about 10^{-3} radians, becoming less as R increases.

Now for the action of the photodetector. (In this term we include not only the sophisticated photomultiplier tube, but also the humble photographic plate.) We will assume firstly that the detector reacts only to a very narrow range of frequencies, of the order of a natural linewidth, at a time, and secondly that there is an intensity threshold, somewhat higher than the zero-point intensity, below which no activation occurs. For a fuller discussion of these two assumptions see Reference[21].

With these assumptions it is possible to model the signal as a plane wave with fixed amplitude β, two polarization variables (θ,ϕ) and a phase variable ψ. These latter three are taken to be random variables with probability density*

$$\rho(\theta,\phi,\psi) = \cos2\phi/(2\pi^2)$$

$$(0 < \theta < \pi, \ -\frac{\pi}{4} < \phi < \frac{\pi}{4}, \ 0 < \psi < 2\pi). \qquad (4.2)$$

A signal with these values of (θ,ϕ,ψ), propagating in the z-direction, has the electric vector

$$E_x = Re[\beta e^{i\omega t+i\psi}(\cos\theta\cos\phi + i\sin\theta\sin\phi)],$$

$$E_y = Re[\beta e^{i\omega t+i\psi}(-\sin\theta\cos\phi + i\cos\theta\sin\phi)],$$

$$E_z = 0. \qquad (4.3)$$

For times less than the natural lifetime ($\sim$ 30 ns) of the atomic transition, the variables (θ, ϕ, ψ) are well defined. We propose that the photo-detector, through a mechanism not yet understood, can be sensitive, at any given time, only to a narrow range of frequencies of the order of the reciprocal of this lifetime. It follows that the relevant part of the zero-point background may also be modelled by an electric vector like (4.3), except that β is different. We shall choose units such that, for the background, $\beta = 1$, and write its electric vector as

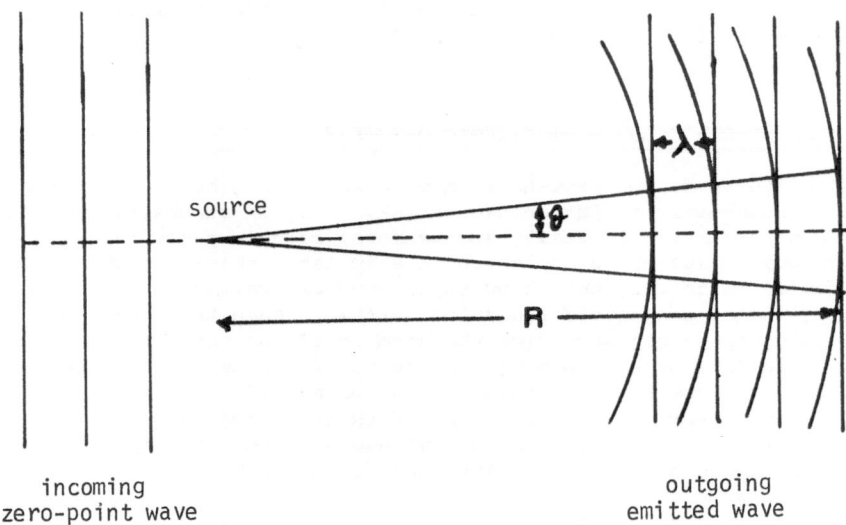

source

incoming
zero-point wave

outgoing
emitted wave

FIG.2: Superposition of an incoming plane
wave and an outgoing spherical wave
to give needle radiation.

---

* The density introduced in (4.2) has been shown[30] to be invariant under the action of half- and quarter-wave plates, or indeed of a plate which, in (4.3), changes the phase of E_y by an arbitrary amount α.

$$E_x^0 = \text{Re}[e^{i\omega t + i\psi'}(\cos \theta' \cos \phi' + i \sin \theta' \sin \phi')],$$

$$E_y^0 = \text{Re}[e^{i\omega t + i\psi'}(-\sin \theta' \cos \phi' + i \cos \theta' \sin \phi')],$$

$$E_z^0 = 0, \tag{4.4}$$

where the set (θ', ϕ', ψ') has the probability density (4.2) and is independent of (θ, ϕ, ψ).

The modelling of a polarization by a pair of variables representing elliptical polarization is, of course, standard in classical optics[29], but has been rare[30] up to this point in discussions of the Bell inequalities. I would say that,in general, the attempt to study realist alternatives to the quantum description of coincidence data for photon counting has been somewhat half-hearted. The quantum description of an atomic light signal does not allow it to have a well-defined relative phase for two perpendicular directions of polarization, but this is by no means the case for the realist alternative. For anyone familiar with macroscopic optics there is nothing "hidden" about the variables (θ, ϕ, ψ).

Now when the signal (4.3) is incident on a photodetector, we will assume that the probability of activation in a time interval w is given by

$$\text{Prob. (activation)} = \alpha(2\overline{E^2} - \gamma)w, \tag{4.5}$$

where the bar denotes time averaging, α is an efficiency parameter, and γ is a threshold level. This latter quantity certainly has to be greater than one, in order to exclude the possibility that the detector is triggered by the zero-point field alone. Then, since our model has a fixed signal amplitude, the activation probability for an unanalyzed signal is

$$\text{Prob (activation)} = \alpha(\beta^2 - \gamma)w \quad \text{(unanalyzed signal)}. \tag{4.6}$$

We now consider what happens when an ideal beam splitter is interposed between the source and the detector. The situation is illustrated by Fig. 3, where it will be seen that the interference of the signal E with the zero-point field E_0 results in a randomness in the amplitudes of the transmitted and reflected signals. The transmitted and reflected signals will be

$$E_t = 2^{-\frac{1}{2}}(E + E_0), \quad E_r = 2^{-\frac{1}{2}}(E - E_0), \tag{4.7}$$

where E and E_0 are given by (4.3) and (4.4).

It is now possible to see how the interference of the signal with the zero-point field causes the characteristic "all-or-nothing" behaviour for the detection rates in the two channels, usually thought to prove the particle nature of light. The signal-to-noise ratio, β, must be greater than one, but if it is not much greater it may result that only one out of the transmitted and reflected signals has an intensity above the threshold γ. To take account of this possibility we should modify (3.5) for the t-channel to read

$$\text{Prob.(activation)} = \alpha(2\overline{E_t^2} - \gamma)_+ w, \tag{4.8}$$

with a corresponding expression in the r-channel. Here the notation $(\ )_+$ means that we put zero if the bracket is negative.

A detailed calculation of the counting rates, N_t and N_r, in the t- and r- channels, and of the coincidence rate, N_c, has been made in Reference[21]. It involves an averaging over the sets (θ, ϕ, ψ) and (θ', ϕ', ψ'), using the weight function (4.2). Here we give only the results of this calculation. Denoting by R_0 the number of signals, assumed well separated, emitted by the source per second, we obtain

$$N_r = N_t = \alpha\beta R_0 w\pi^{-1} \int_0^\pi d\psi_1 \int_0^{\pi/2} \sin\theta' \cos\theta' \, d\theta' (\xi + \cos\theta' \cos\psi_1)_+,$$

(4.9)

$$N_c = \alpha^2\beta^2 R_0 w^2 \pi^{-1} \int_0^\pi d\psi_1 \int_\theta^{\pi/2} \sin\theta' \cos\theta' \, d\theta' \, (\xi^2 - \cos^2\theta' \cos^2\psi_1)_+,$$

(4.10)

where $\qquad \xi = \beta^{-1}(\tfrac{1}{2}\beta^2 + \tfrac{1}{2} - \gamma).$ (4.11)

The quantity measured in a recent experiment[31] to exhibit the particle nature of light is[*]

$$\alpha' = \frac{R_0 N_c}{N_r N_t}.$$

(4.12)

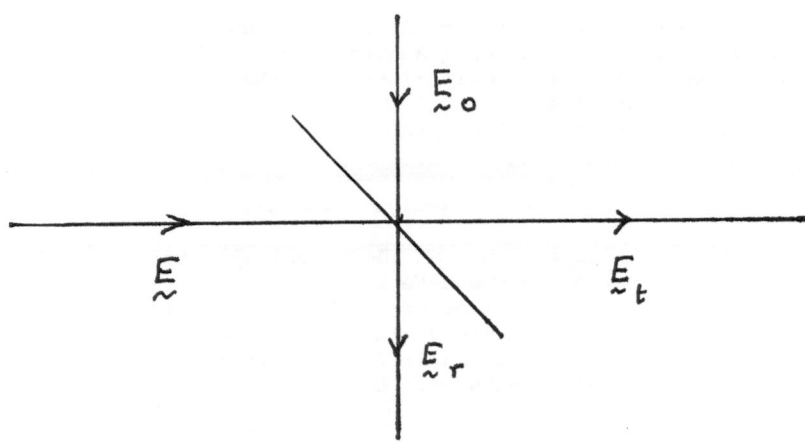

FIG.3: The action of a beam splitter on a signal $\underset{\sim}{E}$. The signal is split and mixed with the noise component $\underset{\sim}{E_0}$ to give the reflected and transmitted signals $\underset{\sim}{E_r}$ and $\underset{\sim}{E_t}$.

---

[*] More precisely α', defined by (4.12), is a function of w, because (4.9) and (4.10) should be corrected to take account of the possiblity that the signals from two or more atoms overlap. When (4.9) and (4.10), in their present, uncorrected form, are substituted in (4.12), both R_0 and w cancel out and we obtain the value of α' for the limit w = 0. For details see reference[21].

Table 1

β	γ	α'
1.511	1.642	0.000
1.5200	1.500	0.013
1.550	1.375	0.072
1.570	1.327	0.111
1.600	1.275	0.164
1.630	1.235	0.212

In Table 1 we give values of α' for a range of values of β. The threshold γ is adjusted to make the value of N_t equal to $1/2\ \alpha\ R_0 w\ (\beta^2-\gamma)$, that is one half of the counting rate in the undivided channel, as given by (4.6). In a purely wave theory of the type we are considering, we would not necessarily expect this to be satisfied exactly, but I suspect that, because it "obviously" is satisfied in the photon description, nobody has yet checked it experimentally. For $\beta \leqslant 1.511$ we obtain the "all-or-nothing" behaviour, that is $\alpha' = 0$, and as $\beta \longrightarrow \infty$ we obtain the value, $\alpha' = 1$, normally described as "classical". We now see that this latter behaviour, which excludes the possibility of a real zero-point fied, is more accurately described as "deterministic classical", and that the model of stochastic optics, which is fully classical, allows for all values of α' between 0 and 1.

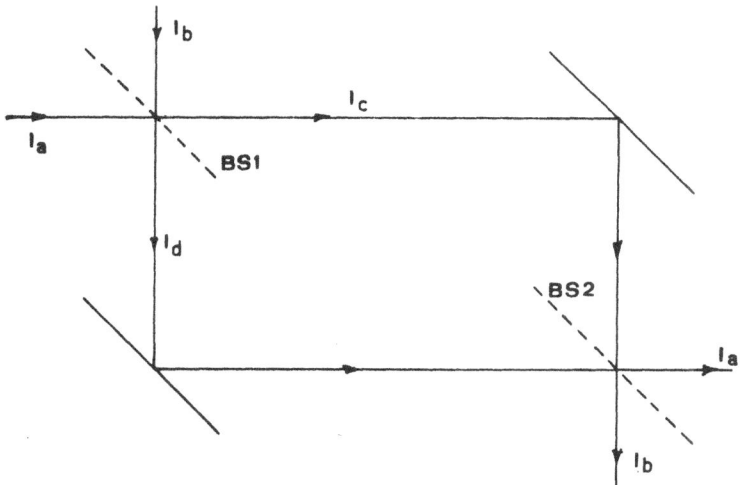

FIG.4: Recombination experiment. From an incoming
signal (I_a) and the relevant part of the
zero point field (I_b), two waves are produced
(I_c and I_d) in a beam splitter (BS1).
Another beam splitter (BS2) recombines the
waves reproducing exactly the initial
intensities if the waves arrive in phase.

Grangier, Roger and Aspect[31], the authors quoted above, also showed the wave nature of light, by recombination of the transmitted and reflected signals (see Fig. 4). In a purely wave theory such as ours, the *existence* of interference fringes is not at all problematic; the major problem is the particle nature discussed above. What may seem remarkable is that these authors obtained a very high (> 98 per cent) fringe visibility. Since a beam splitter causes, as we have seen, some fuzziness in the values of N_r and N_t, which, for $\beta > 1.511$, is manifested in a nonzero value for N_C, it may appear that a second beam splitter would add further fuzziness. Reference to Fig. 4, however, shows that the incoming signal and noise are *perfectly reproduced* in the outgoing channels of the second beam splitter, if the path lengths are equal. This is because all the "relevant" modes of the zero-point field are contained in E and E_0. No additional zero-point noise enters at the second beam splitter. The hundred-per-cent fringe visibility is therefore a consequence of the time-reversible property of Maxwell's equations.

We have explained both the "wavelike" property of recombination and the "particlelike" property of anticorrelation using a purely wave model in both cases. The zero-point, or sub-threshold field plays a crucial role in the explanation of particle behaviour. Now we see, through the analysis of recombination, that, although one of the signals E_t or E_r is, with high probability, sub-threshold, it nevertheless carries "latent order"[32] which becomes manifest when the two signals are recombined at the second beam splitter. A term used by some advocates of wave-particle duality[33] for the sub-threshold signal is that of "empty wave", but we believe that our understanding of the phenomenon is greatly improved by the knowledge that this wave is *not* empty; it carries energy and does not differ qualitatively from the wave said to be carrying the photon in the other channel.

5.   THE MECHANISM OF ENHANCEMENT

As we indicated in Section 2, enhancement is a property of linear polarizers. This means that the detection probability of a signal (θ, ϕ, ψ) may be increased as a result of its interaction with a linear polarizer of orientation θ_A, for certain values of these four variables. We shall now show that the property receives a natural explanation when we take into account the interference of the signal with the zero-point field at the polarizer.

There is some similarity between the actions of a polarizer and a beam splitter. Indeed the latter may also be said to exhibit a form of enhancement. Following the analysis of Section 4, a signal E, having amplitude β, is transmitted as $2^{-1/2}(E + E_0)$, where E_0 has unit amplitude and random phase relative to $\tilde{E}$. Then, for $\tilde{\beta} < 2^{1/2} + 1$, there are some phases for which the amplitude of the transmitted wave exceeds that of the incident wave.

However, in the case just considered there is no enhancement when we average over the noise variables. We shall find that the corresponding interaction with a polarizer can give enhancement even after taking this average. Only after averaging over both noise and signal variables does the transmitted signal have an intensity less than the incident signal.

Referring again to Fig. 3, if the device is a perfect linear polarizer with its axis in the x-direction, then a signal E is converted into a transmitted signal E_t and a reflected signal $\tilde{E_r}$, where

$$\underset{\sim}{E}_t = (E_x, E_{0y}), \quad \underset{\sim}{E}_r = (E_{0x}, E_y). \tag{5.1}$$

In this case it is convenient to use a different representation of the elliptical polarization. We write

$$E_x = \beta \cos \theta' \cos(\omega t + \psi_1), \tag{5.2}$$

$$E_y = \beta \sin \theta' \cos(\omega t + \psi_2), \tag{5.3}$$

$$E_z = 0.$$

This has been shown[21] to be equivalent to (4.3), the corresponding probability density being

$$\rho(\theta', \psi_1, \psi_2) = \sin 2\theta'/(4\pi^2),$$

$$(0 \leqslant \theta \leqslant \pi/2, \ 0 \leqslant \psi_1, \psi_2 < 2\pi). \tag{5.4}$$

Then the counting rates in the transmitted and reflected channels are

$$N_t = N_r = \alpha R_0 w \int_0^{\pi/2} \sin 2\theta' d\theta' \int_0^{\pi/2} \sin 2\theta'' d\theta'' \ .$$

$$. \ (\beta^2 \cos^2\theta' + \sin^2\theta'' - \gamma)_+ \tag{5.5}$$

and the coincidence rate is

$$N_c = \alpha^2 R_0 w^2 \int_0^{\pi/2} \sin 2\theta'' d\theta''$$

$$(\beta^2 \cos^2\theta' + \sin^2\theta'' - \gamma)_+ (\beta^2 \sin^2\theta' + \cos^2\theta'' - \gamma)_+. \tag{5.6}$$

Corresponding to equation (4.12), we may define the "Grangier parameter" for the perfect polarizer

$$\alpha_p' = \frac{R_0 N_c}{N_r N_t} , \tag{5.7}$$

and the efficiency parameter

$$\epsilon_p = \frac{2N_t}{\alpha R_0 w (\beta^2 - \gamma)} . \tag{5.8}$$

In Table 2 we give values of α_p and ϵ_p for the same values of β and γ as were used, for the beam splitter, in Table 1. Recall that, for each β, γ was chosen such as to make the beam splitter efficiency exactly one.

It will be observed that, in the range $1.511 < \beta < 1.63$, the values of ϵ_p and α_p do not differ from those of ϵ and α' by more than 5%. We shall see, in the next section, that the range of values of β allowed by stochastic optics is in fact rather smaller than this, and, for the allowed subrange, the parameters for the two devices agree to within 2%.

The phenomenon of enhancement is exhibited very clearly in equation (5.5). The intensity of the transmitted beam is

$$I_t = \beta^2 \cos^2\theta' + \sin^2\theta''. \tag{5.9}$$

Not only do there exist values of (θ', θ'') for which this is greater than the incident intensity of β^2; an average of I_t over the noise variable θ'' gives

Table 2

β	γ	ϵ_p	α_p	α
1.511	1.642	.947	.000	.000
1.520	1.500	.962	.014	.013
1.550	1.375	.979	.076	.073
1.570	1.327	.986	.115	.111
1.600	1.275	.994	.164	.164
1.630	1.235	1.000	.208	.212

$$\bar{I}_t = \beta^2 \cos^2\theta' + 1/2, \tag{5.10}$$

and this also is greater than β^2 for some values of θ'.

This explanation of enhancement was given, essentially, in my report to the Urbino conference in 1985[28], where it was attributed to a property of the signal called *superpolarization*. This arises because the signal has the same polarization both above and below the "sea" of zero-point radiation. Hence both of these parts carry a Malus law factor of $\cos^2\theta'$ in (5.10), so that for small values of θ', the addition of the noise through the action of the polarizer gives enhancement.

6.COINCIDENCE COUNTS IN ATOMIC CASCADES

With the theory of stochastic optics, developed in the last three sections, it is now possible to see how the pattern of coincidence counts observed in the atomic-cascade experiments receives a natural (local realist) explanation.

Referring to Fig. 1, the set of pairs of signals should, according to stochastic optics, be described by a probability density $\rho(\theta_R, \phi_R, \psi_R, \theta_G, \phi_G, \psi_G)$, where $(\theta_R, \phi_R, \psi_R)$ are the polarization and phase variables of one light signal (which we call "Red") and $(\theta_G, \phi_G, \psi_G)$ are the corresponding variables of the other ("Green') signal. For a 0-1-0 cascade of the type used in most experiments, we shall find that the data obtained are consistent with the assumption that $\theta_R = \theta_G$, $\phi_R = \phi_G$, $\psi_R = \psi_G$, while for the 1-1-0 cascade the corresponding assumption is $\theta_R = \theta_G + \frac{\pi}{2}$, $\phi_R = \phi_G$, $\psi_R = \psi_G$. Thus we may consider that $\lambda \equiv (\theta_R, \phi_R, \psi_R)$, which means that we assume *perfect correlation* for the two emitted signals. The density $\rho(\lambda)$ is now completely determined by the symmetry considerations of Section 4 and is given by (4.2), that is

$$\rho(\lambda)d\lambda = d\theta_R d\phi_R d\psi_R \cos 2\phi_R/(2\pi^2). \tag{6.1}$$

Consider now the detection rates, p_1 and p_2. With the polarizers removed we may use expression (4.6)

$$P_1(\infty,\lambda) = \alpha_1(\beta^2-\gamma)w, \tag{6.2}$$

$$P_2(\infty,\lambda) = \alpha_2(\beta^2-\gamma)w, \tag{6.3}$$

where we are assuming that the parameters β and γ are the same for both signals. With a polarizer in place, P_1 is given essentially by (5.5), but it is now convenient to express it in terms of the (θ, ϕ, ψ) - rather than the $(\theta', \psi_1, \psi_2)$ - representation:

$$P_1(\theta_A, \lambda) = \alpha_1 w \int_0^{\pi/2} \sin 2\theta' d\theta' .$$

$$. \ [\frac{1}{2} \beta^2 \{1 + \cos(2\theta_R - 2\theta_A)\cos 2\phi_R\} + \sin^2\theta' - \gamma]_+ . \quad (6.4)$$

A similar expression is used for $P_2(\theta_B, \lambda)$, the noise variables at the two polarizers being independent random variables. The integration in (6.4) is straightforward, and the double integration which follows when (6.1) and (6.4) are substituted in (2.2) may be performed numerically[21]. Note that we have already calculated the right hand side of (2.1) by the simpler procedure of Section 6, finding, for values of β between 1.511 and 1.63, a result very close to the experimental one, namely (see the values of ϵ_p in Table 2)

$$P_1(\theta_A) \simeq 1/2 \ p_1(\infty). \quad (6.5)$$

The quantity measured in the one-channel coincidence experiments is

$$r_{12}(\theta_A - \theta_B) = \frac{R_{12}(\theta_A, \theta_B)}{R_{12}(\infty, \infty)} , \quad (6.6)$$

and it has been shown[16] that, for models of the type we are considering, the predictions for two-channel experiments[5] may be obtained from this same function by essentially the same arguments as are used in the quantum analysis [2,3]. In Table 3 we give, for various values of $\alpha = (\theta_A - \theta_B)$ and for the same values of β as in Tables 1 and 2, the predicted values of $r_{12}(\alpha)$. We give also, for comparison, the quantum value of r_{12} for ideal polarizers (see (2.5)).

Table 3 - Values of $r_{12}(\alpha)$

$\beta\backslash\alpha$	0	$\pi/8$	$\pi/4$	$3\pi/8$	$\pi/2$
1.511	.537	.422	.201	.049	.009
1.520	.500	.407	.217	.070	.020
1.550	.472	.396	.232	.092	.040
1.570	.462	.391	.238	.102	.050
1.600	.451	.387	.243	.113	.061
1.630	.443	.383	.247	.122	.071
Ideal quantum values	.500	.427	.250	.073	.000

The analysis I have given is for ideal polarizers, so, at the present stage, it is not possible to make a detailed quantitative comparison with experiment. Nevertheless, one striking feature of these predictions is that $r(\pi/8) - r(3\pi/8)$ is greater than 0.25 for all values of β in the range we are considering. It has been established[2,3] that, for any local realist theory without enhancement, this quantity cannot exceed 0.25, and we have therefore shown explicitly how a model with enhancement can violate homogeneous Bell inequalities. The difference between the present model and previous ones with the same property is that it is based on rather general proposals about the nature of the electromagnetic field. This makes it possible to design a very wide variety of experimental tests of the theory, which is what we now consider.

7. SUGGESTIONS FOR FUTURE EXPERIMENTS

As we indicated in Section 2, stochastic optics may explain not only static experiments in which the light signals are analyzed with all kinds of polarizing devices and beam splitters, but also experiments which exhibit time-dependent effects such as photon antibunching. However, there is already enough material on the static phenomena studied in the last three sections to suggest a wide range of experimental tests.

With respect to the attempted realizations of the E.P.R.B. experiment, the results of Section 6, show very conclusively that we now have not only ad hoc enhancement models violating the homogeneous Bell inequalities but at least one such model which is based on a general theory of the nature of the electromagnetic field. Actually there has been one other model of this nature[34,35], based on a wave-particle-duality theory of the electromagnetic field.

The main difference between the predictions of the wave-particle-duality model and that developed here is that the former gives results in agreement with quantum optics for all "single-photon" experiments; its differences from quantum optics can, in principle, *only* be exhibited by experiments of the E.P.R. type. In stochastic optics we take the view that atomic light signals ("photons") can be divided. There is, for example, an energy-carrying wave in both the channels of a beam splitter, as we saw in Section 4. This means that predictions different from quantum optics may be expected in stochastic optics, even in the "single-photon" domain.

One example of this is the different predictions for the intensity of an atomic light signal after passing through two linear polarizers in series, for which stochastic optics gives small but measurable deviations[21] from Malus' law. Rather interestingly the wave-particle-duality model predicts a substantial deviation from Malus' law in the case where two polarizers are used for one of the signals of a cascade. This has led to the setting up of a three-polarizer experimental test at Stirling[36], but the results have so far been inconclusive, owing to the large corrections[37] expected, both in the wave-particle duality theory and in quantum optics, due to quite small polarizer inefficiency. Such effects will also be very important in stochastic optics, as may be appreciated by noting that, for example, the pile-of-plates polarizers used by the Orsay group[4] pass 3% of the intensity of the "wrong" polarization, which means almost 20% of the amplitude. This will obviously leave considerable scope for superposition of the weak and strong polarization components at the second polarizer.

There are two areas where the predictions of stochastic optics are rather more definitely at variance with quantum optics. The first of these may be seen by reference to Table 3, where it will be noted that, for all values of β listed, $r(\pi/4)$ is less than its quantum value of 0.25. Of course, this quantity has already been measured[2-6], but I do not think the statistics are good enough to exclude all values of β listed in Table 3. Nevertheless, I think that the values for $\beta = 1.511$

and $\beta = 1.52$ can be so excluded. This has important implications for the anticorrelation experiment[31] discussed in Section 4, because it would now seem that, according to Table 1, the Grangier parameter α' is not less than 0.07. Emilio Santos and I have examined the evidence from this experiment, especially that part of it for which the accidental coincidences are minimized, and we have concluded[38] that the best estimate for α' is about 0.06 with a 90% confidence interval $0 \leqslant \alpha' \leqslant 0.17$. This would seem to justify us in discarding the value $\beta = 1.63$ and concluding with some confidence that the parameter β lies somewhere between 1.55 and 1.60. Naturally, better statistics would make possible better estimates of both $r(\pi/4)$ and α', with the possibility of eventually refuting either quantum optics or stochastic optics.

8.   CONCLUDING REMARKS

The business of science is to analyze, and to analyze is to separate[12]. To declare something "unanalyzable"[39] is to abdicate responsibility. But it also leaves open a great opportunity for scientific discovery to those who disregard the declaration. Those of us who have insisted on analayzing, that is separating, the action of two devices some 10 metres apart (detectors A and B in Fig. 1), have discovered enhancement.

I have tried to show, in collaboration with Emilio Santos, one possible way to push the analysis further. It involves disregarding at least two more such declarations. Light signals, even if they originate in a single atomic transition, have definite amplitudes and phases for both directions of polarization, that is a full set of Stokes parameters. If, according to the quantum description, that means they are in a simultaneous eigenstate of two non-commuting observables, then so much the worse for the quantum description. Furthermore the analysis I am suggesting indicates quite strongly not only that the "photon" is divisible, but also that two fragments of it can activate, indeed possibly have already been observed to activate, two separate photodetectors.

There is some good experimental work for someone here.

9. REFERENCES

1.   A. Einstein, B. Podolsky and N. Rosen, Can quantum mechanical description of physical reality be considered complete?, Phys. Rev. 47:777 (1935).

2.   J. F. Clauser and A. Shimony, Bell's theorem: experimental tests and implications, Rep. Prog. Phys. 41:1881 (1978).

3.   F. Selleri and G. Tarozzi, Quantum mechanics, reality and separability, Riv. Nuovo Cim. 4:1 (1981).

4.   A. Aspect, J. Dalibard and G. Roger, Experimental test of Bell's inequalities using time-varying analyzers, Phys. Rev. Lett. 49:1804 (1982).

5.   A. Aspect, P. Grangier and G. Roger, Experimental realization of Einstein-Podolsky-Rosen-Bohm Gedanken experiment. A new violation of Bell's inequalities, Phys. Rev. Lett. 49:91 (1982).

6.   W. Perrie, A. J. Duncan, H. J. Beyer and H. Kleinpoppen, Polarization of the two photons emitted by metastable atomic deuterium: a test of Bell's inequality, Phys. Rev. Let. 54:1790 (1985).

7.   E. T. Jaynes, Electrodynamics today, in: "Coherence and Quantum Optics", L. Mandel and E. Wolf, eds., Plenum New York (1978).

8.   T. A. Welton, Some observable effects of the quantum-mechanical

fluctuations of the electromagnetic field, Phys. Rev. 74:1157 (1948).

9.   T. W. Marshall, A classical treatment of blackbody radiation, Nuovo Cim. 38:206 (1964).

10. R. Hanbury-Brown and R. Q. Twiss, Correlation between photons in two coherent beams of light, Nature 177:27 (1956).

11. J. F. Clauser and M. A. Horne, Experimental consequences of objective local theories, Phys. Rev. D10:526 (1974).

12. A. Einstein, Quantum mechanics and reality, in: "The Born-Einstein Letters", Macmillan, London, (1971) pp.168-173. "If this axiom (the Principle of Local Action) were to be completely abolished, the postulation of laws which can be checked empirically would become impossible" (I have changed Born's translation of "Prinzip der Nahewirkung".)

13. M. Planck, Verh. Deutsch. Phys. Ges. 13:138 (1911).

14. W. Nernst, Verh. Deutsch. Phys. Ges. 18:83 (1916).

15. T. W. Marshall, A naive remark concerning the Bohm-Aharonov experiment, Phys. Lett. 79A:147 (1980).

16. D. Home and T. W. Marshall, A stochastic local realist model for the E.P.R. atomic-cascade experiment which reproduces the quantum-mechanical coincidence rates, Phys. Lett. 113A:183 (1985).

17. T. W. Marshall, Testing for reality with atomic cascades, Phys. Lett. 100A:225 (1984).

18. E. Santos, Stochastic electrodynamics and the Bell inequalities, in: "Open Questions in Quantum Physics", G. Tarozzi amd A. van der Merwe, eds., Reidel, Dordrecht (1985).

19. A. Einstein, On the quantum theory of radiation, in: "Sources of Quantum Mechanics", B. L. van der Waerden, ed., Dover, New York (1968).

20. P. L. Knight and L. Allen, "Concepts of quantum optics", Pergamon, Oxford, (1983).

21. T. W. Marshall and E. Santos, Stochastic optics: a classical alternative to quantum optics, University of Santander preprint (1986).

22. M. Dagenais and L. Mandel, Investigation of two-time correlations in photon emissions from a single atom, Phys. Rev. A18, 2217 (1978).

23. T. H. Boyer, Quantum zero-point energy and long-range forces, Ann. Phys. (N.Y.), 56:474 (1970).

24. P. Claverie and S. Diner, Stochastic electrodynamics and quantum theory, Int. J. Quantum Chem., 12 Suppl. 1:41 (1977)

25. L. de la Peña, Stochastic electrodynamics, its development, present situation and perspectives, in: "Stochastic Processes Applied to Physics and Other Related Fields", B. Gomez, S. M. Moore, A. M. Rodriguez-Vargas and A. Rueda, eds., World Scientific, Singapore (1983).

26. T. H. Boyer, A brief survey of stochastic electroydnamics, in: "Foundations of Radiation Theory and Quantum Electrodynamics", A. O. Barut, ed., Plenum, New York (1980).

27. T. H. Boyer, The classical vacuum, Sci. American, August (1985).

28. T. W. Marshall, Towards a realist theory of measurement, in: "Microphysical reality and quantum formalism", Conference proceedings, Urbino (1985).

29. M. Born and E. Wolf, "Principles of Optics", Pergamon, Oxford (1984).

30. T. W. Marshall and E. Santos, Local realist model for the coincidence rates in atomic-cascade experiments, Phys. Lett. 07A:164 (1985).

31. P. Grangier, G. Roger and A. Aspect, Experimental evidence for a photon anticorrelation effect on a beam splitter: a new light on single-photon interference, Europhys. Lett. 1:173 (1986).

32. D. Greenberger, Some new wrinkles on the measurement problem, in: "Microphysical reality and quantum formalism", Conference

proceedings, Urbino (1985).

33. F. Selleri, Gespensterfelder, in: "The Wave-Particle Dualism, S. Diner, D. Fargue, G. Lochak and F. Selleri, eds., Reidel, Dordrecht (1984).

34. A. Garuccio and F. Selleri, Enhanced photon detection in EPR type experiments, Phys. Lett. 103A:99 (1984).

35. F. Selleri, Local realistic photon models and EPR type experiments, Phys. Lett. 108A, 197 (1985).

36. A. J. Duncan, The two-photon decay of metastable atomic deuterium: recent test of Bell's inequality, report to conference "Microphysical reality and quantum formalism", Urbino (1985).

37. A. J. Duncan and F. Selleri, private communications.

38. T. W. Marshall and E. Santos, Statistical analysis of the experimental evidence for a photon anticorrelation effect in a beam splitter, University of Santander preprint (1986).

39. N. Bohr, Discussion with Einstein on epistemological problems in atomic physics, in: "Albert Einstein: Philosopher-Scientist", ed. P. A. Schilpp, Tudor, New York (1957) "any attempt of subdividing the phenomena will demand a change in the experimental arrangement introducing new possibilities of interaction between objects and measuring instruments which in principle cannot be controlled."

NEO-HERTZIAN ELECTROMAGNETISM

Thomas E. Phipps, Jr.

908 South Busey Avenue
Urbana, Illinois 61801 (USA)

In this talk -- aimed at modernizing and rejuvenating electromagnetic theory -- I shall be concerned (a) with new physical interpretations of some century-old mathematics (first published by Heinrich Hertz, but independently rediscovered by several modern investigators), (b) with the surprising implications for photon behavior, and (c) with needed experimental testing.

Investigators interested in possible quantum violations will require no reminding that the most "quantum" process we know -- the one whose characteristics first evidenced historically the existence of energy quanta -- is the detection of light. The principal descriptor we have of light, Maxwell's equations, gives no hint of the existence of quanta (photons) -- but has nevertheless cast its shadow (field theory) upon all we claim to know about light. Perhaps we know less than we think we know. If any minds are opened to new possibilities in an unexpected direction, my talk will have achieved its purpose.

DEFICIENCIES OF MAXWELL'S THEORY

I came to an interest in electromagnetism indirectly, after many years of skeptical thinking about kinematics. As a result of my dissatisfaction with such aspects of the theory as the Ehrenfest paradox, I arrived eventually at the conclusion that special relativity is only half right. That is, it is right in its description of single particles and single worldlines, but wrong in asserting spacetime symmetry, 4-space metric structure, etc. -- in short, false (or at best unproven) in all its relational assertions about worldlines in the plural.

The outcome of my prolonged immersion in kinematics was an hypothesis regarding the invariants of that subject; namely, that those invariants are object length and particle proper time. This mixing of the old and new sacrifices the alleged symmetry of spacetime and neatly divides honors between Newton (to whom length was an invariant) and Einstein (the inventor of proper time). When I had reached this point in my studies I was forced to recognize that one cannot make a move in kinematics without a

matching move in electromagnetism -- for the two subjects dove-
tail precisely. Unsymmetrical length and proper-time invariants
had somehow to be made compatible with electromagnetic theory --
the origin and stronghold of spacetime symmetry. So, about six
years ago, I turned to Maxwell's mathematics and began to look
for soft spots that might reveal alternative theoretical
possibilities.

The first thing one notices about Maxwell's equations is
that they are not -- at the first order of approximation, $O(v/c)$
-- Galilean invariant. (Indeed, they are not invariant at any
order under any known coordinate transformation.) The failure at
first order is a most grievous fault -- for it is fruitless to
seek "corrections" at second or higher orders to mistakes made at
first order. A theory must be perfect at all lower orders before
one is justified in giving any consideration to its higher-order
refinements. This invariance flaw in Maxwell's equations was the
cause of much distress to nineteenth-century experimentalists.
For noninvariance meant (on the face of it) that the theory, as
given, was first-order valid in only one "fundamental" reference
system. Consequently, when one changed inertial systems,
observable $O(v/c)$ effects such as optical fringe shifts were
predicted and calculable, provided one knew the velocity of one's
system with respect to the fundamental system in which Maxwell's
equations were valid. This motivated the great nineteenth-
century snark-hunt for an "ether wind." Though we laugh at them
nowadays, the hunters were not crazy; they simply took Maxwell at
his word, as he surely deserved to be taken.

The fact that no first-order effects were observed meant
that Maxwell's equations, which predicted such effects, were
wrong at first order. Observe that I do not mince words about
this. I state the only logical conclusion at which nineteenth-
century physicists had any historical right to arrive. It was
not until the twentieth century that artful dodges such as Minkowski's "covariance" and the artistic covering-over of first-
order errors with second-order flowered wallpaper made their
advent. Why, then, did not nineteenth-century physicists forth-
rightly draw this negative conclusion about Maxwell's equations
and get to work changing those equations to make them first-order
Galilean invariant? Well, one of those physicists -- Heinrich
Hertz -- did just that. I will proceed to that part of my story
in a moment. As for the rest of them ... well, perhaps in theme
the Watergate cover-up was not altogether the invention of the
twentieth century.

Before turning to Hertz's work, let me mention a second flaw
or clue to be found in Maxwell's equations. This one concerns
parameterization. The basic physical elements of Maxwellian
theory are (i) the field source, (ii) the field detector, and
(iii) the system or "frame" of reference. The latter permits
quantification of the location of the "field point" upon which
the comoving "observer" fixes his attention. Of these the tangi-
ble, material items are the field source and the field sink or
detector. There really is no necessary ether in the theory, nor
any all-pervading ectoplasmic "field," because the "field vari-
ables" $\vec{E}(\vec{r},t)$ and $\vec{B}(\vec{r},t)$ may be construed as simply mathematical
notations for the numerical readings on certain black boxes --
call them field meters -- present at the selected "field point"
$(\vec{r},t)$. In other words the field variables do not describe an
abstract "field," existing independently of all measurement, but

stand for measurement readings on certain instruments. Through this interpretation one becomes aware of the foreshadowing of quantum mechanics by Maxwell's equations.

To be specific, we can fabricate an electric field meter of sorts by placing inside a small box made of insulating material a charged pith ball, under frictionless orthogonal constraints, attached to strain gauges that read out digitally (given suitable calibration) the three components $(\underline{E}_x, \underline{E}_y, \underline{E}_z)$ of the forces needed to prevent the pith ball from moving with respect to the walls of the box. A similar box containing orthogonal current elements can provide us with three magnetic field numbers, $(\underline{B}_x, \underline{B}_y, \underline{B}_z)$. This, in thought, is how the Maxwell field can be operationally defined ... and in science the operational defining of an entity renders more abstract or philosophical definitions redundant. What I am saying, then, is that for a physicist the "field" is nothing more nor less than what a field detector detects.

A field detector is a composition of matter, as Einstein's patent office would put it, and it is a well-known feature of compositions of matter that they possess physical degrees of freedom. Now we come to my point about parameterization: If we examine Maxwell's equations in their simplest or "microscopic" form,

$$\vec{\nabla} \times \vec{B} - \frac{1}{c} \frac{\partial \vec{E}}{\partial t} - \frac{4\pi}{c} \vec{j}_s = \emptyset \qquad (\vec{j}_s = \rho \vec{v}_s) \qquad (1a)$$

$$\vec{\nabla} \times \vec{E} + \frac{1}{c} \frac{\partial \vec{B}}{\partial t} = \emptyset \qquad (1b)$$

$$\vec{\nabla} \cdot \vec{B} = \emptyset \qquad (1c)$$

$$\vec{\nabla} \cdot \vec{E} - 4\pi \rho = \emptyset \ , \qquad (1d)$$

we see that in addition to the field variables descriptive of readings on field meters they contain parameters descriptive of the degrees of freedom of field sources. That is, the mean velocity $\vec{v}_s$ in the laboratory of the electrified particles that produce the field is parameterized by the current (density) vector $\vec{j}_s = \rho \vec{v}_s$. So the motional freedoms of any charged bits of matter responsible for creating the field are explicitly parameterized. But what about that other bit of matter, the absorber, detector, or field sink? I refer to our field meter itself -- the composition of matter that gives operational defin-ability to the "field." Alas, look high, look low, you will find no parameter in Maxwell's equations describing the motional free-doms of the black box that acts as a field meter. Field sources have their motions parameterized, but not field sinks.

So what? Is it not legitimate to pick an arbitrary "field point" of interest and to define the field as what is measured by a field meter at rest there? Generations have thought so, to be sure, but the answer is yes and no. Yes, insofar as choice of an arbitrary field point is concerned and provided we agree to exclude all consideration of relative motion of observers -- i.e., we allow a preferred system of reference. The location of the field point, which is not a composition of matter, is always entirely at the observer's option. But to define the field meter, which is a composition of matter, as at rest with respect to the observer or his field point is another affair entirely. Only absolutists can afford the kind of restrictive definition that denies the facts of physical mobility of matter. The field

meter is a material object whose physical degrees of freedom all
would-be "relativists" deny at their peril. If one seeks any
kind of "relativity theory," the degrees of freedom of all compositions of matter involved in the problem must be parameterized.
The alternative is to create a "preferred system" with respect to
the motional state of some material object in the universe --
which by definition spells death to relativism.

Within the field meter elemental quantum-level influences --
physically unique localized interactions or "events" of field
detection -- are occurring. These events are specific to that
composition of matter -- that piece of "apparatus" -- and are not
interchangeable with similar quantum measurement events occurring
in other field meters in other states of motion, even though the
meters may in thought all be instantaneously present at practically the same field point. Thus the field meter at rest in
inertial system S is in motion in another system S' ... and, if
our electromagnetic field equations lack parameters to describe
detector motions, such equations can never describe in S' the
specific field-detection events that occur within the particular
meter at rest in S. The issue is one of descriptive power:
Maxwell's equations are fine for describing what Maxwell set out
to describe, namely, events occurring in detection instruments at
rest in his laboratory. But he did not have differently-moving
laboratories in mind -- that was Einstein's contribution, which
came much later. And a poorly thought-out contribution it was
... for Maxwell's equations lack the parameters to describe (even
at first order) the motional freedoms of field sinks. To define
any composition of matter as at rest automatically creates a
"preferred system" with respect to that bit of matter. Any such
theory is marked by an indelible stain of absolutism.

Einstein's approach, instead of correcting Maxwell's
parametric deficiency, was to clone it. He made every observer a
"preferred observer" with respect to his own bit of matter -- his
private field meter. Now each observer had his own by-definition
immovable detector. So, the lying about natural degrees of freedom was symmetrized among observers and made a universal feature
of the world. How georgeously symbolic and utterly appropriate
to the twentieth century! The cream of the jest is that this
symmetrizing of "preferences" among observers and of lies about
degrees of freedom, this building of the castle of relativism
upon the rock of absolutism, was done in the name of the relativity principle -- the one principle that must never be admixed in
any way with Maxwell's equations because of their fundamental
first-order nonrelativism.

This genuine mess, unwittingly fathered by Maxwell, cannot
in any way be cleared up until an electromagnetic theory is
adduced that parameterizes the motions of all relevant compositions of matter with respect to some one observer. We have to
get the representation of nature right (meaning parametrically
complete) in one system. Given a theory that is right in one
system, its parameterization will automatically suffice to
describe the mobility of all relevant matter with respect to all
observers -- thus providing the basis for a true "relativity
theory" from which "preferred observers" are permanently banished. In mathematical terms what is needed in the theory is some
parameter $\vec{v}_L$ that describes field detector velocity with respect
to the observer or his field point. Such a sink-velocity parameter is needed to match the source-velocity parameter $\vec{v}_s$ already

present. Until electromagnetic theory contains such a parameter, the stain of Maxwell's absolutism will never be removed.

While we are seeking such a theory -- <u>i.e.</u>, once we gather the courage to tamper in any way with the sacred equations of Maxwell -- we might as well aim to correct also the first deficiency I mentioned by making our improved theory first-order (Galilean) invariant. Such invariance will ensure the satisfaction of the relativity principle and eliminate the need for fancy footwork, fast talk about "covariance" (the second-order band-aid on a first-order hemorrhaging wound), and self-contradictory verbalisms such as "Lorentz invariance." Of course we shall not dare to lay rude hands on such an elegant and successful theory as Maxwell's unless we can be assured of recovering all his predictive results through discovery of what is called a "covering theory." Mathematically, then, what we must seek (to put this all together) is an <u>invariant covering theory</u> of Maxwell's electromagnetism ·-- one that is more richly parameterized through possession of an extra velocity-dimensioned parameter $\vec{v_d}$ describing detector velocity. This brings us to the work of Hertz.

HERTZ'S INVARIANT COVERING THEORY

Hertz is known primarily as the experimentalist who assured Maxwell's lasting fame by confirming his prediction of electromagnetic waves, but consultation of Hertz's book[1] further reveals him as a theoretician of stature. Unfortunately, he was a man of few words, so from his book we know little of what motivated him to put down the equations he did, or what he thought of them. One can only speculate that he perceived the fallacy of trying to describe electromagnetic phenomena with first-order noninvariant equations, so he touched-up Maxwell's equations to make them invariant under first-order inertial (Galilean) transformations. There was really nothing to it -- just a matter of replacing Maxwell's partial time derivatives with what we now call total or convective time derivatives:

$$\frac{\partial}{\partial t} \rightarrow \frac{d}{dt} = \frac{\partial}{\partial t} + \vec{v_d} \cdot \vec{\nabla} \ . \tag{2}$$

Hertz remarked in passing, without proof, that his equations were invariant, but did not emphasize that they differed significantly from Maxwell's. In fact, he mentioned the matter in such an offhand way that nobody seems particularly to have noticed the difference -- and Einstein in his 1905 paper refers to "Maxwell-Hertz equations" that have nothing to do with Hertz. The distinction between Hertz and Maxwell was forgotten in this century until the modern historian of science, Arthur I. Miller, pointed it out in his book[2] on Einstein. (Miller unfortunately muddied the waters by "correcting" Hertz, saying that he really meant "covariance," not "invariance." Not so. His equations, being unsymmetrical between space and time, are not guilty of covariance. Hertz said and meant invariance, as we shall soon verify.)

Hertz's use of a total time derivative brought into the equations of electromagnetism for the first time a velocity-dimensioned parameter, written by him in component form as (α, β, γ) and written above as $\vec{v_d}$. In view of my preamble I need not tell you that I interpret $\vec{v_d}$ as detector velocity in the laboratory. Hertz, however, fell into a deep interpretational trap -- in that he saw $\vec{v_d}$ as "ether velocity." Having made that basic slip, which was probably fore-ordained by the ether fixation of practically all physicists of his time (including Maxwell), he proceeded to tumble all the way for an old idea of Stokes's that the ether is 100% convected by material objects. Thus Hertz called his theory an "electrodynamics of moving bodies" -- by which he meant that $\vec{v}_{l}$ parameterized the motion of just any "body" that happened to move in the laboratory. From this misapprehension it could be deduced that a dielectric moving in the lab should produce a magnetic field detectable there. A few years later, after Hertz's untimely death, the experiment was done [3,4] with negative results, so Hertz's theory was discredited and forgotten -- until the present.

In our era I rediscovered the invariance prescription, Eq. (2), as did S. Kosowski of the Polish Academy of Sciences, F. D. Tombe of Belfast, Northern Ireland, and perhaps others I know not of. All worked independently, of course, and none suspected Hertz's priority until I stumbled over the Miller reference and perceived that only an archaic notation (and Miller's "covariance" fixation) prevented Hertz' mathematical achievement from being recognized. Hertz himself pioneered the movement to disregard Maxwell's "physics" through his famous dictum that "Maxwell's theory is Maxwell's equations." I should like here and now to return the favor by stating a parallel dictum, that "Hertz's theory is Hertz's equations" -- for his ether interpretation, like Maxwell's before him, is without basis in today's physical thought ... but his mathematics is more relevant than ever. I hope the truism conveys itself that the physics resides in the interpetation, not in the mathematics -- and that (to judge from the breakdown of consensus) the hard part is the physics, not the mathematics. For, to complete my story, I have to report that no two of the modern rediscoverers of Hertz's mathematics agree as to physical interpretation either with each other or with Hertz. Again, therefore, what I offer here is a minority report of one.

As a curious sociological fact I call it to your attention that when Maxwell's equations, coupled to a false ether interpretation, were refuted by observation his equations were retained, his interpretation was discarded, and he himself was elevated to the pinnacle of the pantheon of theoretical physics. But when Hertz's equations, coupled to a false ether interpretation, were refuted by observation his equations and interpretation were discarded together and he himself (as theoretician) was relegated to the ashbin of history of physics. In the modern lexicon "Hertz" is only a trendy misnomer for "cycles per second." There is no mystery to this as long as history is written by people like Professor Miller, who are open partisans of Einstein. But there is a great pity to it, and an injustice to science, since it is Hertz's equations -- representing an invariant covering theory of Maxwell's equations -- that are the nobler and more worthy scientifically to endure.

The "covering theory" aspect follows at once from the fact that in Eq. (2) the partial and total time derivatives become identical in the special case $\vec{v}_{l} = 0$. That is, Hertz's equations reduce to Maxwell's equations in the special case $\vec{v}_{l} = 0$, which by my interpretation means the case in which the field detector is at rest in the observer's laboratory. That, as we know, is exactly the special case of the "immobile field meter" assumed by Maxwell and the only case in which his equations can be expected to possess physical validity. Hertz's equations, being more richly parameterized, "cover" additional moving-detector cases not described by Maxwell's more restricted formalism. They possess therefore additional physical implications about which I shall speak presently.

As for the invariance of Hertz's equations at first order, let us pause to verify this for only one of them, since time is pressing. Confining attention to first-order considerations, we see that the Galilean transformation for the case of inertial system S' (primed coordinates) moving with velocity $\vec{v}$ with respect to S (unprimed), namely,

$$\vec{r}' = \vec{r} - \vec{v}t \quad , \qquad t' = t \tag{3}$$

implies that

$$\frac{\partial}{\partial t} = \frac{\partial x'}{\partial t} \cdot \frac{\partial}{\partial x'} + \frac{\partial y'}{\partial t} \cdot \frac{\partial}{\partial y'} + \frac{\partial z'}{\partial t} \cdot \frac{\partial}{\partial z'} + \frac{\partial t'}{\partial t} \cdot \frac{\partial}{\partial t'}$$

or

$$\frac{\partial}{\partial t} = \frac{\partial}{\partial t'} - \vec{v} \cdot \vec{\nabla}' = \frac{\partial}{\partial t'} + \vec{v}' \cdot \nabla' \quad , \tag{4a}$$

where $\vec{v}' = -\vec{v}$ is the velocity of S with respect to S'. Similarly,

$$\frac{\partial}{\partial x} = \frac{\partial x'}{\partial x} \cdot \frac{\partial}{\partial x'} + \frac{\partial y'}{\partial x} \cdot \frac{\partial}{\partial y'} + \frac{\partial z'}{\partial x} \cdot \frac{\partial}{\partial z'} + \frac{\partial t'}{\partial x} \cdot \frac{\partial}{\partial t'} = \frac{\partial}{\partial x'}$$

etc., so

$$\vec{\nabla}' = \vec{\nabla} \quad . \tag{4b}$$

Under this transformation an arbitrary velocity $\vec{V}$ obeys the first-order velocity addition law,

$$\vec{V}' = \vec{V} - \vec{v} \quad \text{or} \quad \vec{V} = \vec{V}' - \vec{v}' \quad . \tag{5}$$

The fact that our field meters can have digital readouts which must display the same numbers in the view of observers in arbitrary states of relative motion implies numerical or scalar invariance of the field variables:

$$\vec{E}'(x',y',z',t') = \vec{E}(x,y,z,t) \tag{6a}$$

$$\vec{B}'(x',y',z',t') = \vec{B}(x,y,z,t) \tag{6b}$$

The arguments represent coordinates of the same physical point (occupied by the field meter), as measured in the primed and unprimed systems. Note the sharp contrast of this scalar invariance to Lorentz covariance. Similarly, scalar invariance of charge density, $\rho' = \rho$, may be assumed, so that

$$\vec{u_s'} = (\rho \vec{v_s})' = \rho' \vec{v_s'} = \rho (\vec{v_s} - \vec{v}) = \vec{u_s} - \rho \vec{v}$$

or

$$\vec{u_s} = \vec{u_s'} + \rho \vec{v} = \vec{u_s'} - \rho' \vec{v}' \quad . \tag{7}$$

Let us consider the Hertz equation analogous to Eq. (1a). It contains a source term whose transformation properties are readily established: The current $\vec{u}_m$ measured by a meter that comoves with our field meter is related to the current $\vec{u}_s$ measured by a meter at rest in the laboratory by

$$\vec{u}_m = \vec{u}_s - \rho\,\vec{v}_d \quad . \tag{8}$$

Applying Eqs. (2) and (4)-(8), together with c' = c, we see that this field equation transforms invariantly:

$$\vec{\nabla}\times\vec{B} - \tfrac{1}{c}\frac{d\vec{E}}{dt} - \tfrac{4\pi}{c}u_m = \vec{\nabla}\times\vec{B} - \tfrac{1}{c}\left(\tfrac{\partial}{\partial t} + \vec{v}_d\cdot\vec{\nabla}\right)\vec{E} - \tfrac{4\pi}{c}\left(\vec{u}_s - \rho\vec{v}_d\right)$$

$$= \vec{\nabla}'\times\vec{B}' - \tfrac{1}{c}\left[\tfrac{\partial}{\partial t'} + \vec{v}'\cdot\vec{\nabla}' + (\vec{v}_d' - \vec{v}')\cdot\vec{\nabla}'\right]\vec{E}' - \tfrac{4\pi}{c}\left[(\vec{u}_s' - \rho'\vec{v}') - \rho'(\vec{v}_d' - \vec{v}')\right]$$

$$= \vec{\nabla}'\times\vec{B}' - \tfrac{1}{c}\left(\tfrac{\partial}{\partial t'} + \vec{v}_d'\cdot\vec{\nabla}'\right)\vec{E}' - \tfrac{4\pi}{c}\left(\vec{u}_s - \rho\vec{v}_d\right)'$$

$$= \vec{\nabla}'\times\vec{B}' - \tfrac{1}{c}\frac{d\vec{E}'}{dt'} - \tfrac{4\pi}{c}u_m' = 0 \quad . \tag{9}$$

Note that to obtain the invariant Hertz equation we have had to make two changes in the corresponding Maxwell equation:

(i)   The partial time derivative $\partial/\partial t$ in Eq. (1a) is replaced by the total derivative, $d/dt = \partial/\partial t + \vec{v}_d\cdot\vec{\nabla}$.

(ii)   The source current $\vec{u}_s$ is replaced by $\vec{u}_m = \vec{u}_s - \rho\vec{v}_d = \rho(\vec{v}_s - \vec{v}_d)$, which is seen to involve only the relative velocity of source charge and field detector.

Invariance of the other Hertz equations, obtained by similar modifications of their Maxwell counterparts, is easier to prove and will be omitted here, since it has been demonstrated else-where[5,6]. (We say that an equation is "invariant" if each term in it transforms invariantly. In the present instance the invariance is "manifest," because it can be shown that each symbol in each term transforms invariantly -- e.g., (d/dt)' = (d/dt), etc.) In summary, the first-order invariant Hertz equations are

$$\vec{\nabla}\times\vec{B} - \tfrac{1}{c}\frac{d\vec{E}}{dt} - \tfrac{4\pi}{c}\vec{u}_m = 0 \tag{10a}$$

$$\vec{\nabla}\times\vec{E} + \tfrac{1}{c}\frac{d\vec{B}}{dt} = 0 \tag{10b}$$

$$\vec{\nabla}\cdot\vec{B} = 0 \tag{10c}$$

$$\vec{\nabla}\cdot\vec{E} - 4\pi\rho = 0 \quad . \tag{10d}$$

Because of this invariance there exists no preferred inertial system, a relativity principle is automatically satisfied, and all first-order attempts to detect motion (via optical fringe shifts, etc.) with respect to a "fundamental" system are fore-doomed. This fits with Potier's principle [7] (the nineteenth-century form of the relativity principle, which states that no first-order effect of motion with respect to an ether can be detected by optical means), the proof of which depends on Fermat's principle. The implication is that Fermat's principle is in agreement with Hertz's equations: Both predict no fringe shifts affecting interference patterns under changes of inertial system.

What it also means -- which I think nobody has recognized before -- is that Maxwell's equations are in first-order disagreement with both Fermat's and Potier's principles, and with all observations that confirm them. For at the first order in v/c Maxwell's equations are noninvariant under inertial transformations, hence predict first-order fringe shifts that are denied by observation, by Fermat-Potier, and by Hertz. I hope that nothing more needs to be said against Maxwell's equations and consequently against Einstein's special relativity theory -- which in essence is not a relativity theory but a theory of preferred systems comoving with electromagnetic field meters.

Let me try to clarify in operational terms the basic difference between Hertzian and Maxwellian electromagnetism: The "field," as operationally defined, is a different object in the two theories. That is the vital point to grasp. Both theories share the same black-box detectors, but there the resemblance ends. Hertz, as I have reinterpreted his mathematics, gives a complicated definition of field involving an extra parameter, the velocity of a black box in the laboratory. Since the field detectors move and the field point is considered at rest, it is necessary to picture a linear stream of detectors passing continually through the field point in order to follow evolution of the Hertzian field in time at a fixed place in the lab. In consequence of the Hertzian field being thus a more complicated object, it has simpler transformation properties, given by Eq. (6). That is, the Hertz field is a scalar invariant under first-order inertial transformations. Electric and magnetic fields, though complicated in definition, retain their separate identities and never get mixed together.

The Maxwell field, on the other hand, is operationally defined in a very simple way, by the readings on a black box at rest in the laboratory. But this simplicity of definition entails complicated transformation properties, known as covariance, whereby electric and magnetic field components lose their identities and get "scrambled" by linear combination via a complicated Lorentz transformation. The latter is alleged to have something to do with inertial motion, but mysteriously must invoke second-order considerations in order to answer first-order questions. The "electro-magnetic" scrambling has been supposed from the earliest days to represent real physics; but in the Hertzian view the scrambling is just one of the penalties for a noninvariant mathematical formulation. In summary, there is a trade-off of complexities between the two theories: Hertz offers simple mathematical transformation properties at the price of a complicated operational definition of "field." Maxwell offers a simple operational definition of "field" and complicated transformation properties. The aesthetic advantages clearly are with Hertz's theory -- mathematically because it is an invariant covering theory, physically because (i) it treats field sources and sinks (parametrically) on a par, (ii) it employs only relative velocities among physically relevant objects (with no necessary employment of "schesic" velocities relative to frames).

So far I have merely recapitulated Hertz's mathematics and offered a new interpretation of his extra velocity-dimensioned parameter. Validity of the whole discussion has been confined to first-order considerations. In order to bring in the "Neo-" of my title I must say a word about higher-order refinements. this is a big subject, about which I must refer anyone really interested to a forthcoming book [6] of mine. Here I shall merely say that to harmonize electromagnetic theory with the kinematics I mentioned at the outset, of which the invariants are length and proper time, it is necessary to identify which object concerned in the electromagnetic descriptive problem is the one whose proper time will be used to parameterize the description. To make a short story shorter, the object in question is obviously the field detector. This is the central actor in all field theory. So, we merely have to introduce τ_d, the proper time of the detector, as the time parameter of the field-descriptive problem and substitute it for the total frame-time derivatives appearing in Eq. (10); that is, $d/dt \rightarrow d/d\tau_d$. The only tricky part concerns the source term in (10a), which must employ a (non-Einsteinian and non-Galilean) higher-order velocity composition law in determining relative velocities between field meter (or ρ-meter) and field source particles. I will not take time for this here.

As for what happens to the Lorentz force law, the first thing is that the velocity appearing in $\vec{v} \times \vec{B}$, which is normally interpreted as charge velocity relative to the observer's reference frame, is reinterpreted as charge velocity relative to the $\vec{B}$-field detector. Beyond that I ought not to speculate, though I cannot resist noting that the Lorentz force law, because it derives from the Biot-Savart law, has problems about balancing action-reaction in the analysis of forces on portions of conductive circuits. To deal with this, an interesting generalization of the Lorentz force law, derived from Ampere's original law of force between current elements and involving measurable contributions from the so-called "gauge" quantities, such as $\vec{\nabla} \cdot \vec{A}$ (which thus play a new, operationally quantifiable, physical role), has been proposed by Wesley[8]. The superiority of the Ampere law for describing forces upon parts of circuits has been confirmed by experimental studies of Graneau[9]. According to Wesley, Maxwell-Lorentz theory is limited in validity to the case in which all currents vanish on the boundary of the space considered -- hence to the Coulomb gauge, $\vec{\nabla} \cdot \vec{A} = 0$. When one cuts into a circuit, to examine forces on part of it, the currents do not vanish on the boundary, $\vec{\nabla} \cdot \vec{A} \neq 0$, and any amendments to the force law involving $\vec{\nabla} \cdot \vec{A}$ become significant.

Let me say no more on this, but leave you with the suggestion that this sub-area of nineteenth-century physics is ripe for reappraisal -- which becomes both possible and desirable once the intellectual stranglehold of covariance has been broken. It delights me to think that the famous "gauge invariance" might be an incidental casualty of any struggle to get the force law of classical electrodynamics right. I feel sure the last word has not been said on this subject -- and, mind you, I am not talking about higher-order refinements, but about first-order effects and understandings. The building of cloud-castles of theory upon gauge invariance, or upon vague analogies in general, serves to remind us how much of modern theoretical physics represents forms of sympathetic magic.

CRUCIAL EXPERIMENTATION

Testing the predictions of Hertzian electromagnetism, insofar as they differ from Maxwellian predictions, proves to be not an entirely trivial task. The conceptual distinctions are clear, but since I am running out of time I must be brief about them and about the experimental situation. Suffice it to say that I wasted five years of my own time in basement optical experimentation, until I realized that the equivalent of Potier's principle applies to the detector-convective effects I sought, just as to the ether-convective effects of the nineteenth century. The upshot is that no cheap optical experiment in the spatial domain (interference, diffraction, etc.) can decide the issue, and only a more expensive experiment in the time domain can hope to do so.

From the neo-Hertzian field equations one derives at once the corresponding wave equation,

$$-\nabla^2 \vec{E} + \frac{1}{c^2}\frac{d^2}{d\tau_d^2}\vec{E} = 0 , \qquad (11a)$$

where τ_d is field detector proper time. In one dimension this reduces to

$$-\frac{d^2 E}{dx^2} + \frac{1}{c^2}\gamma_d^2\left(\frac{\partial^2}{\partial t^2} + 2v_d\frac{\partial^2}{\partial x \partial t} + v_d^2\frac{\partial^2}{\partial x^2}\right)E = 0 \qquad (11b)$$

in view of

$$\frac{d}{d\tau_d} = \gamma_d\frac{d}{dt} = \gamma_d\left(\frac{\partial}{\partial t} + \vec{v}_d\cdot\vec{\nabla}\right), \qquad \gamma_d \equiv \left(1 - \frac{v_d^2}{c^2}\right)^{-\frac{1}{2}}. \qquad (11c)$$

On trying a solution of the form E = E(x + αt), we find for α a quadratic equation, $\alpha^2 + 2v_d\alpha - c^2 + 2v_d^2 = 0$. A d'Alembert type of solution for the E-field, employing the two roots, is then

$$E = E_1\left(x - \left[\sqrt{c^2 - v_d^2} - v_d\right]t\right) + E_2\left(x - \left[\sqrt{c^2 - v_d^2} + v_d\right]t\right) \qquad (12)$$

where, to first order, E_1 represents a wave of arbitrary shape traveling along the x-axis to the left at speed c - v_d , and E_2 is another arbitrary wave traveling to the right at speed c + v_d . If the detector moves to the right ($v_d > 0$) the wave traveling to the left is slowed down and the wave traveling to the right is speeded up. If the detector moves to the left ($v_d < 0$) the reverse is true. In all cases there is a first-order convection or pulling-along of the wave field by the detector's motion relative to the observer's laboratory. This convective effect on phase velocity vanishes only if the detector is at rest in the laboratory -- the usual case in actual laboratory practice.

On solving the correponding wave equation in two or more dimensions one finds a general expression for the phase velocity of the form

$$u = \sqrt{c^2 - v_d^2} + \frac{\vec{v}_d\cdot\vec{k}}{|\vec{k}|} \qquad (13)$$

where $\vec{k}$ is the propagation vector of the light and all velocities are measured in the laboratory. It is this dot-product nature of the convection effect that coincides with the situation dealt with by Potier's principle. That is, the mathematical problem of ether-convection of light is isomorphic to the problem of detector-convection of light specified by Eq. (13).

Thus according to neo-Hertzian theory each photon destined for detection has its phase velocity of propagation linearly altered by the component of detector velocity parallel to propagation direction. Each photon, in other words, "knows" where it is going to be absorbed, knows the state of motion of the absorber before it gets there, and adjusts its speed of propagation accordingly. If this seems spooky, remember our initial warning that the photon, contrary to Einstein's simplistic view, is the most "quantum" -- and the most nonlocal -- of all known quantum processes. It is therefore entitled to be the most mysterious and the most counterintuitive in its behavior. That it has not hitherto been seen in that light is testimony to the weight of Einstein's influence on the perceptions of his fellow physicists. Before bowing to this influence it would be well to recollect the words of another physics guru, Bohr, who said that "The apparatus as a whole makes the measurement." Is there a better word for that than spooky? That happens to be the word Einstein chose to describe the EPR situation. As a post-Einsteinian, I do not say that a thing has to be spooky to be right; I merely say that spooky things are not necessarily wrong.

How would one test this experimentally? I have examined rather carefully (a) the idea that the issue may already "accidentally" have been tested, (b) the possibility of astronomical tests, etc. My conclusion on all counts is negative: I have been able to discover no adventitious data that might settle the matter. (Remember -- we are dealing with a covering theory, which means that everything Maxwell predicts Hertz predicts as well. Covering theories of validated theories are seldom trivial to refute.) There seems to be no substitute for controlled laboratory experimentation, and no evading the need to have a radiation absorber move in the laboratory.

To fix our ideas, let us think about a specific setup that could be realized in a practical experiment, using it as an illustration of possibilities. Suppose we have in our laboratory a horizontal disk of radius R that rotates with rim speed v. At opposite ends of one diameter of this disk we mount two fast-responding photodiodes that are to be our radiation detectors. In the plane of the disk at a large distance D from the disk center we place a pulsed light source, say, a laser capable of rapid modulation -- meaning large intensity changes occurring within a short time ϵ. We may suppose the photodiodes capable of response in a time comparable to ϵ.

The detectors are so oriented that both "look" toward the source at the moment the disk diameter on which they are located is perpendicular to the line from source to disk center. At that moment during each cycle of disk rotation one detector is advancing toward the source with speed v and the other is retreating from it with the same speed, as measured in the laboratory. By the photon convection hypothesis we wish to test -- as embodied in Eq. (13) -- the advancing detector causes photons it will absorb to travel in the laboratory at phase velocity $u = c - v$, to first order, and the retreating detector causes the photons it will absorb to travel at speed $u = c + v$. If, as we may suppose, the propagation medium is nondispersive, these phase velocities are also signal or energy transport velocities. This will be the case if the experiment is done in vacuum.

By choosing parameters such that v << (R/t) << c, where t = D/c is the mean time of photon flight, we assure that the light arrives at the disk as practically a plane wave, and that the distance moved by either detector during photon propagation is small compared to the disk radius. Suppose that at the exact instant when the light modulation or "flash" reaches the center of the disk the inter-detector line is perpendicular to the propagation direction. Then according to Maxwell-Einstein the flash, consisting of a single wavefront, will simultaneously reach the two detectors and the first-order time difference between signals detected by the photodiodes will be $(\Delta t)_E = 0$. According to the Hertzian hypothesis to be tested, on the other hand, the retreating detector will have speeded-up its photons to c + v, with the result that this detector "sees" the flash slightly earlier, by an amount Dv/c^2. By the same hypothesis the advancing detector will slow its photons to c - v, with detection delayed, as a result, by an interval of the same magnitude; so the predicted time difference of signal arrivals is $(\Delta t)_H = 2Dv/c^2$. This is a first-order (linear) effect in v, but it is numerically small because of c^2 in the denominator.

To avoid having to observe one-time events, it is expedient to consider emitting repeated flashes at the disk rotation frequency. The desired experimental information then becomes encoded as phase information, and we require a way of establishing a reliable phase reference. It occurs to me that this might be accomplished by attaching to the disk surface a plane mirror along the inter-detector line, with its reflecting surface normal to the disk surface -- i.e., lying in a vertical plane. Then at the instant the normal to the mirror coincides with the propagation direction the condition mentioned above will be satisfied, and a flash reaching the center of the mirror at this time will be reflected back to the point of its origin. This is true according to both Maxwell and Hertz, because the center of the disk or mirror is essentially at rest in the laboratory, and the two theories always tell the same story about that special case. So, we simply have to observe on a screen behind the laser source the location of the flash reflection, and to adjust the phase of flash emission (which we suppose to be at the experimenter's control) in such a way that the flash reflection comes straight back to the source.

If the flash-phasing condition just mentioned is satisfied, it then becomes simply a question of choosing parameters such that the difference between the Einsteinian and Hertzian predictions is resolvable; that is, the hypothesized signal reception time difference between the two detectors, $2Dv/c^2$, must exceed the resolving time ϵ of the system. To put in some numbers, consider v = 10^{-6}c. Then D must be greater than $10''\epsilon$ kilometers. Thus if ϵ, the flash or modulation resolving time of the system (source plus detectors) is 10^{-10}second, then a 10-kilometer vacuum chamber will be required. Obviously this is not for basement experimenters like myself. Nevertheless, it seems a rather easy enterprise compared to some of the heroic experiments in which large amounts of government physics money are being invested these days. The necessary crucial experiment -- of whose feasibility it seems to me there can be little doubt -- could be done either on earth or in orbit. It is not impossible that the latter option might give it more likelihood of attracting funding.

I wish to conclude with an appeal to this conference: If a list of approved or endorsed experiments should happen to be one of the outputs of our collective wisdom, I ask that a test of neo-Hertzian vs. Maxwellian electromagnetism be included.

REFERENCES

1. H. R. Hertz, "Electric Waves," MacMillan, London and New York (1900).
2. A. I. Miller, "Albert Einstein's Special Theory of Relativity: Emergence (1905) and Early Interpretation (1905-1911)," Addison-Wesley, Reading, MA (1981).
3. A. Eichenwald, Ann. d. Phys. 11:1 and 421 (1903).
4. H. A. Wilson, Phil. Trans. 204:121 (1904).
5. T. E. Phipps, Jr., Ann. Fond. L. de Broglie 8:325 (1983) and 9:41 (1984).
6. T. E. Phipps, Jr., "Heretical Verities: Mathematical Themes in Physical Description," Classic Non-Fiction Library, POB 926, Urbana, IL 61801 (forthcoming).
7. A. Potier, J. de Physique (Paris) 3:201 (1874).
8. J. P. Wesley, "Causal Quantum Theory," Benjamin Wesley, 7712 Blumberg, W. Germany (1983).
9. P. Graneau, "Ampere-Neumann Electrodynamics of Metals," Hadronic Press, Nonantum, MA (1985).

EFFECTIVE PHOTON HYPOTHESIS VS. QUANTUM POTENTIAL THEORY: THEORETICAL

PREDICTIONS AND EXPERIMENTAL VERIFICATION*

E. Panarella

National Research Council
Ottawa, Canada K1A 0R6

ABSTRACT

This paper will review the basic formulation of effective photon theory, the experimental results of photoemission from a laser-irradiated metal that led to the concept of effective photon, and the experiments of laser-induced gas ionization which can be interpreted with the effective photon hypothesis. Then, it will review alternate theories that infer the existence of effective photons from different premises. It will be shown that a distinct difference exists between the alternate theories, and in particular quantum potential theory, and effective photon theory. The first predict that higher-than-normal energy photons in a laser beam are a consequence of the geometrical manipulation of the beam. The latter postulates that the geometry has no effect, and that the intensity of light is the important parameter leading to the effective photons. Such difference brings testable effects and an experiment has been performed to discriminate between effective photon theory and the alternate theories. Although the results of the experiment do not rule out the alternate theories, they do, however, provide strong evidence that higher-than-normal energy photons are a light intensity rather than a beam geometry effect. Effective photon theory is then extended. Starting from some theoretical considerations based on an interacting photon model proposed by the present author in another paper in these Proceedings, the coefficients α and β_ν of the effective photon energy relation $\varepsilon = h\nu/[1-\beta_\nu I^\alpha]$ for $\lambda = 10600$ Å are derived. Finally, the validity of this relation in predicting the minimum intensities of light necessary to ionize the noble gases, at the foregoing wavelength, is demonstrated.

---

*Accounts of parts of this work have appeared in the literature and are abstracted here: Lett. Nuovo Cim. 3, 417 (1972), Found. Phys. 4, 227 (1974).

TABLE OF CONTENTS

1. INTRODUCTION

The effective photon concept is a novel theoretical hypothesis which has been advanced in the 1970's as a consequence of the results of an experiment of laser-irradiated metal[1] and of the large number of experimental results available at that time on laser-induced gas ionization phenomena[2-16], experiments which could be explained in a unified and consistent way with the assistance of such novel concept.

The logical basis upon which the effective photon hypothesis rests is that, as long as one can safely adopt fundamental formulas of physics in increasingly wider range of parameters, one shouldn't refrain from doing so. However, when the range of parameters is so far away from that in which the validity of a particular formula has been clearly demonstrated, and a degree of divergence occurs between the expected experimental results and the actual results, then it is legitimate to inquire whether the formula in question is universal, i.e. applicable to all situations, or not.

238

This is the case for one of the most fundamental formulas of physics

$$\varepsilon = h\nu \tag{1}$$

expressing the energy ε of photon of frequency ν. This formula is independent of light intensity and its validity was verified right at the time when it was postulated, namely when it was inserted by Planck in his blackbody radiation law, as well as in countless other times.

The experiments, however, in which Eq. (1) applied so well were all "low light intensity" experiments, meaning by this term "low number of photons per unit area per unit time". But, when in the 1960's the optical laser was invented, and intensities of light of the order of $10^{12} - 10^{15}$ W/cm^2 (corresponding to ~$10^{30} - 10^{33}$ photons/cm^2.sec) were routinely obtained in the laboratory with focussed laser beams, new phenomena started to appear, such as ionization of gases by laser beams and photoemission from laser-irradiated metal surfaces. These phenomena were not expected, because the photon energy of the laser light used was invariably well below the ionization potential of the irradiated gas, or the work function of the metal, respectively.

Concomitantly, some of the difficulties faced by those classical theories which were immediately advanced in order to try to explain these novel phenomena[17-22] led to the notion that perhaps, at the very high light intensities provided by laser beams, formula (1) breaks down. A novel photon energy expression was then postulated "ad hoc", simply as a calculational tool, in order to see if the experimental results of irradiation of gases and of metal surfaces could be explained in a consistent way with such novel theoretical approach. The postulated photon energy expression was

$$\varepsilon \simeq \frac{h\nu}{1 - \beta_\nu f(I)} \tag{2}$$

where $h\nu$ is the normal photon energy, β_ν is a positive coefficient, and $f(I)$ is a function of the light intensity. Clearly, the term $\beta_\nu f(I)$ had to significantly differ from zero only in the case of very high intensity laser light, so that, at normal light intensities, Eq. (1) was retrieved. The photons of enhanced energy that obeyed Eq. (2) where then called "effective photons".

Eq. (2) has been successful in interpreting the experiments mentioned above[23-25] as well as others[26]. In this review, we shall first re-examine in detail the experiment of metal irradiation with a high power ruby laser beam which led first to the idea that higher-than-normal energy photons in the beam were responsible for the photoelectric emission from the metal. We shall then review the experimental results of laser-induced ionization of gases which can also be interpreted through relation (2). Then, since in recent years at least three theories have appeared which explain classically, i.e. without resorting to an "ad hoc" hypothesis, the existence of higher-than-normal energy photons[27-31] in laser beams, these theories will be reviewed and their predictions will be compared with those of effective photon theory. Since the alternate theories predict the existence of higher-than-normal energy photons even at very low light intensities, albeit with very small probability, whereas for effective photon theory these photons can exist only at very high light intensities, an experiment has been carried out to discriminate between the former and the latter theories. The results show that, with high probability, the effective photons are an intensity effect.

In recent years, the effective photon hypothesis has gradually been losing its original character of an "ad hoc" postulate and acquiring more and more a theoretical basis. Elementary considerations on the very nature of photons has led to a justification of the original postulate[32]. Moreover, some considerations on the possibility that photons might interact at intensities of light much below those predicted by QED has led to an interaction relation for photon[33-34, 43] which predicts that a photon cannot occupy a volume smaller than $\sim\lambda^3$. This means that the number density of photons N has to be: $N \simeq \lambda^{-3}$. In essence, this expression relates the intensity of light I to the frequency of photons ν, because $I \propto N$ and $\nu \propto \lambda^{-1}$. Hence, the effective photon hypothesis, which indeed rests on the notion that frequency and energy of a photon are not independent of light intensity, receives in this way theoretical support. By applying these concepts, we shall see that the parameters contained in the expression $\beta_\nu f(I)$ of Eq. (2) can be determined and a new verifications of the validity of that formula in the interpretation of laser-induced gas ionization phenomena will be provided.

2. EXPERIMENT ON ELECTRON EMISSION FROM A LASER IRRADIATED METAL SURFACE[1]

This experiment was designed to discriminate between the two intensity-dependent theories that could explain photoelectric emission from a laser-irradiated metal surface, namely multiphoton theory[17-21] and effective photon theory[23]. The predictions of the two theories are:

I - <u>Multiphoton theory</u>

1) $$i \propto I^n \qquad (3)$$

where i is the electron current released from a metal by a purely photoelectric effect, I is the light intensity, n is the integer part of $\frac{W}{h\nu} + 1$, where W is the work function of the irradiated metal and $h\nu$ is the photon energy.

2) $$E_{max} = nh\nu - W \qquad (4)$$

where E_{max} is the maximum kinetic energy of the emitted electrons.

For multiphoton theory therefore the electron current is proportional to the n-power of the light intensity and the maximum energy of the emitted electrons is independent of light intensity.

II - <u>Effective photon theory</u>

1) $$i \propto I \qquad (5)$$

i.e. the electron emission is a single-photon process[35], and therefore linear with light intensity.

2) $$E_{max} = f(I) \qquad (6)$$

i.e. the maximum energy of the emitted electrons is a function of light

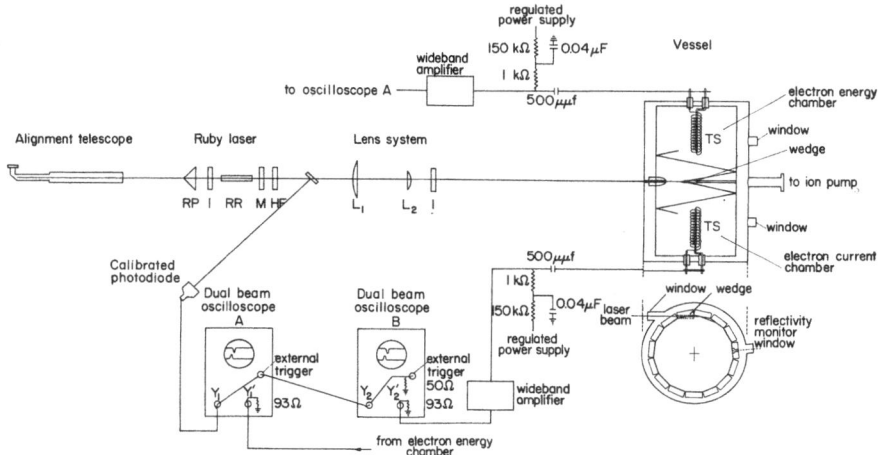

Fig. 1   Schematic drawing of the apparatus used to discriminate
between effective photon theory and multiphoton theory
of electron emission from a laser irradiated metal
surface. RP: rotating prism; I: iris; RR: ruby rod;
M: front mirror; HF: high-pass filter; L_1, L_2:
lenses; TS: tungsten electrode.

intensity, because the photon energy, according to Eq. (1), is a function
of the light intensity.

The experimental apparatus used to discriminate between the two
theories is shown in Fig. 1. In essence, laser light is allowed to
interact with a set of metal surfaces in a highly evacuated chamber and a
simultaneous determination of electron current and (relative) electron
energy is obtained. Since the most important requirement of the
experiment was to reduce as much as possible any spurious electron
emission of thermionic origin, the following steps were taken: a) the
angle of incidence of the laser light with the metal surface was reduced
to a small value = 12°, b) the surface of the metal was polished flat to
$\frac{1}{10}\lambda$, and c) the laser power was never above 1 MW. Moreover, the laser
light entering the chamber through a small hole interacted with a set of
identical consecutive surfaces arranged in poligonal order until complete
absorption occured.

In detail, the experimental apparatus is made up of three parts:
the optical system for the generation and control of the intensity of
light, the vacuum vessel where the interaction light-metal takes place,
and the electronic detection system.

The optical system is composed of an alignment telescope, a
Q-spoiled ruby laser and a lens system. The Q-spoiled ruby laser is a
conventional rotating prism laser made up of a Czochralski rod
7 cm long × 1 cm diameter, an iris inserted into the optical cavity in
order to restrict the emission to the homogeneous central zone 0.3 cm in
diameter, and a front mirror. A high-pass filter in front of the laser
prevents UV radiation from the flash lamp from reaching the chambers.
The laser power is monitored by means of a calibrated photodiode ITT

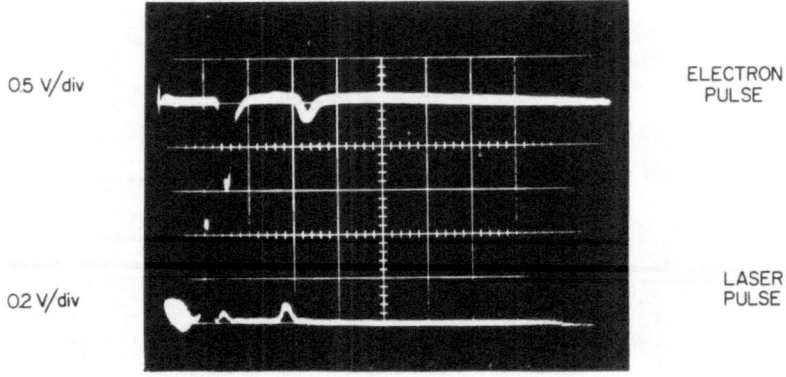

0.5 V/div

ELECTRON
PULSE

0.2 V/div

LASER
PULSE

TIME SCALE: 0.1 μ sec/div

Fig. 2   Oscilloscope record of the electron current and of the
laser pulse when the laser delivers two pulses. It is
clear from this record that the electron emission from
the metal is proportional to light intensity.

F-4000 irradiated by 4% of laser light. The lens system has the purpose
of increasing the light intensity, not by focussing the beam, but by
simply reducing its cross section. The lens system is made up of two
plano-convex lenses L_1 and L_2 separated by a distance equal to the sum of
their focal lengths. During the experimentation, lens L_1 had a constant
focal length 23.6 cm, while lens L_2 was allowed to change focal length
from 23.6 cm to 14.3 cm to 10.0 cm.

The vacuum vessel (which contains an inner chamber) is a cylindrical
container properly sealed and connected to an ion pump. After bake out,
a vacuum of $4 \cdot 10^{-9}$ torr could be achieved. The inner chamber is made of
15 flat hardened steel bars 25.95 cm long, polished on the internal faces
to $\frac{1}{10}\lambda$. The reflectivity of the polished faces is 58.5% at the ruby
wavelength λ = 6943 Å. The bars are arranged so that the polished faces
are in polygonal order tangent to a circle 7.14 cm radius and are mounted
on two polished end plates and a centre plate in order to form two
identical chambers, the electron current being measured in one chamber
and the electron energy in the other. The laser beam enters the vessel
through a window and is split in two parts by a 12° polished steel wedge
located in the interior of the chamber. After reflection at the wedge,
the beams are directed along a helical path in their respective chamber
and are not able to re-emerge through the 0.9 cm diameter entrance hole.
The collection of electrons is accomplished by tungsten wire electrodes
having spiral form with the terminal leads coming out of the vessel
through high vacuum feed-through electrical connections. This particular
form was chosen so that the electrodes can perform two functions: 1) the
collection of electrons, 2) act as heaters for deposition on the internal
surfaces of the chambers of different metals, such as aluminum, silver,
gold etc.

The electronic detection system connected to each chamber is
composed of a storage capacitor (C = 500 $\mu\mu$F), a regulated D.C. power
supply, a wideband amplifier and a Tektronix 555 dual-beam oscilloscope.

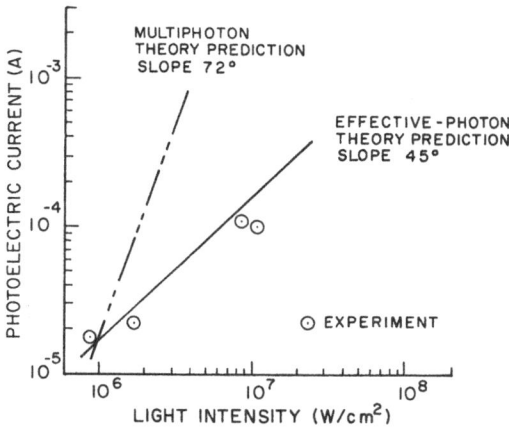

Fig. 3   The electron emission from a metal is directly
proportional to the light intensity, in agreement with
the effective photon theory, but in disagreement with
multiphoton theory, which calls for a slope of 72° of
the current vs. light intensity plot.

The wideband amplifier has a bandpass from 200 Hz to 50 MHz, input and output impedance 93 Ω, gain 40 db, rise time 10 ns and pulse delay time 40 ns.

The tests have been done with a constant positive voltage of 2600 V at the electron current electrode. At the electron energy electrode a positive voltage is applied and reduced at each laser shot until the signal almost disappears. At this time the positive voltage of the electrode, combined with the negative voltage of the space charge, is just enough for the most energetic electrons to reach the electrode. The electrode voltage is then kept constant and the light intensity is increased. The signal amplitudes are then compared and an estimate of the (relative) electron energy, as a function of the light intensity, can be obtained.

Fig. 2 shows an oscilloscope record of the electron pulses (upper trace) and laser pulses (lower trace) when the laser delivers two pulses of different intensities. It is apparent from this record that the amplitude of the electron pulse is proportional to the light intensity. To make a better comparison, we plotted (Fig. 3) the peak electron current vs. laser intensity obtained from different records and we found that the slope of the line was ~45° as predicted by effective photon theory. This result has to be contrasted with the prediction of multiphoton theory, which calls for a slope of 72°, because n = 3 for the irradiated metal (steel of work function W = 5 ÷ 6 eV).

It remains now to verify if the maximum energy E_{max} of the emitted electrons is a constant, as predicted by multiphoton theory, or a function of the light intensity, as predicted by effective photon theory. From top to bottom, Fig. 4 shows three sets of signals from the electron energy chamber (left pictures) and electron current chambers (right pictures), taken under increasing values of light intensities from set No. 1 to set No. 3. The light intensity was increased by leaving the beam collimated and changing the lenses, as indicated on the left of each

+300 V   +2,600 V

LENS L₁
FOCAL LENGTH 23.6cm

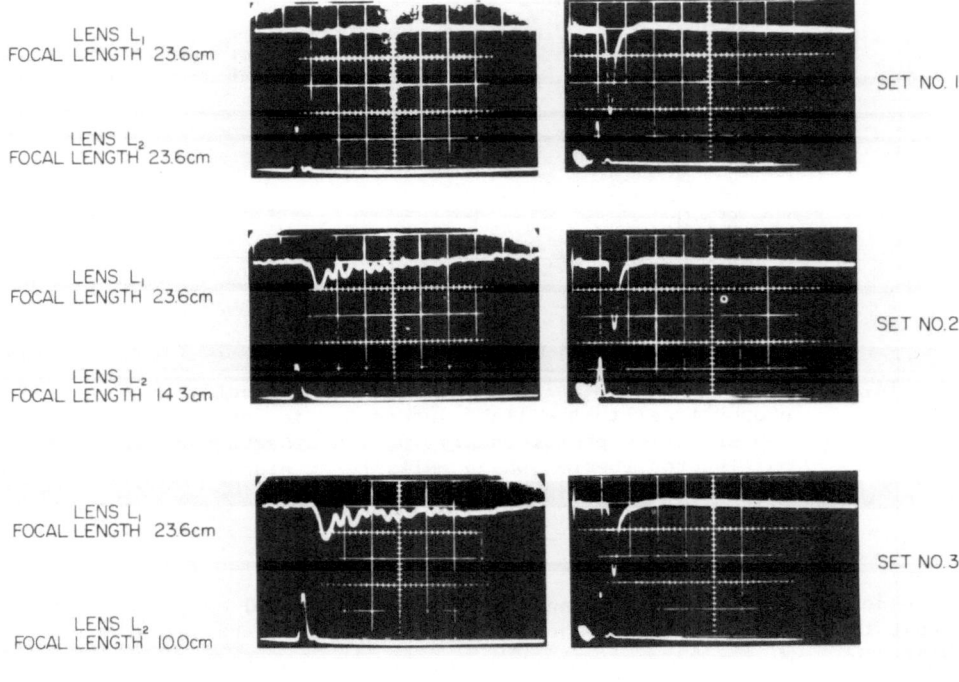

LENS L₂
FOCAL LENGTH 23.6cm   SET NO. I

LENS L₁
FOCAL LENGTH 23.6cm

LENS L₂
FOCAL LENGTH 14.3cm   SET NO.2

LENS L₁
FOCAL LENGTH 23.6cm

LENS L₂
FOCAL LENGTH 10.0cm   SET NO.3

UPPER TRACE 0.05 V/div   UPPER TRACE 0.5 V/div
LOWER TRACES 0.5 V/div
TIME SCALE 0.1 μ sec/div

Fig. 4   Oscilloscope records of simultaneous determination of
         (relative) electron energy and electron current for
         different laser intensities. The signals on the left
         records (upper traces) clearly show that the electron
         energy increases with light intensity.

set of photographs. The lower trace in each photograph shows the laser
pulse, as monitored with the calibrated photodiode. A positive voltage
of 300 V was applied to the collector of the electron energy chamber. At
this voltage, and at the lowest of the possible light intensities, the
electrons are not energetic enough to overcome the voltage barrier given
by the space charge and the electron signal almost disappears (Set No. 1,
left figure, upper trace). When the laser intensity is increased by the
action of the combined lenses L_1 and L_2, the electron signal increases.
This could mean that the electrons are more energetic and can more easily
overcome the space charge barrier, or that there is a larger production
of electrons in the chamber. From an inspection of the photographs at
the right, one has to rule out the latter assumption, because the
electron pulses are of almost always constant amplitude in all three
situations. Hence, the electrons are more energetic when the intensity
is increased. This is again in agreement with the prediction of
effective photon theory and contrary to that of multiphoton theory.

244

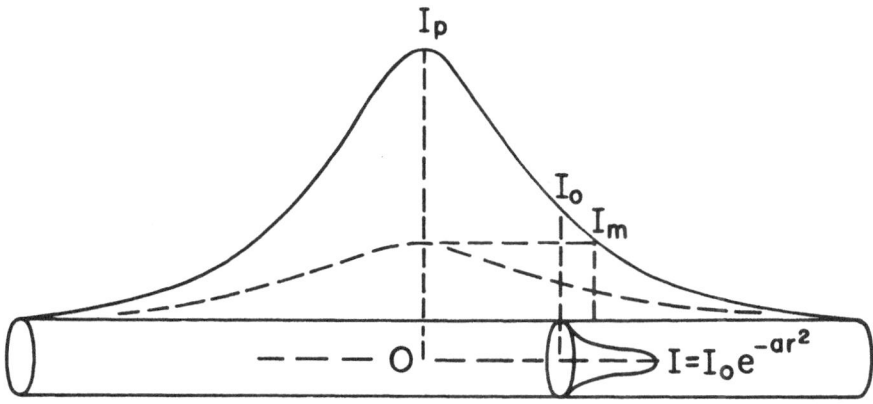

Fig. 5 Distribution of light intensity in a laser pulse.

3.   EFFECTIVE PHOTON THEORY

     The approach of effective photon theory to the solution of the
laser-induced gas ionization and breakdown problem follows a pattern
quite different from that of either multiphoton[17-21] or cascade theory[22].
Effective photon theory assumes that, at the extremely high intensities
of light available from modern lasers, as used in the ionization of
gases, photon-photon interaction begins to play a role with the consequent effect of subjecting the energy of photons to a variation as a
function of the density of the photon sea in which the light particles
are immersed. The postulated variation is expressed as

$$\varepsilon = \frac{h\nu}{1 - \beta_\nu f(I)} \qquad (2)$$

where β_ν is a constant positive coefficient, $f(I)$ a function of the light
intensity* and the product $\beta_\nu f(I) \to 0$ for low intensity light. Starting
from (2), the theory proceeds by deducing how many photons, in a laser
pulse of given temporal and spatial intensity distribution, are "effective", i.e. have energy equal to or greater than the ionization potential
of the gas atoms. The ionization process is then considered to follow a
single-photon mechanism and the number of ions generated to be proportional to the number of effective photons in the laser pulse.

     In the following sections, the theory will be formally developed and
its predictions will be verified against available experimental results.

3.1 Mathematical Formulation

     The problem to be solved is the following. Given a laser pulse
(Fig. 5) whose temporal intensity distribution is assumed for simplicity
the be   triangular and with a gaussian spatial distribution

---

*In order to easily compare the theory with the experimental results, the
 light intensity in (2) conserves the conventional definition. In other
 words, it represents a quantity proportional to photon number density.
 This is possible because intensity measurements in high power laser
 beams are usually carried out by scaling up measurements done on an
 attenuated beam where $\varepsilon = h\nu$.

$I(r) = I_o e^{-ar^2}$, where $I(r)$ is the intensity at a distance r from the beam center and I_o is the intensity at the centre $r = 0$, deduce how many photons are effective and calculate the number of ions produced by simple photoelectric effect. The analysis proceeds as follows. First, assume that $f(I) = I^\alpha$ where α is an undetermined exponent. In a cross section of the beam the effective photons are confined within a circle of radius r_p which can be deduced from the equation

$$W = \frac{h\nu}{1 - \beta_\nu I^\alpha \exp(-\alpha a r_p^2)} \tag{7}$$

where W is the ionization potential of the gas investigated. The coefficient β_ν is found by considering another parameter I_m, defined as the peak intensity of a laser pulse containing only one effective photon. Necessarily, this photon is located at the point 0 where $r = 0$ and

$$\beta_\nu = \frac{W - h\nu}{W I_m^\alpha}. \tag{8}$$

Substitution of (8) into (7) yields

$$r_p = \left(\frac{1}{a} \ln \frac{I_o}{I_m}\right)^{\frac{1}{2}}. \tag{9}$$

The total number of effective photons N_p crossing a section of the beam in unit time is

$$N_p = \frac{2\pi}{h\nu} \int_o^{r_p} I(r)\ r dr = \frac{\pi}{h\nu a} (I_o - I_m). \tag{10}$$

The volume where the interaction effective photons–gas particles takes place is considered a cylinder of length $\Delta\ell$ (which depends on the focal length of the lens used to focus the beam) and radius $r_p(\max) = \left(\frac{1}{a} \ln \frac{I_p}{I_m}\right)^{\frac{1}{2}}$ where I_p is the peak intensity of the laser pulse. The interaction of these effective photons with the gas atoms within the focal volume creates the following number of ions:

$$dN_i = \sigma (N_a - N_i)\ N_p\ dt \tag{11}$$

where σ is a proportionality constant which depends on the nature of the gas investigated and the frequency of light used. Eq. (11) can be written in this way

$$\frac{dN_i}{N_a - N_i} = \frac{\sigma\pi}{h\nu a} (I_o - I_m) \frac{dt}{dI_o}\ dI_o \tag{12}$$

For a triangular laser pulse $\frac{dt}{dI_o} = \frac{\Delta t}{I_p}$. Two cases can be considered. In one, the ratio $\frac{\Delta t}{I_p}$ is assumed to vary from one laser pulse to another. In the second $\frac{\Delta t}{I_p} = K$, i.e. the ratio is constant. The integration of (12) between I_m and I_p yields the ion density N_i for the two cases, respectively:

$$N_i = N_a \left[1 - e^{-K_o I_p \left(1 - \frac{I_m}{I_p}\right)^2} \right] \qquad (13)$$

$$N_i = N_a \left[1 - e^{-K_1 I_p^2 \left(1 - \frac{I_m}{I_p}\right)^2} \right] \qquad (14)$$

where $K_o = \dfrac{\sigma \pi \Delta t}{h \nu a}$ and $K_1 = \dfrac{\sigma \pi K}{h \nu a}$.

Formulae (13) and (14) are the general relations to be obeyed by the laser induced ionization phenomenon. Specifically, they allow the calculation of the number of ions N_i generated by a laser pulse of peak intensity I_p or, in the case of gas breakdown, the calculation of the breakdown intensity threshold I_{th} as a function of gas pressure N_a, lens focal length f and pulse-width Δt. This is possible because $N_i = N_b =$ const at breakdown and (13) and (14) yield respectively:

$$I_{th} = \frac{1}{K_o} \ln \frac{N_a}{N_a - N_b} \qquad (15)$$

$$I_{th} = \left(\frac{1}{K_1} \ln \frac{N_a}{N_a - N_b} \right)^{\frac{1}{2}}. \qquad (16)$$

3.2 Comparison of Effective Photon Theory with the Experimental Results

The four relations (13) to (16) are the required formulae to verify the experimental results. The first two relations (13) and (14) have to be used in the pre-breakdown phase, the last two (15) and (16) are to be employed when the parameters of the phenomenon have been measured at breakdown. As to the choice between (13) or (14) and (15) or (16), one has to look at the information supplied with the experimental results. If the laser pulse-width has been maintained constant during the course of the experimentation, (13) or (15) will be used. If the pulse-width has been varied, which implies that presumably $\frac{\Delta t}{I_p} \simeq$ const from shot to shot*, then (14) or (16) will be used.

The verification of the theory with the experimental results will be a qualitative one because the parameters K_0 and K_1 appearing in (13) to (16) contain the coefficient σ which is unknown at present and a normalization procedure will be adopted. The other parameter I_m that appears in the theory can be deduced from the experiments and means to find I_m will be provided.

3.2.1 Intensity dependence of the ionization process. The experimental results in this case seem to exhibit a common feature. If the laser intensity is just sufficient for the detection of a few ions, then any slight increase of light intensity yields a very large increase in the number of ions generated. By contrast, when the laser intensity is high enough to generate a large number of ions, then a further increase of laser intensity yields a linear to quadratic ion production. Moreover, the trend of the ionization curves does not appear to be affected neither by the wavelength of the light used nor by the nature of the gas investigated.

---

*This is because, when higher intensities are extracted from the laser, the pulse-width increases and the ratio $\frac{\Delta t}{I_p}$ remains approximately constant.

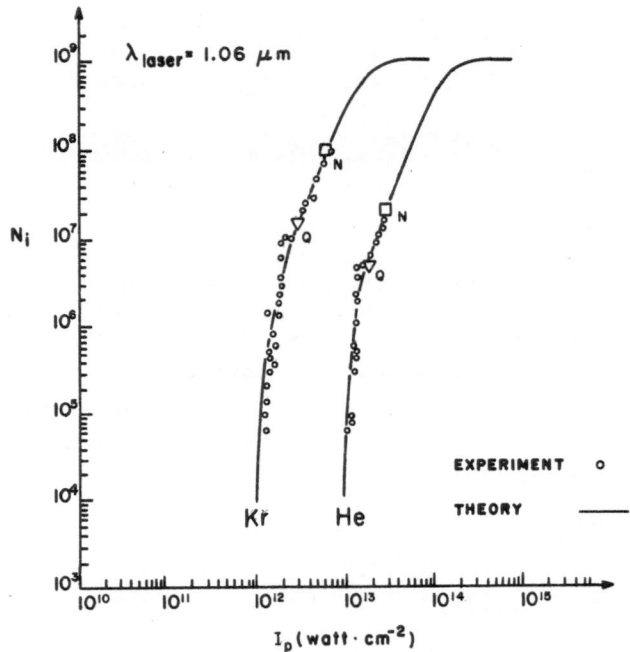

Fig. 6 Variation of total number N_i of ions of krypton and
helium as a function of laser intensity I_p in the
focussing region. The experimental points are taken
from Ref. 11. N is the normalization point. Q is the
point where the curve begins to have a quadratic
behaviour.

These results are predicted by Eqs. (13) and (14). Let us write an
approximate expression for these equations

$$N_i \simeq K_o N_a I_p \left(1 - \frac{I_m}{I_p}\right)^2 \tag{13a}$$

$$N_i \simeq K_1 N_a I_p^2 \left(1 - \frac{I_m}{I_p}\right)^2 \tag{14a}$$

obtained after expansion of the exponential function up to the second
term. From (13a, 14a) it appears that, when I_p is slightly larger than
I_m, the ion density N_i is a rapidly increasing function of I_p, because of
the quadratic term $\left(1 - \frac{I_m}{I_p}\right)^2$. When $I_p \gg I_m$ the term $\left(1 - \frac{I_m}{I_p}\right)^2$ reduces to
1 and ion production is linear (13a) or quadratic (14a) with the light
intensity. On the other hand, the nature of the gas and the wavelength
of the light used do not appear in Eqs. (13a, 14a), which then have to be
obeyed by all gases independently of light wavelength.

A comparison of theory with the experiments can be done only if we
know I_m, the minimum peak intensity of a laser pulse <u>necessary</u> (but not

sufficient) to ionize only one atom*. However, we can deduce a value of I_m from the intersection of the experimental points with the line corresponding to $N_i = 1$.

Let us proceed with the verification of the theory. In Fig. 6 the variation of the total number of ions is reported, as calculated from relation (14), as a function of laser intensity I_p, in the case of ionization of Krypton and Helium with a Neodymium laser ($\lambda = 1.06$ μm). The experimental points have been taken from Ref. 11. Relation (14), rather than (13) was used because the laser pulse-width has not been maintained constant but varied somewhat during the course of the experimentation. The normalization point has been chosen at N and I_m has been assumed equal to 10^{12} for Krypton and 9×10^{12} for Helium. As can be seen from the figure, the agreement between theory and experiment is excellent and the same agreement can be found when we consider other gases and different wavelengths[11].

When the analysis does not require the knowledge of I_m, because the experiment has been carried out for values of I_p much larger than I_m, the agreement between theory and experimentation is also very good. Fig. 7 reports the ionization of Cesium at $\lambda = 0.53$ μm. The experimental points[15] lie on the quadratic portion of the curve and insertion of two arbitrary values of I_m (= 10^4 and 10^5) does not affect the trend of the upper part of the theoretical line.

3.2.2. <u>Time dependence of ionization before breakdown.</u> In a crucial experiment by Chalmeton and Papoular[2] it was found that $\dfrac{d \ln N_e(t)}{dt}$, where $N_e(t)$ is the electron density evolution in the focal region before breakdown, was independent of laser peak intensity I_p and gas density N_a and was only a function of time t.

In order to see if this result is predicted by the present theory, we need to rewrite Eq. (13) in terms of time t

$$N_i(t) = N_e(t) = N_a \left[1 - e^{-\frac{\sigma \pi I_p}{2h\nu a \Delta t} t^2 \left(1 - \frac{t_m}{t}\right)^2} \right] \qquad (17)$$

for $t \leqslant \Delta t$, where $t_m = \dfrac{\Delta t \cdot I_m}{I_p}$ is the time at which $I_0 = I_m$ and single ionization has been assumed to occur in the laser-gas interaction region.

For the range of times t in which Chalmeton and Papoular detected light emission, i.e. for times t in which $N_i \ll N_a$ and $t \gg t_m$, Eq. (17) is well approximated by

$$N_i(t) = N_e(t) \cong \frac{\sigma \pi I_p}{2h\nu a \Delta t} N_a t^2 \qquad (18)$$

_____

*I_m is the intensity necessary for the photon energy to attain the gas ionization potential W and for one effective photon to be found in the laser pulse. I_m cannot however be sufficient to ionize one atom because the quantum yield is never equal to 1.

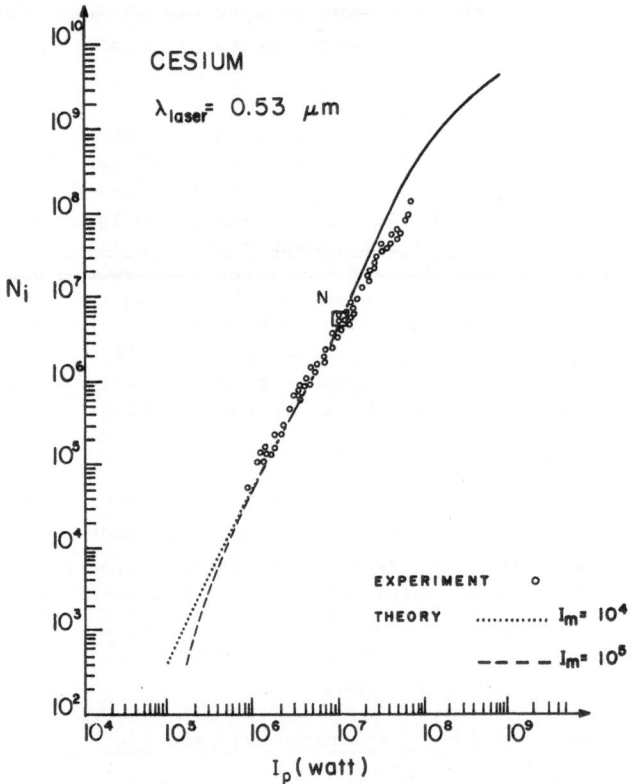

Fig. 7 Variation of total number of ions N_i of cesium at
$\lambda = 5300$ Å. I_m is the minimum laser intensity required
to ionize one atom. The experimental points are taken
from Ref. 15.

obtained after a series expansion of the exponential function limited to
the second term. Taking now the logarithm of both sides of (18) and
differentiating with respect to time t, we get

$$\frac{d \ln N_e(t)}{dt} = \frac{2}{t} . \tag{19}$$

This relation shows that the variation of the logarithm of the light
emission with respect to time t is only a function of time t, in
agreement with the experiment.

3.2.3. <u>Pressure dependence of the gas breakdown laser threshold
intensity</u>. We proceed now with the verification of the theory in the
case of breakdown intensity threshold measurements as a function of
pressure (or gas density). We shall compare the theory with the
experimental results of Okuda et al[7-9].

From the context of the papers by these authors, it seems that the
laser pulse-width has not been maintained constant but varied somewhat in
the course of the experimentation between 50 and 100 nsec. Hence, Eq.
(16) has to be applied. We first write Eq. (16) in the following way:

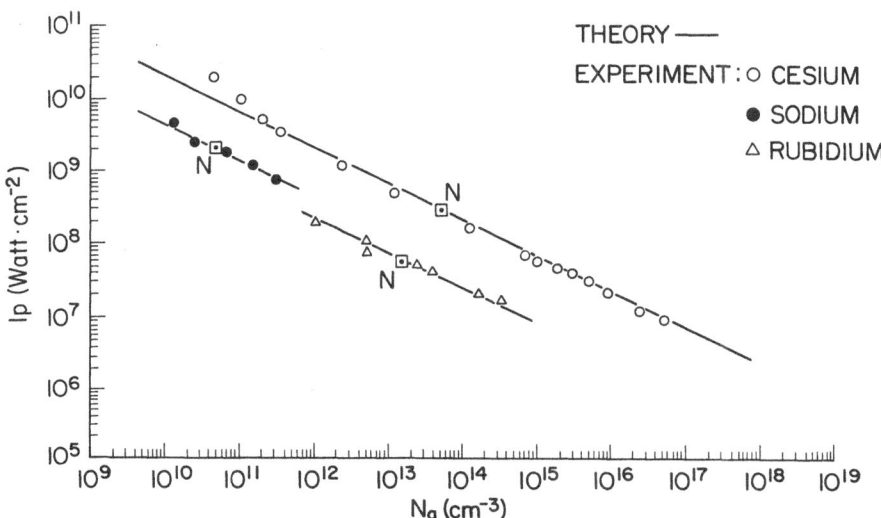

Fig. 8 Gas density dependence of breakdown threshold intensity in cesium, sodium, and rubidium. The experimental points are taken from Refs. 7-9. N is the normalization point.

$$(\sigma\pi K/h\nu a)\ I_{th}^{2} = \ln[N_{a}/(N_{a} - N_{b})] \tag{20}$$

from which

$$N_{b} = N_{a}\{1 - \exp[-(\sigma\pi K/h\nu a)\ I_{th}^{2}]\} \tag{21}$$

After a series expansion of the exponential function limited to the second term we get

$$N_{b} \simeq (\sigma\pi K/h\nu a)\ N_{a}\ I_{th}^{2} \tag{22}$$

from which we get

$$I_{th} = (N_{b}h\nu a/\sigma\pi K)^{\frac{1}{2}}\ N_{a}^{-0.5} = \text{const} \times N_{a}^{-0.5} \tag{23}$$

By plotting relation (23) in Fig. 8 for the case of breakdown of cesium, sodium, and rubidium[7-9], we find excellent agreement between theory and experiment. The experimental results of Ref. 10 are also in agreement with the theory. In fact, the two limiting relations (15) and (16) say that the slope of the lines of the breakdown intensity threshold versus pressure varies between -0.5 and -1. This is what has been reported in Ref. 10.

3.2.4. Focal volume dependence of the breakdown threshold intensity. Relations (15) and (16), when explicitly written in the following way

$$I_{th} = \frac{2h\nu a}{\sigma\pi\Delta t}\ \ell n\ \frac{N_{a}}{N_{a} - N_{b}} = \text{const.}\cdot a \tag{15b}$$

$$I_{th} = \left(\frac{2h\nu a}{\sigma\pi K}\ \ell n\ \frac{N_{a}}{N_{a} - N_{b}}\right)^{\frac{1}{2}} = \text{const.}\ (a)^{\frac{1}{2}} \tag{16b}$$

show that in an experiment in which all parameters but the gaussian factor a are held constant, the threshold intensity for gas breakdown is either proportional to a or to $(a)^{\frac{1}{2}}$. Now, the factor a is related to the lens focal length f and divergence θ of the laser beam through the following expression

$$a = \frac{4 \ln 2}{d^2} = \frac{4 \ln 2}{f^2 \theta^2} \qquad (24)$$

as one deduces from the knowledge that, by definition, the diameter d of the focal spot is measured between points where the laser intensity drops to one half the value at r = 0. Hence, relations (15b) and (16b) state that the slope of the lines of I_{th} versus f ranges between −2 and −1 or, in terms of the rms electric field E in the focal spot, between −1 and −0.5. This is what has been reported in the literature[3,5-6].

3.2.5. <u>Temporal dependence of the breakdown threshold intensity.</u> The analysis, in this case, is very simple. The time evolution of the ionization before breakdown is given by the approximate expression (18). At breakdown $N_i = N_b$ = const and (18) becomes

$$I_{th} = const \cdot t_b^{-2} \qquad (18a)$$

which, when written in terms of the rms E field in the focal spot, becomes

$$rms \; E_{th} \propto t_b^{-1}. \qquad (18b)$$

This expression shows that a plot of the rms E field necessary for breakdown vs time t has a slope equal to −1, in agreement with the experimental results[36].

3.2.6. <u>Concluding remarks.</u> In conclusion, effective photon theory is indeed able to predict, in a unified and consistent way, the experimental results of laser-induced gas ionization and breakdown. The theory has been formulated from the "ad hoc" assumption (2). In Sec. (6.1) we shall provide a theoretical background for that assumption which will lead to a determination of the minimum intensity I_m of light necessary to create one effective photon in a laser pulse.

4. ALTERNATE THEORIES OF PHOTON ENERGY ENHANCEMENT

In recent years, at least three models[27-31] have been proposed in order to explain photon energy enhancement in focussed laser beams. The first model[27] is based on standard quantum mechanics formalism. It rests on the notion that focussing decreases the position uncertainty of the photons, thereby increasing their momentum uncertainty. This, in turn, is related to photon energy, which must therefore have a lower bound determined by the geometrical focussing parameters, and independent of light intensity. The second model[28-29] is based on classical electromagnetic theory. It shows that the frequency spectrum of an electromagnetic wave can be related to the properties of the wave-vector spectrum, which is a function of the focussing geometry of the light beam under consideration. The stronger the focussing, the broader the frequency or energy spectrum and, here too, the light intensity plays no role in such broadening. The third model[30-31] is based on the postulated existence of a quantum force $F_\mu = -\partial_\mu Q$, where Q is the de Broglie-Bohm quantum potential, which accelerates the photons. Such acceleration persists even at low beam intensity, with small but non-vanishing probability, and the only requirement in order to have these energetic photons is that the phase relation between neighbouring photon lines of flow be disturbed, a fact that can be produced by focussing the light beam, or can be found in the interfering zone of intersecting beams. In other words, without focussing or other means of disturbing the phase relation, no energetic photons are produced.

The above three models of photon energy enhancement are therefore beam geometry-dependent and light intensity-independent. By contrast, the effective photon model is geometry-independent and intensity-dependent. Such difference leads to testable effects. In the following, we shall first analyze more in depth the three intensity-independent models, and then describe an experiment designed to discriminate between these theories, and verify whether or not higher-than-normal energy photons exist at low intensity, in a range where the effective photon model predicts that such photons should not exist.

4.1 Model Based on Quantum Formalism[27]

This model, based on an application of the Heisenberg uncertainty principle to a focussed laser beam, derives a formula for the lower bound of the energy E of photons in a focal circle of radius r

$$E > hc/2r \sin \theta = h\nu\lambda/2r \sin \theta = \varepsilon \lambda/2r \sin \theta \qquad (25)$$

where θ is the lens half-cone angle, and ε is the original photon energy, before focussing. This formula shows that, in order to have some photons of energy E higher than the original energy ε, the laser beam must be focussed into a circle of radius

$$r < \lambda/2r \sin \theta. \qquad (26)$$

Since Ineq. (25) and (26) are intensity-independent, higher-than-normal energy photons should appear at any light intensity, provided Ineq. (26) is satisfied. The question of whether or not they will appear remains probabilistic, but obviously the probability will increase with the number of photons present in the focal volume.

4.2 Model Based on Classical Electromagnetic Wave Theory of Laser Line Broadening[28-29]

This model derives a relation between the line broadening $\Delta\omega$ and the lens focussing angle 2θ

$$\Delta\omega/\omega_0 = \left[\tfrac{1}{2}(\chi - 1)\right]^{\frac{1}{2}} \sin^2 2\theta \qquad (27)$$

where ω_0 is the center frequency of the laser used, and χ is a constant. This broadening is then used to calculate the probability $Q(W)$ of having photon energies above the ionization potential of a gas or the work function of a metal W, assuming as laser line shape a Lorentzian distribution of damping factor $\varepsilon = 0.2$. It is found $Q(W) = \left[2\varepsilon X^2 f(\varepsilon)\right]^{-1} \exp(-\varepsilon X)$, where $X = (W - \hbar\omega_0)/\hbar\Delta\omega$, and $f(\varepsilon) = 1.08$.

The prediction of this model is that, if a laser pulse contains N photons, ionization of a gas or electron emission from a metal irradiated with the laser will occur whenever $N \times Q(W) > 1$.

4.3 Model Based on Quantum Potential Theory[30-31]

In this model the photon is considered as an oscillating localized particle with nonzero mass which moves along the lines of flow of a four-vector wave field $A_\mu = \exp[P + i(S/\hbar)]\, a_\mu$, where P and S are real functions of the coordinates X_μ, and a_μ is a real four-vector with $a^\mu a^\mu = 1$. The motions can be derived from the spin-1 Lagrangian[37-38].

$$L = -\tfrac{1}{4} F^*_{\mu\nu} F_{\mu\nu} - \tfrac{1}{2} \mu_\gamma^2 A^*_\mu A^\mu \qquad (28)$$

where a mass term ($\mu_T = mc^4/\hbar$) is added to the Maxwell term $-\tfrac{1}{4} F^*_{\mu\nu} F_{\mu\nu}$. In this stochastic interpretation of quantum mechanics the real part of the wave equation

$$\partial_\mu F^{\mu\nu} = \mu_\gamma^2 A^\nu \qquad (29)$$

yields the relativistic Hamilton-Jacobi equation

$$\partial_\mu S\, \partial^\mu S + m_\gamma^2 c^2 + \hbar^2\, \partial_\mu a_\nu \cdot \partial_\mu a_\nu - \hbar^2 (\Box P + \partial^\mu P) = 0 \qquad (30)$$

i.e. $P_\mu P^\mu + m_\gamma c^2 + Q + T = 0$, where $Q = -(\partial_\mu P \partial_\mu P + \Box P)$ represents the usual de Broglie-Bohm quantum potential and $T = \partial_\mu a_\nu \cdot \partial_\mu a_\nu$ a spin-quantum torque, with $a_\mu = $ const. Relation (30) shows that the energy $E = h\nu = \partial S/\partial T$ is not, in general, a constant of the motion along a photon path, except when Q and T vanish as it occurs in the case of parts of linearly polarized beams where neighbouring current lines are both straight and parallel. Therefore, if one disturbs the phase relation in a laser beam, which can be obtained by focussing, and/or uses unpolarized beams, the quantum potential is no longer zero and the photon energy is no more a constant of the motion but depends on the value of the quantum potential which oscillates violently in the region of the intersecting beams. Accordingly, the quantum potential can enhance the photon energy and, if $h\nu > W$, photoelectric emission can occur at the position of the minimum of the quantum potential, namely at the focal spot. As in the case of the previous models, this enhancement is not an intensity effect, but it would survive even at low intensities, albeit with reduced probability.

5. EXPERIMENT DISCRIMINATING BETWEEN EFFECTIVE PHOTON THEORY AND THE ALTERNATE THEORIES

We proceed now with the description of the experimental setup used to detect the possible presence of higher-than-normal energy photons in low intensity laser beams (Fig. 9). The setup essentially consists of a He-Ne cw laser model Melles Griot 05 LLR 851 and of a chamber containing two electrodes. The output from the laser is a 3 mW randomly polarized beam of 7.7×10^{-2} cm diameter at $1/e^2$ points and divergence 1 mrad. The beam enters the chamber through a quartz window and is sharply focussed onto the surface of the negatively charged electrode by means of a simple plano-convex lens of 0.34 cm focal length. The ground electrode, in the center of which is mounted the lens by means of a thin collar, is of a grid type structure so as to allow UV light from a Philips 93109 high pressure mercury lamp, when positioned in front of the quartz window, to illuminate the negative electrode and release electrons from it by photoelectric emission. The negatively charged electrode is made of a 3.5 cm diameter massive block of aluminum, the irradiated area being polished to a reflectivity of 96% at the laser wavelength ($\lambda = 6328$ Å). The reflectivity at the UV line wavelength used ($\lambda = 2537$ Å), as selected

254

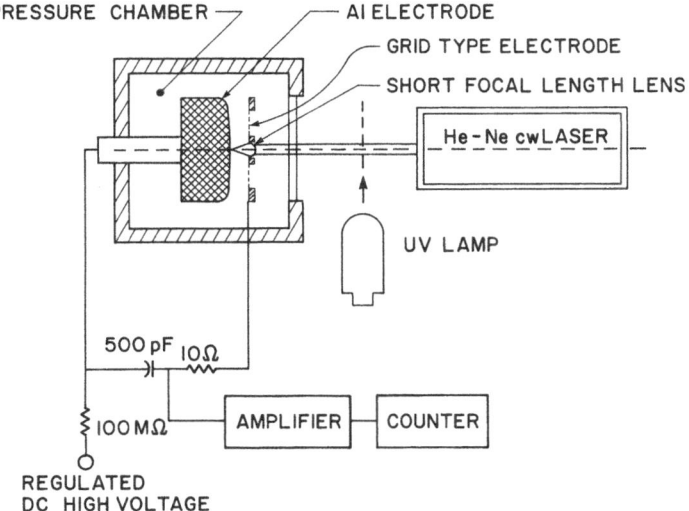

PRESSURE CHAMBER ⎯⎯ ⎯⎯ Al ELECTRODE
⎯⎯ GRID TYPE ELECTRODE
⎯⎯ SHORT FOCAL LENGTH LENS
He-Ne cwLASER

UV LAMP

500 pF 10Ω
100 MΩ
AMPLIFIER   COUNTER
REGULATED
DC HIGH VOLTAGE

Fig. 9 Experimental setup to reveal higher-than-normal energy
photons in a low intensity focussed He-Ne laser beam.

with a narrow band interference filter, is 30%. The chamber was slightly
pressurized at 1 psi above atmospheric pressure with Nitrogen in order to
have a constant environment around the electrodes. Finally, the Al
electrode was connected to a 500 pF capacitor, which was charged with a
regulated DC high voltage power supply to 6950 V through a 100 MΩ
resistor. The other electrode was grounded. Since any electron emission
from the Al electrode was amplified by electron avalanche through the gas
and produced a minute spark between the two electrodes, the number of
sparks was counted with a counter and an amplifier through a 10 Ω
resistance in series with the capacitor.

The objective of the experiment was the following. If one charges
to a sufficiently high voltage one of the two electrodes inside the
pressure chamber, self-breakdown of the gas occurs. This is due to the
fact that, by field emission effect, an electron is born at that time
from the cathode material and an avalanche electron multiplication within
the gas sets in. The current from the cathode material therefore reaches
at that time a value such that the gas, from a state of almost infinite
resistance, becomes an almost perfect conductor[39], i.e. it breaks down.
A spark is therefore a means of detecting the detachment of an electron
from the cathode surface. Such electron, on the other hand, can be
created by other means, for instance by photoelectric effect, with the
same result as before. A UV lamp can be used for this purpose, supplying
photons of sufficiently high energy to overcome the work function of the
cathode material and release an electron. If this is the case, however,
we have to keep the voltage across the electrodes just slightly below the
self-breakdown voltage, so as to avoid field-emission electron
generation. Once it is proven that electrons can be released with a UV
lamp, a low intensity He-Ne laser beam is then used to irradiate the same

cathode material. Normally no breakdown will occur in this case because the photon energy is well below the work function of the material. However if, by focussing, higher-than-normal energy photons are generated within the focal area on the surface of the cathode, as predicted by the above theories, they will be able to release electrons from it by simple photoelectric effect and lead to breakdown in the same manner as UV photons do. Hence, any observed spark will be a means of revealing the presence in the laser beam of higher-than-normal energy photons.

All three theories predict that the appearance of energetic photons is a probabilistic effect. The probability therefore increases with the number of photons present on the cathode surface at any one time. For this reason, the experiment was run with the full available laser intensity I on the cathode material, calculated as follows. The diameter of the focal circle is given, to a good geometric approximation[40], by $d = f\theta$, where θ is the divergence of the laser beam. For a lens of focal length of 0.34 cm, beam divergence of 10^{-3} rad, and laser power 3 mW, we find $I = 3.30 \times 10^4$ W/cm^2.

This high intensity cold give rise to some competing thermionic emission from the irradiated metal. To be sure that this is not the case, we calculate first the temperature rise on the surface of the cathode caused by I. In the simple case of the heating of a semi-infinite solid by a cylindrical beam of diameter 2a, the temperature T obeys the Laplace equation

$$\frac{\partial^2 T}{\partial r^2} + \frac{1}{r}\frac{\partial T}{\partial r} + \frac{\partial^2 T}{\partial z^2} = 0 \tag{31}$$

which has the general solution $T = \int_o^\infty f(k)\, \partial_o(rk)\, e^{-kz}\, dk$, where $f(k)$ is an unknown function, and ∂_o is the Bessel function. The boundary conditions are

$$T \to 0 \qquad\qquad \text{for } z \to \infty$$

$$-\varkappa\frac{\partial T}{\partial z} = P \qquad\qquad \text{for } z = 0 \text{ and } r < a$$

$$-\varkappa\frac{\partial T}{\partial z} = 0 \qquad\qquad \text{for } z = 0 \text{ and } r > a,$$

where $\varkappa$ is the thermal conductivity of the solid. We use the Weber-Schafheitlin integral[41]

$$\int_o^\infty \partial_o(rk) \cdot \partial_1(ak)\, dk = \begin{cases} 1/a & \text{for } r < a \\ 0 & \text{for } r > a \end{cases}$$

from which follows that

$$f(k) = \frac{Pa}{\varkappa k}\, \partial_1(ak)$$

and

$$T = \frac{Pa}{\varkappa} \int_o^\infty \frac{\partial_1(ak)}{k}\, (rk)\, e^{-kz}\, dk. \tag{32}$$

This integral cannot be expressed in terms of tabulated functions, but the maximum temperatures is achieved at the point $z = r = 0$, hence[42]

256

$$T_{max} = \frac{Pa}{\mathcal{K}} \int_{p}^{\infty} \frac{\partial_1(ak)}{k} \, dk. \tag{33}$$

Since the reflectivity of the aluminum cathode material is 96%, the beam power density absorbed is 1.32×10^3 W/cm^2, which corresponds to 3.15×10^2 cal/sec/cm^2. For aluminum $\mathcal{K} \cong 0.5$ cal/sec/cm and $a = 1.7 \times 10^{-4}$. We have from (33)

$$T_{max} = \frac{3.15 \times 10^2 \cdot 1.7 \times 10^{-4}}{0.5} = 0.11°C,$$

an insignificant temperature rise.

We are now in a position to verify if the experimental conditions are such to satisfy the requirements of the foregoing three models for enhanced energy photon production.

Model 1 - For the parameters of the laser focussing geometry indicated above, the lens half-cone angle is $\theta = 6.46°$ and $\lambda/2\sin\theta = 2.81 \times 10^{-4}$ cm. Since the radius of the focal spot is 1.7×10^{-4} cm, Ineq. (26) is satisfied and higher-than-normal energy photons should appear in the focal volume, according to this model.

Model 2 - For the focussing angle $2\theta = 2 \times 6.46° = 12.92°$, original laser photon energy 1.96 eV, and W = 4.28 eV (work function of Aluminum), we find that the line broadening is

$$\hbar\Delta\omega = 0.29$$

and X = 7.93. Hence

$$Q(W) = 7.51 \times 10^{-3}.$$

In order to have $N \times Q(W) > 1$, the number of photons in the laser pulse must be $> 1.33 \times 10^2$. This condition has been greatly exceeded in the experiment.

Model 3 - Three conditions are required in order to increase the probability of having enhanced energy photons:

1. strong focussing,
2. random polarization of the laser beam, and
3. irradiated metal positioned at the focus of the laser beam.

All three conditions have been satisfied in the experiment.

The experiment was carried out as follows. Before each run, the chamber was flushed several times with Nitrogen and the pressure was then set at 1 psi. The voltage across the electrodes was gradually raised until self-breakdown of the gas occurred, which was observed through the presence of sparking. The voltage at which self-breakdown occurred was 7000 V. This voltage was then lowered to 6950 V, after which self-breakdown disappeared. The UV lamp and interference filter were then placed in front of the chamber, so that photons from the $\lambda = 2537$ Å line going through the filter illuminated the negatively charged electrode. At that time breakdown of the gas resumed. The counter was then set for a counting time of 0.3 sec, during which one spark was invariably counted. The UV lamp was then replaced by the He-Ne laser beam and the counting time, from one experiment to the other, was increased from a few seconds to several hours. The last experiment that we performed ran for 9 hours. Finally, at the end of the irradiation with the He-Ne laser, the UV lamp was put back. Again, we observed one spark in the same 0.3 sec time period.

Even with the longest irradiation time of 9 hours, we never observed any spark with the He-Ne laser beam. This result allows us to deduce a lower bound for the probability of the appearance of higher-than-normal energy photons in the focussed laser beam according to the foregoing three theories. The probability will be calculated as a ratio of the number of UV photons that are capable of releasing an electron from the irradiated surface during the 0.3 sec counting time to the number of He-Ne laser photons arriving on the same surface during 9 hours irradiation time.

The UV lamp was calibrated and was known to deliver a photon flux of 4.18×10^{12} sec^{-1} at $\lambda = 2537$ Å (= 4.89 eV). Since sparks were intermittently produced which discharged the 500 pF condenser (Fig. 9), it is necessary to calculate first how many of these photons, during the 0.3 sec counting time, are able to find the condenser recharged to at least 6930 V, which is the minimum voltage for gap breakdown to occur with UV light. The time required to recharge the condenser to 6930 V through the 100 MΩ resistance is 0.29 sec. Hence, during (0.30 − 0.29) sec = 0.01 sec the voltage across the gap is sufficient to yield breakdown. The number of UV photons arriving during such time is 4.18×10^{10} and, taking into account that 30% of them are reflected, 2.5×10^{10} are the useful UV photons. As far as He-Ne laser photons at $\lambda = 6943$ Å (= 1.96 eV), the total number arriving during 9 hours of irradiation exposure is 1.23×10^{19}, after taking into account that their reflectivity at the electrode surface is 96%. The probability is therefore

$$\frac{2.5 \times 10^{10}}{1.23 \times 10^{19}} = 2.04 \times 10^{-9}.$$

Clearly, this probability refers only to the presence of 4.89 eV photons during a time period of 9 hours. It does not exclude that photons of energy higher than the original laser photon energy (= 1.96 eV) but well below 4.89 eV be present in the focal volume and that their probability of being present there be much greater than the one just calculated.

We would like now to say only a few words on the prediction of the effective photon hypothesis in terms of the threshold intensity for enhanced energy photon production because this subject will be thoroughly discussed in the next Section. It will be shown that, for effective photon theory, anomalous effects arise only at photon number densities much greater than $6 \, \pi^{-1} \, \lambda^{-3}$ cm^{-3}, i.e., when more than one photon occupy a volume of diameter equal to λ. In the case of the He-Ne laser light, such photon number density is 7.54×10^{12} cm^{-3}, which corresponds to a photon flux of 2.26×10^{23} cm$^{-2} \cdot$sec^{-1} or to a light intensity of 7.07×10^{4} W$\cdot$cm^{-2}. We were below these figures in the experiment.

In conclusion, of all the theories advanced, it seems that the effective-photon model of intensity-dependent photon energy enhancement is still the most successful in explaining the anomalous photoelectric effect both in metals and gases.

6. EXTENSION OF THE EFFECTIVE PHOTON ANALYSIS

At the time of the proposal of effective photon theory[23] the existence of effective photons was postulated and formalized through the following photon energy expression:

$$\varepsilon = h\nu \cdot \exp(\beta_\nu I^\alpha) \simeq \frac{h\nu}{1 - \beta_\nu I^\alpha} \tag{2a}$$

where $h\nu$ is the normal photon energy, α and β_ν are positive coefficients, and I is the light intensity above a threshold value I_{th} for the effect to appear. The above expression applies to all cases of ionization of gases with very high intensity laser beams, because $I \gg I_{th}$ in this case. Whenever I is not much larger that I_{th} (see the following section for details), the correct expression is

$$\varepsilon = h\nu \cdot \exp\left[\beta_\nu (I - I_{th})^\alpha\right] \simeq \frac{h\nu}{1 - \beta_\nu (I - I_{th})^\alpha}. \tag{2b}$$

Eq. (2a) provides for some effective photons to exist whenever $\beta_\nu I^\alpha$ significantly differs from zero, a fact which normally occurs at the focus of high intensity laser beams, without precluding that effective photons might even exist in unfocussed beams. Moreover, since conservation of energy and photon number in a light pulse is obviously maintained, eq. (2a) implies that the pulse must have photons not only of higher-than-normal energy but also of lower-than-normal energy. Specifically, the conservation of light pulse energy and photon number requires that

$$h\nu = \langle\varepsilon\rangle = \langle h\nu \exp\left[\beta_\nu(I - I_{th})\left|I - I_{th}\right|^{(\alpha - 1)}\right]\rangle. \tag{2c}$$

where $\langle\ \rangle$ indicates averages over the photons in the region of beam intensity I. For $I > I_{th}$, we retrieve the effective photon expression (2a). For $I < I_{th}$, we have "tired" photons. In other words, in a small region of beam intensity I_{av} there is a distribution of photon energies ε and intensities I. Where locally $I > I_{th}$ there are effective photons; where $I < I_{th}$ there are "tired" photons.

In essence, therefore, Eq. (2a), and even more so Eq. (2c) are postulated expressions of a line broadening with light intensity. Moreover, one can also interpret Eq. (2a) as postulating that a variation of photon energy with light intensity leads to a light frequency change, this interpretation being in agreement with the experimental results of Ref. (35). Hence Eq. (2a) can also be written as

$$\varepsilon(I) \simeq h\left(\frac{\nu}{1 - \beta_\nu I^\alpha}\right) = h\nu_1(I). \tag{2a}$$

This expression states that the energy of a photon or the frequency of light are not independent of light intensity but are a function of it. Eq. (2a) therefore goes beyond the conventional interpretation of the fundamental Planck-Einstein relation $\varepsilon = h\nu$ by introducing a (continuous) functional relation between frequency ν and light intensity I. Allen's remark[37] that a contradiction exists between Eq. (2a) and Planck-Einstein relation is therefore valid only up to the point that, every time the photon energy expression is required, one should inquire whether the intensity of light is high enough to introduce a correction factor in the light frequency. Once the correction factor is introduced and the modified frequency ν_1 is calculated, all other fundamental formulas of physics, such as the de Broglie formula relating photon momentum and wavelength, retain their structure.

The fundamental postulate (2a) clearly implies the existence of a hitherto unknown photon-photon interaction force acting at intensities of light much below those predicted by quantum electrodynamics. In Refs. (34-35, 43) it was demonstrated that an interaction law for photons can be derived from a dispersion-free version of the Heisenberg principle, in which equality replaces inequality and Δ-quantities become sharp. Such interaction law yields the prediction that photons do not travel as particles in a stream but, because they tend to stick together, they tend to form bunches or clumps. This implies that the detection of photons at very low intensities cannot be linear with intensity, because it is not a single-particle phenomenon. This prediction has been recently verified in a number of precise experiments[43]. Moreover, the photon bunching effect is not new in the literature and actually has been reported on several occasions, starting with the pioneer work by Hanbury-Brown and Twiss[44] and best summarized in a recent book by Loudon[45]. We would like now to examine if the same interaction law which has led to the foregoing results is capable of correctly predicting a characteristic feature of the gas ionization experiments, namely the minimum intensity of light for ionization effects to take place. Once the theory is shown to be able to predict this feature, then it will be a matter of routine to derive the coefficients α and β_ν of Eq. (2a).

6.1 Minimum Intensity of Light Necessary to Ionize a Gas

In the elementary analysis of Refs. (34-35, 43), an interaction relation for photons was derived from plausible arguments as

$$p = \frac{const}{r} \qquad (34)$$

where r is the distance between two photons and p is the momentum transferred along the r-direction from one photon to the other. The analysis showed that a photon cannot approach another one any closer than a characteristic distance r_o, which was assumed to be the equivalent of the wavelength λ in the classical wave theory of light. This implies that a photon occupies a volume of space equal to or greater than $\sim\lambda^3$. In terms of photon number density N, the maximum allowed value is therefore

$$N_{th} = \frac{1}{(4\pi/3)(\lambda/2)^3} \qquad (35)$$

or, in terms of photon flux F

$$F_{th} = N \cdot c = 6c/(\pi\lambda^3) \qquad (36)$$

or light intensity I

$$I_{th} = F \cdot h\nu = (6hc^2)/(\pi\lambda^4) \qquad (37)$$

where the notation I_{th} indicates that this is the maximum intensity at which individual photons of energy $h\nu$ can avoid producing effective photons; in other words, it is the threshold intensity required to produce effective photons. However, it is well known that, in focussed high intensity laser beams, these values are exceeded, in the sense that the number density of photons is much larger than $(\lambda)^{-3}$, where λ is the original laser wavelength. Since photons cannot come any closer than λ, when this happens a mechanism must be operative, perhaps photon-photon inelastic scattering, the consequence being that the photon wavelength is reduced and the frequency increased, or the line broadened. In a small region of average light intensity I_{av}, there will be a distribution of intensities such that

$$I_{loc} = N_{loc} \cdot c \cdot h\nu_{loc} \tag{37a}$$

where the subscript loc indicates local values. A laser beam subjected to such line broadening will be able to ionize a gas of ionization potential W, provided the modified frequency ν_1 of some photons obeys the following relation

$$h\nu_1 = hc(\lambda_1)^{-1} = W \tag{38}$$

from which

$$\lambda_1 = hc/W. \tag{39}$$

Insering (39) into (37), one gets

$$I_m = (6\pi\, h^3 c^2)\, W^4 \tag{40}$$

where I_m is the minimum light intensity required in order to ionize a gas of ionization potential W. Eq. (40) is remarkable in predicting that the minimum local light intensity I_m required to produce an ionization of energy W is independent of the wavelength of the incident radiation.

Clearly, in the interaction of laser light with a gas, such minimum intensity has to be exceeded, especially in the case of a rarified gas. In other words, because of the rather small cross section for collisions between the effective photons and gas atoms, not all effective photons will be able to produce ionization of the gas atoms. This is the same effect as in the interaction of normal light and a metal surface, in which the quantum efficiency[46] is of the order of 10^{-3}. In comparing therefore the theoretical minimum intensity of light I_m capable of producing ionization effects with the minimum intensity I_{exp} which experimentally can produce one ion, one should observe if the cross section is constant for different gases. If it is so, one can conclude that the theory is capable of predicting the experimental results.

We have carried out such comparison. In Table I are reported the theoretical values of I_m derived from Eq. (40) for the five noble gases and compared with the experimental I_{exp} for $\lambda = 10600$ Å, as reported in Ref. (24). Considering the uncertainty of the data and the approximate nature of the theoretical analysis, the ratio of the two intensities is indeed constant.

6.2 <u>Derivation of the Coefficients α and β_ν in the Approximate Expression for the Effective Photon Energy</u>

Since the values so found of I_m seem to be correct, we can proceed with the derivation of the values of α and β_ν of relation (2a). Let us write I_m in terms of the modified photon energy expression (1a)

$$I_m = N \cdot c \cdot h\nu_1 = \frac{6c}{\pi\lambda_1^3[1-\beta_\nu(I_m-I_{th})^\alpha]}\, h\nu = \frac{6hc^2}{\pi\left\{\lambda[1-\beta_\nu(I_m-I_{th})^\alpha]^4\right\}}. \tag{41}$$

We get:

TABLE I

Gas	Ionization potential	I_m	I_{exp}	Ratio
Helium	24.58 eV	1.75×10^9 W/cm^2	1.1×10^{13} W/cm^2	1.59×10^{-4}
Neon	21.56 eV	1.04×10^9 W/cm^2	8.0×10^{12} W/cm^2	1.30×10^{-4}
Argon	15.76 eV	2.96×10^8 W/cm^2	2.5×10^{12} W/cm^2	1.18×10^{-4}
Krypton	14.00 eV	1.84×10^8 W/cm^2	1.2×10^{12} W/cm^2	1.54×10^{-4}
Xenon	12.13 eV	1.04×10^8 W/cm^2	7.0×10^{11} W/cm^2	1.48×10^{-4}

Comparison between the theoretical values of the minimum light intensity I_m necessary to ionize the noble gases with Neodymium laser light ($\lambda = 10600$ Å) and the experimental values I_{exp}. Since the ratio between these values is approximately constant (within the uncertainties of the data) for such diverse gases and range of light intensities as the above, it appears that the theoretical analysis is correct.

$$I_m \, \lambda^4 \left[1 - \beta_\nu (I_m - I_{th})^\alpha\right]^4 = \frac{6hc^2}{\pi}. \qquad (42)$$

In this expression the unknowns are α and β_ν, because I_m can be taken from Table I. From (37) we derive the value of I_{th} for $\lambda = 10600$ Å. It is: $I_{th} = 9.009 \times 10^3$ W·cm^{-2}. Inserting this value of I_{th} and taking from Table I the values of I_m for the pairs of gases (Helium, Xenon) and (Neon, Krypton), and averaging the α and β_ν values, we get

$$\alpha = 0.01846$$

$$\beta_\nu = 0.64384$$

for $\lambda = 10600$ Å, and the relation (2a) for laser beams of $\lambda = 10600$ Å is therefore

$$\varepsilon(I) = \frac{1.17}{1 - 0.64384(I - 9.009 \times 10^3)^{0.01846}} \text{ eV} \qquad (43)$$

where 1.17 eV is the original photon energy. When $I \gg 9.009 \times 10^3$ W·cm^{-2}, the following simplified expression can be used

$$\varepsilon(I) = \frac{1.17}{1 - 0.64384 \, I^{0.01846}} \text{ eV.} \qquad (44)$$

6.3 Derivation of the Coefficients α and β_ν in the Exact Expression for the Effective Photon Energy

In the previous section, we have assumed that expression (2a) was valid and have added the theoretical notion of a threshold intensity I_{th} for photons to gain energy, which enabled us to derive the parameters α and β_ν.

The origin of the formula (2a) was left undefined. It might have as origin, for instance, the compositness structure of the photon itself as indicated in Ref. 32, or it might be a first approximation of an exponential, as suggested here.

If we maintain the latter hypothesis, we can calculate the parameters α and β_ν from the exact expression

$$\varepsilon = h\nu \cdot \exp \left[\beta_\nu (I - I_{th})^\alpha\right] = h\nu_1. \tag{45}$$

The procedure is the same as before. We insert first the above expression in $I_m = N \cdot c \cdot h\nu_1$

$$I_m = N \cdot c \cdot h\nu_1 = N \cdot c \cdot h\nu \cdot \exp \left[\beta_\nu (I - I_{th})^\alpha\right] \tag{46}$$

from which we get

$$I_m \lambda^4 \left\{\exp\left[-\beta_\nu (I_m - I_{th})^\alpha\right]\right\}^4 = 6\pi^{-1} hc^2 \tag{47}$$

$$\ln \beta_\nu + \alpha \ln (I_m - I_{th}) = \ln \left\{-[\ln(I_{th}/I_m)]/4\right\}. \tag{48}$$

This is a linear (straight line) relation $y = mx + c$, if we make the identifications

$$y = \ln \left\{-[\ln(I_{th}/I_m)]/4\right\}$$

$$x = \ln(I_m - I_{th})$$

$$m = \alpha$$

$$c = \ln\beta_\nu.$$

The values of I_m (Table I) are at least 4 orders of magnitude larger than the value of I_{th}, so the subtraction of I_{th} in the computation of x is insignificant. The least squares straight line through the 5 x, y values computed from the data in Table I yields the following values of α and β_ν, and also standard errors in these computed quantities

$$\alpha = 0.09308 \pm 0.00088$$

$$\beta_\nu = 0.4213 \pm 0.0035.$$

These are perhaps better values of α and β_ν than those provided in the previous section. However, it is believed that one should now wait until the experimentalists come up with some results which, either confirm the validity of the approximate expression (2a) or of the improved expression (45), or require an altogether new theoretical hypothesis leading to a different effective photon formula.

6.4 Total Number of Effective Photons Contained in a Laser Pulse and their Spectral Distribution

We shall use the exact photon energy expression (45) in order to calculate the total number of effective photons contained in a laser pulse, as well as their spectral distribution.

The analysis proceeds as follows. Let us write Eq. (45) in terms of wavelength λ rather than frequency ν

$$h\nu_1 = h\nu \, e^{\beta_\nu (I - I_{th})^\alpha} \tag{46}$$

$$\nu_1 = \nu \, e^{\beta_\nu (I - I_{th})^\alpha}$$

$$\lambda_1 = \lambda \, e^{-\beta_\nu (I - I_{th})^\alpha} . \tag{47}$$

This is the fundamental relation between intensity of light and wavelength obeyed by the effective photons.

Given a typical laser pulse of peak intensity I_p, having a triangular temporal distribution and guassian radial distribution of intensity

$$I = \frac{I_p}{\Delta t} \, t \cdot e^{-ar^2} \tag{48}$$

where Δt is the pulse duration at half-power and $a = \dfrac{4 \ell n 2}{f^2 \theta^2}$ is the gaussian factor (f = focal length of the lens used for focussing; θ = laser beam divergence), the total number of effective photons contained in such pulse is given by

$$N_{eff} = \frac{2\pi}{h\nu} \frac{I_p}{\Delta t} \int_{t_{th}}^{\Delta t} dt \int_o^{r_{th}} t \, e^{-ar^2} \, r dr \tag{49}$$

where t_{th} and r_{th} are the threshold time and radius, respectively, for effective photons to appear in the laser pulse. Eq. (49) calculates the number of effective photons, i.e., higher-than-normal energy photons, the "tired" photons being of equal in number. We can derive r_{th}, which is the radius where $\lambda_1 = \lambda$. We have in this case

$$I = I_{th}$$

$$\frac{I_p}{\Delta t} \, t \cdot e^{-ar_{th}^2} = I_{th}$$

$$r_{th} = \left[-\frac{1}{a} \ell n \left(\frac{\Delta t}{I_p} \frac{I_{th}}{t} \right) \right]^{\frac{1}{2}} . \tag{50}$$

t_{th} is given by

$$t_{th} = \frac{I_{th}}{I_p} \Delta t \tag{51}$$

because $r = 0$ in this case. Integration of (49) yields

$$N_{eff} = -\frac{\pi}{h\nu} \frac{I_p}{\Delta t \, a} \int_{t_{th}}^{\Delta t} t \, dt \, \left[e^{-ar^2} \right]_o^{r_{th}}$$

$$= -\frac{\pi}{h\nu} \frac{I_p}{\Delta t \, a} \int_{t_{th}}^{\Delta t} t \left(\frac{I_{th}}{I_p} \frac{\Delta t}{t} - 1 \right) dt$$

$$= \frac{\pi}{h\nu} \frac{I_p}{\Delta t \, a} \int_{t_{th}}^{\Delta t} (t - t_{th}) \, dt$$

$$= \frac{\pi}{h\nu} \frac{I_p}{\Delta t} \frac{1}{a} \left[\frac{t^2}{2} - t_{th} \, t \right]_{t_{th}}^{\Delta t}$$

$$= \frac{\pi}{2} \frac{I_p}{h\nu} \frac{1}{a} \left[\Delta t - 2t_{th} + \frac{t_{th}^2}{\Delta t} \right]. \tag{52}$$

This expression can be written in terms of the intensity threshold I_{th}

$$N_{eff} = \frac{\pi}{2} \frac{I_p}{h\nu} \frac{\Delta t}{a} \left[1 - 2 \frac{I_{th}}{I_p} + \left(\frac{I_{th}}{I_p} \right)^2 \right]. \tag{53}$$

On the other hand, the total number of original photons N_{orig} contained in the laser pulse is given by

$$N_{orig} = \frac{4\pi}{h\nu} \frac{I_p}{\Delta t} \int_o^{\Delta t} t \, dt \int_o^\infty e^{-ar^2} r dr \tag{54}$$

Integration of (54) yields

$$N_{orig} = \frac{\pi \, I_p \, \Delta t}{h\nu \, a}. \tag{55}$$

The number of photons in a laser pulse that have not changed wavelength is therefore

$$N_{unpert} = N_{orig} - 2N_{eff} \tag{56}$$

the number 2 in this expression appearing because the number of "tired" photons is equal to the number of effective photons.

In order now to derive the spectral distribution of the effective photons, we shall consider a typical wavelength $\lambda_c < \lambda$ and calculate the number of photons at this wavelength within a typical bandwith $\Delta\lambda = 100$ Å, i.e., between $\lambda' = \lambda_c - 50$ and $\lambda'' = \lambda_c + 50$. Starting from (47) we get

$$\left(\ell n \frac{\lambda}{\lambda_c} \right)^{1/\alpha} = \beta_\nu^{1/\alpha} \left(\frac{I_p}{\Delta t} t \, e^{-ar^2} - I_{th} \right) \tag{57}$$

from which

$$\frac{I_p}{\Delta t} t \, e^{-ar^2} = \frac{1}{\beta_\nu^{1/\alpha}} \left(\ell n \frac{\lambda}{\lambda_c} \right)^{1/\alpha} + I_{th}. \tag{58}$$

The total number of effective photons of $\lambda = \lambda_c$ and $\Delta\lambda = 100$ Å is given by

$$N_{\lambda_c} = \frac{2\pi}{h\nu} \frac{I_p}{\Delta t} \int_{t_{\lambda_c}}^{\Delta t} t dt \int_{r_{\lambda_c-50}}^{r_{\lambda_c+50}} e^{-ar^2} r dr$$

$$= \frac{2\pi}{h\nu} \frac{I_p}{\Delta t} \int_{t_{\lambda_c}}^{\Delta t} t dt \left[- \frac{e^{-ar^2}}{2a} \right]_{r_{\lambda_c-50}}^{r_{\lambda_c+50}}$$

$$= \frac{\pi}{\beta_\nu^{1/\alpha} \, h\nu \, a} \left[\left(\ell n \frac{\lambda}{\lambda_c-50} \right)^{1/\alpha} - \left(\ell n \frac{\lambda}{\lambda_c+50} \right)^{1/\alpha} \right] \int_{t_{\lambda_c}}^{\Delta t} dt \tag{59}$$

because

$$\left[-e^{-ar^2}\right]_{r_{\lambda_c}-50}^{r_{\lambda_c}+50} = \frac{\Delta t}{I_p \ t \ \beta_\nu^{1/\alpha}} \left[\left(\ell n \ \frac{\lambda}{\lambda_c-50}\right)^{1/\alpha} - \left(\ell n \ \frac{\lambda}{\lambda_c+50}\right)^{1/\alpha}\right] \qquad (60)$$

as derived from (58). Carrying the integration on the right side of Eq. (59) one gets

$$N_{\lambda_c} = \frac{\pi(\Delta t - t_{\lambda_c})}{\beta_\nu^{1/\alpha} \ h\nu \ a} \left[\left(\ell n \ \frac{\lambda}{\lambda_c-50}\right)^{1/\alpha} - \left(\ell n \ \frac{\lambda}{\lambda_c+50}\right)^{1/\alpha}\right] \qquad (61)$$

t_{λ_c} is derived from (58) for r = 0

$$t_{\lambda_c} = \frac{\Delta t}{I_p} \left[\frac{1}{\beta_\nu^{1/\alpha}} \left(\ell n \ \frac{\lambda}{\lambda_c}\right)^{1/\alpha} + I_{th}\right]. \qquad (62)$$

Hence

$$N_{\lambda_c} = \frac{\pi \ \Delta t}{\beta_\nu^{1/\alpha} \ h\nu \ a} \left\{1 - \frac{1}{I_p} \left[\frac{1}{\beta_\nu^{1/\alpha}} \left(\ell n \ \frac{\lambda}{\lambda_c}\right)^{1/\alpha} + I_{th}\right]\right\}$$

$$\cdot \left[\left(\ell n \ \frac{\lambda}{\lambda_c-50}\right)^{1/\alpha} - \left(\ell n \ \frac{\lambda}{\lambda_c+50}\right)^{1/\alpha}\right]. \qquad (63)$$

This is the expression to be used to calculate the number of effective photons of $\lambda = \lambda_c$ and $\Delta\lambda = 100$ Å contained in a laser pulse of peak intensity I_p, original wavelength λ, original photon energy $h\nu$, pulse-width Δt, and gaussian factor a. The threshold intensity I_{th} is given by (37).

If one wants to know the shortest possible wavelength λ_{cutoff} of the effective photons contained in the laser pulse, it can be derived from (58) for r = 0 and t = Δt. One gets:

$$\lambda_{cutoff} = \lambda \ e^{-\beta_\nu (I_p - I_{th})^\alpha}. \qquad (64)$$

We have used Eqs. (53), (55), (56), (63) and (64) to compile Table II, which reports the values of N_{orig}, N_{eff}, N_{unpert}, λ_{cutoff} and N_{λ_c} for selected wavelengths from $\lambda_c = 10000$ Å to λ_{cutoff}, for a Neodymium laser pulse ($\lambda = 10600$ Å) of peak intensity I_p, divergence $\theta = 1$ mrad, and pulse-width 10 nsec focussed with a lens of focal length f = 1 cm. The intensity threshold for such laser is $I_{th} = 9.009 \times 10^3$ W·cm^{-2}, $\alpha = 0.09308$ and $\beta_\nu = 0.4213$. If one wants the values of N_{orig}, N_{eff}, N_{unpert}, and N_{λ_c} for different Δt and different f from the ones tabulated, they can be easily obtained by knowing that they are proportional to Δt and to the square of f.

TABLE II

	I_p (W·cm^{-2})			
	10^5	10^8	10^{11}	10^{14}
N_{orig}	6.04×10^9	6.04×10^{12}	6.04×10^{15}	6.04×10^{18}
N_{eff}	2.50×10^9	3.02×10^{12}	3.02×10^{15}	3.02×10^{18}
N_{unpert}	1.04×10^9	1.09×10^9	1.09×10^9	1.00×10^9
N_{λ_c} 8000Å	3.46×10^2	3.81×10^2	3.81×10^2	3.81×10^2
6000Å	4.39×10^5	4.83×10^5	4.83×10^5	4.83×10^5
4000Å	1.13×10^8	1.37×10^8	1.37×10^8	1.37×10^8
3000Å	$\emptyset$	2.26×10^9	2.27×10^9	2.27×10^9
1000Å	$\emptyset$	$\emptyset$	3.06×10^{12}	3.06×10^{12}
500 Å	$\emptyset$	$\emptyset$	7.55×10^{13}	7.69×10^{13}
100 Å	$\emptyset$	$\emptyset$	$\emptyset$	4.04×10^{16}
60 Å	$\emptyset$	$\emptyset$	$\emptyset$	7.31×10^{17}
λ_{cutoff}	3131 Å	1021 Å	124 Å	2 Å

Values of N_{orig}, N_{eff}, N_{unpert}, λ_{cutoff} and N_{λ_c} for selected wavelengths λ_c, for a Neodymium laser pulse ($\lambda = 10600$ Å) of peak intensity I_{th}, divergence $\theta = 1$ mrad, 10 nsec duration, focussed with a lens of focal length f = 1 cm.

7. CONCLUSIONS

This review paper on the effective photon theory has covered work in this field which started at the beginning of the 1970's, when the first results of ionization of gases with high intensity laser beams were made available, and has continued since then.

Since the original "ad hoc" hypothesis, namely that the energy of a photon might be a function of light intensity, has not been contradicted so far neither by the experiments of laser-induced gas ionization nor by those of photoemission from laser irradiated metals, this has led, in recent times, to inquire whether these photons might have a classical origin. Three theories have been then put forward that explain "classically" the existence of effective photons. Of these theories, the Quantum Potential Theory is the most complete. It predicts, however, that higher-than-normal energy photons are a beam geometry, rather than a light intensity effect. An experiment reported in this paper, which was designed mainly to test the validity of the Quantum Potential Theory claim, has not ruled it out completely and more refined tests need to be done. However, the experiment has provided initial evidence that, if the effect is not an intensity effect, the probability for the existence of intensity-independent effective photons is rather small.

By contrast, the confidence acquired in the validity of the effective photon formula has led in recent years to inquire about its theoretical origin. Clearly, effective photons are a consequence of a hitherto unknown interaction relation between photons. From plausible arguments this relation has now been derived as a dispersion free version of the Heisenberg principle where equality replaces inequality and Δ-quantities become sharp. This interaction relation leads naturally to the notion that one cannot have more than one photon in a volume of space $\sim\lambda^3$, which is tantamount to saying that intensity and frequency of light are related, as propounded by effective photon theory.

Starting from this fundamental notion, effective photon theory has been extended and the minimum intensities of light necessary to ionize the noble gases have been derived. Also, the parameters α and β_ν of the effective photon formula have been calculated.

Much work still needs to be done on the verification of effective photon theory in different experimental situation. In particular, it is believed that the most important requirement at present is the direct and unequivocal experimental confirmation of the the existence of effective photons in high intensity focussed laser beams. This can only be achieved through advanced experimentation of the photon-photon scattering type. If scattered photons of shifted wavelength will be found in such experiments, then indeed the prediction that photons can be subjected to an energy change as a function of light intensity will have received its ultimate test.

ACKNOWLEDGEMENT

We would like to thank P. Savic, A. Hsu, R.R. Haria and G.P. van Blockland for assistance in the extension of the mathematical analysis of effective photon theory.

REFERENCES
1. E. Panarella, Lett. Nuovo Cim. 3, 417 (1972).
2. V. Chalmeton and R. Papoular, Compt. Rend. 264B, 213 (1967).
3. D.C. Smith and A.F. Haught, Phys. Rev. Lett 16, 1085 (1966).
4. P. Belland, C. DeMichelis, and M. Mattioli, Opt. Comm. 4, 50 (1971).
5. A.F. Haught, R.G. Meyerand, and D.C. Smith, in "Physics of Quantum Electronics," P.L. Kelley, B. Lax, and P.E. Tannenwald, eds. (1966), p. 509.
6. R.W. Minck, and W.G. Rado, in "Physics of Quantum Electronics," P.L. Kelley, B. Lax, and P.E. Tannenwald, eds. (1966), p. 527.
7. T. Okuda, K. Kishi, and K. Sawada, Appl. Phys. Lett. 15, 181 (1969).
8. K. Kishi, K. Sawada, T. Okuda, and Y. Matsuoka, J. Phys. Soc. Japan 29, 1053 (1970).
9. K. Kishi, and T. Okuda, J. Phys. Soc. Japan 31, 1289 (1971).
10. R.G. Tomlinson, E.K. Damon, and H.T. Buscher, in "Physics of Quantum Electronics," P.L. Kelley, B. Lax, and P.E. Tannenwald, eds. (1966), p. 520.
11. P. Agostini, G. Bariot, G. Mainfray, C. Manus, and J. Thebault, IEEE J. Quantum Electr. QE-6, 782 (1970).
12. G.S. Voronov, G.A. Delone, and N.B. Delone, Soviet Phys. - JETP 24, 1122 (1967).

13. M. Louis-Jacquet, Compt. Rend. 270B, 548 (1970).
14. V. Chalmeton, J. de Phys. 30, 687 (1969).
15. B. Held, G. Mainfray, C. Manus, and J. Morellec, Phys. Lett. 35A, 257 (1971).
16. G.A. Delone and N.B. Delone, JETP Lett 10, 265 (1969).
17. L.V. Keldish, Soviet Phys. JETP 20, 1307 (1965).
18. B.A. Tozer, Phys. Rev. 137A, 1665 (1965).
19. H.B. Bebb and A. Gold, Phys. Rev. 143, 1 (1966).
20. Y. Gontier and M. Trahin, Phys. Rev. 172, 83 (1968).
21. H.R. Reiss, Phys. Rev. 1A, 803 (1970).
22. F. Morgan, L.R. Evans, and C. Grey Morgan, J. Phys. D: Appl. Phys. 4, 225 (1971).
23. E. Panarella, Found Phys. 4, 227 (1974).
24. E. Panarella, Can. j. Phys. 54, 1815 (1976).
25. E. Panarella, Phys. Rev. Lett. 33, 950 (1974).
26. N.S. Nogar, Spectr. Lett. 11, 243 (1978).
27. A.D. Allen, Found Phys. 7, 609 (1977).
28. A.L. de Brito and A. Jabs, Can. J. Phys. 62, 661 (1984).
29. A.L. de Brito, Can. J. Phys. 62, 1010 (1984).
30. C. Dewdney, A. Garuccio, A. Kyprianidis, and J.P. Vigier, Phys. Lett. 105, 15 (1984).
31. C. Dewdney, A. Kyprianidis, J.P. Vigier, and M.A. Dubois, Lett. Nuovo Cim. 41, 177 (1984).
32. P. Raychaudhuri, paper in these Proceedings.
33. E. Panarella, Ann. Fond. Louis de Broglie 6, 197 (1981).
34. E. Panarella, Ann. Fond. Louis de Broglie 10, 1 (1985).
35. E. Panarella, Phys. Rev. 16A, 672 (1977).
36. A.V. Phelps, in "Physics of Quantum Electronics," P.L. Kelley, B. Lax, and P.E. Tannenwald, eds (1966), p. 538.
37. M. Moles and J.P. Vigier, Compt. Rend. 276, 697 (1973).
38. A. Garuccio and J.P. Vigier, Lett. Nuovo Cimmento 30, 57 (1981).
39. A.L. Donaldson, M.O. Hagler, M. Kristiansen, L.L. Hatfield, and R.M. Ness, J. Appl. Phys. 57, 4981 (1985).
40. M.V. Klein, "Optics" (Wiley, New York 1970) p. 136.
41. I.S. Gradshteyn and I.M. Ryzhik, "Tables of Integrals, Series and Products," (Academic Press, New York 1980) p. 692, No. 6.575(1).
42. I.S. Gradshteyn and I.M. Ryzhik, ibid. p. 684, No. 6.561(14).
43. E. Panarella, other paper in these Proceedings.
44. R. Hanbury-Brown and R.Q. Twiss, Proc. Royal Soc. A242, 300 (1957); ibid 291 (1958).
45. R. Loudon, "The Quantum Theory of Light" (Clarendon Press, Oxford 1973) p. 227.
46. "RCA Phototubes and Photocells - Technical Manual PT-60," 1963, Radio Corporation of America, p. 5.

COMPOSITENESS OF PHOTONS AND ITS IMPLICATIONS

Probhas Raychaudhuri

Department of Applied Mathematics
Calcutta University
Calcutta - 700009, India

ABSTRACT

It is suggested that the photon may be taken as a composite state of
a neutrino-antineutrino pair when the composite character is described by
the vanishing of the wave-function renormalisation constant and the
nonvanishing of a certain composite coupling constant. There must be two
different kinds of neutrinos, one associated with left circular
polarization and another with right circular polarization. In this
formalism these polarizations would have to transform into one another
under inversion and, therefore, correspond to the same representation of
the Lorentz group. To construct a photon of spin 1, we shall need
neutrinos of both helicities. The two neutrinos associated with a single
photon are both either β-neutrinos (ν_L) or μ-neutrinos (ν_R); it is
impossible to say which without reference to the vacuum state. The field
representing neutrinos would have to be quantized in accordance with the
parastatistics to allow the existence of neutrinos in the same dynamical
state. The neutrinos (ν_L, ν_R) transform into one another under inversion
and therefore correspond to the same representation of the full Lorentz
group. In the above formalism neutrinos are of opposite spin and thus
neutrinos are parafermions (satisfying parastatistics of order 2).
Therefore, the neutrino theory of light can be reformulated without
rotational invariance. In this formalism the photon may oscillate
between a left-handed and a right-handed photon. It is expected that the
system of photons interacts differently at very low, intermediate and
very high photon densities. Thus it is expected that at very high photon
densities the usual form $E = h\nu$ may not be valid. Implications of the
above photonic structure will be discussed in connection with the
anomalous photoelectric effect, wave-particle duality, replication of
photons, superluminal transmission etc. Considering the composite
coupling constant of the photon we will discuss the possibility of a
stochastic interpretation of quantum mechanics.

1. INTRODUCTION

The possibility that the photon may be a composite state of neutrino
and antineutrino pair was first suggested by de Broglie[1] in connection
with the specific properties of the photoelectric effect. The composite

271

nature of the photon was inferred by Jordan[2] on the basis of statistical arguments. However, this type of simple picture of the photon was found not to obey Bose statistics because of the underlying Fermi statistics of its components. In fact, if two such photons were in the state with momentum $\vec{p}$, two components ν's and ($\bar{\nu}$'s) of these photons would be in the state with the same momentum $\vec{k}$. Kronig[3] succeeded in constructing the photon field out of these neutrinos. Pryce[4] has shown that the Kronig theory is not invariant under the group of spatial rotations about the direction of photon momentum as an axis. Barbour, Bietty and Touschek[5] argued that a photon in neutrino theory is always longitudinally polarized. Ferretti[6] suggested that the photon can be considered as a limiting state of a bound state of two non-zero mass particles with a given angular momentum when the binding energy as well as the mass $\to 0$. Perkins[7] attempted to solve the problem and has shown in his formalism also that the photon operators do not satisfy Bose commutation relations. Again, the impossibility of constructing linearly polarized photons seems to be a very serious defect of his theory. But in 1972 Perkins[8] constructed linearly polarized photon operators and the theory was shown to describe truly neutral photons. A composite photon distribution is obtained which is similar enough to Planck's distribution to satisfy the experimental results. To form a composite photon one must give up the exact Bose statistics. Bandyopadhyay and Raychaudhuri[9] showed that the photon may be taken as a composite of neutrino-antineutrino pair where the composite character is described by the vanishing of a certain wave function remormalisation constant and the nonvanishing of a certain composite coupling constant. Green[10] showed how the neutrino theory of light can be reformulated, without violating rotational invariance, provided the neutrinos satisfy parastatistics of order 2.

We will discuss in Sec. 2 the photon as a composite particle in quantum field theory (QFT). In Sec. 3 we will discuss the parastatistics and the compositeness of a photon. In Sec. 4-7, respectively, we will discuss the photonic structure, the anomalous photoelectric effect, the wave-particle duality, replication of photons and superluminal transmission. In Sec. 8 we will discuss the possibility of a stochastic interpretation of quantum mechanics.

2. THE PHOTON AS A COMPOSITE PARTICLE IN QUANTUM FIELD THEORY (QFT)

The general features of composite particles have been developed by different authors on the vanishing of the wave function renormalisation constant and the nonvanishing of a certain composite coupling constant.

Broido[11] argued that a composite particle in local Lagrangian QFT is obtained by taking an elementary particle and letting its wave function renormalisation constant go to zero, $Z_3 = 0$ in such a way that

$$\lim_{Z_3 \to 0} \frac{g'Z_1}{Z_3 \delta m^2} \tag{1}$$

is finite and nonzero, where g' is the renormalised coupling constant and δm^2 is the mass shift describing the composite particle. In quantum electrodynamics (QED), within the framework of the perturbation theory (QEDP), Z_3^{-1} diverges to all orders. The divergence of Z_3^{-1} is an ultraviolet divergence which will have to be removed by cut-offs. Thus the $Z_3 \to 0$ transition process is accomplished by the transition $\Lambda^2 \to \infty$, where Λ is the cut-off factor. Hence, in QEDP the divergence of Z_3^{-1} has nothing to do with the zero rest mass of the photon and would persist also for small non-zero values of the renormalised photon mass. Therefore, zero is not an isolated point of the spectrum of p_μ^2 for the photon field; if this is regarded as a composite, there is no particle interpretation for it. Pointing out these drawbacks Broido[11] argued that in QEDP the photon cannot be taken as a composite state of an electron-positron pair. Broido[11] also reached the same conclusion from an analysis outside the framework of the perturbation theory.

Bandyopadhyay[12] suggested that a photon can interact weakly only with two-component neutrinos having an interaction Lagrangian of the form

$$L = ig\bar{\phi}\gamma_\mu\phi A_\mu = ig\bar{\psi}\gamma_\mu(1+\gamma_5)\psi A_\mu \qquad (2)$$

Here ϕ is the two-component neutrino wave function and ψ is the four component function defined as

$$\phi = \frac{1}{2}(1+\gamma_5)\psi \qquad (3)$$

where g is the photon-neutrino weak coupling constant ($g = 10^{-10}e$). It cannot behave as an electric charge due to the equivalence of two-component neutrinos with four-component majorana neutrinos. For a majorana neutrino,

$$\chi(x) = e^{-1}\bar{\chi}(x), \qquad (4)$$

and therefore $\chi(x)$ does not allow the gauge transformation of the first kind,

$$\chi(x) \to e^{i\alpha}\chi(x). \qquad (5)$$

Thus, the majorana neutrino does not admit of gauge transformations of the first kind, but it does admit of a γ_5 gauge transformation. Let us take the Hamiltonian for the two-component neutrino (Weyl) field interacting with an electromagnetic field:

$$H = \frac{1}{4}\left[\bar{\phi}\gamma_\mu\frac{\partial\phi}{\partial x_\mu} - \frac{\partial\bar{\phi}}{\partial x_\mu}\gamma_\mu\phi - \frac{\partial\phi}{\partial x_\mu}T\bar{\phi} + \phi\gamma_\mu T\frac{\partial\bar{\phi}}{\partial x_\mu}\right] - \frac{ig}{2}A_\mu$$

$$\cdot\left[\bar{\phi}\gamma_\mu\phi - \phi\gamma_\mu T\bar{\phi}\right] + \text{h.c.} \qquad (6)$$

This is invariant under the gauge transformation

$$\phi \to e^{ig\Lambda(x)}\phi(x) \qquad (7)$$

$$A_\mu \to A_\mu + \frac{\partial\Lambda}{\partial x_\mu} \qquad (8)$$

but the transformation (7) does not allow the neutrino mass to be strictly zero. However, the Hamiltonian is invariant under

$$\phi \to e^{i\gamma_5\alpha}\phi \qquad (9)$$

and ensures that the mass of the neutrino before and after interaction is always zero.

The associated conservation law:

$$\partial_\mu J_\mu = \partial_\mu(\bar{\phi} \gamma_\mu \phi) = 0 \tag{10}$$

Now, following Ryan and Okubo[13]:

$$V\phi V^{-1} = \frac{1}{2} (1+\gamma_5) \chi(x) \tag{11}$$

where $\chi(x)$ is a majorana field,

$$\chi(x) = \frac{1}{2} \left[\chi(x) + c^{-1} \bar{\chi}(x)\right] \tag{12}$$

$$\phi(x) = \frac{1}{2} (1+\gamma_5) \chi(x). \tag{13}$$

We can write:

$$VHV^{-1} = H' = \frac{1}{8} \left[\bar{\chi} \gamma_\mu \frac{\partial \chi}{\partial x_\mu} - \frac{\partial \bar{\chi}}{\partial x_\mu} \gamma_\mu \chi - \frac{\partial \bar{\chi}}{\partial x_\mu} \gamma_\mu^T \bar{\chi} + \chi \gamma_\mu^T \frac{\partial \bar{\chi}}{\partial x_\mu}\right]$$
$$- \frac{ig}{4} \left[\bar{\chi}\gamma_\mu\gamma_5 \chi - \chi\gamma_5^T\gamma_\mu^T\bar{\chi}\right] A_\mu + h.c. \tag{14}$$

Other parts vanish because of the majorana condition on $\chi(x)$. Equation (14) is clearly invariant under the gauge transformation,

$$\chi(x) \rightarrow e^{ig\gamma_5\Lambda(x)} \chi(x) \tag{15}$$

$$\Lambda_u \rightarrow \Lambda_u + \frac{\partial \Lambda}{\partial x_u} ,$$

and from the gauge invariance equations (8) and (15), it is clear that the photon mass is zero.

The associated conservation law is

$$\partial_\mu J_{\mu 5} = \partial_\mu(\bar{\chi}\gamma_\mu\gamma_5\chi) = 0 \tag{16}$$

Clearly, the theory is renormalisable.

The majorana neutrino, however, allows the formation of the pseudovector $\bar{\chi}\gamma_\mu\gamma_5\chi$ and thus may allow the pseudo charge (chiral charge)[14,15],

$$q = \int d^3 \vec{x} \chi^+ \gamma_5 \chi, \tag{17}$$

in terms of Weyl operators,

$$q' = -\int d^3 \vec{x} \phi^+ \phi. \tag{18}$$

Therefore, the coupling constant in photon-neutrino weak coupling cannot behave as an electric charge of a neutrino.

Now, with an analysis similar to that in QEDP, we can use the following spectral representation to calculate Z_3^{-1} for neutrino dynamics:

$$Z_3^{-1} = \int_o^\infty d\sigma^2 \rho_R(\sigma^2) \tag{19}$$

where

$$\rho_R(\sigma^2) = \delta(\sigma^2) + g^2/12\pi^2 \theta(\sigma^2) \frac{1}{\sigma^2}, \tag{20}$$

with

$$\theta(\sigma^2) = 1, \quad \sigma^2 > 0$$
$$\qquad = 0, \quad \sigma^2 < 0$$

It is to be noted that here we come across the same divergence difficulties as in QEDP which has to be removed by cut off. Apart from the dependence of Z_3 on the mass and coupling constant, it is also dependent on the cut-off factor. We can write in a general form,

$$Z_3^{-1} = 1 + \int_0^{\Lambda^2} f(M^2, g^2, m_p) \frac{dM^2}{M^2}, \qquad (21)$$

where m_p is the renormalized photon mass. However, it is observed that for photons and two-component neutrinos the renormalized mass as well as the bare mass is zero due to the above-discussed gauge transformation. Thus zero is essentially an isolated point of the spectrum of p_μ^2 for the photon field. This shows that, in neutrino dynamics, the vanishing of Z_3 arises from its functional dependence on the external parameters. Thus, the main difficulty which crops up in using the $Z_3 = 0$ condition in QEDP is removed in neutrino dynamics. In view of the above conclusion, we have seriously considered condition (1). We write explicitly the renormalised constant,

$$\phi = Z_2^{-\frac{1}{2}} \phi^{(u)}$$
$$J_\mu = Z_1 \bar{\phi}_\nu \gamma_\mu \phi_\nu$$
$$\quad = Z_1 Z_2^{-1} \bar{\phi}_\nu^{(u)} \gamma_\mu \phi_\nu^{(u)} \qquad (22)$$
$$A_\mu = Z_3^{-\frac{1}{2}} A_\mu^{(u)}$$
$$g = Z_1^{-1} Z_2 Z_3^{\frac{1}{2}} g^{(u)},$$

where ϕ is the two-component neutrino wave function. The superscript (u) stands for unrenormalized. Here the two-component neutrino current $\bar{\phi}_\nu \gamma_\mu \phi_\nu$ is a conserved quantity and therefore the Ward-Takahashi identity holds for neutrino dynamics. As a result we get $Z_1 = Z_2$ and $Z_3 = 0$ for vanishing bare spinor mass[16]. Maris et al[16] have shown that, in this special case, $Z_1 = Z_2 = 0$ in all physical gauges. Evidently this result is also valid in neutrino dynamics. Following Kang and Land[17] we have found $Z_3 m^2 = \lim_{s \to \infty} sZ(s)$. Again we have $Z_1 = \lim_{s \to \infty} \Gamma(s)$. Now, for $Z_3 = 0$ and $Z_1 = 0$, we find that the asymptotic behaviour for the field theoretic quantities in the approximation of elastic unitarity $\lim_{Z_3 \to 0} \frac{yZ_1}{Z_3 \delta m^2}$ is finite and non-zero[9]. Thus, the condition of compositeness is satisfied in neutrino dynamics.

Raychaudhuri[18] suggested that the weak-interaction photon does admit of the usual gauge transformation and, apart from electromagnetic inter-action, the photon can also take part in the weak interaction with all other particles, with a coupling constant $g = 3 \times 10^{-6} e$ which is different from Bandyopadhyay's coupling constant ($10^{-10} e$). A parity-violating direct coupling to a hadron is expected to exist with a coupling constant $10^{-5} e$ to $10^{-6} e$ as a consequence of weak interactions[19]. The effect on P_γ

of such a weak electromagnetic coupling is typically included among exchange effects or off mass-shell effects and has been calculated to make a contribution of 10^{-9} to 10^{-10} to P_γ[20]. The effect of the coupling constant proposed by Raychaudhuri[18] is to change the rate of astrophysical neutrino processes by 10^{-2} or 10^{-3} order less[18]. If we take this coupling constant, then all the results calculated by Raychaudhuri during 1971-1972 provide evidence for the existence of a weak interaction of photons[22-23].

3. PARASTATISTICS AND COMPOSITENESS OF THE PHOTON

Ken-ichi Ono[24] ascribed the origin of nonconservation of parity to the asymmetry of the internal structure of particles rather than to the asymmetry of space. Raychaudhuri[25-26] suggested that the neutrino in any process may occur in the forward and backward direction. The total energy emitted for the decay process may be actually shared by both neutrinos (i.e., ν_L and ν_R) in the forward and backward direction. The asymmetry in the β-decay process appears to be due to the asymmetry of the pair of ($\bar{e}\ \bar{\nu}_L$), ($\bar{e}\ \bar{\nu}_R$) occurring in the process. Raychaudhuri[25-26] suggested that the solar neutrino flux discrepancy may be accounted for if neutrinos undergo oscillations so that the beta decay energy is shared equally between $\nu_L(\nu_\varepsilon)$ and $\nu_R(\nu_\mu)$, thus considerably reducing the flux and mean energy of the ν_ε to which the ^{37}Cl reaction of Davis experiment is sensitive.

A well known theorem of quantum mechanics shows that, if the interchange of two similar particles does not produce a new state, the particle must satisfy either Fermi statistics or Bose statistics. After the discovery of two neutrinos the situation changed. The fact that all the neutrinos are not identical, although they are dynamically indistinguishable, has shown that elementary particles are not completely characterized by mass and spin, but that some other invariant is needed. However, it is conceivable that the two particles should be dynamically similar, i.e., they correspond to the same irreducible representation of the Lorentz group, but aren't identical. If this possibility is admitted (e.g. for neutrinos), then the particle concerned may satisfy statistics more general than Fermi or Bose statistics. Green[10] has shown that the neutrinos are parafermions (they satisfy parastatistics of order 2). This not only accounts for the existence of distinct particles but also provides a suitable invariant by means of which the particle may be distinguished. Green[10] showed that if neutrinos are parafermions, photons can be composite pairs of neutrinos and antineutrinos. According to Green, there must be two different kinds of neutrinos, one associated with left circular polarization and another with right circular polarization of the photons. In this formalism, they would have to transform into one another under inversion and therefore correspond to the same representation of the Lorentz group. Again, since the two neutrinos are associated with the electron and muon respectively, the application of parastatistics to the neutrinos has consequences for the leptons in general. According to Green[10], a natural extension allows us to infer that the electron and muon are also parafermions. They are represented by the same field variable and Dirac's equation can be formulated in a way which allows for the prediction of light and heavy particles.

Thus the neutrino theory of light can be reformulated and there is no difficulty in thinking of the photon as a composite state of a neutrino and antineutrino pair.

4. POSSIBLE STRUCTURE OF THE PHOTON

A non-zero mass for photons destroys the gauge invariance of electromagnetic theory. This gives rise to a longitudinal photon besides the transverse one. According to Bass and Schrödinger[27] the interaction of longitudinal photons with matter is so feeble as to make them irrelevant for the attainment of thermal equilibrium in a cavity. Its wall would be essentially transparent to them. The reason for this is the tremendous Lorentz contraction of longitudinal fields when going from the rest frame of a photon mass to the cavity rest frame. Hence, one cannot eliminate the photon mass only on the basis of the absence of the $\frac{3}{2}$ factor in the Stefan-Boltzmann law. Again, according to Schwinger[28], gauge invariance has nothing to do with the vanishing mass of the photon. We also suggest that γ_5 invariance does not necessarily imply that the mass of the neutrino is strictly zero. There is some evidence that the neutrino mass cannot be zero[29]. At absolute zero temperature only the combinations of $\nu_\mu + \bar\nu_e(\nu_R + \nu_L)$ and $\nu_e + \bar\nu_\mu(\nu_L + \bar\nu_R)$ are stable. But, at slightly higher temperatures than zero, the left-handed photon is a combination of $\nu_L + \bar\nu_L$ and $\nu_L + \bar\nu_R$ (or $\nu_R + \nu_L$). Similarly, the right-handed photon is a combination of $\bar\nu_R + \nu_R$ and $\nu_R + \bar\nu_L$(or $\nu_L + \bar\nu_R$). These are stable. For conservation of parity the above combination for the structure of the photon is proposed. Thus, the system of the photon is a combination of a right-handed photon and a left-handed photon. In this formalism a photon is oscillating between left-handed photon and right-handed photons. The structure $\nu_L + \bar\nu_R$ (or $\nu_R + \bar\nu_L$) will be between $\nu_L + \bar\nu_L$ and $\nu_R + \bar\nu_R$ in this photon system. That is, the photon system may be both circularly polarized and plane polarized simultaneously. Therefore, the photons in a beam may be arranged in a geometrical pattern. An interesting point is that the photon is an oscillating particle which moves on the average along the lines of a continuous wave vector field A_μ.

We suggest that the number of longitudinal photons ($\nu_L + \bar\nu_R$, $\nu_R + \bar\nu_L$) in the photon system will depend on the number density of photons. As the number density increases, the proportion of longitudinal photons decreases and the average photon separation decreases. This is due to the clustering effect of the mutual interactions of photons, (i.e., neutrinos and antineutrinos) so as to form a type of lattice structure.

When the photon system is placed in an atom, the $\nu_R + \bar\nu_L$ and $\nu_L + \bar\nu_R$ portion of the photon changes to $\nu_L + \bar\nu_L$ and $\nu_R + \bar\nu_R$. When the spin direction of the neutrino is changed (i.e., $\nu_L \to \nu_R$ and $\bar\nu_L \to \bar\nu_R$ in $\nu_R + \bar\nu_L$ and $\nu_L + \bar\nu_R$) the energy will be increased more than that of the original structure; they must therefore come out from the system to ionize the atom. In our structure the longitudinal photon ($\nu_L + \bar\nu_R$ and $\nu_R + \bar\nu_L$ pair) will be in the middle of the system and, therefore, the energy will be increased in the middle of the pulse. The photon energy varies with the decrease of the number of longitudinal photons. When a photon whose energy has increased relative to the energy of the surrounding photons will get out and ionize an atom, the reshuffling of the system of photons induces the exit of $\nu_R + \bar{\nu}_R$ and $\nu_L + \bar{\nu}_L$ pairs from the system of photons. Energy gain by the photon (in order to conserve energy) can only be obtained at the expenses of energy loss from surrounding photons; this effect is similar to Panarella's[30] effective photon hypothesis but unlike the quantum potential hypothesis[31]. During the superposition of wave packets a wide range of photon frequencies will appear due to the reshuffling of the system of photons.

5. THE ANOMALOUS PHOTOELECTRIC EFFECT

Panarella[30-32] has pointed out that ionization of gases induced by laser radiation has been observed where it should not because the energy of the incident light quanta is supposed to be a factor of 5 or more below the ionization potential of the gases investigated. He pointed out that the classical theories, e.g. multiphoton and cascade theories, cannot explain the experimental results. He therefore advanced the effective photon hypothesis which states that the expression $E = h\nu$ should be modified into the form,

$$E = \frac{h\nu}{[1-\beta_\nu f(I)]} = h\nu \cdot \exp\left[\beta_\nu f(I)\right] , \qquad (23)$$

where E is the photon energy, h is the Planck constant, ν is the light frequency, I is the light intensity, and β_ν is a coefficient whose product with $f(I)$ significantly differs from zero only in the case of high intensity laser light. The above expression implies that the photon energy may reach values above the ionization energy of the gas atoms.

We wish now to derive how the photon energy enhancement occurs in our photon structure. A particle in a many-particle system constantly affects the potential it feels from the other particles. If the photon–neutrino weak coupling exists, then a neutrino and an antineutrino in the photon sea can be considered to be bound by the force corresponding to the exchange of a photon. It may be assumed that the potential corresponding to the binding force between a neutrino and an antineutrino helps to keep the neutrinos and antineutrinos bound quasielastically to their equilibrium positions, just as the electromagnetic interaction can be considered to keep the oppositely charged particles in a dielectric bound quasielastically to their rest positions. Eventually, any external electromagnetic field, by its weak interaction with the neutrinos, will displace them from their equilibrium position. This will give rise to a polarization effect, similar to the polarization of a dielectric. In a photon, the neutrino and the antineutrino constitute a single dynamical system. Any interaction on a single neutrino is an interaction on the system which affects the state of the whole system. Therefore, we can take the effective energy of a photon in a beam of photons (i.e., the system of neutrino and antineutrino pair) as

$$E = \frac{h\nu}{\left[1 - \frac{4\pi N g^2 c^2}{m(\nu^2 - \bar{\nu}_o^2)}\right]} , \qquad (24)$$

where m is the mass of the neutrino, $\bar{\nu}_o$ is the characteristic oscillation

278

frequency of the neutrinos in the sea, g is the coupling constant $3 \times 10^{-6} e$, and N is the number of neutrinos per unit volume. If $\bar{\nu}_0 \ll \nu$ for any natural photon, we get

$$E = \frac{h\nu}{[1 - \frac{4\pi N g^2 c^2}{m\nu^2}]} . \qquad (25)$$

This relation actually corresponds to the energy of a photon propagating in a sea of (free) neutrinos and antineutrinos (i.e., in a sea of photons). Thus the interaction force between neutrino and antineutrino (i.e., between photon and photon) is responsible for the energy variation with the neutrino density (photon density). The enhancement of photon energy depends on the photon mass, number density and composite coupling constant. Taking $g = 3 \times 10^{-6} e$ $m = 10^{-4} eV$, $\nu = 10^{15}/sec$ and $N = 2.4 \times 10^{21}$, we get $E = 10 \, h\nu$. In this way it is possible to explain the observed photoelectric effect when extremely high intensity laser beams are focussed in a gas[30]. Here we have taken N as the neutrino density which is almost equal to half the photon density due to the compositeness of the photon. We will get the enhancement of photon energy when $2.64 \times 10^6 < N/m\nu^2 < 2.6455 \times 10^8$ is satisfied. If $\nu_0 = \nu$ we must take into account the damping term in equation (25).

6. WAVE-PARTICLE DUALITY

It has been suggested by Panarella[33] that the single particle concept of the photon should be modified in order to explain some results obtained by him in diffraction experiments. In our photon model, the $\nu_L + \bar{\nu}_R$ and $\nu_R + \bar{\nu}_L$ pair is in middle of the system of photons. We can expect therefore that diffraction and interference phenomena are not single particle phenomena as the photon is made from the combination of $\nu_L + \bar{\nu}_L$, $\nu_R + \bar{\nu}_L$, $\nu_L + \bar{\nu}_R$ and $\nu_R + \bar{\nu}_R$. Photons are clumped together due to the mutual attractive forces, i.e., due to the attraction of neutrinos. Photons are arranged in a geometrical shape. In our model, a right-handed photon and a left-handed photon constitute a single dynamical system. Any interaction on a single photon is an interaction on the system and affects the whole system. Before entering an interferometer, the system is split in two parts: one part is $\nu_L + \bar{\nu}_L$ and $\nu_R + \bar{\nu}_L$ ($\nu_L + \bar{\nu}_R$), the other is $\nu_R + \bar{\nu}_R$ and $\nu_L + \bar{\nu}_R$ ($\nu_R + \bar{\nu}_L$), which recombine out of the interferometer and produce the observed diffraction phenomena. If they come from the same source we will find the correlation observed by Hanbury-Brown and Twiss[34]. It is observed that the probability for finding two photons right on top of each other is twice the value for finding two photons at a large separation. Interference is already there within the system of photons. If the photons come from the two different sources and enter an interferometer, they will still interfere only with themselves. Since the structure of the system of photons depends on the number density of photons, the diffraction pattern also depend on e n mber density of photons. When the photon density is high, interference occurs between $\nu_L + \bar{\nu}_L$ and $\nu_R + \bar{\nu}_R$ as there are very few $\nu_R + \bar{\nu}_L$ and $\nu_L + \bar{\nu}_R$ pairs in the system. But when the number density is low, interference will be small due to the reasonable proportion of $\nu_R + \bar{\nu}_L$ and $\nu_L + \bar{\nu}_R$ pair in the photon system. We believe that, when the number density is high, the $\nu_L + \bar{\nu}_L$ and $\nu_R + \bar{\nu}_R$ grow in proportion to light intensity. If we gradually reduce the intensity of light, then, in the photon structure, the $\nu_L + \bar{\nu}_R$ and $\nu_R + \bar{\nu}_L$ pair gradually increases; thus the photon clump will expand and the photon flux per unit area will decrease. In this case each clump will be well separated but still its shape will be geometrical. By this means we can explain Panarella's[33] observation that, at low intensity (low density), the photon flux per unit area decreases faster than at high intensity (high density). In conclusion, we suggest that the diffraction and interference phenomena can be described by the wave nature of clumped photons.

7. REPLICATION OF PHOTONS

It is well known that for certain nonseparably correlated EPR pairs of photons, once an observer has made a polarization measurement on one member of the pair, the other one, which may be far away, can be for all purposes of prediction regarded as having the same polarization. If this second photon could be replicated, and its precise polarization measured as above, it would be possible to ascertain whether the first photon had been subjected to a measurement of linear or circular polarization. In this way the first observer would be able to transmit information faster than light by encoding his message into his choice of measurement. If photons can be cloned, a plausible argument can be made for the possibility of faster than light communications.

Three possible ways to replicate photons are:

a) Conservative replication: The original photon (mixture of right-handed and left-handed photons) remains intact and a new photon is formed from an amplifier. With this mechanism it is not possible to produce several copies of a photon system, each having the same state as the original.

b) Semiconservative replication: When the original photon interacts with the amplifier, we will get two photons at the first stage which contain one new neutrino (antineutrino) and one original antineutrino (neutrino), as the photon is a composite state, a neutrino-antineutrino pair. Ultimately, we will get only two photons (which contain one new neutrino (antineutrino) and one original antineutrino (neutrino)) with many photons. In this mechanism, too, there is no possibility of producing several copies of a photon system, each having the same state as the original one.

c) Dispersion replication: This mechanism involves the distribution of the original neutrino-antineutrino pair ($\nu_L + \bar{\nu}_L$, $\nu_R + \bar{\nu}_R$, $\nu_L + \bar{\nu}_R$, $\nu_R + \bar{\nu}_L$) among the four (neutrino-antineutrino) chains of two new photons. In our model one photon contains two chains. If this type of replication is observed, then it is possible that the amplifier can produce photon systems with arbitrary polarizations. But the linear character of the quantum mechanical evolution does not allow arbitrary states of polarization[35]. In all these mechanisms, angular momentum should be conserved.

Thus it is not possible to produce several copies of a photon, each having the same state as the original. Perfect and certain replication of any single photon is impossible and information faster then light is not possible.

8. QUANTUM MECHANICS AS A STOCHASTIC PROCESS

From the very beginning of quantum mechanics there were attempts to interpret it as a stochastic process. The stochastic quantum mechanics is able to reproduce all the predictions of nonrelativistic quantum mechanics, but it has the following feature: the diffusion drift of the random motion of a particle in the medium is not preassigned, but depends on the wave function which characterizes an ensemble as a whole. Therefore, the procedure of preparation of the quantum ensemble strongly influences the properties of the underlying medium.

Guth[36] has derived the classical uncertainty relations by prescribing uncertainty for long times and certainty for short times. The complementary quantities will then be the co-ordinate and an average momentum. The physically simplest and clearest example for these new uncertainty relations comes from the theory of Brownian motion, particularly Rayleigh's problem (one dimensional). Guth[36] considered the slowing down of a large particle of a given velocity by multiple collisions with a gas of small molecules. For short times before any collision occurs, there is no statistical description and therefore we have certainty. For long times, the multiple collisions introduce statistics and we have uncertainty. Here we will examine the hypothesis that the particles in a neutrino (degenerate) sea are subjected to Brownian motion. The motions of the Brownian particles are maintained by fluctuations in the collisions with neutrinos of the neutrino sea. Here we are considering the slowing down of a particle (say an electron) of a given velocity by multiple collisions with a sea of neutrinos. The slowing down of a particle in a given medium can be described by Kramer's equation[36,37],

$$\frac{\partial f}{\partial t} + \frac{p}{m} \frac{\partial f}{\partial q} - \frac{1}{\tau} \frac{\partial}{\partial p} (pf) = 0 , \qquad (26)$$

where p and m are the momentum and mass of the particle, $\tau = m/\mu$, and μ is the frictional constant. The third term in equation (26) is due to frictional forces. According to Stokes' law,

$$\frac{m}{\mu} = \frac{1}{\tau} = \frac{m}{6\pi r \eta} , \qquad (27)$$

where r is the radius of the particle and η is the coefficient of viscosity of the surrounding medium. Taking the first two moments with respect to p of equation (26), we obtain a system of two first order differential equations:

$$\frac{\partial n}{\partial t} + \frac{\partial j}{\partial q} = 0 \qquad (28)$$

$$\tau \frac{\partial j}{\partial t} + D \frac{\partial n}{\partial q} + j = 0 \qquad (29)$$

where

$$n(q,t) = \int f \, dp, \quad j(q,t) = \int \frac{pf \, dp}{m}, \quad D = \tau/m^2 \int \frac{p^2 f}{n} \, dp ,$$

and the approximate assumption is made that D is independent of q. Both n and j satisfy the telegrapher's equation in the form

$$\left[\tau \frac{\partial^2}{\partial t^2} + \frac{\partial}{\partial t} - D \frac{\partial^2}{\partial x^2}\right] \begin{bmatrix} n \\ j \end{bmatrix} = 0 \ . \tag{30}$$

Guth[36] pointed out that the telegrapher's equation (30) with $\tau = \frac{i\hbar}{mc^2}$ satisfies the Klein-Gordon (also Dirac) wave equation for a free particle if the factor $\exp\left[-(i/\hbar \ mc^2 t\right]$ is split from it. But, in this way, the well-known formal transition from the classical diffusion equation to the Schrödinger equation, obtained by introducing an imaginary diffusion coefficient $D = i\hbar/2m$, persists. The time, according to relativity, can be expressed by an imaginary relaxation time τ, and all this is true also for the three-dimensional case. It was suggested that there exists a sea of degenerate neutrinos and antineutrinos in the universe which can be held together in an equilibrium position by the binding force generated by the weak interaction of photons. We consider that the medium is in the form of a neutrino sea. Although the neutrino-antineutrino interaction is very small, we can think of some frictional effect in the neutrino sea. This means that the Brownian motion of a particle cannot be very smooth and velocities will not exist. We know that the 3°K cosmic microwave background corresponds to an energy of 3×10^{-4}eV and since we have taken the photon as a composite state of a neutrino-antineutrino pair, we can assume that the temperature of the neutrino is also 3°K. With this consideration in mind we shall calculate the value of τ.

For the case of a weak interaction of photons with coupling constant $g(=10^{-6}e)$ the cross section for the $\bar{\nu}\nu \to \nu\bar{\nu}$ interaction can be written as

$$\sigma_{\nu\nu} = \frac{10^{-22}}{E^2} g^4 \ cm^2 \ , \tag{31}$$

where E is in Mev. The coefficient of viscosity is defined as[37]

$$\eta_{\nu\nu} = \frac{1}{3} \rho_\nu \ t_{\nu\nu} = \frac{1}{3} \rho_\nu \ \frac{4\pi}{(E/c\hbar)} \cdot \frac{1}{g^4} \ . \tag{32}$$

If we consider that the universe is finite and its radius R is 10^{28} cm, then the mean free path $ct_{\nu\nu}$ of a neutrino should be within the radius of the universe to get the effect of the neutrino sea. Since

$$ct_{\nu\nu} \simeq \frac{1}{\sigma_{\nu\nu} \ n_\nu} \simeq \frac{4\pi}{(E/c\hbar)} \frac{1}{g^4} \simeq 10^{28} \ cm \tag{33}$$

we know that

$$\rho_{\nu\nu} \simeq 3 \times 10^{-21} \ E^4 \ g/cm^3 \tag{34}$$

where E is in eV. Therefore:

$$\eta_{\nu\nu} \simeq \frac{3 \times 10^{-25}}{g^4} \ . \tag{35}$$

Thus $\tau \simeq 10^7 \ g^4 \ sec = 10^{-21} \ sec$ for $r = 2\times10^{-11}$ cm, which is almost the Compton radius of the electron. If the motion of the electron is described by a Markov process in co-ordinate space as in the Einstein-Smoluchowski theory[39], then, in order to calculate τ we will consider the second restriction on the validity of Stokes' law which is

282

$$\tau = \frac{\alpha \ell m}{6\pi\eta r^2} \tag{36}$$

where $\alpha > 1$. We take $\alpha = 1.21$ and $\ell = 3\times10^{-11}$ cm $= \tau c$. For $r = 2.4\times10^{-11}$ cm, $\tau = 10^{-21}$ sec. The result $\tau \simeq 10^{-21}$ and $r = 2\times10^{-11}$ cm by both of these methods suggests that the neutrino sea connected with the weak interaction of photons with coupling constant $10^{-6}e$ may be the sub-quantum medium of Bohm, Vigier and de Broglie in the universe.

9. DISCUSSION

Photons usually satisfy Bose statistics and the compositeness of the photon can be thought of as a fermion pair. Therefore, it is natural to inquire about the relationship existing between these two descriptions. Obviously the fermion description must be deeper and also more complete because the system is a fermion pair. The boson description, however, is not and cannot be a complete description. The boson description can only apply to the behaviour of the system as a whole; it cannot give any information about the particle inside. In other words, in the boson description, the internal structure must be frozen. There must be a fundamental interaction between a neutrino and an antineutrino to form a composite photon with coupling constant g. In the photon structure which we have proposed, it is possible that a wave packet of a photon (or of photons) can be divided into two parts and can be separated in a macroscopic distance.

The influence of the neutrino sea on the radiative corrections in quantum electrodynamics may be calculated. This was studied by Aurela[40]. Such corrections should exhibit space and time variations due to variations in the neutrino sea. We assume that the distribution of neutrino matter in the universe follows the same pattern as the distribution of matter. Therefore, in the region near massive bodies, the density of neutrinos and antineutrinos will be greater than in remote regions. We also suggest that a neutrino sea (finite medium) exists which may be capable of transmitting electromagnetic radiation[41].

It should be mentioned here that, if we consider the neutrino sea as a fluid, then it is possible to show that the quantum potential $- \frac{\hbar^2}{2m}\frac{\Delta\alpha}{\alpha}$ may be conceived as originating from internal stresses in the fluid, even though this stress depends on the derivatives of the fluid density[42]. It may be possible to explain the enhancement of photon energy due to the above quantum potential which arises from the internal stress in the fluid. Thus, when photons that emerge from an atomic cascade with well-defined intensities interact with optical devices (such as a linear polarizer or a quarter wave plate), they undergo stochastic changes of intensity in addition to the change in polarization.

The BCS theory of superconductivity can be applied to the cosmic neutrino sea[43] in the weak interaction theory of photons and may have interesting implications.

It may be mentioned here that Pecker et al[44] claimed that some redshifts can be explained by considering a new type of process in which the incident photon scatters inelastically on 3°K radiation to give four less energetic photons. But Adrovandi et al[45] showed that such a process does not lead to strong peaking in the forward direction which is necessary in order to obtain a point-like image of the source. The shift of the spectral line corresponds to the decrease of energy of a quantum, i.e. a loss of part of the energy of a quantum on its way from remote objects to the terrestrial observer. On its way to the observer the quantum interacts with the 3°K photon and gives up a portion of its energy to it. Since the photon is taken as a composite of a neutrino and antineutrino pair, we suggest that the photon is scattered on a neutrino (i.e., 3°K radiation) by

$$\gamma + \nu \rightarrow \gamma + \nu \,, \tag{37}$$

and loses part of its energy on its way to the observer, where:

$$\frac{d\sigma}{d\Omega} = \frac{g^4}{(4\pi)^2} \frac{1}{E \ E_F} \cdot \frac{1}{(1-\cos\theta)} \,, \tag{38}$$

in which E is the incident energy of the photon and E_F is the energy of the neutrinos. The process $\gamma + \nu \rightarrow \gamma + \nu$ may explain the observed spectral shift as there will be strong peaking (see equation (38)) in the forward direction, which is necessary in order to get a point-like image of the source.

REFERENCES

1. L. de Broglie, Matter and Light, George Allen and Union Ltd., London (1939).
2. P. Jordan, Z. Physik 93, 464 (1935).
3. R.d.L. Kronig, Physica 3, 1120 (1936).
4. M.H.L. Pryce, Proc. Roy. Soc. (London) 165, 247 (1938).
5. I.M. Barbour, A. Bietti and B.F. Touschek, Nuovo Cimento 28, 453 (1963).
6. B. Ferretti, Nuovo Cimento 33, 265 (1964).
7. W.A. Perkins, Phys. Rev. B137, 1291 (1965).
8. W.A. Perkins, Phys. Rev. D5, 1375 (1972).
9. P. Bandyopadhyay and P. Raychaudhuri, Phys. Rev. 3, 1378 (1971).
10. H.S. Green, Progr. Theor. Phys. 47, 1400 (1972).
11. M.M. Broido, Phys. Rev. 157, 1444 (1967).
12. P. Bandyopadhyay, Phys. Rev. 173, 1481 (1968).
13. C. Ryan and S. Okubo, Suppl. Nuovo Cimento 2, 234 (1964).
14. L.A. Radicati and B. Touschek, Nuovo Cimento 5, 1693 (1957).
15. W. Pauli, Nuovo Cimento 6, 204 (1957).
16. Th. A.J. Maris, D. Dillenburg and S. Jacob, Nucl. Phys. B18, 366 (1980).
17. K. Kang and D.J. Land, Nuovo Cimento 63A, 1053 (1969).
18. P. Raychaudhuri, unpublished (1975).
19. E.M. Henley, A.H. Huffman and D.U.L. Yu, Phys. Rev. D7, 943 (1973).
20. E.M. Henley, Phys. Rev. C7, 1344 (1973).
21. P. Raychaudhuri, Canad. J. Phys. 48, 935 (1970).
22. R.B. Stothers, Phys. Lett. 36A, 5 (1971).
23. B. Kuchowicz. Rep. Progr. Phys. 39, 291 (1976).
24. Kenichi Ono, Scientific papers of the College of General Education, University of Tokyo 13, No. 1 (1963).
25. P. Raychaudhuri, Spec. Sci. and Tech. 5, 167 (1982).
26. P. Raychaudhuri, Proc. 18th ICRC 7, 100 (1983). Spec. Sci. and Tech. 4, 267 (1981).
27. L. Bass and E. Schrödinger, Proc. Roy. Soc. (London) A232, 1 (1955).
28. J. Schwinger, Phys. Rev. 125, 397 (1962).
29. V.A. Lubimov, E.G. Novikov, V.Z. Nozik, E.F. Tretyakov and V.S. Kosik, Phys. Lett. 94B, 266 (1980).

LOGICAL MEANINGS IN QUANTUM MECHANICS FOR AXIOMS AND FOR IMAGINARY AND TRANSFINITE NUMBERS AND EXPONENTIALS

William M. Honig

Western Australian Institute of Technology
Perth, Bentley, 6102
Western Australia

ABSTRACT

It is shown how the logical status of axioms and of the imaginary number i are isomorphic. Logical meanings for various number forms are discussed. Exponentials of 2 and of e can be shown to refer to discontinuous and continuous designations for their exponents, respectively. The equal symbol in equations can be mapped to transfinite ordinals, so that exponentials of such transfinite quantities labelled with i, can refer to the set of all sets of physical situations to which an axiom, in the form of an equation, applies. This may be useful in interpreting and unifying Psi and Electromagnetic wave functions.

I. INTRODUCTION

The purpose of this note is to amplify previous remarks on this subject.[1,2,3]

Section II is a discussion of the logical status of axioms and of i (the square root of minus one). In a logical sense they both have the same meaning: undecidable; but via the recent logical methods of Cohen[4] they are "forced" to function as terms which have the status: true. Thus the mathematical definition (the logical status) of a scientific axiom is that it is an undecidable statement which is forced (and not the canonical: assumed) to be true. The isomorphism between the logical meaning of i and the logical status of axioms permits that i be used to label number symbols which designate axioms.

The set theoretic meaning for exponentials of base 2, which comes from the Power Set Theorem of Axiomatic Set Theory[5,6] can be extended to exponentials of base e. In Section III, a discussion is given which tries to show that 2 and e in the forms 2^a and e^b can be given a meaning which denotes that a and b apply to discontinuous and continuous quantities (or sets), respectively. It then follows that these exponential forms refer to the set of all sets (discontinuous or continuous, respectively) represented by the number symbols a and b.

Section IV suggests a meaning and use for transfinite ordinals as number symbols designating the equal symbol in equations. If such an equation as for instance, E = hf, is an axiom, then a transfinite ordinal labelled with the coefficient i can stand for the complete set of (E, f) number pairs of the axiom equation. This results in the conclusion that such a transfinite number can stand for the complete set of physical values that the original axiom (E = hf) can give and thus it can stand for the axiom directly.

Section V then suggests that the imaginary exponential form $e^{i\theta}$ stands for the set of all sets of physical values (situations) which the axiom θ can refer to. This section concludes with a procedural and epistomological discussion of the Euler form:

$$e^{i\theta} = \cos \theta + i \sin \theta.$$

This procedure results in a logical interpretation for the combination of the forms $e^{i\omega t}$ and $e^{iEt/\hbar}$.

Much of the above profits from the insight offered by the Non-Standard numbers[7,8] and from Frege[9] who linked the logical meaning of a number (zero) to the concept of paradox.

II. THE LOGICAL STATUS OF SCIENTIFIC AXIOMS AND A LOGICAL MEANING FOR i

An axiom of a theory cannot be true because it cannot be deduced from any more basic statement which we already know is true. Neither can it be false because it cannot be shown to be undeducible from any such more basic statement. This falsity, in addition, cannot exist because no one would work on the physical predictions of a theory which is established from a false axiom since it would yield results which conflict with physical reality. Thus an axiom is not true and is not false. If we apply the Law of the Excluded Middle (modus ponens) to both parts of the previous statement we get that an axiom is both false and true. Both of the two previous statements provide a very clear definition for undecidability and this quality is one component of an axiom. Furthermore, an axiom has the additional crucial quality, which is, that we, the people who consider and use theories, assume, at least tentatively or even more strongly, that the axiom is true. We arrive thus at a definition for an axiom: it is a statement (mathematical or otherwise) which in a formal sense is undecidable, and which is simultaneously assumed to be true.

The most striking feature of axioms; that they are assumed to be true, appears to be a tentative and social act. It is tentative because, a la Popper, all those who deal with an axiom and its deductions will accept the truth of the axiom as long as the physical results predicted by the theory based on the axiom appears to be in congruence with measurements. It is social because it is a necessity for the dissemination of a theory that its axioms be clearly communicated to anyone wishing to use the theory. This emphasizes that axiom statements are arbitrary and not deducible from anything else.

Logical (both Boolean and non-Boolean) expressions may be treated in quite an algebraic manner. This is built on the work of Stone and many others, see Kiss, Sikorski and their references[10,11] Kiss has shown that such treatments can be extended to undecidable statements.

Let an axiom (statement or equation) be symbolized by β and its negation or the assertion of its falsity by $-\beta$. We let the simultaneous existence of both these statements be represented by logical multiplication, resulting in the expression, which is algebraic:

$$\beta \times (-\beta) \qquad (1)$$

which is, therefore, a mathematical expression for undecidability, (the parentheses delineate the second symbol).

An axiom, per the previous discussion, must not only be undecidable in the above formal sense, but it should also be possible to construct a logical algebraic form which will also include its most striking feature: that it is assumed to be true. This means that the undecidability of the axiom as given in the expression above should be assumed or rather "forced" to be true. Forcing the above expression to be true in the P.J. Cohen[4] sense can be effected by forcing (equating) it to the truth value: true, which can be taken as 1 (one) using the principles of Algebraic Logic and Boolean Algebra as discussed by Kiss[10] and Sikorski[11], thus:

$$\beta \times (-\beta) \equiv 1 \qquad (2)$$

The triple line symbol standing usually for definition is used here to mean forcing. Subsequent algebraic operations with Eq. (2) will be carried out using the = symbol. Solving algebraically for β :

$$-\beta^2 = 1 \qquad (3)$$

$$\text{or} \qquad \beta^2 = -1 \qquad (4)$$

$$\text{or} \qquad \beta = +i, -i \qquad (5)$$

$$\text{or} \quad \beta = i \quad \text{and also} \quad -\beta = i \qquad (6)$$

Statements, equations, and symbols may now be labelled with i (as a coefficient) in order to identify them as axioms or as possessing the qualities of being undecidable and simultaneously true. Some additional remarks are made here.

1. The logical multiplication, expression Eq.(1) above, which is used as a definition for undecidability is not a Boolean logical expression. In the Boolean case one always knows that logical multiplication (or set intersection)cannot exist for two items which are defined via modus ponens and are thus the negation of each other. Undecidability is an explicit tentative acceptance of ignorance. It is our human condition that permits us to see that although ignorance defined in this way is a non-Boolean concept which violates modus ponens it confers a useful and clear meaning upon $\beta \times (-\beta)$: the human conception, undecidability.

2. The Eqs.(6) above show that the i formalism treats equally well an axiom and its negation and shows that the negation of an axiom is also an axiom. This feature has been shown to be useful in Quantum Mechanics and in the Special Theory of Relativity.[12]

3. A comment is made here on the meaning of zero with reference to axioms. The meaning of zero has usually been taken as denoting any and all nullities (null elements). For example, if the set of six apples has those six apples removed from the set, then that set still contains the null element, or zero apples. A similar set of oranges will have the null element also, or zero oranges. Of course, one may take (a la Frege) the number zero as consisting of the set of all null elements. For these two cases the single symbol 0 (zero) is consistently applied because apples and oranges enjoy the same logical status. They are objects of our contemplation which can be defined in terms of other qualities or relationships but they do not have the

same logical status as axioms. This is the reason for our belief that objects of our contemplation which have a different logical status from each other should have their null elements labelled differently. Thus 0 and i0 should be the symbols for deductive and axiomatic null elements, respectively. This is useful in what follows and in some previous work.[12]

III. A SET THEORETIC MEANING FOR THE FORMS, 2^a AND e^b

The Power Set Axiom states that the set of all subsets, [PS], of a set [A] with discrete members (including the null element) which may be finite or enumerably infinite and which has the cardinality expressed by a, is the set with the cardinality 2^a. Thus, e.g., a set with 3 non-null members will have its [PS] (set of all subsets) consit of 2^3 members[5,6].

It is well known the [PS] can be concieved of as the insertion set [P|A], where [p] is the two member set [0,1] and where [A] is the discrete set with the cardinality a, so that:

$$[PS] = [p|A] = 2^a. \qquad (7)$$

The set [p|A] is the set of all insertion of [p] into [A] and can be concieved for an arbitrary subset [A_0] of [A] as the insertion into [A] of the set [1,0] with 1 applying to all the members of [A_0] and 0 applying to the rest of the set [A]. The word "applying" is taken as a multiplicative operation between one or zero on the other hand and a particular set member [A_i] on the other hand. Thus 1 preserves the existence of an [A_i] and 0 denies that existence. The set of all such insertions of [p] into all the subsets of [A] results in the [PS] of [A] with the cardinality 2^a. Directing ones attention to [p]; this set with members 1 and 0 can be concieved of as a definition for the quality: discreteness. Paring the 1, 0 set members of [p] with the qualities existence and non-existence, respectively, does not permit any representation for continuous variation.

Thus 1 corresponds to the presence of a member of the set [A] and 0 to the absence of a set member, i.e., in this way every set member of [A] can be displayed or not depending on the 0,1 assignment. It is important to note that every member of [A] can be quite explicitly picked out in this way. If, however, the set [A] were continuous then this procedure using [p] = [0,1] would be useless, which again emphasizes how the set [p] defines discreteness.

Associating the number 1 or the number 0 with each element of a set [A] is really equivalent to defining a function, f, where

$f(a_i) = 1$ if a_i is to be a subset of [A]
and $f(a_i) = 0$ if a_i is not to be a subset of [A].

Such functions, called characteristic functions (or ch. f.) are closely connected with the operations of Boolean Algebra. The ch. f. for the example given above consists of the set [1,0]. Two useful theorums for such ch. f sets are described in logical terms:

a) The ch. f. set corresponding to the intersection of two or more general sets are the (Cartesian) product set of the ch. f. sets of each general set.

b) The ch. f. set corresponding to the union of two disjoint general sets is the sum of the ch. f. sets of each general set.

Intersection and union of sets corresponds to logical multiplication and logical addition. Thus for 2 sets [A] and [B],

288

logical multiplication corresponds to those elements of [A] and [B] that lie in both [A] and [B]; whereas logical union corresponds to elements of [A] and [B] which lie in [A] or in [B] or in both.

The definition for discreteness which has been given, together with the Power Set axiom lead to a conclusion about the set theoretic meaning for 2 in the cardinal 2^a of the set of all sets of [A]. The meaning for the cardinal 2 must here connote that it is the cardinality of the set [1,0] and that the set [A] is discrete. Thus the meaning for 2 (a la Frege) stands for the set of all sets which are discrete. The quality discreteness must be clear, operative, and especially unique for both logical multiplication and logical addition, as presented in the previous paragraph, because both of these logical operations should apply in normal procedures associated with theories and experiments. The use of the theorums a) and b) above can be used to construct a set of progressions to a cardinality number for all sets which represent discreteness. Since these progressions for sums and for sets should give the same cardinality number (2) we see that this occurs for discreteness ("all" is approached when n approaches infinity).

Thus: $\qquad$ $[1,0] \to [1,0]^m \to [1,0] \to {}_2 1 \to 2$

or: [Single Discreteness Set] → [Cartesian Product of n Discreteness Sets]
$\qquad \to {}_2 1 \to 2$,

and for the logical addition case:

$$(1 + 0) \to (1 + 0)^m \to (1 + 0) \to [1,0] \to {}_2 1 \to 2$$

or: [Discreteness Sum] → [Product of n Discreteness Sums] $\to {}_2 1 \to 2$, where n approaches infinity in both cases.

These detailed explanations appear to be trivial with respect to the result (that 2 represents the cardinality of a number that means discreteness for exponents of this number).

In view, however, of the analysis and development of non-standard numbers and literal infinitesmals[7] it is tempting to view continuous sets in a similar manner to the discussion just given for discrete sets. Instead of the discreteness difinition set [1,0] we take the set [1,δ] where δ is a literal infinitesmal with a non-standard meaning and which is defines as:

$$\delta = \left| \text{im } 1/n \text{ AS } n \to \infty \right. \qquad (8)$$

so that: $[1,{}^\delta]$ $\qquad$ is [1,1/m] for $m \to \infty$

and this set has no null element, by definition as per Robinson's methods[7]. This set [1,δ] is thus meant to represent the qualities associated with continuousness. Now 1 and δ are, for example, a point (a real number) on the continuous interval between 0 and 1; and an infinitesmal increment beyond that point respectively. In an analogous manner to that given above for the discreteness sets, one gets now for $n \to \infty$:

$$(1 + 1/n) \to (1 + 1/n)^m \to [e] \to e$$

$$[1,1/n] \to [1, 1/n]^m \to [e] \to e$$

where e is a non-standard set with no null element.

The progression from $[1, 1/n]^n$ to e must be the same result as that for the addition expressions above it for the reasons given previously, and is also thus consistent with the similar expressions for the dicreteness case. The conclusion, therefore, is that e is number which has a non-standard set theoretic meaning corresponding to continuous (or continuity) for arguments in its exponent.

These comments identify the exponential forms of 2 and of e, say 2^a and e^b, each of which is a set of all sets whose exponents a and b refer to discrete and continuous variables, respectively. They can also, with this meaning represent the set of all physical situations to which an exponent number symbol which stands for an axiom can apply.

IV. A MEANING AND USE FOR TRANSFINITE ORDINALS

Generally the laws of physics are written as equations in terms of symbols describing physical reality. Thus symbols are given to concepts like force, pressure, velocity, energy, and the like. Their numerical magnitudes are coupled with dimensions which define their meaning. These numerical magnitudes lie in the real number field and a normalization can be set up, of course, which maps this number range to the 0,1 interval, which is necessary in what follows.

The human procedure for ascertaining some details of physical reality consists of the use of both a theoretical framework of statements and an experimental setup to find, say, the magnitude of a physical varaible. Suppose one is ignorant of the magnitude of a physical varaible. Upon the manipulation of theory and experiment a value for the magnitude of the physical variable is found; this with a specified accuracy. Such a procedure, however, may be incomplete because the person doing it may be ignorant of significant information affecting the outcome of the experiment. Even so, the performance of the procedure a large number of times can be used to determine the magnitude of an average value of a physical variable. Thus where a particular outcome occurs with a fraction f of the total number of repetitions, that fraction is defined, as by von Mises and Reichenbach, as the probability of the outcome. The meaning for f equal to zero or one would then correspond to impossibility or certainty of an outcome. This gives meaning to the real number range zero to one, and it refers to the relative frquency of the outcome of a well defined physical variable.

Although probabiliy in this way can be associated with a subjective ignorance of the parameters defining a physical variable, it should rather be called an objective ignorance and this probability should be called an objective probability. This is because all observers using the identically almost complete set of concepts and axioms would get the same probable values for the outcome of experiments on such well defined aspects of physical reality.   All this depends, of course, on identical extended sets of definitions and a logical theory containing the deductions and theorums formed from the axioms of the theory.

On the other hand, the axioms of any theory are statements that have a peculiar logical significance. Axioms must necessarily always be statements which cannot be logically deduced from other statements. If they could be so deduced then those other statements would be called the axioms. Of course the ongoing progress in our scientific theories may reduce the axioms of a previously accepted theory to deductions in a new theory. The new theory, however, would still be based on axioms (newer, deeper axioms). Thus each in the sequence of deeper theories will always be based

on axioms. If one wishes to preserve a Boolean logic in the theories it may
be useful to consider an axiom as a statement whose "mechanism" is unknown.
This idea is supplementery to that given previously.

Our ignorance of the magnitudes of well defined physical variables
and of the "mechanism" for axioms can be distinguished clearly by dividing
our ignorance into two parts. The first is that ignorance which is associated with the occurrence of different magnitudes of well defined physical
variables and thus where probability lies in the zero to one number range.
This is objective because acceptance of a reigning theory carries with it
acceptance of the axioms of the theory (and therefore also acceptance of
the ignorance of the mechanism of the theory's axioms).

The second type of ignorance is the ignorance of the mechanism
of an axiom. This type of ignorance cannot be ascertained via any test
which gives numerical answers. This type of ignorance, however, could
lie in imaginary or transfinite number fields as will be discussed. When
Frege mapped operations with one and with the plus sign to a Boolean
formalism he also mapped operations with zero to the concept of paradox
or inconsistency. This is an idea which can also be applied to axioms by
postulating that the fact that one must be ignorant of the mechanism of an
axiom can be mapped to the real number value of i (the square root of minus
one). Thus the fact that the real number value for i is unknown can be identified with an axiom statement whose mechanism is unknown. This is
in a literal sense a subjective ignorance because there is no agreement
on the future axioms of deeper theories which could someday be used to
deduce a present day axiom. This is now expressed via a mathematical
formalism.

Using the symbols + and x where:

+ is the union of the sets (say, [A] and [B]) standing on either side of
+ ; in the logical sense it stands for an element in either [A] or [B], or both.

x is the intersection of [A] and [B] ; logically it is an element that is
in both [A] and [B] simultaneously. Boole adopted the designations:

$$1 = \text{Universe of Discourse set}$$

$$0 = \text{Null set.}$$

We adopt the restricted designations:

> $\underline{1}$ = Universe of physical phenomena (well defined variables)
> set: represented by their symbolic designation, i.e.,
> force, pressure, etc., which obey scientific laws and
> thus are based on the axioms of a theory.
>
> (10)
>
> $\underline{0}$ = The absence of the above, or the physical phenomena
> null set.

Using these designations we set:

$$\sim \underline{0} = \underline{1} \, , \quad \sim \underline{1} = \underline{0} \qquad (11)$$

where $\sim$ is negation, and modus ponens is being used.

Now $[\underline{1} + \underline{0}]$ is the universe of well defined physical variables or their null
set. This includes all the theorems and deductions of a theory, but it does
not include the axioms of the theory. The symbol $\underline{1}$ can be decomposed into
a dual set, say [A], and its complement [~A]:

$$[A]+[\sim A] = \underline{1} \qquad (12)$$

Consistent with the discussion up to now we let:

$$\sim[\underline{1} + \underline{0}] \equiv i[M + \sim M] \qquad (13)$$

where i designates the axiom(s) of a theory, M, or the set of its complements [~M]; which can be combined in two different ways:

$$[M] + [\sim M] \qquad \text{and} \qquad [M] \times [\sim M]. \qquad (14)$$

The second expression has already been used to express undecidability and forced to the truth value: true (or one) in the earlier discussion of the logical status of axioms. The first expression is ambiguous; it can be written two different ways and equated to zero and to one in the field of imaginary number;

$$M \sim M = i0 \qquad (15)$$

and
$$[M] + [\sim M] = i \qquad (16)$$

Eq. (15) is the assertion of an axiom followed by its denial and results in the null axiom element. Eq. (16) has the truth value: one and lies in the i number field which contains all axioms (and their negations, which are also axioms). This has been treated elsewhere.[13]

It remains to suggest a logical meaning for the form $e^{i\theta}$. From the preceeding discussion it can be taken as the set of all sets to which the axiom statement or number symbol θ applies.

One may take the Planck energy relation:

$$E = \hbar \qquad (17)$$

which can be considered as an axiom of Quantum Mechanics (as an axiom statement). The set which can represent the above relation is the set of all the number pairs $[E,\omega]$ which is the continuous set of all such pairs which satisfy Eq. (17).

It can be converted into the axiom statement θ, but only after the terms on both sides of the equal sign are made into pure numbers with no dimensional designation. This is necessary because in Axiomatic Set theory the only meaning which can be given to exponentials is as the number of the set of all sets represented by the exponent (which is a discrete set) and which is thus a pure number. Rearranging Eq. (17) slightly and mult-iplying by t (the time variable) is necessary to make each side into a pure number;

$$Et/\hbar = \omega t \qquad (18)$$

The meaning for an exponential in e (the form $e^{i\theta}$) was shown in a prev-ious section to designate θ as a continuous set. We take the set of all continuous sets to which the Planck energy relation in the form of Eq. (18) applies as an axiom as:

$$e^{i\theta} \quad \text{or} \quad e^{i[Et/h = \omega t]} \quad \text{or} \quad e^{iEt/\hbar} = i\omega t \qquad (19)$$

The bizarre appeerence of the equal sign in the exponent can be made some-what more palatable if it can be mapped to a number. With one additional supposition the last expression above can be written as:

$$e^{iEt/\hbar} + (=) + (i\omega t) \tag{20}$$

where the supposition is that + means lies next to. One may then modify this to become:

$$e^{iEt/\hbar} \ e^{=} \ e^{i\omega t} \tag{21}$$

It now appears that in Eq. (21) the first and last factors can be the Psi function of Quantum Mechanics and the wave function of Electromagnetic theory, respectively. Since the arguments of each of these two factors are equal by Eq. (17), one gets:

$$e^{iEt/\hbar} = e^{i\omega t} \tag{22}$$

If we replace the equal symbols in Eqs. (21) and (22) by say, a symbol, α, and set these terms equal to each other then we get:

$$e^{\alpha} = \alpha \tag{23}$$

where the equal symbol in Eq. (23) above stands also for equality; but it must be different, possibly deeper then the meaning of α, see remarks below.

There have been a number of discussions of the numerical value of which satisfies Eq. (23) above. These discussions suggest that Eq. (23) can be interpreted by means of transfinite ordinals where:

$$\lim_{\alpha \to \infty} e^{\alpha} = \alpha \tag{24}$$

and that α can be the transfinite ordinal which is the ordinal number for the well ordered set (1, 2, 3, 4, ...). [5,6,14,15] There are minor differences between these authors in that Fraenckel and Kamke appear to restrict the base number, which we have as e, to integers; whereas Sierpinski permits that it be any real number. The discussion given in the last section of the forms 2^a and e^b, however, can be taken as a justification for the replacement of 2 as a base by e as a base. Additional discussions by the above authors also appear to give additional meaning to the equal symbol in Eq. (23). It is the second step in a hierarchy of the transfinite ordinal numbers which start from the well ordered integer set above. We retain the designation α for the first transfinite ordinal, the solution of Eq. (23), although it is usually referred to in the above discussions as ω (possibly $\underline{\omega}$ would be better) but in order not to cause confusion with the radian frequency ω of the Planck energy relation the α is retained.

The equation $E = \hbar\omega$ can now be written as

$$\alpha \quad E \mid \hbar_\omega \tag{25}$$

In order to be consistent with the multiplicative operations which have been defined for the first transfinite ordinal α and with the requirement that pure numbers be the only terms in the above equation when it is symbolized by θ when it is to be used in the form $e^{i\theta}$, the above is converted to:

$$(E/\hbar\omega) \ \alpha \ 1 \tag{26}$$

which according to the manipulative rules for transfinite ordinals can be represented by α alone, where α is the ordinal of the sequence:

$$1, 2, 3, 4, \ .... \tag{27}$$

The infinite set of pairs E, ω can be matched to the sequence above, and i α can also according to the manipulative rules for transfinite ordinals stand for the forms in Eq. (20), (21), or (22).

V. LOGICAL MEANINGS FOR IMAGINARY EXPONENTIALS

Taking the set of all sets (of all the physical situations) to which an axiom θ applies, as $e^{i\theta}$, we postulate that this form should correspond to a probability. The logarithm of that probability would then logically correspond to the recovery or discovery of the underlying axiom from the set of all physical situations to which the axiom applies. Obviously the logarithm of the above imaginary exponential should recover the axiom symbol θ.

The only other occasions in science when the logarithms of probabilities are of use is when the Boltzmann definition of Entropy and the Shannon definition of Information are being employed.[16,17]

Such a procedure cannot be of any help in finding the unknown mechanism of an axiom but rather it may help find if specific axiomatic statements will explain the data. This will not be a mechanical way of finding new axioms, but may permit a clearer evaluation of the physical effects that new axioms predict. Thus, according to the ideas discussed here axioms must always be created to fit physical situations and not the reverse.

All of the previous discussion leads naturally to the 2 questions:

1. What is the logical significance of the first and last factors of Equation (21), which appear to be the Psi and Electromagnetic Wave functions, respectively?

2. What is the logical significance of the Euler relation:
$$e^{i\theta} = \cos \theta + i \sin \theta$$

The first question has been discussed in some detail.[3,12]

The second question must be broken into 2 parts; first, for transfinite representations of θ, and second, for finite representations of θ, this latter corresponding to the wave functions above. For the transfinite representation of θ as an axiom, the Euler relation appears to be a mapping from the infinite θ range of the axiom to the range: plus to minus one (real) and also the range plus to minus i (imaginary). All that can be said about this at the moment is that it might represent a normalized physical variable and a normalized axiomatic condition of some kind.

The second part of the second question has the same answer as the first question and the reader is referred to the same references as in that answer. The referenced discussions consider the wave functions in both Quantum Mechanics and Special Relativity to be transformations between situations or rest frames which have axioms sets which are the negative or logical inverse of each other.[3,12]

REFERENCES

1.   Honig W.M., Transfinite Ordinals as Axiom Number Symbols For The Unific-
          ation of Quantum & Electromagnetic Wave Functions, IJTP,
          15, 87-90 (1977).
2.   Honig, W.M., On the Logical Status of Axioms-As Applied in QM and STR,
          Bull.A.P.S., 31, 844, March (1986).
3.   Honig, W.M., Chap. I, VI, VII in "The Quantum and Beyond", Philo-
          sophical Library, New York (1986).
4.   Cohen, P.J., pp107-129 in "Set Theory and the Continuum Hypothesis",
          W.A. Benjamin, San Francisco (1965).
5.   Fraenckel, A., Chap. II, III, and pp202-203 in "Abstract Set Theory",
          North-Holland, Leiden (1961).
6.   Fraenckel, A., Chap. II, III, and 158-159 in "Abstract Set Theory",
          North-Holland, Leiden (1966).
7.   Robinson A., Chap. I, II, III in Non-Standard Analysis,
          North-Holland, Leiden (1966).
8.   Davis, J. & Hersh, A., pp45-90,237-250 in "The Mathematical Experience",
          Penguin, London (1983).
9.   Frege, G., pp 30-60 in "Foundations of Arithmetic" (translation),
          Blackwells, Oxford (1950).
10.   Kiss, A.,"An Introduction to Algebraic Logic" Westport, Conn. (1961).
11.   Sikorski, R.,"Boolean Algebras" Springer Verlag, Berlin (1969).
12.   Honig, W.M., Godel Axiom Mappings in Special Relativity and in
          Quantum-Electromagnetic Theory, FOP, 6, 37-57, (1976).
13.   Chapter VII, Reference 3.
14.   Sierpinski, W., "Cardinal and Ordinal Numbers", Warsaw (1958).
15.   Kamke, E., Chap. III and IV in "Theory of Sets", Dover New York (1950).
16.   Kennard E., pp 367-372 in "Kinetic Theory of Gases", McGraw-Hill,
          New York (1949).
17.   Shannon, C., "Mathematic Theory of Communication", University of
          Illinois Press, Urbana (1949).

THE AHARONOV-BOHM EFFECT AND THE

QUANTUM INTERFERENCE PHASE SHIFT DUE TO AN ELECTROSTATIC FIELD

Giorgio Matteucci and Giulio Pozzi

Center for Electron Microscopy, CISM and GNSM-CNR
Department of Physics, University of Bologna
via Irnerio 46, 40126 Bologna, Italia

INTRODUCTION

Diffraction and interference effects of charged particles are well-known and accepted wave-particle phenomena in quantum physics. However the phase shift, observable in an interference pattern, introduced between two coherent electron beams propagating around a magnetic field through a field free region (Aharonov and Bohm 1959), has been the subject of an extensive debate both from theoretical and experimental viewpoints. The significance of this experiment, which concerns the physical meaning of fields and potentials in electromagnetism and quantum theory, has been recently emphasized by Olariu and Popescu (1985). The authors compare this significance with that which Michelson's experiment had in demonstrating the limitations of the traditional notions of space and time. This quite surprising interference effect has been verified and confirmed by a number of experiments (for a review see Olariu and Popescu (1985)) the last of which regards the investigation of completely shielded superconducting toruses by means of electron holography (Tonomura et al., 1986). This experiment definitely demonstrates that electrons do not directly experience any magnetic field, therefore the phase difference cannot be attributed to an external leakage field or to Lorentz-force effects on the portion of the electron beam going through the magnet, as some authors have defended.

The attention given to this kind of experiment has not been devoted to the case involving the scalar potential. In fact the time dependent experiment, also suggested by Aharonov and Bohm (1959), is beyond the present experimental capabilities while the time independent version, proposed by Boyer (1973), does not possess the same striking and paradoxical aspect as the magnetic case. Although the results of the interference experiments are very similar, in the electrostatic case electrons experience the effect of the field. Starting from Boyer's proposal, we have recently realized an equivalent interference experiment in which a constant phase difference is introduced between the beams by the electrostatic field associated with a bimetallic wire (Matteucci et al., 1984; Matteucci and Pozzi, 1985). The aim of this paper is to consider the basic features of this experiment in relation to the magnetic one, and to present the preliminary results concerning a new experimental set-up by which it is possible to increase and to obtain a continuous control of the phase difference.

THEORY

In electron optics, the interaction of the electron beam with a specimen can be conveniently described by solving the non-relativistic time-dependent Schroedinger equation in the high energy approximation (Messiah, 1970). The following points should be considered: a) the static electromagnetic field of the specimen is described by the scalar potential $V(\bar{r})$ and by the vector potential $\bar{A}(\bar{r})$; b) in the standard operating conditions of the experiments carried out by a transmission electron microscope, the kinetic energy eE of the beam is large when compared to the potential energy $eV(\bar{r})$ of the specimen, so that the electrons propagate almost parallel to the optical axis both above and below the specimen. Under these assumptions, the effect of the specimen on the wave-function associated with the electron beam is described by a transmission function $\Psi(x_o, y_o)$ given by:

$$\Psi(x_o, y_o) = \exp[i\,\varphi(x_o, y_o)] \qquad (1)$$

where x_o and y_o are the co-ordinates of a point P_o in the specimen plane and the phase shift $\varphi(x_o, y_o)$ is given by:

$$\varphi(x_o, y_o) = \frac{\pi}{\lambda E}\int V(x_o, y_o, z)\,dz - \frac{2\pi e}{h}\int A_z(x_o, y_o, z)\,dz \qquad (2)$$

where λ and e are the electron wavelength and the electron charge respectively, and h is the Planck constant. The integrals are taken along an electron path parallel to the optic axis z inside and outside the specimen to include stray fields.

Let us now consider the experiment, suggested by Aharonov and Bohm (1959), by introducing, between two interfering beams, a long solenoid (Bayh, 1962) or better a superconducting hollow cylinder with trapped magnetic flux (Lischke, 1969), as sketched in Figure 1 (a). With this set-up, electrons are not locally influenced by the magnetic field which is zero outside the solenoid. The z component of the vector potential around the solenoid can be written as:

$$A_z(x, y, z) = \frac{F\,x}{2\pi(x^2 + z^2)} \qquad (3)$$

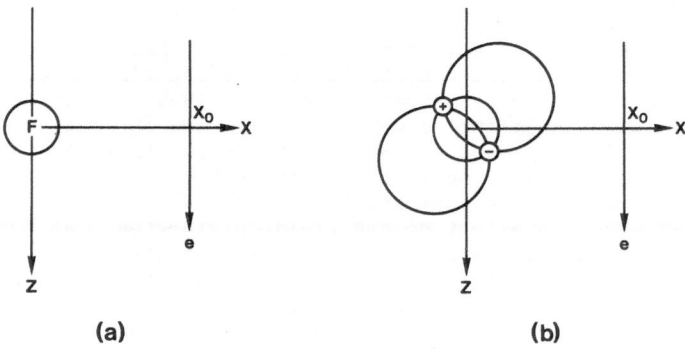

(a)   **(b)**

Fig. 1. Coordinate system and geometry for: (a) the AB effect involving an enclosed magnetic flux F, and (b) the electrostatic field due to two lines having opposite charges.

where F is the enclosed magnetic flux. It turns out that the phase shift suffered by the electrons is:

$$\varphi(x_0, y_0) = \pi e\ F\ segn(x_0)/h \tag{4}$$

and the observable phase difference $\Delta\varphi$ is given by:

$$\Delta\varphi = 2\pi e\ F\ /h \tag{5}$$

The same result can be directly obtained if the phase difference is written using the Stokes theorem.

Let us consider the electrostatic case: the field associated to a bimetallic wire having a zero net charge is equivalent, at long distances, to that of a dipole line, Figure 1 (b), having a strength proportional to $R\ \Delta V$, where R is the wire radius and ΔV the contact potential difference. More precisely:

$$V(x,z) = \frac{2R\ \Delta V\ (x\ \sin\vartheta + z\ \cos\vartheta)}{\pi\ (x^2 + z^2)} \tag{6}$$

where ϑ is the angle between the dipole and the electron beam directions. Introducing Equation (6) in (2) to calculate the phase shift, there results:

$$\varphi(x_0, y_0) = 2\pi R\ \Delta V\ segn(x_0)\ \sin\vartheta/(\lambda\ E) \tag{7}$$

whereas the observable constant phase difference is given by:

$$\Delta\varphi = 4\pi R\ \Delta V\ \sin\vartheta/(\lambda\ E) \tag{8}$$

Therefore, by rotating the wire, it is possible to have a phase difference varying continuously between zero and its maximum value, which is a few π for wires about 1 μm in diameter, with a contact potential difference of $\Delta V = 0.5$ V, observed at 100 kV accelerating voltage.

Another point deserves to be recalled: the electrons interact with the electrostatic field and the phase difference could depend on different lateral deflections suffered by electrons passing on either side of the wire. In the small angle approximation, the deflection is given by:

$$\alpha = \frac{1}{E} \int \frac{\delta V}{\delta x}(x_0, z)\ dz \tag{9}$$

which can be shown to be zero for the field topography given by Equation (6). This means that the diffraction envelope is not affected by the field and can be used as a reference to reveal the interference fringe shift.

Following Boyer's (1973) analysis, it is interesting to see how the phase shift can be interpreted as a classical electrostatic lag effect. The influence of the z component of the electric field varies the electron velocity, and causes a spatial lag Δl given by:

$$\Delta l = 2\ R\ \Delta V\ \sin\vartheta/E \tag{10}$$

between electrons travelling along each side of the dipole. The corresponding phase difference is therefore:

$$\Delta\varphi = 2\pi\ \Delta l/\lambda \tag{11}$$

which is exactly the same as that of Equation (8).

Let us finally remark analogies and differences between the magnetic and electric case. In both cases, the space is multiple-connected owing to the presence of the solenoid or the bimetallic wire, the lateral deflection of the electron beam vanishes and the effect is detectable as a displacement of the interference fringe pattern with respect to the unperturbed diffraction envelope. However, in the electric case, the effect depends on the energy eE of the electrons, i.e. is a dispersive phase shift. As pointed out by Boyer (1973) in his analysis using wavepackets, this means that in the electric case the lag effect can be made larger than the wavepacket length, thus destroying the interference pattern (Schmid, 1984) whereas such effect is not expected in the magnetic case, where the phase shift is independent of the electron energy.

EXPERIMENTAL SET-UP

The analysis of the foregoing section points out that the constant phase difference in the electrostatic case is essentially due to a dipole line field. In our former set-up (Matteucci and Pozzi, 1985) this field was realized by means of a bimetallic wire, so that the experimenter could only vary the angle between the electron beam and the dipole or the accelerating voltage, having no control on the dipole strength which is proportional to the wire radius and to the contact potential difference.

This drawback can be overcome if one realizes a macroscopic dipole line, by means of two parallel conducting wires, held at opposite potentials, Figure 1 (b). In this case, the phase difference should increase from the few π of the bimetallic wire to several hundreds or thousands of π depending on the potential applied to the wires and on their distance. Moreover, it should also be possible to investigate how the phase shift varies between the wires as this region becomes accessible. These advantages are made at the expense of a more complicated experimental realization.

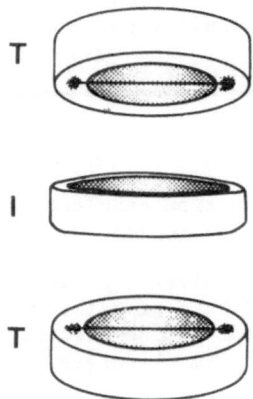

Fig. 2. Exploded view illustrating the elements of the macroscopic dipole. T: platinum apertures on which the platinum wire is soldered with conducting cement. I: electrical insulator.

The first device was realized by soldering onto two standard platinum apertures (300 μm in diameter) a commercial Wollaston platinum wire after its silver sleeve had been removed by chemical etching. The two apertures T were then superimposed and electrically isolated as shown in Figure 2. The two wires were then oriented parallelly. This "sandwich-like" device was then mounted on a specimen holder of an electron microscope equipped with an electrical contact and connected to an external voltage supply. In this way it was possible to bias the upper wire while the lower one was earthed. In addition, a goniometer stage allowed the variation of the orientation of the electric dipole by $\pm$ 24° around the rod axis. The observations were carried out with a Philips EM400T electron microscope equipped with a field emission source and a Möllenstedt-Düker (1956) electron biprism inserted at the level of the selected area aperture plane.

A schematic drawing of the whole set-up is shown in Figure 3. The coherent electron beam coming from the source S first illuminates the two wires (i.e. the macroscopic dipole D), and then propagates as far as the biprism plane whose axis is oriented not parallelly to the two wires. The electron biprism is a thin conducting wire W which can be charged by means of an external voltage supply with respect to two earthed plates P. The wire W splits the wavefront of the incoming electron beam, and the electrostatic field produces a deflection and a subsequent overlapping in a plane OP below the wire. A system of interference fringes will be observed if the coherence conditions are satisfied (Missiroli et al., 1981).

In the case of Figure 3 the interference region in the OP plane is divided into three parts by the projected shadows H of the two wires. In the regions M two wavefronts with the same phase shift overlap, and therefore a symmetric fringe system is expected, equal to that observable with the biprism alone. In the region N between the two shadows, the waves coming from the opposite sides of the dipole overlap. Thus the electrons experience different phase shifts and a resulting phase difference, given by Equation (8). The net effect is a lateral displacement of the interference fringes with respect to the unperturbed system or to the diffraction envelope, as roughly sketched in the lower part of Figure 3. It should be remarked that only the non-integral fraction $\Delta\varphi = 2\pi\varepsilon$ ($\varepsilon < 1$) of the phase difference can be estimated from the asymmetry of the intensity distribution. This drawback is not present when the experiment is carried out and recorded dynamically since the lateral shift of the fringes can be monitored continuously when the phase difference is varied.

EXPERIMENTAL RESULTS

The results described in this section were obtained with the following electron-optical conditions: the microscope was operated in the diffraction mode with the second condenser and objective lens switched off, whereas the diffraction lens was used to conjugate the specimen plane with the final viewing screen or the photographic plate. This operating condition is slightly different from that depicted in Figure 3, as the images of the wires are observed instead of their projected shadows and a virtual instead of a real system of interference fringes is formed on the OP plane.

The experiment was carried out by rotating the "sandwich" in order to obtain the narrowest cross-section of the two wires (as seen from the incident electron beam direction), that can be considered as a single impenetrable region for the electrons.

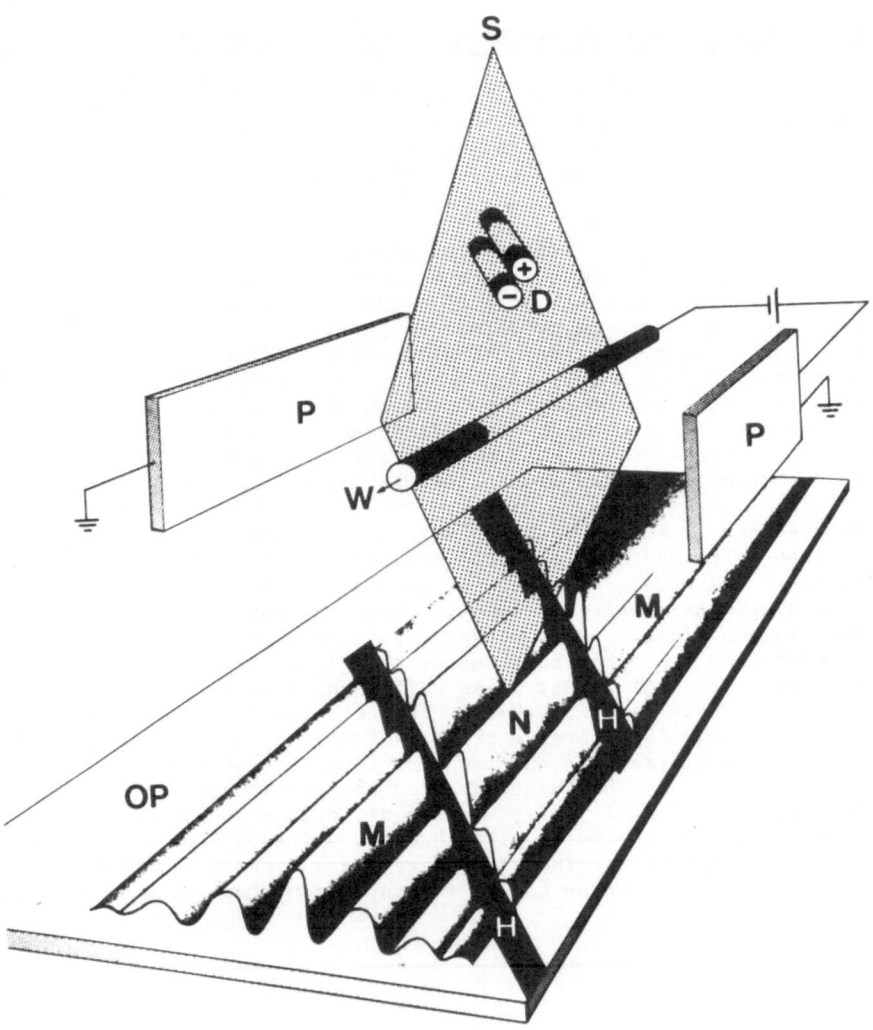

Fig. 3. Schematic set-up for the interference pattern formation. S:
field emission source. D: macroscopic dipole placed at the
standard specimen level. W and P: wire and ground plates
respectively of the electron biprism, which is inserted at the
selected area aperture plane. OP: observation plane, where the
phase difference can be detected in the region N placed between
the dipole shadows H, with respect to the regions M where the
interference pattern remains unperturbed.

By varying the potential applied to the upper wire, the fringe systems in the outer regions M of the interferogram are displaced laterally in opposite directions, whereas the fringes between the two images of the wires are tilted, Figure 4. This effect is related to the presence of an external field due to the biased wire and not compensated by the earthed one. The two wires behave essentially like a single charged wire, i.e. an electron biprism.

By increasing the potential applied to the upper wire, it was possible to compensate this field and make the axis of symmetry of the interference fringe system on the right and on the left of the image of the two wires coincide again, within the accuracy of observations carried on the fluorescent screen. In this condition the two wires behave as a macroscopic dipole, although with a fixed bias. Therefore, a phase difference between the two interfering beams was introduced by slowly tilting the macroscopic dipole of an angle so as to prevent electrons passing through the two wires. In our experimental conditions for angles within 1° it was possible to observe, in a very striking way, directly on the viewing screen, the dynamical change of the phase in the region between the shadows (compare region N of Figure 3) whereas the outer fringes (regions M) remained unperturbed.

The effect is less remarkable when recorded on the plate. In the micrograph of Figure 5 (a) in correspondence with the region N it is possible to reveal a phase difference with respect to the regions M. This photograph was recorded at a slightly lower biprism voltage in order to obtain a stronger diffraction modulation of the interference fringes. The phase shift is further evidenced by the microdensitometer traces of Figure 5 (b) and (c) that correspond respectively to the outer region of the interferogram in which the fringes are centrosymmetric with a central maximum, whereas in the inner region they are asymmetric.

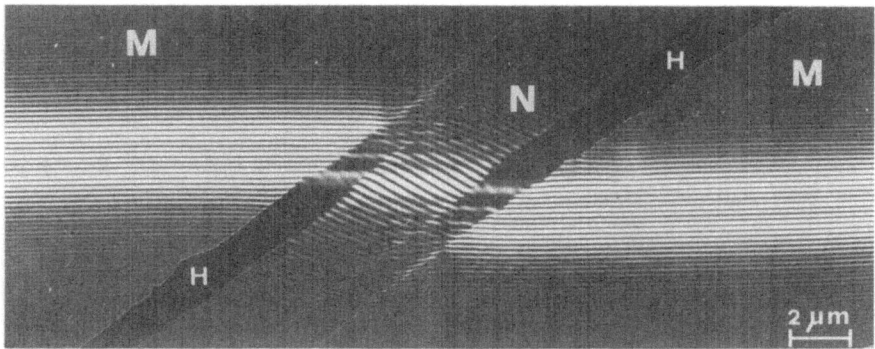

Fig. 4. Micrograph showing the interference pattern in the regions M and N as sketched in Figure 3, where the images of the two wires of the dipole are overlapped. The lateral displacement of the outer interference system is due to an uncompensated field originated by the "sandwich" of Figure 3. The biprism voltage is $V_F=12V$, the dipole voltage is $V_D=18V$.

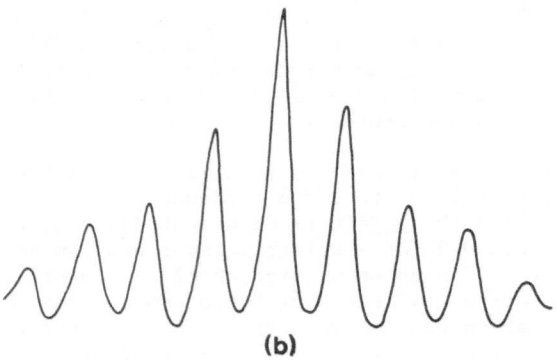

(b)

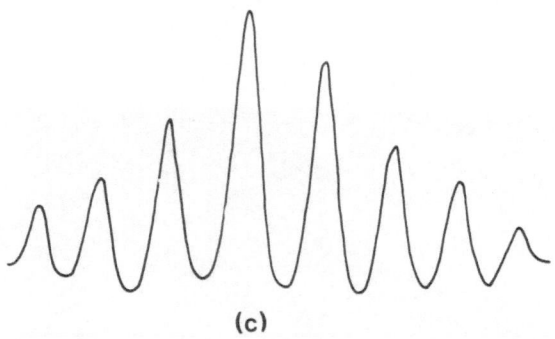

(c)

Fig. 5.　(a) Interference pattern obtained in the same experimental
conditions of Figure 4 but with V_F=8V. In this case the
effect of the biprism field present in Figure 4 has been
compensated by applying a higher external voltage to the
dipole: V_D=36V.　(b) and (c) are the microdensitometer
traces corresponding respectively to the regions M and N of
(a).

Figure 6 (a) and (b) have been recorded at larger positive and negative angles, in order to show the trend of the fringes between the two wires. It can be ascertained that the fringes are parallel and inclined with respect to the outer fringes, thus indicating a linear trend of the phase difference between the wires. However, the most striking effect is represented by the deformation of the overlapping region due to the deflecting field between the wires.

It may be noted that the most relevant features of the observed patterns, i.e. the deformation of the overlapping region (shadow effect) and the trend of the interference fringe system, can be accounted for by modelling the dipole as made of two parallel electron biprisms of opposite strength and by means of the asymptotic approximation of the image wavefunction in interference electron microscopy (Missiroli et al., 1981).

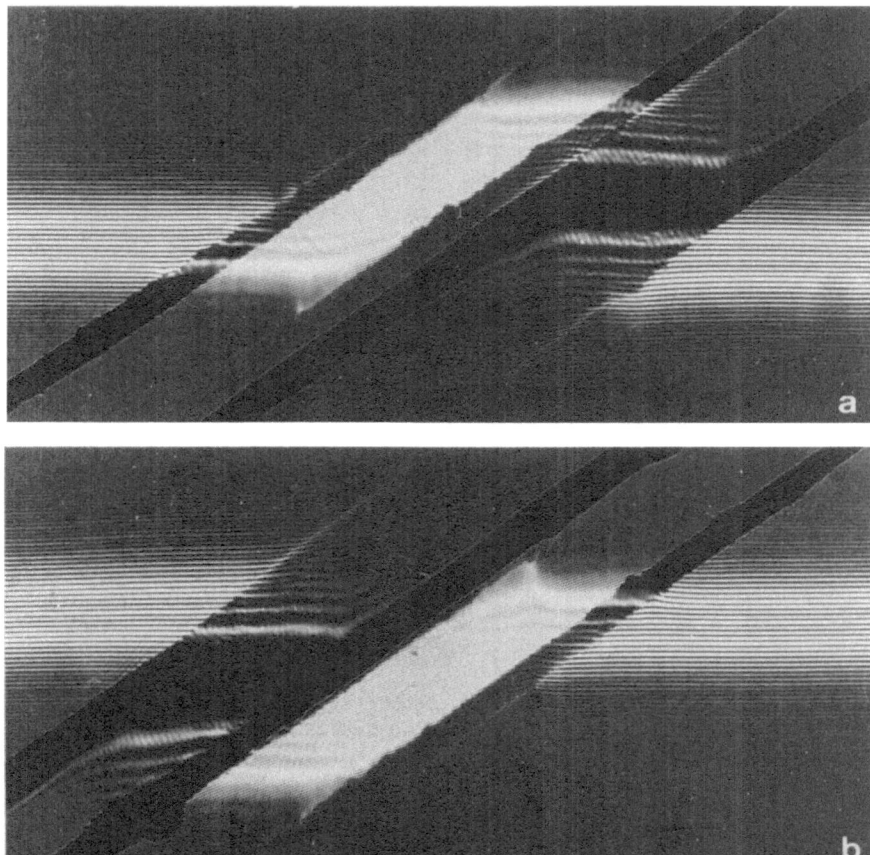

Fig. 6. (a) Interference pattern taken by rotating the "sandwich". The images of the two wires are easily distinguishable. The interference fringes show that the trend of the phase variation is linear. (b) Same as (a), but with the dipole rotated in the opposite direction. In this case the central images of the wires are superimposed. V_F=12V, V_D=36V.

CONCLUSIONS

The preliminary results reported in this work show that a quantum phase shift can be introduced by a macroscopic dipole and detected in an interference experiment. This phase shift arises through a local interaction of electrons with the electrostatic field of the dipole and can be interpreted either as a classical lag effect or as a local effect of the scalar potential. By improving the shifting device, for example by applying an opposite potential to the two wires, it will be possible to vary in a continuous way the phase difference up to values of thousands of π. By increasing the electrostatic field of the dipole it should be possible, at least in principle, to introduce a phase difference larger than that corresponding to the longitudinal coherence of the electron wavepacket and to obtain an estimate of the wavepacket length.

AKNOWLEDGEMENTS

Thanks are due to Dr G.F. Missiroli for useful discussions and for the critical reading of the manuscript, and to Miss Rossana Prola for her help with the microdensitometry. This work has been supported by funds from Ministero della Pubblica Istruzione, Italia.

REFERENCES

Aharonov, Y., and Bohm, D., 1959, Significance of electromagnetic potentials in the quantum theory, Phys. Rev. , 115:485.
Bayh, W., 1962, Messung der kontinuierlichen Phaseschiebung von Elektronenwellen im kraftfeldfreien Raum durch das magnetische Vektorpotential einer Wolfram-Wendel, Z. Phys. , 169:492.
Boyer, T.H., 1973, Classical electromagnetic deflections and lag effects associated with quantum interference pattern shifts: considerations related to the Aharonov-Bohm effect, Phys. Rev. D , 8:1679.
Durand, E., 1964, "Electrostatique", Masson et C ,Paris.
Lischke, B., 1969, Direct observation of quantized magnetic flux in a superconducting hollow cylinder with an electron interferometer, Phys. Rev. Lett. , 22:1366.
Matteucci, G., and Pozzi, G., 1985, New diffraction experiment on the electrostatic Aharonov-Bohm effect, Phys. Rev. Lett. , 54:2469.
Matteucci, G., Missiroli, G.F., Porrini, M., and Pozzi, G., 1984, Recent observations on a new electrostatic phase-shifting effect, in "Proceedings of the International Symposium on Foundations of Quantum Mechanics", S. Kamefuchi, ed., Physical Society of Japan, Tokyo, 39.
Messiah, A., 1970, "Quantum Mechanics", North Holland, Amsterdam.
Missiroli, G.F., Pozzi, G., and Valdrè, U., 1981, Electron interferometry and interference electron microscopy, J. Phys. E:Sci. Instrum. , 14:649.
Möllenstedt, G., and Düker, H., 1956, Beobachtungen und Messungen an Biprisma-Interferenzen mit Elektronenwellen, Z. Phys. , 145:377.
Olariu, S., and Popescu, I.I., 1985, The quantum effects of electromagnetic fluxes, Rev. Mod. Phys. , 57:339.
Schmid, H., 1984, Coherence length measurement by producing extremely high phase shifts, in "Proceedings of the Eighth European Congress on Electron Microscopy", Á. Csanády, P. Röhlich and D.Szabó, ed., Programme Committee of the Congress, Budapest, 285.
Tonomura, A., Osakabe, N., Matsuda, T., Kawasaki, T., Endo, J., Yano, S., and Yamada, H., 1986, Evidence for Aharonov-Bohm effect with magnetic field completely shielded from electron wave, Phys. Rev. Lett. , 56:792.

DISCUSSION III

Compiled and interpreted by D.W. Kraft and E. Panarella

De Martini. I have a question for Matteucci. In the Bohm-Aharonov (B-A) experiment, it is important to separate the effects of the field and the potential. What is the magnitude of the effect of the electrical potential with respect to the real effect of the electric field? Is it 0.1? Is it 0.001? This is important.

Matteucci. If you want to calculate the total phase difference, you have to calculate the scalar potential.

Boyer. There are two different points of view expressed. Aharonov and Bohm tried to show that it's just the potentials and not the field that they entered at all. They had a magnetic case and an electrostatic case when they turned the potentials on and off while the charges were inside the tubes. When the charges were inside these tubes, they were screened from any field, but they claimed there would still be an effect. Now, I was looking at it from an entirely different point of view, contrary to what Aharonov and Bohm had, and I suggested this electrostatic effect to show that the electrostatic fields could give a lag effect showing the same interference pattern as what had arisen from the B-A effect for a magnetic case. What I was proposing is that the energy conservation suggested that maybe the magnetic case was also exactly a lag effect, but I didn't say that the electrostatic case in any way involves only the potentials rather than the fields. It really involves fields fully in the form that's being treated at the moment. Now if they can screen those fields and confine the charges inside some tubes, the way Aharonov and Bohm actually proposed, then they are seeing effects on the phase difference which are not related to electrostatic effects. But this is related to an electrostatic effect, comes out nicely, giving the same answer either from the semi-classical or quantum point of view. Now I didn't get your last comments on proposed experiments and I would be interested to hear what you are proposing for the future.

Matteucci. We want to measure the wave packet length.

Boyer. Right, and did you mention also possible shielding in the way that Aharonov and Bohm had proposed? I think it would be very interesting to be able to measure how many wavelengths difference and what sort of coherence length you get.

Matteucci. This is possible if you apply a potential to the wires of the macrodipole so that you can obtain a phase shift, that is you have to destroy the interference pattern. At this point you can infer that you can obtain an evaluation of the coherence length.

Selleri. This is a very interesting set of experiments on the B-A effect because they point to the existence of zero energy effects. In other words, in a situation in which the electric and magnetic fields are both zero, classically you do not have any energy and so if you are left only, in one case, with the vector potential and in the other case, with the scalar potential, you think that there is no energy density. In spite of this you see an effect in the intereference figures which is also a systematic effect in that there is a systematic shift in one direction even though the envelope remains the same. All the fringes shift, which means that the ensemble of electrons has received a lateral kick.

Matteucci. No. You do not have lateral shift. The electrons do not suffer any lateral deflection. They suffer only a change of velocity in this case.

Selleri. But the fringes have shifted.

Matteucci. Yes, you introduce a phase difference.

Edmonds. Isn't there an electric field in the region that these electron waves are passing through around the wires?

Matteucci. Yes, of course there is a fringing field due to the magnetic field of the current in the wires.

Edmonds. They are actually then traveling in an electric field. It is not analogous to the B-A effect?

Matteucci. No. It is not a B-A effect. Those experiments can be considered as a demonstration of the experiments suggested by Boyer, the experiment with the two tubes held at different potentials for the duration of the entire experiment. The geometry of course is different. Here are two wires placed inside the two electron beams. In the two-tube experiments, of course, the electrons pass through the tubes, but they, in both cases, feel the fringing fields when entering and leaving the two tubes, or they feel the fringing field due to the two wires.

Boyer. I would like to make one more comment on the question of energy. Even the ordinary B-A effect really involves substantial energy changes. When the electron is passing this tube, unless you change the currents in the solenoid, the overlap of the fields means there is a substantial magnetic energy and if you assign that energy change back to the velocities of the passing charges, then you get exactly this lag effect that one sees in the electrostatic case. And so I was simply proposing that the actually observed B-A effect for the magnetic field is not that at all; it is not an effect of potential but a collective reaction of whatever is producing the magnetic field of the solenoid back on the electron. It is really a classical electromagnet lag effect, but the final step, which produces the intereference pattern clearly is not handled within the classical theory.

Selleri. Which way? Nonlocally or locally?

Boyer. Local. Purely local classical electromagnetics.

Selleri. Are there leaking magnetic fields?

Boyer. Not leaking. When the electron is passing the solenoid, the fields of the electron act on the currents. They must act on the currents. If you pretend they don't act on the currents, then you have a magnetic field in the inside which involves an interference term between the magnetic field of the solenoid and the magnetic field of the passing charge. And that's ignored in the B-A calculations but that's a substantial magnetic energy change If one assigns that energy change to the passing particle, in other words to say that the energy balance is provided by changing the electric and magnetic fields which produce an energy change back on the passing electron, then you get a lag effect which would give exactly the phase shift to the final packets, corresponding to the B-A effect. So I propose the electrostatic case simply to undercut the ordinary interpretation of the magnetic B-A effect and say possibly that this is a current. Now I did computer simulations of classical charged particles going around in a circle and charged particles passing on either side, and then just seing exactly what the classical electromagentism would say. If you find those particles, they speed up as the passing particle passes. You do get the effect qualitatively but at that time I didn't have enough computer power to make a realistic solenoid. But you do, qualitatively, see the slowing and speeding of the charged particles, depending on which side of the solenoid the electron passes.

Edmonds. If the electrons travel farther out from the solenoid, is the effect the same, provided the solenoid length is infinite?

Boyer. If it's a finite solenoid, the effect falls off. But if it's an infinite solenoid there's no change. As long as your solenoid is long enough. But it's a fascinating thing. You can easily do the classical calculation and they were published in Physical Review. I separated the two parts because ordinary physicists accepted all the classical calculations but then when I wanted to relate them to the B-A effect, I could find no physicist who agreed with me and therefore I published this as a second piece.

Edmonds. The electron affects the magnetic fields in the solenoid which affects the electrons producing the magnetic field in the wires. Right?

Boyer. Right.

Edmonds. Why must that do anything to the electrons?

Boyer. Oh, because you see if you change the currents in the solenoid, then the fields outside must change. The only reason there is zero magnetic field outside is because of a very specific pattern of currents inside the solenoid.

Aspden. I haven't seen Boyer's paper on this subject but I agree with him for the following reasons. First of all, in the magnetic version, if you have a current in the infinite solenoid, you certainly do not get a magnetic field outside because you know that a closed path integration goes to zero. But what happens is that the electron comes along passing the solenoid, the electron itself induces an interaction with the current in the solenoid, but this interaction produces an electric displacement of charge in the solenoid which has the effects of slowing the electron that passes on one side and speeding up the electron on the other side. It's an electrostatic effect. This can be related to classical electromagnetic theory, if you don't take the Lorentz version of the electromagnetic interaction. But, if you take the version that has an additional term which might allow for induction, I think this touches the B-A effect in the magnetic version, and can go back to the fundamental interaction between current elements and can actually determine which law is true, whether it's the Ampere law or one of the other laws, but it certainly doesn't tell us anything about the Lorentz law.

Dewdney. If the electron goes on only one side of the coil, what does it interfere with?

Boyer. The interference is something different. All I want to argue about is the classical lag. On one side of the solenoid, the charged particle will be slowed first as it's approaching and then speeded up when it's leaving; on the other side the order is reversed. When two electrons which started together are past the solenoid, one is ahead of the other and therefore there is a relative displacement of the wave packets. Now that is exactly what the electrostatic experiment here confirms, namely that relative displacement will lead to an interference pattern shift analogous to that of the B-A effect. Therefore, conceivably, the B-A effect is also really a classical electromagnetic lag effect.

De Martini. I tend to agree with many of Marshall's ideas and I don't see them as revolutionary; actually I see them as very close to QED. For example, it's been demonstrated that the zero point fluctuations are actually responsible for spontaneous emission. This comes out of making standard quantum mechanical assumptions about the reality of an experiment and the hermiticity of an operator. So what Marshall said this morning was that spontaneous emission is stimulated by the zero point energy. Secondly, I agree with Marshall's reluctance to talk about particles. My only point of disagreement with Marshall is that he had to introduce a nonlinearity to guarantee the stability of the field. In fact the idea of Einstein and de Broglie was to build up a soliton solution that is stable in time and space. Your solution is not stable. Quantum field theory suggests that the quantum numbers may be related to soliton stability. So, I agree with Marshall that the particle is not a stone. Certainly, what Einstein had in mind was that the particle is a soliton solution, i.e. a perturbance with an associated energy to a field.

Vigier. I have two remarks to make: With respect to the beginning of Marshall's talk and the general nature of the discussions here, I am concerned that "Quantum Violations" may not be an appropriate title for this Conference. Although we can reinterpret quantum mechanics, I have, so far, seen no evidence of quantum violations. Accordingly, I suggest that the Proceedings of this Conference be published under the title, "Quantum Uncertainties." This title reflects more closely the nature of these discussions and would draw less antagonism from the community in which those who question the Copenhagen interpretation are still very much in the minority.

My second remark deals with Marshall's models. First, I must say that what De Martini has been saying about the soliton is exactly de Broglie's point of view. In 1927, he proposed the pilot wave, saying that the photon is a point and the Maxwell field surrounds and guides it. However, he himself knew very well that this was a simplification for the convenience of the calculations. He and Einstein wanted to have something like nadelstrahlen and this is physically absolutely necessary because any linear solution of Maxwell's equations is bound to spread with time. Now you can do exactly the same experiments on photons originating in distant galaxies as you can with photons produced 50 cm away; you can do the same photoelectric effect, the same double scattering effect, etc., so those nadelstrahlen just won't do. You have to add something to Maxwell's equations and you have to find a soliton-like solution if you want to defend that point of view, which I also approve. I am sympathetic to your argument about the background zero point field. If those neutron experiments justify the wave and particle model, then something must carry the de Broglie waves, and I see

310

no other possibility saying that the de Broglie waves are collective
motions, like a sound wave, carried by a random vacuum. This is where,
in my opinion, SED has failed, by dropping the particle aspect, for as
soon as you say that the particle is just something, extended if you
like, moving in a chaotic field, you cannot recover the quantized orbits
correctly. Quantum mechanics has been a very successful theory and any
reinterpretation must recover the known, experimentally tested
predictions. That is why I favor the quantum potential approach because
it goes beyond quantum mechanics while at the same time recovering the
predictions of quantum mechanics. If you don't do that, if you just
ignore some aspect such as the hydrogen atom orbits or the double
scattering procedure or the photoelectric effect, then you are in trouble
with precisely the points where quantum theory has scored its biggest
successes. Therefore, any attempt to re-establish causality and realism
in quantum theory must start with a model where one knows one will
recover the known observations.

Selleri. I want to ask Marshall about the photoelectric effect for if
his theory cannot explain it, and I belive it cannot, then this theory is
untenable.

Marshall. I agree with Vigier that we shold rename the Conference and I
would like to use a name similar to that of the Bari Conference, which
was Open Questions in Quantum Physics.

The closeness of SED to QED is natural because of the similarity of
the correlation algebra to the commutation algebra. For the free
radiation field it is rather striking and the explanations it provides
for the Casimir effect is almost the same. To that extent one would
expect a much closer and quicker approach to QED from this direction than
one would from the double solution approach. However, I would not say
that we have anything like solitons and indeed if we did have something
like solitons, we would be much closer to what the de Broglie school
thinks we should have.

As far as the electromagnetic field is concerned, all my instincts
are to disagree with the de Broglie school on this point. The
nadelstrahlung which I described is sort of halfway between a soliton and
a well-defined bunch. The latter can travel large distances while
retaining its shape. What the de Broglie school wants me to have is a
soliton, and a soliton would keep its size all the way to distant stars.
For a really disastrously spreading thing, the size of the bundle would
vary linearly with the distance, while the thing I'm describing varies
with the square root of the distance. Therefore, it can go quite a long
way without deformation.

Now as for obtaining Bohr orbits, this is an area where SED has not
done as well as I had hoped when I began to work in this area, and to
that extent I agree with Vigier.

With respect to te photoelectric effect, don't forget that there was
a semiclassical theory in the 60's and 70's that did actually do the
photoeffect; that's the Jaynes theory. The Jaynes theory treated atoms
quantum mechanically, but it did the radiation field classically and it
did the photoeffect. It failed when it came to the kind of things that I
described this morning, because Jaynes rejected the idea of a real zero
point field. What I am suggesting is that one may be able to put
together the successful ideas of Jaynes with respect to photoelectricity
and the successful ideas of Santos and myself with respect to these,
previously-considered, particle aspects of the photon. We have got
aspects of the photon without any actual particles, any little dots in

the middle of the nadelstrahlung. The nadelstrahlung is a highly
localized solution of Maxwell's equations and it's spread out a long way
in length, of the order of meters, and it's spread only a little way in
width, of the order of half a millimeter.

De Martini. You should explain the phenomenology of the photoelectric
effect.

Marshall. Jaynes showed that if you put a classical electrical field
into the nonrelativistic Schrödinger equation, you get the photoelectric
effect.

Vigier. Not at huge distances.

Marshall. I think it does at huge distances.

De Martini. No. It doesn't have enough energy.

Honig. I would like to ask Selleri about enhancement and if he has a
mechanism for it.

Selleri. I agree with the idea of enhancement and I think it contains a
lot of physics, from which can come great developments. I am glad that
Marshall is studying enhancement; we are also studying enhancement at
Bari, although we adopt a dualistic approach. We believe that photons
are particles and waves at the same time, and we study enhancement within
this framework rather than with SED.

Santos. It is perfectly possible to have solutions of Maxwell's
equations which propagate without deformation, because the equations are
for a motion in a nondispersive medium, such as a vacuum. Therefore, a
wave packet from a distant star should arrive there with the same shape.

I agree that SED up to now has not been able to provide a good
explanation for the hydrogen atom or for the photoelectric effect.
However, SED has succeeded in explaining the Casimir effect and, in
general, for those problems in which the essential ingredient is the
radiation by itself, not the interaction of the radiation with matter.
What Marshall and I have initiated is the study of the propagation of the
radiation, with ad hoc assumptions for the emission and absorption of
radiation. Of course, this is only tentative; it is not the final
theory, but a step in that direction.

CRITICAL ANALYSIS OF THE TESTS OF LOCAL REALISM

VERSUS QUANTUM MECHANICS

Emilio Santos

Departmento de Física Teórica
Universidad de Santander
39005 Santander, Spain

1. INTRODUCTION

P. A. M. Dirac [1] emphasized the fact that several dramatic advances in physics have come from the solution of a previous conflict between two well established theories (which is just an example of the usefulness of the dialectic method). For instance, the contradiction between Newton's mechanics and Maxwell's electromagnetism was solved by Einstein with the introduction of special relativity. Similarly, the conflict between special relativity and Newtonian gravitation gave rise to general relativity. A conflict similar to the latter is the one existing at present between local realism and quantum mechanics. In fact, in both cases the conflict arises because it is predicted a violation of the principle of local action: by gravitational forces at a distance in the first case, and by the influence of a measurement on another one performed at a distant place, in quantum mechanics. As the principle of local action (signals cannot travel faster than light) is at the roots of relativity theory, I guess that a dramatic advance in physics will soon take place, when this conflict is solved.

Local realism stands for the conjunction of a general philosophical principle (material systems have properties independently that they are measured) and the physical principle of local action. The discovery that some predictions of quantum mechanics are not compatible with local realism was made by J. S. Bell [2] in 1965. More than 20 years have elapsed since and it is not yet clear whether the incompatibility of principle can be tested empirically. My personal conjecture is that it cannot. Of course, this opinion departs from the current state of opinion that not only can the contradiction be shown by experiments, but that local realism has been already refuted empirically. I am in strong disagreement with the last assertion, and the main purpose of this talk is to show that it is false. The second aim is to indicate what experiments may likely show a violation of quantum mechanics.

In Sections 2 and 3 I will briefly review the theoretical basis for the conflict between local realism and quantum mechanics. The main criticism to the experiments performed is made in Sections 5 and 6. Finally, in Section 7 I shall indicate optical photon experiments that may likely show a violation of quantum theory.

2. LOCAL HIDDEN-VARIABLE THEORIES

Proofs of the contradiction between local realism and quantum mechanics have been considered with different levels of generality. In my opinion, as the generality of the proofs increases, their clarity decreases. A consequence of this fact is that several people, including eminent authors [3], have arrived at the conclusion that the said contradiction does not exist. Therefore, it seems to me convenient to start analyzing the conflict between quantum mechanics and a restricted form of local realism, represented by deterministic local hidden-variable (LHV) theories. It is interesting to point out that the original proof by J. S. Bell [2] was given for this restricted class of theories.

The experiments giving rise to a contradiction between the predictions of quantum mechanics and those of any deterministic LHV theory are the following. Some event takes place in a region S (the source) at a time t_0, and signals (maybe particles) are sent from the source to two separated regions of space, A and B. A measurement is performed at time $t_1 > t_0$ in region A and another measurement at $t_2 > t_0$ in region B. It is assumed that $|t_1-t_2| < c/R_{12}$, where R_{12} is the distance between regions A and B and c is the speed of light. For the sake of brevity I shall call EPR experiments those of the kind just described. The hypothesis that the measurements so performed cannot have any influence on each other is called locality, although in other contexts it is called relativistic causality. This hypothesis is a particular case of the principle of local action. The contradiction arises because, for some EPR [4] experiments, quantum mechanics predicts correlations between the results of the measurements which are stronger than any possible correlation established by the common past of the signals arriving at A and B (i.e., established in the source).

If an EPR experiment is performed N times, a sequence of pairs of real numbers is obtained, $\{\alpha_i, \beta_i\}$, i = 1, ..., N. The empirical correlation is defined by

$$<\alpha\beta> = 1/N \sum_{i=1}^{N} \alpha_i\beta_i. \tag{1}$$

It must be pointed out that the usual definition of correlation in mathematical statistics is not (1), but rather,

$$<\alpha\beta> \text{ stat} = \frac{<\alpha\beta>-<\alpha><\beta>}{(<\alpha^2>-<\alpha>^2)^{1/2}(<\beta^2>-<\beta>^2)^{1/2}}. \tag{2}$$

This fact has given rise to some confusion. For instance, Eq. (2) has been used to show counterexamples to theorems established for (1). Although either (1) or (2) could be used for the proof of impossibility of deterministic LHV theories, practically all published proofs involve (1).

The prediction of any deterministic LHV model for (1) can be obtained from the probability $P^{\alpha\beta}$ (a, b) of getting the pair of numbers $\{\alpha, \beta\}$ in the measurements made in A and B respectively. This probability is

$$P^{\alpha\beta}(a, b) = \int d\mu(\lambda) \, p^{\alpha}[\lambda(t), a] \, p^{\beta}[\lambda(t'), b],$$

$$t < t_1-c/R_1, \quad t' < t_2-c/R_2, \tag{3}$$

where $R_1(R_2)$ is the distance from the source S to region A(B), c the speed of light, $p^{\alpha}[\lambda, a]$ the probability of getting the result α in the region A with a specified measuring apparatus a, $\lambda(t)$ being a set of parameters describing the state of the source at time t, and similarly for $p^\beta[\lambda, b]$. An appropriate probability measure in the space of samples of (the sto-chastic process) $\lambda(t)$ is used Eq. (3) means that the probability of getting the result α, for a given measuring apparatus, a, set in A, depends on the history of the source, specified by $\lambda(t)$. Then $p^\alpha[\lambda, a]$ is a functional of $\lambda(t)$ and similarly for $p^\beta[\lambda, b]$. The fact that a product of probabil-ities appears in (3) means that any correlation between a and b was es-tablished in the common past (the source).

Equation (3) is the general framework for deterministic LHV theories if p^α and p^β represent certainties (i.e., they can only take the values 1 or 0). On the other hand, Eq. (3) is a framework for stochastic LHV the-ories if p^α and p^β may take other values in [0, 1]. In view of this fact, it seems absurd to insist on determinism, as I have done above. The point is the following: Eq. (3) is a necessary and sufficient conditon for de-terministic LHV theories, but only a sufficient conditon for LHV theories. The lack of necessity of (3) for general LHV theories has been recently stressed by Popper [3] but it should be clear long ago. In fact, it is rather trivial [5] to exhibit stochastic LHV models in which Eq. (3) is not used (in particular there is no factorization within the integral). It is enough to average over one of the parameters in (3) in order to de-stroy the factorized form of the integrand. On the other hand, such an averaging transforms a deterministic model into a stochastic one. For this reason it is not controversial that any deterministic LHV model of an EPR experiment involves (3), to the point that (3) can be used as a definition of deterministic local realism. In contrast, as said above, it is controversial whether (3) represents adequately stochastic (probabil-istic) LHV models. Even more controversial is whether (3) is necessary for local realism, a not well defined concept supposedly more general than hidden-variable theory. I do not want to enter in the polemic and I avoid it simply by giving the following classification:

Class I. People arguing that (3) does not necessarily apply to all EPR experiments.

Class II. People considering that any EPR experiment can be described by (3).

Class I includes two quite different subclasses: a) People supporting the Copenhagen interpretation of quantum mechanics, and b) some critics of the orthodox view which, nevertheless, do not attach any relevance to the Bell inequalities [3, 6]. It should be pointed out that many orthodox people do attach a relevance to the Bell theorem as having shown the use-lessness of hidden-variable theories. I belong to class II, so I should give arguments in favor of the necessity of (3). However, my experience is that I will not succeed in changing a single person from class I to class II with such arguments, so I will not insist any more. It must be realized that any argument for the necessity of (3) will involve at some point a formal definition of LHV theory, definition which will not be accepted by people in class Ib. The remainder of the present paper concerns the analy-sis of the contradiction, if any, between the predictions of (3) and those of quantum mechanics.

3. HOMOGENEOUS AND INHOMOGENEOUS BELL'S INEQUALITIES

The proof of the contradiction between local realism and quantum mechan-ics usually goes through two steps. Firstly, a Bell's inequality is derived from (3) for an EPR experiment. Then, it is shown that the quantum mechani-cal prediction violates the inequality. (There are derivations of Bell's inequalities [7, 8] not using explicitly (3), but hypotheses apparently more general. However, I doubt that they are really more general.) Actually, the Bell inequalities cannot be derived directly from (3), but an approximation is needed. In fact, it is necessary to replace the time-dependent expression (3) by a time independent one writing, instead of (3),

$$p^{\alpha\beta}(a, b) = \int p^{\alpha}(\lambda, a) \; p^{\beta}(\lambda, b)\rho(\lambda) \; d\lambda, \; \lambda \in \Lambda, \qquad (4)$$

where Λ is a set of (hidden) variables with probability density $\rho(\lambda)$. This is the usual starting point for the derivation of the Bell inequalities. The change from (3) to (4) involves neglecting some memory and collective effects, which will be considered in more detail in Section 5.

A simple derivation of the Bell inequality from (4) goes as follows [9]. It is considered the repetition of four similar EPR experiments all with the same source prepared in identical conditions, but using either apparatus a or a' in A and either apparatus b or b' in B. The variables α and β are dichotomic observables taking the possible values ± 1 (or 1 and 0) and it is considered the case $+1$ only (which corresponds usually to the record of a count in a detector). Then, four primary probabilities are involved fulfilling the obvious inequalities

$$0 \leq p(\lambda, a), \; p(\lambda, a'), \; p(\lambda, b), \; p(\lambda, b') \leq 1, \; \lambda \in \Lambda \qquad (5)$$

where the superscript $+1$ has been omitted for simplicity. From (5) it follows easily

$$p(\lambda, a)p(\lambda, b) - p(\lambda, a)p(\lambda, b') + p(\lambda, a')p(\lambda, b) + p(\lambda, a')p(\lambda,b') \leq$$

$$p(\lambda, a') + p(\lambda, b). \qquad (6)$$

If (6) is averaged over λ (i.e., multiplied by $\rho(\lambda)d\lambda$ and integrated) one gets

$$P(a, b) - P(a, b') + P(a', b) + P(a', b') \leq P(a') + P(b). \qquad (7)$$

This inequality involves, not only the coincidence probabilities considered in (3) and (4), but also the single probabilities

$$P(a') = \int \rho(\lambda)p(\lambda, a')d\lambda, \; P(b) = \int \rho(\lambda)p(\lambda, b)d\lambda. \qquad (8).$$

In consequence, I have proposed to call (7) <u>inhomogeneous</u> Bell's inequality. The important point is that its derivation involves the inhomogeneous second inequality (5) (besides the homogeneous first inequality (5)).

Inequality (7) is difficult to test empirically due to the fact that only a fraction of the signals produced in the source can be analyzed with the apparata placed at A and B. In the first place, the signals may travel in the wrong direction not arriving at regions A or B. Secondly, the measuring apparata have a finite efficiency, so that not all signals arriving are detected. If f is the fraction of the signals emitted which are actually detected, P(a') and P(b) are of order f whilest P(a, b), etc., are of order f^2. Therefore, the inequality (7) cannot be violated except if f is close to unity. More or less plausible estimates could be made about the fraction of all signals which arrive at the measuring apparata, so estimating the different probabilities involved in (7) from the counting rates, but the procedure is uncertain. Even more dangerous is to extrapolate the results actually measured with low-efficiency detectors in order to estimate the results with ideal (100% efficient) detectors. I shall return to this point in Sec. 4.
A procedure suggested to avoid the problem of the finite efficiency of detectors is to test <u>homogeneous</u> inequalities, involving only coin-

cidence probabilities. Such inequalities cannot be derived from (5), but it is necessary to use homogeneous inequalities instead, like

$$0 \leq p(\lambda, a), p(\lambda, a'), p(\lambda, b), p(\lambda, b') \leq p(\lambda, \infty), \qquad (9)$$

where $p(\lambda, \infty)$ will be defined below. Hence, it is easy to derive

$$0 \leq P(a', \infty) + P(\infty, b) - P(a, b) + P(a, b') - P(a', b) - P(a', b') \leq$$

$$P(\infty, \infty), \qquad (10)$$

which are <u>homogeneous</u> inequalities, involving only coincidence counts. Inequalities (9) have been postulated on the basis that the apparatus placed at A(B) usually involves a selector (polarized, Stern-Gerlach, etc.) and a detector, and $p(\lambda, \infty)$ is the detection probability when the selector is not present. (The labels a, a', b, b' usually represent the orientations of the selector.) Then, it is assumed as plausible that the detection probability with the selector in place can be written

$$p(\lambda, a) = q(\lambda, a)p(\lambda, \infty) \qquad (11)$$

where $q(\lambda, a)$ is the probability that a signal crosses the selector. From the fact that $q(\lambda, a) \leq 1$, inequality (9) follows. What is not obvious, however, is Eq. (11), which presumes that crossing the selector does not influence the detection probability of the signal. In fact, if this possibility is taken into account we should write $p(\lambda, a, \infty)$ instead of $p(\lambda, \infty)$ in (11) and the inequalities (9) and (10) cannot be derived. The derivation of (10) involves therefore the contradiction of assuming that <u>local</u> influences between selector and detector in region A <u>do not</u> exist in order to prove that <u>nonlocal</u> influences between apparata in regions A and B <u>do</u> exist if (10) is empirically violated [10].

In summary, homogeneous inequalities like (10) do not derive from local realism (represented by Eqs. (4) and (5)) but from particular assumptions like (9). Note that (9) cannot be tested directly because it should be true for any value of the hidden variables λ. On the other hand, inhomogeneous inequalities like (7) could only be tested with very efficient detectors.

4. EXPERIMENTAL TESTS OF THE INEQUALITIES

Many empirical tests of Bell's inqualities have been proposed till now and several have been performed. Practically all tests involve the measurement of the correlation between the spin projections (or polarizations) of pairs of particles (or photons) prepared in a pure quantum state (e.g., a spin singlet). I shall not attempt to review all proposed tests but to discuss only two most important kinds: atomic cascade (or related ones) and molecular tests. (A review of the empirical tests previous to 1978 was given by Clauser and Shimony [7] and only the Orsay [11] and Stirling [12] groups have made experiments after.) A common feature of these two kinds of tests is that they involve low energies (of the order of electron-volts), whilst energies thousand times larger are involved in most other tests (e.g., pairs of gamma rays produced in the decay of positronium or proton-proton scattering) [7]. Low energy tests have the advantage that the selector used in the EPR experiment (polarizers in the atomic cascade tests or Stern-Gerlach analyzers in the molecular tests) can be described in classical terms. In contrast, the spin projection (or polarization) of a high energy particle (or photon) can be only measured by the interaction with another particle, and this process must be analyzed using quantum concepts. The recourse to quantum theory for the analysis of the experiment invalidates the test of local realistic theories in most high energy tests. (From now on I shall use the word photon only as a shorthand for light signal, without necessarily attaching quantum properties to it.)

In atomic cascade experiments [7], the inhomogeneous inequality (7) is very well fulfilled, the left hand side being about one thousand times greater than the right hand. This is due to the combination of two facts. In the first place, the pair of photons to be analyzed appear in a three-body decay with the consequence that only scarcely the two photons propagate in opposite directions. The probability that a photon enters the lenses system (covering an angle not greater than 60°) is about 10% and the probability that both photons of a pair enter is of 1%. It is possible, in principle to derive the inhomogeneous inequality (7) only for pairs of photons such that both enter the lenses systems (it is enough to average (6) over the appropriate subensemble of values of λ). It is not so easy, however, to estimate the probabilities involved in (7) from the measured rates because the ratio between coincidence and single counting rates is not the same as the ratio between coincidence and single probabilities. But even if the fraction of single counts corresponding to pairs both entering the measuring apparata could be accurately estimated a problem appears with the low efficiencies of photon detectors. In fact, the efficiencies are of order 15%, so that the left hand of (7) is about one hundred times higher than the right hand even for the restricted subensemble of pairs just discussed.

The standard procedure to circumvent the problem is to assume that the ensemble of pairs actually detected is a representative sample of the pairs arriving at the detectors. But this implictly uses an assumption of indistinguishability for the photons which is typically quantal. It is obvious that this assumption means nothing, for instance, for classical wave packets. In the experiments performed, the hypothesis of indistinguishability is incorporated through Eq. (11) or inequality (9), and the inequalities tested have been always homogeneous like (10). It is not necessary to repeat here the arguments given in the previous section showing that homogeneous inequalities are not genuine Bell's inequalities, derived only from local realism. In conclusion, atomic-cascade experiments are inadequate for the test of local realism, a fact already recognized 12 years ago [9], but not seriously taken into account until recently [13].

This conclusion does not mean that the experiments performed have been useless, because, if we believe in local realism, they have shown the existence of a new physical phenomenon: the violation of the inequalities (9) by light signals. That is, the detection probability of a light signal can be enhanced in crossing a polarizer. The discovery of enhancement opens a new line of research in optics whose consequences may be far reaching. My colleugue T. W. Marshall speaks in this conference [14] about our joint work on this line.

In view of the impossibility of testing local realism with atomic-cascade experiments, due to the low efficiency of optical photon detectors, Lo and Shimony [15] have proposed a molecular test. In it, a sodium molecule in a singlet state is dissociated, by laser light, in two atoms whose spin components along chosen axes can be measured with Stern-Gerlach analyzers. This kind of test solves, in principle, the two difficulties discussed above for the atomic-cascade tests . In fact, the molecular dissociation is a two-body problem so that the resulting atoms travel in opposite directions with the consequence that if one enters the aperture of the first measuring system, also the other atom of the pair likely enters the corresponding aperture in the second system. In the second place, very efficient detectors are available and also the Stern-Gerlach analyzers have a high efficiency. A careful analysis, however, has shown that the experiment, as it was initially proposed, does not provide a reliable test of LHV models [16].

Then the present status of the empirical test of Bell's inequalities can be summarized as follows. <u>No experiment performed till now has provided a valid test of local realism</u>. In fact, high energy experiments should be interpreted using quantum theory if the results are to be considered as violations of Bell's inequalities, which invalidates them as tests of (classical-like) LHV models. Atomic-cascade experiments are useless due to the low efficiency of optical photon detectors. Also, no suitable test seems to be in preparation for the near future. The lack of a true empirical test of local realism versus quantum mechanics 20 years after Bell's discovery suggests that the contradiction of principle may not exist in practice.

5. TIME-DEPENDENT AND COLLECTIVE EFFECTS IN THE SOURCE

As said in Section 3, the Bell inequalities are not derived from (3), which I use a representation of local realism, but from (4). In going from (3) to (4) an approximation is made with the result of neglecting a number of possible time-dependent and collective effects in the source. (Even (3) is not general enough for the analysis of the actual experiments, which use an electronic coincidence circuit producing an asymmetry between the apparata in regions A and B, asymmetry not reflected in (3) [17].) In practice, the source consists of a set of particles which emit pairs of signals when they are suitably excited by an external field. For instance, in atomic cascade experiments, the source consists of atoms excited by laser light (or other means) that emit pairs of photons. Now, using Eq. (4) is equivalent to consider that the behavior of the source can be described as a set of instantaneous events (e.g., the decay of each atom) and that all types of correlations between such events can be neglected. Then, the single time-dependent experiment is analyzed as if it consisted of a large number of independent experiments. (It is paradoxical that the analysis uses the most naive form of local realism.) The neglected correlations, however, may give rise to several effects that I analyze in this section and the following one. They can be classified in three groups:

1) <u>Time-dependent effects</u>, due to correlations between events at different times

2) <u>accidental coincidences</u>, due to events arising at the same time in different places;

3) <u>rescattering</u> of a signal by a particle in the source (e.g., a photon by an atom).

As an example of the relevance of these effects, I shall analyze in detail the problem of the accidental coincidences in the following section. Here I consider briefly time-dependent effects and rescattering.

The first kind of time-dependent effect is the possibility of communication amongst the measuring apparata at regions A and B, or between these apparata and the source, which is possible in all static experiments. This communication may give rise to violations of the Bell inequalities without any conflict with the principle of local action. The experiment of Aspect, Dalibard, and Roger [11] with time-varying analyzers has been performed in order to block this loophole in the refutation of LHV theories. The problem of possible communication between the measuring apparata in static experiments received a great deal of attention in the past decade, but now we have realized that the problem is not very relevant. In the first place already the violation of genuine Bell's inequalities in static experiments would be very strange, because any LHV theory able to explain this violation should involve interaction not decreasing with distance. Secondly, and more important, the inequalities violated in atomic cascade experiments are not genuine (inhomogeneous) Bell's inequalities as discussed in Sections 3 and 4. In consequence, I shall not discuss the point any more.

Another kind of time-dependent effect is related to the practical way of dealing with coincidence counts by means of an electronic circuit. The fact that this establishes correlations between signals emitted at different times and may produce violations of the Bell inequalities has been noticed by several authors in the last few years [17-20]. In particular, Pascazio has pointed out that the coincidence counting electronics used in atomic cascade experiments can act as a source of non-locality [20]. The effect would not arise if there were no correlations between pairs of signals emitted by the source at different times. In practice, weak sources are used in order that each pair of emitted signals is well separated from other pairs, so minimizing the correlations (for instance, in atomic cascade experiments, the time interval between the decays of two different atoms is usually larger than the lifetime of one atom in the middle state of the cascade). However, the hypothesis that signals are well separated in time is not at all classical and should be avoided if LHV theories are to be refuted. For instance, in a pure wave theory of light the source is assumed to emit a continuous wave rather than instantaneous signals. In spite of these criticisms, my opinion is that only short time correlations are likely to be important and these are discussed in detail in the next section. Long time correlations (e.g., involving times much larger than the lifetime of the intermediate state in atomic cascade experiments) are probably less important.

The rescattering of photons (trapping of light) in the atomic source was already noticed in the early (1972) experiment of Freedman and Clauser [7]. Subequent atomic cascade experiments used lower density sources and the problem did not arise until the Orsay experiments [11], which again used intense sources in order to have better statistics. A controversy exists about the relevance of rescattering in these experiments. Aspect et al. [11] recognized the use of more intense sources than Freedman and Clauser, but their beam had smaller transverse dimensions. In these conditions they estimated that less than 1% of the detected light could have experienced a rescattering. However, Selleri [21] estimated that the fraction is much larger and Pascazio [22] obtained a value as high as 33%. In a later paper, Aspect and Grangier [11] have criticized these estimates.

Maybe more important than the exact amount of rescattering is to predict the modification that this fact could produce in the counting rates. In principle, three possibilities should be taken into account: (i) photons initially moving in the right direction to enter the lenses system which are scattered outside them; (ii) photons initially in the wrong direction which are scattered inside; and (iii) photons initially in the right direction scattered so that they yet enter the apertures. The second effect is likely more important than the first one, for reasons to be explained below, so the actual counting rate should be smaller than the one without rescattering. This is, however, irrelevant because only ratios of coincidence counting rates are used in the tests of the (homogeneous) Bell's inequalities. The only effect that remains is a depolarizing effect which would result in correlations between polarization smaller than predicted by quantum theory. No such effect is found either in Freedman and Clauser's or in the Orsay experiments. An accurate estimation of the depolarizing effect is not easy, but it is likely small. This is because the Doppler effect prevents rescattering except for photons moving at right angles with the atomic beam. In fact, the longitudinal dispersion in speeds of the atoms in the beam is large whilst the transverse speeds are very small. As the axis of the lenses systems is perpendicular to the atomic beam, the scattering is more likely at small angles where the depolarization is smaller The argument, however, is rough and a correct estimate of the depolarizing effect should be welcome. The possibility that a photon is coherently scattered by several atoms should also be taken into account.

As a summary of this section, I conclude that time-dependent effects and rescattering of signals by particles in the source are likely not very important for the tests of Bell's inequalities.

6. THE CORRECTION FOR ACCIDENTAL COINCIDENCES

In any experimental test of the Bell inequalities the source contains many particles emitting pairs of signals. When two signals are detected in coincidence, it is impossible to know whether they come from the same particle or from two different ones or if they have been emitted collectively by several particles. Actually, in the derivation of the Bell inequalities given in Section 3 it is irrelevant whether a pair of signals is emitted by one particle or by the whole source. Therefore any analysis of the way in which pairs of signals are emitted is not necessary. However, it is a fact that most experiments performed until now do not show a violation of the Bell inequalities by the raw empirical data, but only after a background subtraction of the "accidental coincidences." I made this criticism at the Bari workshop [23] held in 1983, on the basis that any background subtraction increases the correlation. At that time only the early experiment by Freedman and Clauser showed a violation of the (homogeneous) Bell inequalities by the raw empirical data [7]. In particular, none of the most accurate experiments by the Orsay group [11] showed that violation. The criticism led Grangier and Aspect to perform a new experiment in 1985 which showed a significant violation of a (homogeneous) Bell's inequality without making any correlation [11]. As a consequence, any criticism resting upon the correction for accidental coincidences in atomic cascade experiments is now obsolete. (Of course, it is also irrelevant because these experiments do not test genuine, i.e., inhomogeneous, Bell inequalities as said in Section 4.) However, the criticism is relevant for the proposed molecular experiments [15, 16]. Then, it is worth to show a specific local realistic model which violates the Bell inequality after the usual correction for accidental coincidences. For the sake of clarity, I construct the model for atomic cascade experiments although, as said above, the criticism is not important in this case.

In a typical experiment, atoms in a source are excited by incoming light (e.g., a laser). It is assumed that, from time to time, one of the excited atoms decays with the emission of two photons, say a "green" photon and a "blue" photon (these colors correspond to the wavelengths of the photons emitted in the cascade $4p^2\,^1S \rightarrow 4s4p\,^1P_1 \rightarrow 4s^2\,^1S_0$ of the calcium atom, used in four of the experiments [7, 11]). After crossing a lenses system, a filter and sometimes a polarizer, each photon can be detected by an appropriate phototube. A coincidence circuit exists such that a count is recorded whenever a green photon is detected at a time, say t_0 and a blue photon is also detected, by other detectors, within the time interval ($t_0 + t$, $t_0 + t + t_w$). The coincidence window, t_w, is usually chosen to be larger than the lifetime, τ, of the intermediate state of the atomic cascade. Two rates, $R_{exp}(\phi, t)$ and $R_{exp}(\infty, t)$, can be measured as functions of the delay time, t. $R_{exp}(\phi, t)$ is the coincidence rate with two polarizers inserted, one for each photon, at a relative angle ϕ. $R_{exp}(\infty, t)$ is the coincidence rate when both polarizers are removed. (In some cases it is also recorded the coincidence rate with only one polarizer inserted.)

The single rates may be measured too, both with and without polarizers. We shall label $R_1(R_2)$ the single rates of green (blue) photons at the corresponding detector with the polarizer inserted and $R_1'(R_2')$ the single rate without polarizer.

The relevant data reported in a typical experiment can be fitted by the following expressions:

$$R_{exp}(\phi, t) = R(\phi) \exp(-t/\tau) + A; \quad R_{exp}(\infty, t) = R_0 \exp(-t/\tau) + B; \quad (12)$$

$$R_1 R_2 = A/t_w; \quad R_1' R_2' = B/t_w; \quad R(\phi)/R_0 = C \pm D \cos 2\phi; \quad (13)$$

where the constants A, B, C, D and R_0 are positive. The (±) sign in Eq. (13) corresponds to the 0-1-0(1-1-0) cascade of the calcium (mercury) atom used in four [7, 11] (three [7]) experiments. The Stirling group [12] uses the two-photon decay of atomic deuterium, which is not a cascade, but is similar to the 0-1-0 case with respect to the polarization correlation of photons. As seen in (12), the time variation of the coincidence rates consists of two parts: a constant background (A or B), and an exponentially decreasing term ($R(\phi)$, or R_0, times $\exp(-t/\tau)$). It is standard practice to assume that the background term (A or B) corresponds to accidental coincidences and the exponentially decreasing term corresponds to true coincidences. The main purpose of the present section is to show that this assumption is questionable if one attempts to refute all LHV models.

Actually, not all the parameters involved in Eqs. (12) and (13) are given in the published reports of the experiments, but usually these values can be obtained from the reported data. In general, it has been reported that the background rates, A and B, are proportional to the square of the cascade rate, R, whilst $R(\phi)$ and R_0 are just proportional to R. This is interpreted as an additional argument for ascribing the background to accidental coincidences. In order to show that neither the dependence on the time delay given by Eqs. (12) nor the dependence on the cascade rate found empirically are sufficient justification for the background subtraction practice, I construct a simple model which reproduces Eqs. (12) and (13) and gives also the empirical dependence of the parameters on the cascade rate for most experiments.

In the first five [7] experiments the quantity more accurately measured is the Freedman parameter

$$\delta = |R(3\pi/8) - R(\pi/8)|/R_0 - 1/4. \quad (14)$$

The refutation of LHV theories rests upon the evidence of a positive value for δ because it is claimed that all LHV enhancement-free (see Section 4 above) models predict $\delta < 0$, this being the Bell inequality tested. That inequality can be easily derived from (10). However, this claim involves making specific assumptions about the mechanism of production of accidental coincidences. Without these assumptions the parameter to be tested is not δ, but the following one:

$$\delta' = |R_{exp}(3\pi/8, 0) - R_{exp}(\pi/8, 0)|/R_{exp}(\infty, 0) - 1/4 =$$

$$[\delta - 1/4(B/R_0)] (1 + B/R_0)^{-1}. \quad (15)$$

In order to make this point clear, we shall show that there are enchancement-free models violating $\delta < 0$, but not $\delta' < 0$.

The model assumes that each photon is characterized by a single parameter λ and that this parameter has the same value for the two photons (green and blue) emitted from the same atom. The probability of a true

322

coincidence count in a time-delay channel, conditional to the decay of an atom is assumed to be

$$P(a, b; t) = \exp(-t/\tau)\int p(\lambda, a)p(\lambda, b)\rho(\lambda)d\lambda, \qquad (16)$$

where I use a notation similar as in (4). The factor $\exp(-t/\tau)$ is the probability that the blue photon is emitted within t and $t + t_w$ after the emission of the green one (assuming $t_w \gg \tau$). Similarly, if both polarizers are removed, the coincidence probability is assumed to be

$$P(\infty, \infty; t) = \exp(-t/\lambda)\int p(\lambda, \infty)p(\lambda, \infty)\rho(\lambda)d\lambda, \qquad (17)$$

where ∞ represents the absence of polarizer. The model will be enhancement-free if (9) is fulfilled.

In order to calculate the probability of an accidental coincidence, I define $W(\lambda'/\lambda; t)d\lambda'$ to be the probability that there is one blue photon with parameter in $d\lambda'$ of λ', emitted in the time interval $(t_0 + t, t_0 + t + t_w)$, conditional to the emission of a green photon with parameter λ at time t_0, excluding the case that both photons come from the same atom. Then, the probability of an accidental coincidence in the time-delay channel conditional to the decay of an atom at time t_0 will be

$$P_{ac}(a, b; t) = \int d\lambda\rho(\lambda)p(\lambda, a)\int W(\lambda'/\lambda; t)p(\lambda', b)d\lambda'. \qquad (18)$$

The function p is assumed to be the same in (18) and (16) because the lenses systems, polarizers and detectors "do not know" whether the arriving photons come from different atoms or from the same one. In analogy with (18), if the polarizers are removed we will have

$$P_{ac}(\infty, \infty; t) = \int d\lambda\rho(\lambda)p(\lambda, \infty)\int W(\lambda'/\lambda; t)p(\lambda', \infty)d\lambda'. \qquad (19)$$

Finally the total coincidence counting rates are

$$R_{exp}(\phi, t) = R[P(\phi, t) + P_{ac}(\phi, t)];$$

$$R_{exp}(\infty, t) = R[P(\infty, t) + P_{ac}(\infty, t)], \qquad (20)$$

where R is the rate of decay of atoms in the source. We have simplified the notation by writing ϕ instead of (a, b) and (∞) instead (∞, ∞). For single rates, similar arguments to those leading to (20) give

$$R_1 = R\int d\lambda\rho(\lambda)p(\lambda, a); \quad R_1' = R\int d\lambda\rho(\lambda)p(\lambda, \infty); \quad 1 \leftrightarrow 2; \quad a \leftrightarrow b. \qquad (21)$$

The model for accidental coincidences depends essentially on the choice of the function $W(\lambda/\lambda'; t)$. In order to get a suitable expression for this function, we shall consider its long-time and short-time behavior. It seems obvious that polarization of a blue photon emitted at time t could not possibly be correlated with the polarization of a green photon emitted by another atom at time 0 if t is very large, so that W cannot depend on λ for large t. Also, the integral of W over λ' (which is the expected number of blue photons emitted between t and $t + t_w$) should be equal to Rt_w. Therefore, we must assume

$$W(\lambda'/\lambda; t) \rightarrow Rt_w\rho(\lambda') \text{ if } t \rightarrow \infty,$$

ρ being the same density function used in (16) to (19). In contrast, there are no restrictions in principle for atoms decaying at the same, or almost, the same time. These emissions might be uncorrelated in polarization, so that again $W \propto \rho(\lambda')$, or completely correlated, so that $W \propto \delta(\lambda - \lambda')$. For instance, quantum mechanics predicts complete polarization correlation between a photon produced by stimulated emission and the photon inducing that decay. (Note that if the green photon of atom 1 is correlated with the green photon of atom 2, it is also correlatd with the blue photon emitted by atom 2 subsequently.) Such correlations cannot be forbidden in a general LHV theory. Consequently, we choose the simple, but general enough, expression

$$W(\lambda'/\lambda; t) = Rt_w\{[1 - \alpha \exp(-t/\tau)]\rho(\lambda') + \alpha \exp(-t/\tau)\delta(\lambda - \lambda')\},$$

$$0 \leq \alpha \leq 1, \tag{22}$$

that fulfills the asymptotic condition stated above. This expression involves the same exponential appearing in (16) and (17), that corresponds to the decay law of the intermediate state in the atomic cascade. Equation (22) is not the most general expression that may be used. In particular, we might have written β instead of α in front of $\delta(\lambda - \lambda')$ with the condition $\beta \geq 0$. This choice, however, will make the model more complex without much profit. I obtain an enhancement-free model by choosing λ to be uniformly distributed in the interval $[0, 2\pi]$ and the functions p to be

$$p(\lambda, a) = p(\lambda, \infty)[^1/_2 + \beta \cos(2\lambda - 2a) - \gamma \cos(6\lambda - 6a)],$$

$$P(\lambda, b) = p(\lambda, \infty)[^1/_2 \pm \beta \cos(2\lambda - 2b) \mp \gamma \cos(6\lambda - 6b)]. \tag{23}$$

For the sake of simplicity, we assume also that $p(\lambda, \infty)$ is independent of λ. The positivity of p and the no-enhancement condition (9) put the following constraints on the values of β and γ

$$\beta \leq \sqrt{3}/3; \quad (2\gamma)^{1/3} - 2\gamma \geq 2\beta/3 \text{ if } 9/16 \leq \beta \leq \sqrt{3}/3. \tag{24}$$

Values of β smaller than 9/16 have no interest in practice. The model parameters α, β, and γ should be fitted to the data of each experiment.

From the model Eqs. (16) to (24) it is straightforward to derive Eqs. (12) and (13) with the parameters given by

$$A = B/4; \quad B = R^2 t_w(\infty)^2; \quad R_0 = Rp(\infty)^2; \quad R_1 = R_2 = \tfrac{1}{4} Rp(\infty);$$

$$R_1' = R_2' = Rp(\infty); \tag{25}$$

Note that the model predicts correctly the empirical dependence of the different parameters on the cascade rate, R, and also the empirical time dependence in Eqs. (12) and (13). The model does not predict Eq. (13) exactly, but the approximate expression

$$R(\phi) = (R_0/4)\{1 + 2(1 + \alpha Rt_w)[\beta^2 \cos 2\phi \pm \gamma^2 \cos 6\phi]\}. \tag{26}$$

The Freedman parameter (14) is easily calculated from (26) to be

$$\delta \text{ model} = (\sqrt{2}/2)(\beta^2 - \gamma^2)(1 + \alpha Rt_w) - \tfrac{1}{4} \tag{27}$$

Taking (22), (24), and (26) into account, it is possible to choose values of α, β, and γ giving a positive Freedman parameter for experiments such that the product Rt_w is higher than 0.12. This shows that experiments with a cascade rate above this limit are unreliable as tests of enhancement-free objective local models. In contrast with the irrelevance of the parameter δ (16), any violation of $\delta' < 0$ (δ' defined by (15)) refutes our model. In fact, from (15) and (27) it follows

$$\delta' \text{ model} = (\sqrt{2}/2)(\beta^2 - \gamma^2)(1 + \alpha Rt_w)(1 + Rt_w)^{-1} - \tfrac{1}{4} \leq$$

$$(\sqrt{2}/2)(\beta^2 - \gamma^2) - {}^1/_4 \leq -0.027, \tag{28}$$

where the first inequality follows from the condition $\alpha < 1$ (see (22) and the second one from (24). This proves our thesis, namely, that Bell's inequalities must be tested with the raw empirical rates (Eq. (15)) and not with the rates corrected by a background subtraction (Eq. (14)).

As said above, the recent experiment by Aspect and Grangier [11] shows a very significant violation of a (homogeneous) Bell inequality, so that enhancement-free LHV theories can be considered as refuted.

7. THE SEARCH FOR QUANTUM VIOLATIONS

The history of physics shows that every theory is sooner or later superseded by another theory which is correct in a wider domain. Then, we should expect that quantum theory has a limited validity and some future experiment will violate it. This is not, however, the common belief. In fact, quantum theory is not presently considered just a theory, but a general framework for physical theories which has replaced the classical framework. Then, a large fraction of the scientific community sees the future progress of physics in the replacement of one particular quantum theory by another one. This kind of progress took place, for instance, in going from quantum electrodynamics to electroweak theory. There is another fraction of the scientific community which finds serious conceptual problems in the usual quantum theory, and is searching for either a new formalism or a new interpretation of the present formalism (most times in terms of LHV theories). In consequence, when Bell proved [2] the incompatibility between LHV theories and the quantum formalism, some people said, even before any experiment, "LHV theories are dead" whilst others said "quantum theory is dead." For the latter it was just a question of time to find an empirical violation of quantum theory.

There was a surprise to realize that no experiment performed before Bell's work was able to discriminate between quantum and LHV theories. Almost a decade elapsed until the first experiment considered reliable was made (by Freedman and Clauser [7] in 1972). An important loophoole, due to the possibility of communication between detectors in static experiments, was blocked a decade later (Aspect et al. [11] in 1982). It is remarkable that another loophole due to the low efficiency of optical photon detectors, although well known after the paper by Clauser and Horne [9], was not taken so seriously as it should. Now, after, a critical analysis of the performed experiments, the state of the art can be summarized by the following three conclusions:

1) No experiment yet performed has shown a violation of LHV theories.

2) If LHV theories are to be maintained, a new physical phenomenon must be admitted: the enhanced detection probability of some light signals that have crossed a polarizer.

3) The domain of phenomena where a real conflict exists between quantum and LHV theories is narrower than previously assumed.

My personal conjecture is that the said domain may not exist [17]. If this is so, the hope for a violation of quantum theory decreases very much. Indeed, in that case it is not necessary to change the quantum formalism in order to be able to interpret it for every real (as opposed to ideal) experiment in terms of LHV models. A part of my conjecture is that the Heisenberg uncertainty relations may play some role in avoiding the conflict between LHV and quantum theories. I am not in a position to prove this conjecture, but it is easy to see that the Heisenberg relations give rise to some constraints on the observability of violations of the Bell inequalities.

For instance, in molecular experiments, a molecule is divided in two atoms with mass M each. Assume that the measuring apparata are situated at a distance 2R, and the source midway between them. Initially, the molecule has uncertainties in position and velocity constrained by the Heisenberg relation

$$\Delta x \Delta v \leq 4M\hbar. \tag{29}$$

In consequence each atom arrives at the corresponding detector with a time uncertainty

$$\Delta t \simeq \Delta x/v + (R/v^2)\Delta v \geq \Delta x/v + R\hbar/(4Mv^2\Delta x) \geq (R\hbar/Mv^3)^{1/2}, \tag{30}$$

where the second inequality comes from (29) and the last one is a straightforward mathematical relation. Now, the test of locality (relativistic causality) is only possible if

$$\Delta t \ll 2R/c, \quad c = \text{speed of light}, \tag{31}$$

which gives

$$R \gg c^2/\hbar(4Mv^3). \tag{32}$$

The distance so found is not very large (although macroscopic) for typical molecular experiments, so the constraint is not important. The result, however, is enough to show that the Heisenberg relations play some role in making more difficult the conflict between LHV and quantum theories.

The Bell theorem does not give, for the moment, a clue for finding violations of quantum theory. However, the critical analysis of the performed experiments suggest several possible new experiments where it is more likely to find such violations. In the domain of optics, where most experience has accumulated as a result of the recent experiments, I think that it is convenient to search in two areas:

1) Photon anticorrelation effect in a beam splitter.

2) Two photon polarization correlation in atomic cascade experiments.

The photon anticorrelation in a beam splitter is discussed in the talk by T. W. Marshall. Therefore, I will consider only the possible quantum violations in the coincidence counting rate, $R(\phi)$, for atomic cascade experiments with two polarizers at a relative angle ϕ. As discussed in Section 3, there are LHV models with enhancement which fully agree for $R(\phi)$ with the quantum predictions [9, 25]. These models, however, have the inconvenient feature that the probabilities $p(\lambda, a)$ and $p(\lambda, b)$ (see (4)) are very different functions of their arguments. The need for such asymmetry was proved by Caser [24]. Symmetric LHV models cannot agree completely with the quantum prediction Eq. (13) but predict the existence of terms of the form $\cos 2n\phi$ $(n > 1)$ in the coincidence rate $R(\phi)$ [13]. Marshall [26] has investigated the statistics necessary for the discovery of these terms and it is quite larger than that available till now. There is, however, the hope of getting a theorem that gives a bound for some relevant quantity measurable with lower statistics. Such a theorem would provide a test between symmetrical LHV theories and quantum theory, which will be more useful than Bell's theorem, useless in the domain of atomic cascade experiments as said in Section 4. My feeling is that this is one of the most likely places to find quantum violations.

The test of symmetric LHV theories against quantum theory is easier using calcite polarizers. It is quite remarkable that only one [27] atomic cascade experiment has been made using calcite polarizers and this experiment is the only one showing a violation of the quantum prediction. In spite of this fact, none of the subsequent experiments have used this type of polarizer. The relevant property of calcite polarizers is the very small transmittance, ε_m, for photons with polarization perpendicular to the polarizer's axis ($\varepsilon_m < 10^{-4}$). The polarizers used in all other experiments have ε_m of order of a few percent (which for the transmitted amplitude corresponds to 10-20%). In this sense, calcite polarizers are close to the ideal ones. (They have a relatively low transmission $\varepsilon_m \simeq 0.9$, for photons with parallel polarization, but this departure from ideal behavior is not relevant for our purposes.)

In experiments with calcite polarizers, the constraints posed by symmetric LHV theories are rather stringent. This can be seen if we compare the quantum prediction (with the ideal value $\varepsilon_m = 0$)

$$[R(\phi)/R_0]_Q = \tfrac{1}{4}\ \varepsilon_M^2\ (1 + \cos 2\phi), \tag{33}$$

with the prediction of symmetric LHV theories

$$[R(\phi)/R_0]_{LHV} = 1/\pi \int_{-\pi/2}^{\pi/2} d\theta\ p(\theta)\ p(\theta + \phi),\ p \geq 0. \tag{34}$$

This is not the most general expression because it involves a single hidden variable θ, but it is general enough to show the trend. Even for this simplified form, it is a difficult mathematical problem to find the function $p(\theta)$ giving for (34) the closest fit to (33) (and it is ambiguous to define "closest"). An estimate of the distance between (33) and (34) can be obtained searching for the $p(\theta)$ that gives agreement for the values $\phi = 0$, $\phi = \pi/4$, and $\phi = \pi/2$. It is remarkable that there is a unique solution

$$p(\theta) = \varepsilon_M \text{ if } |\theta| < \pi/4, \text{ 0 otherwise}, \tag{35}$$

which gives

$$R(\phi)/R_0 = {}^1/_2\ \varepsilon_M^2\ (1 - 2\phi/\pi),\ 0 < \phi < \pi/2. \tag{36}$$

This is quite different from (33), as it can be seen by calculating the value

$$F_{LHV} = |R(\pi/8) - \dot{R}(3\pi/8)|/R_0 = {}^1/_4\ \varepsilon_M^2 \tag{37}$$

to be compared with the quantum prediction

$$F_Q = |R(\pi/8) - R(3\pi/8)|/R_0 = {}^1/_4\ \varepsilon_M^2 \sqrt{2}. \tag{38}$$

Alternatively, we may choose to fit the value (38) plus the values for $\phi = 0$ and $\phi = \pi/2$. Again the problem has the unique solution

$$p(\theta) = 2\ \varepsilon_M \cos 2\theta \text{ if } |\theta| < \pi/4, \text{ 0 otherwise, which gives} \tag{39}$$

$$R(\phi)/R_0 = (\varepsilon_M^2/\pi)[(\pi/2 - \phi) \cos 2\phi + {}^1/_2 \sin 2\phi]. \tag{40}$$

But now there is disagreement for $\phi = \pi/4$:

$$[R(\pi/4)/R_0]_Q = \varepsilon_M^2/4,\ [R(\pi/4)/R_0]_{LHV} = \varepsilon_M^2/2. \tag{41}$$

Both with choice (35) and with the choice (39) we obtain perfect agreement for three parameters but a large deviation for the fourth one. This leads us to estimate that the <u>best choice</u> could give some deviation for most values of ϕ, but these deviations should likely be of the order of 1/3 or 1/4 of the maximum one, found above. If this guess is correct, it is likely to have a relative deviation for the parameter F (see (37) and (38))

$$(F_Q - F_{LHV})/F_Q > 0.10 - 0.14. \tag{42}$$

This parameter is important because it is the one most frequently measured (see Section 6, Eq. (14) and comments below it). In the Holt and Pipkin experiment [27], the measured deviation was

$$(F_Q - F_{EXP})/F_Q = 0.19 \pm 0.05, \tag{43}$$

which is remarkably close to the estimate (42); note that both are positive.

In all other atomic cascade experiments [7, 11, 12], the polarizers have a minimum transmittance ranging between 0.01 and 0.04. Then, symmetric LHV models are possible much closer to quantum mechanics. For instance, the pioneering model of this type [13], leads to

$$\frac{[R(\phi)/R_0]_{LHV} - [R(\phi)/R_0]_Q}{[R(\phi)/R_0]_Q} = \frac{\beta \cos 4\phi}{1 + \alpha \cos 2\phi} \tag{44}$$

with typical values $\alpha \simeq 0.84$, $\beta \simeq 0.04$, giving relative deviations of order 0.02, i.e., much smaller than (42). These deviations cannot be measured with the statistics of the performed experiments and so it is not strange that a perfect agreement with quantum theory has been reported.

This discussion suggests the following question: Is the quantum violation observed 12 years ago by Holt and Pipkin a common feature of atomic cascade experiments with calcite polarizers? The importance of the answer shows the urgent need for new experiments of this kind.

REFERENCES

1. P. A. M. Dirac, Methods in theoretical physics, in: "From a life of physics," A. Salam, Ed., Trieste Lectures, International Energy Agency, Wien (1969).
2. J. S. Bell, Physics, New York, 1, 195 (1965).
3. K. Popper, Bell's theorem: A note on locality, in: "Microphysical reality and quantum formalism," G. Tarozzi, Ed., Reidel Publishing Co., Dordrecht (in press).
4. A. Einstein, B. Podolsky, and N. Rosen, Phys. Rev., 47, 777 (1935).
5. F. Selleri, Found. Phys., 12, 645 (1982).
6. L. de la Peña, A. M. Cetto, and T. A. Brody, Lett. Nuovo Cimento, 5 177 (1972).
7. J. F. Clauser and A. Shimony, Rep. Progr. Phys., 41, 1881 (1978).
8. E. Santos, Phys. Lett. A, 115, 363 (1986).
9. J. F. Clauser and M. A. Horne, Phys. Rev., D10, 526 (1974).
10. T. W. Marshall, private communication.
11. A. Aspect, P. Grangier, and C. Roger, Phys. Rev. Lett., 47, 460 (1981); 49, 91 (1981); A. Aspect, J. Daliberd, and G. Roger, Phys. Rev. Lett., 49, 1804 (1982); A. Aspect and P. Grangier, Lett. Nuovo Cimento, 45, 435 (1985).
12. W. Perrie , A. J. Duncan, H. J. Beyer, and H. Kleinpoppen, Phys. Rev. Lett. 54, 1790 (1985).
13. T. W. Marshall, E. Santos, and F. Selleri, Phys. Lett., 98A, 4 (1983).
14. T. W. Marshall, Towards a realistic theory of measurement, in: "Microphysical reality and quantum formalism," G. Tarozzi, Ed., Reidel Publishing Co., Dordrecht (in press).

15. T. K. Lo and A. Shimony, Phys. Rev., A23, 3003 (1981).
16. E. Santos, Phys. Rev., A30, 2128 (1984); A. Shimony, Phys. Rev., A30, 2130 (1984).
17. E. Santos, Phys. Lett., 101A, 379 (1984).
18. G. C. Scalera, Lett. Nuovo Cimento, 38, 16 (1983); 40, 353 (1984).
19. S. Notarrigo, Nuovo Cimento, 83 B, 173 (1984).
20. S. Pascazio, Phys. Lett., 111A, 339 (1985).
21. F. Selleri, Lett. Nuovo Cimento, 39, 252 (1984).
22. S. Pascazio, Nuovo Cimento, 5, 23 (1985).
23. E. Santos, Stochastic electrodynamics and the Bell inequalities, in: "Open questions in quantum physics," G. Tarozzi and A. van der Merwe, Eds., Reidel Publishing Co., Dordrecht (1985).
24. S. Caser, Phys. Lett., 102A, 152 (1984).
25. D. Home and T. W. Marshall, Phys. Lett., 113A, 182 (1985).
26. T. W. Marshall, Phys. Lett., 99A, 163 (1983); 100A, 225 (1984).
27. R. A. Holt and F. M. Pipkin, Preprint Harvard University (1974).

PHYSICAL PHOTONS: THEORY, EXPERIMENT AND IMPLICATIONS

Geoffrey Hunter and Robert L. P. Wadlinger

Department of Chemistry, York University
4700 Keele Street, Toronto, Ontario, Canada   M3J 1P3

Abstract

A physical model of the photon is presented.   It is a solution of Maxwell's equations confined within a finite region of space–time along the photon's axis of propagation.   The rotating/oscillating electromagnetic field has the observed photon–eigenvalues of linear and intrinsic (spin) angular momentum.   The model predicts two angular momentum eigenstates having either positive or negative helicity (left or right circular polarization).

The finite region containing the field is defined by the relativistic principle that congruent events within it are causally connected (separated by timelike intervals).   The region is a circular ellipsoid whose major axis and cross–sectional circumference are both one wavelength long.   Excited states of the field containing two or more quanta of energy within the same ellipsoidal volume represent multiphotons.

This physical model of the photon is consistent with experimental properties of electromagnetic radiation, including photon bunching and anti–bunching, multiphoton absorption, and the transmission of microwaves through apertures.   A microwave experiment designed to measure the photon's diameter is reported; the measured value accords with the theoretical model's prediction within the experimental error of half a percent.

The finite–field model of the photon is both a particle and a wave, and hence we refer to it by Eddington's name "wavicle".   That the wavicle's position is essentially uncertain within the size of its finite domain leads to the idea that the minimum quantum of action arises because the particle cannot transfer its momentum in less time than it takes to traverse the length of its own domain.   Its minimum action (equal to Planck's constant) is the product of its length and its momentum.

Thus the dichotomy of the wave–particle duality of light is replaced by a unity, and the schism between the Copenhagen philosophy of fundamental indeterminacy and the contrary, determinist, view that indeterminacy is only an experimental limitation, is resolved in favour of the latter, because internally the wavicle is classical and causal, but its interactions necessarily involve non–causal, space–like intervals, and hence in interactions (i.e. measurements) the indeterminacy of established quantum mechanics prevails.

The Wave–Particle Duality and Physical Photons

It is well established experimentally that light and matter exhibit both particle–like and wave–like properties. Quantitatively these two properties are related by the De Broglie relation:

$$p \lambda = h \qquad (1)$$

p being momentum, λ wavelength, and h Planck's constant. The optical properties of light (reflection, refraction, diffraction, dispersion, interference) are its wave–like properties. Blackbody radiation, line–spectra, the photo–electric effect, and Compton scattering, are its particle–like properties.

The conceptual distinction between waves and particles is that a wave extends indefinitely in space, whereas a particle is a localised, point–like object. The term "duality" or "dualism" expresses the view that matter and light exhibit two irreconcilable properties, in some experiments behaving as an extended wave, and in others as a collection of point–like particles. This dichotomy is the prevalent viewpoint of most theoretical physicists: that the wave–particle duality is a dilemma that we must simply accept as one of the quirks of nature that defies a deeper understanding.

In the current mainstream of Quantum Electrodynamics (Loudon, 1983; Craig and Thirunamachandran, 1984) the photon is regarded simply as an occupied mode of the electrodynamic field contributing an energy $h\nu$ to the total field–energy; the concept of the photon as a physical entity is implicitly disregarded. We believe that this conventional view of the photon as simply a quantum of energy, is incomplete. Such a view of the hydrogen atom is known to be incomplete; one needs the wavefunctions in order to predict all the properties of the hydrogen atom. By analogy we believe that a complete description of the photon requires photon–wavefunctions as well as energy eigenvalues.

Since the inception of quantum mechanics in the first part of this century, the concept of the photon as a physical entity has been held by several distinguished scientists, including the founding fathers of the wave–particle duality, Albert Einstein and Louis de Broglie. The following quotation from Einstein expresses the concept (Diner, Fargue, Lochak, and Selleri, 1984):

> ".. the energy of a light ray spreading out from a point source is not continuously distributed in space, but consists of a finite number of energy quanta which are localised at points in space, which move without dividing, and which can only be produced and absorbed as complete units".

The term "photon" was coined by G. N. Lewis in a letter to Nature (1926). Like Einstein, he expressed the view that the photon is a physical entity — a localised packet of electromagnetic energy. He conjectured that when an isolated atom in an excited state spontaneously decays to its ground state, the light that is emitted must be a single photon having a well defined energy and frequency, and the process must take place within a time τ that is shorter than the average lifetime of the excited state. Hence the length and radial breadth of the photon can be no larger than τc. More recent studies of multi–photon processes (Chin and Lambropoulos, 1984) indicate that the time τ is comparable with the period of oscillation of the emitted photon, and hence the spatial dimensions of the photon must be of the order of its wavelength $\lambda = \tau c$.

The Dynamics of the Finite Photon

Several models of the photon have been proposed dating back to the work of J. J. Thomson (1924,1925,1936; Bandyopadhyay and Raychaudhuri, 1971; Honig, 1974; Nishiyama, 1977; Levitt, 1978). Thomson's and Honig's models are based upon the field radiated by an oscillating dipole. Like Lewis' concept of the spontaneously radiating atom, they predict that the photon is a localised packet of electromagnetic energy, its size being of the order of its wavelength. The Thomson and Honig models differ from each other and from our model derived below. Our model arose from a units analysis that predicted that the photon is one wavelength long (Wadlinger, 1983).

Our approach was to model a freely propagating photon as an electromagnetic field moving at the speed of light, within a volume that is limited in extent both along, and perpendicular to its axis of propagation. Complete details of our mathematical analysis will be presented elsewhere. In summary, our solutions of Maxwell's equations (Fewkes and Yarwood, 1956) appropriate to a freely propagating photon are:

$$E_x = (\alpha r + \beta/r) \times [AP(\varphi) + BQ(\varphi)] \times S(u) = \mu_0 c \; H_y \; , \; E_z = H_z = 0 \; ,$$

$$E_y = i \times (\alpha r - \beta/r) \times [AP(\varphi) - BQ(\varphi)] \times S(u) = - \mu_0 c \; H_x \qquad (2),$$

where $P(\varphi) = \exp(i\varphi)$, $Q(\varphi) = \exp(-i\varphi)$, $S(u=z-ct) = \exp(2\pi i u/\lambda)$, r is the radial distance from the axis of propagation (the z axis), φ the polar angle around the axis, c the velocity of light, μ_0 the permeability of free space, and α, β, A and B are the arbitrary constants of the general solution.

The components of the electromagnetic field (E_x, H_x, etc) are field amplitudes that are operationally similar to Schrödinger wavefunctions (Green and Wolf, 1953), and hence it is legitimate to operate upon the field components with quantum mechanical operators. The factor S(u) is an eigenfunction of the operator for the z–component of linear momentum ($P_z = (h/2\pi i)\partial/\partial z$) with eigenvalue $+h/\lambda$. This is the De Broglie momentum expected for a photon of wavelength λ moving in the positive z direction.

Similarly the factor $P(\varphi)$ is an eigenfunction of the angular momentum about the z axis ($L_z = (h/2\pi i)\partial/\partial\varphi$) with eigenvalue $+h/2\pi$, and $Q(\varphi)$ is an eigenfunction of L_z with eigenvalue $-h/2\pi$. This angular momentum is intrinsic to the rotating field; i.e. it is spin angular momentum. These eigenvalues accord with the experimental observation that photons have a spin of one unit of $h/2\pi$ (Swartz, 1965).

The function $P(\varphi)$ has positive helicity and $Q(\varphi)$ negative helicity (Craig and Thirunamachandran, 1984). In optics these two states correspond respectively to left and right circularly polarized light. A single photon will have a definite helicity, either positive (B=0) or negative (A=0). Thus circularly polarized light consists entirely of photons with the same helicity. Linearly polarized light corresponds to A=B or A=−B, and elliptically polarized light to A≠B, A≠0 and B≠0.

The Domain of the Wavicle

One of our basic precepts about a wavicle is that it is a causally related entity; i.e. the interval between events within the wavicle must be timelike (Lawden, 1962). The infinitesimal interval ds between a pair of contiguous events is expressed by:

$$ds^2 = c^2 dt^2 - dz^2 - dx^2 - dy^2 \qquad (3)$$

and in the cylindrical coordinates used in (2) by:

$$ds^2 = c^2dt^2 - dz^2 - dr^2 - r^2d\varphi^2 \tag{4}.$$

In view of the rotational motion of the field, the finite interval between a pair of non—contiguous events, Δs^2, must be computed along the geodesics of the non—rectilinear motion. Congruent events have the same phase, the same amplitude, and are on cylinders where r is constant. The phase δ, of the wave (2) is given by:

$$\delta = 2\pi(z-ct)/\lambda \pm \varphi \tag{5},$$

and for a constant phase, differentiation of (5) yields:

$$c\, dt = \mp (\lambda/2\pi)\, d\varphi \tag{6}.$$

Substitution of (6) into (4) together with the constraint r=constant (dr=0) yields the following expression for a finite interval along a geodesic of the rotating field is:

$$\Delta s^2 = [(\lambda/2\pi)^2 - r^2]\, \Delta\varphi^2 - \Delta z^2 \tag{7}.$$

For this interval to be timelike, $\Delta s^2 \geq 0$, and the domain of the wavicle is bounded by the null geodesics $\Delta s^2 = 0$.

Considering a plane z = constant ($\Delta z^2 = 0$), the radius r must satisfy the relation:

$$r \leq \lambda/2\pi \tag{8},$$

which implies that the wavicle's maximum transverse radius r_{max} is:

$$r_{max} = \lambda/2\pi \tag{9}.$$

At this radius the causally related events unite into a single event because the proper time between them is zero. The tangential velocity of the rotating field at this radius, $2\pi\nu r_{max} = (2\pi c/\lambda)r_{max}$, is equal to the velocity of light.

The wave ·motion repeats itself when $\Delta t = \lambda/c$, the period of the oscillation, and this period is equivalent to $\Delta\varphi = 2\pi$. Thus the interval between identical, repeating, points of the wave is:

$$\Delta s^2 = \lambda^2 - (2\pi r)^2 - \Delta z^2 \tag{10}.$$

For a null geodesic ($\Delta s^2=0$) this pair of points coalesces into a single relativistic event, and hence the repeating unit of the wave is bounded by a null geodesic surface defined by:

$$(2\pi r/\lambda)^2 + (\Delta z/\lambda)^2 = 1 \tag{11}.$$

For r = constant, this is the equation of a pair of points on an ellipse. If the origin (z=0) is chosen to be at the centre of the ellipse, then the locus of all such pairs of points is:

$$(2\pi r/\lambda)^2 + (2z/\lambda)^2 = 1 \tag{12}.$$

This is the equation of an ellipse with major axis (along the z—axis) of length λ, and minor axis (in the x—y plane) of length λ/π.

The domain of the wavicle's field is the ellipsoid obtained by revolving the ellipse of eqn.(12) about the z—axis. It is a prolate spheroid (cigar shape) 3.14 times as long as its diameter, its long axis being the axis of propagation. Its length and circumference are both equal to λ, the wavelength. The ellipsoidal surface is the locus of all null geodesics ($\Delta s^2 = 0$) between pairs of equi-phase, equal amplitude points of the field (2).

Waves in free space are normally considered to extend indefinitely; the theory of such waves is usually couched in terms of infinite wave trains. However, waves with discontinuous boundaries were shown to be mathematical tractable by Love (1903). He showed, in particular that "... the velocity of advance of a wave-boundary, involving discontinuity of this type, is the same as that of a wave which does not involve discontinuity.".

This finite domain of the wavicle's field is a quantitative realisation of Einstein's conjecture that events could be causally related "within a local region of space-time".

Energy Eigenstates: Multiphotons

Confinement of the field within the ellipsoid (12) implies that the field must be zero on the surface of the ellipsoid. The solution (2) of Maxwell's equations must be modified to conform with this boundary condition. This can be accomplished by transforming from the coordinates r and z to confocal elliptical coordinates (Stratton, et al, 1956). While the mathematics is somewhat tedious because it involves a pair of simultaneous eigenvalue problems (Hunter and Pritchard, 1967), its essential results are the allowed eigenvalues for the frequency c/λ and angular momentum of the confined field. We have thus far not completed this mathematical imposition of the ellipsoidal boundary conditions. However, the eigenvalues of the energy are obtained as follows.

The eigenvalue equation for the energy eigenvalues is the time-independent Klein-Gordon equation (Bethe, 1964). This covariant eigenvalue equation is derived from D'Alembert's classical, covariant wave equation by differentiating the wavefunction (2) with respect to time to eliminate the time, followed by use of the De Broglie relation (1), the Energy-inertia equivalence relation:

$$E = mc^2 \qquad (13),$$

and the expression for the relativistic mass:

$$m = m_0/[1-(v/c)^2)]^{1/2} \qquad (14)$$

to eliminate the velocity v and the relativistic mass m (Hunter, 1986). It has the form:

$$- (\hbar^2/m_0) \ \nabla^2\psi = [E^2/(m_0c^2) - m_0c^2]\psi \qquad (15).$$

Since the rest-mass, m_0 of the photon is zero, equation (15) simplifies to:

$$- (\hbar^2/m) \ \nabla^2\psi = [E^2/(mc^2)]\psi = E\psi \qquad (16).$$

This covariant eigenvalue equation is for a free photon; i.e. there is no potential energy in the hamiltonian.

The energy eigenstates of the photon-field are obtained by solving equation (16) for a "particle" (really a field) confined within the ellipsoidal box that is the photon's domain. The energy eigenvalues of a particle in a box are (McQuarrie, 1983):

$$E_n = g \ (nh/L)^2/m \qquad (17)$$

where n is the quantum number taking any positive integral value, L is the linear size of the box, m is the particle's mass, and g is a geometrical factor determined by the shape of the box.

Elimination of the relativistic mass m between (13) and (17) for a box of length L=λ (the wavelength) produces, on taking the positive square root:

$$E_n = gnhc/\lambda = gnh\nu \qquad (18).$$

From the Einstein relation:

$$E = h\nu \qquad (19)$$

we deduce that the geometrical factor g=1, and equation (18) thus indicates that the photon energy may be any integral multiple of the quantum hν:

$$E_n = nh\nu \qquad (20).$$

This is a generalisation of the Einstein relation (19). The energy $E_1 = h\nu$ is the ground state (a single—photon), and for n>1 the excited states of the electromagnetic field correspond to what are experimentally known as multiphotons. A multiphoton has the same frequency as the ground state, but it has n times as much angular momentum and energy.

We believe that Multiphotons are commonly produced by the process of stimulated emission. An ellipsoidal photon of energy nhν interacts with an atom in an excited state with energy hν above a lower stationary state. The photon emerges from the interaction with (n+1)hν of energy (in the same ellipsoidal volume) leaving the atom in the lower state. Multiphotons will also be produced whenever a large number of single photons are packed together; e.g. in a light beam focussed to produce a high intensity. Experimentally this is achieved in focussed laser beams (see below).

Like most excited states multiphotons will spontaneously decay to the ground state, in this case producing n spatially separated, but coherent single photons. Such a flock of photons is observed experimentally in the phenomenon of "photon bunching" (Loudon, 1982). "Photon anti—bunching" is simply a way of describing the fact that the energy of a light beam is localised (in the ellipsoidal volumes of our model) rather than being evenly distributed throughout the volume of the beam.

The mathematical representation of multiphoton decay and formation is something that we would like to work out. It is probably related to the theory of photon—photon interference; when 2 ellipsoidal photons interact, what determines whether they produce a multiphoton or interfere destructively? In particular we expect to provide an explanation for the phenomenon of specular interference commonly observed in laser beams. The ellipsoidal photon model seems to imply that interference phenomena arise from interactions between members of a flock of coherent photons; i.e. it is a photon—photon phenomenon rather than a single—photon phenomenon as asserted by Dirac (1967).

Multiphoton Phenomena

Multiphoton absorption by atoms is considered to take place within "one optical period"; i.e. within $\tau = 1/\nu$ (Mainfray and Manus, in Chin and Lambropoulos, 1984, pp.9—10). This is the transit time of the photon wavicle. Mainfray and Manus express the need for having a large number of photons concentrated within 10^{-15} sec (i.e. one optical period). This need is satisfied by our multiphoton model, in which the N photons are congruent in space and time within the same ellipsoidal volume as that of the single photon.

Multiphoton absorption of visible light is observed to take place at intensities above about a megawatt/cm^2 (Panarella, 1974; Held, Mainfray, Manus, and Morellec, 1971). Since the photon wavicle has a cross−sectional area of $\lambda^2/4\pi$, and an intrinsic power of $h\nu/\tau = h\nu^2$, its intrinsic intensity, I_P (power/area) is $4\pi hc^2/\lambda^4$, which is one megawatt/cm^2 for $\lambda=523$ nanometers. At higher intensities than this (disregarding a geometrical packing factor) single ($1h\nu$) photons necessarily overlap to form multiphotons.

For randomly distributed photons the probability of finding a multiphoton with energy $Nh\nu$ is proportional to the ratio of the beam intensity I_B to the photon intrinsic intensity I_P raised to the Nth power. When $I_B<I_P$ this factor $(I_B/I_P)^N$ is much smaller than unity, and when $I_B>I_P$ it is much larger than unity. This explains both the threshold at about $I_B = I_P$, and the observed N^{th} power dependence of the ionization rate upon the intensity of the focussed beam (Chin and Lambropoulos, 1984: p.17 Mainfray and Manus, p.51 Grontier and Trahin, p.68 Crance).

This finite photon overlapping explanation of the Nth power law of multiphoton absorption is simple compared with the complexity of Nth order perturbation theory; the latter is regarded, even by its proponents, as conceptually unsatisfactory (Chin and Lambropoulos, 1984). Our model of multiphoton processes differs from the "effective−photon" model (Panarella, 1977), and from the "quantum potential" theory (Dewdney, Garuccio, Kyprianidis and Vigier 1984). Further analytical work, and possibly experiments, is required to determine which of the four theories best fits the experimental data.

Microwave Generation and Transmission

The axial mode of a helical microwave antenna has a wavelength equal to the circumference of the cylindrical helix, with maximum radiation along the axis of the helix in a well−defined, circularly polarized beam (Kraus, 1950). The axis of propagation and circumference of this antenna correspond precisely with the axis and circumference of the finite photon model. The radiation produced by this antenna consists of wavicles all propagating along the axis of the antenna, and having the same helicity as that of the antenna. In view of the low intrinsic intensity of the microwave wavicle the radiation will consist of multiphoton wavicles containing many quanta ($nh\nu$, $n>>1$).

The transmission of microwaves through a circular aperture is predicted to be strongly attenuated at diameters smaller than λ/π (Andrews, 1960). This is the cross−sectional diameter of the finite photon model. The prediction is based upon the classical electromagnetic theory of a continuous wave. The attenuation is consistent with the simple mechanical notion that the photon cannot pass through an aperture that is smaller than its own diameter of λ/π. The photon wavicle is not, of course a rigid body, and its induced currents in the screen may cause some radiation to appear on the far side of the screen, and in fact the classically derived curve of Andrejewski (King and Wu, 1959) predicts a small finite transmission coefficient at aperture diameters smaller than λ/π. The available experimental evidence is confined to localised field intensity measurements, and while it is broadly in agreement with classical electromagnetic theory and with our photon model, it is inconclusive as evidence for/against the finite photon model. This led us to undertake experiments designed to resolve the issue.

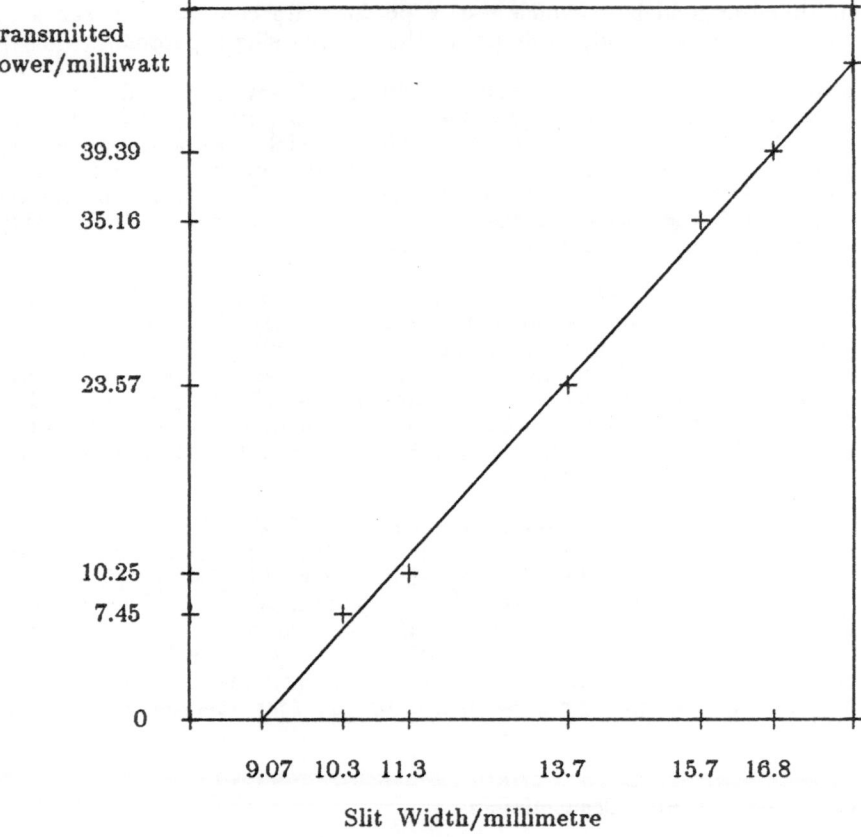

FIGURE 1. Experiment by Fritz Engler, March 29, 1986: X–band microwaves (wavelength = 28.5 mm), rectangular slit (constant length) in thin (0.5mm) aluminum plates. The measured power was corrected by subtraction of the 1st Harmonic content of the beam computed by the formula H = 2.5×(SW−4.54)/(8.1−4.54) based upon the signal at a slit width of 8.1 mm. The least–squares straight line (solid line in the graph) through the 5 experimental points (shown by + , SW = 10.3–16.8) yields an intercept on the slit–width axis of 9.07 ± 0.04 mm (0.5% error). The theoretically expected result is 28.5/π = 9.07 mm.

Measurement of the Photon's Diameter

We have investigated the phenomenon of microwave transmission through apertures in a series of experiments designed to measure the transmitted power as a function of aperture size; we used both circular apertures (size=diameter) and rectangular apertures (size=width). A linear plot of transmitted intensity vs aperture area has an intercept on the aperture axis close to the photon–wavicle's intrinsic diameter of λ/π; the transmitted power is proportional to the difference between the aperture and wavicle areas.

Complete details of these experiments will be reported elsewhere; there were naturally several experimental factors that had to be taken into account in order to obtain an accurate measurement. Our latest result is summarised in Figure 1. The measured cut—off diameter is equal to the value λ/π predicted by our ellipsoidal model within the experimental error of 0.5%. This is direct evidence for the validity of our model of the photon.

Diffraction and Interference at Low Light Intensities

Some experiments on the diffraction of light at low intensities have re—opened the question of whether interference is a single—photon, or a photon—photon, phenomenon (Dontsov and Baz, 1967; Panarella, 1985). The latter explanation accords with our model of the photon as an indivisible, physical entity. In our photon model interference typically occurs between the coherent photons of a photon—bunch.

Light sources often exhibit the property of photon bunching (Loudon, 1983). Stimulated emission adds one quantum of energy ($h\nu$) to the incident photon; e.g. an incident single ($1h\nu$) photon stimulates the atom to give it an additional $h\nu$ of energy, emerging as a double photon having $2h\nu$ of energy. This subsequently decays into 2 spatially proximate, coherent photons. Repetition of this process thus produces bunches of coherent photons. For this physical concept of photon bunching the term "flock" is very appropriate, for the photons of the bunch are travelling together just like a flock of migrating geese. Stimulated emission producing photon bunches is believed to be of common occurence; it is of course dominant in lasers, but is also believed to be important in thermal light sources.

In our photon model interference effects are typically produced by photon bunches. In a diffractometer the members of a bunch will be deflected at different angles depending upon the impact parameter of each photon with the edge of the slit; photons that miss the slit walls are transmitted undeflected. This model of diffraction explains why fringe intensity is a maximum for slits of about a wavelength wide (the intrinsic width of a photon wavicle being λ/π). Interference is produced when a pair of coherent photons have different path lengths to the plane of observation thereby interfering constructively or destructively.

As the incident light intensity decreases the number of photons in bunches may be expected to decrease. However, it should not be assumed that a reduction in intensity necessarily reduces the extent of photon bunching; the actual physical mechanism used to absorb or deflect some of the photons may, or may not, reduce the proportion of photons in bunches.

In a beam of statistically independent photons diffraction still takes place; it results from a single photon's interaction with the walls of the diffracting aperture. However the interference fringes should disappear in a low intensity beam of statistically independent photons. This effect has been observed in a series of experiments by Panarella (1985), and in an earlier experiment by Dontsov and Baz (1967). However similar, recent experiments by Jeffers (1986) do not show the effect; the relative intensity of the side lobes of the diffraction pattern is independent of the beam intensity. In a 1978 review of 5 non—photographic detection experiments (Pipkin, 1978), the only one showing a decrease in fringe intensity with decreasing incident beam intensity was the Dontsov and Baz experiment (1967).

It is now known that the original low—intensity interference experiments of G. I. Taylor (1909) did not display single—photon interference because photographic detectors (as used by Taylor and in three subsequent experiments reviewed by Pipkin, 1978) are now known to be essentially photon—coincidence detectors; i.e. a silver halide grain must absorb at least 4 coincident (i.e. within one optical period) photons to become developable (Panarella and Phipps, 1982).

Despite the low incident intensity used by Taylor his interference patterns must have been produced by photon bunches of at least 4 photons. It was these early experiments whose results were presumed to display single photon interference, that apparently led Dirac into making his well known pronouncement (Dirac, 1967) that interference is a single−photon phenomenon. In the light of modern photographic knowledge the correct inference from Taylor's (and the other photographic) experiments is that photon−bunching is commonplace, even in low intensity beams originating from non−coherent sources such as gas discharge lamps.

Hence the key to resolving the experimental issue is to produce a light source of statistically independent photons; i.e. one with hardly any bunched photons. One way of detecting statistical independence would be to use photographic, and non−photograhic detectors alternately in the same experiment. If the diffraction pattern is not observed with the photographic detector, but is observed with the photo−multiplier, or other non−photographic detector, then the inference would be that the observed pattern was produced by a beam with fewer than 4 photons in any bunches in it. On the other hand if the diffraction pattern is also not observed with the non−photographic detector, then the inference would be that the interference is a photon−photon phenomenon.

A recent photon−counting experiment (Grangier, Roget, and Aspect, 1986) was designed to unequivocally determine whether diffractive interference is a single−photon, or a photon−photon phenomenon. The experimental results support the single−photon hypothesis. The conventional rationalisation of single−photon interference is the dichotomy of the wave−particle duality; the single photon's wave traverses both paths of the interferometer simultaneously, interfering with itself and then manifesting its particle−like character at the detector. Explanations for single−photon interference based upon passage of each photon along only one of the two paths are under consideration by Buonomano (1986) and Surdin (1986).

Experiments with two independent light sources have shown that a photon from one source can interfere with a photon from the other source (Pfleegor and Mandel, 1967; Paul, 1986). Thus photon−photon interference does occur contrary to Dirac's pronouncement (Dirac, 1967). The Grangier (1986) experiment indicates that a single photon can also interfere with itself. However, the interpretation of this experiment is currently the subject of controversy (the Discussion in these Proceedings), and hence the question of whether a photon can interfere with itself as well as with other photons remains unresolved.

Photon Wavicles and the Minimum Quantum of Action

The wavicle's energy is $h\nu = pc$ and its transit time $= 1/\nu$, or alternatively its momentum is $p = h\nu/c = h/\lambda$ and its length is λ. Hence its action = energy × time = momentum × length = h, Planck's constant. Thus the wavicle model of the photon is consistent with the idea that upon absorption, emission, or scattering, the action involved in the process is at least h; the action involved has a minimum because of the photon's finite transit time (or finite length).

A fascinating aspect of the wavicle model is that its linear motion is intimately related to its rotational motion. In terms of its rotational motion the minimum action h is the angular momentum per complete cycle; i.e. for a mass $m = h\nu/c^2$ rotating with angular velocity $\omega = 2\pi\nu$ at a radius $r = r_{max} = \lambda/2\pi = c/\omega$ the angular momentum is $mr^2\omega = h/2\pi$ per radian, and hence the angular momentum per cycle is h. This relationship between linear and rotational motion may eventually lead to an explanation for the universal equivalence of inertial and gravitational masses (Jammer, 1961), for rectilinear motion is related to inertial mass, and accelerated, rotational motion is equivalent to a gravitational field.

The finite wavicle model of the photon supercedes the wave—particle duality theory of light. The wavicle is neither an indefinitely extended wave nor a point—like particle; it is a wave with a well—defined finite extent. Its oscillating/rotating electromagnetic field is consistent with its wave—like properties (particularly interference) and its finite extent explains its particle—like properties (especially line—spectra and the photo—electric effect). The photon—wavicle is, as Einstein anticipated (Diner et al, 1984), a localised, indivisible physical entity that moves without dividing and is only produced or absorbed as a complete unit. It is the atom of light. The wavicle model unifies the wave and particle properties of light in a single physical entity. Thus the dichotomy of light exhibiting both wave—like and particle—like properties (the wave—particle duality paradox) is resolved in the unity of the finite wavicle.

The wavicle model of the photon also provides a resolution of the philosophical dispute between proponents of the Copenhagen interpretation of quantum mechanics (Bohr, Heisenberg, Dirac, etc) and the persistent determinists (Einstein, De Broglie, Schrödinger, etc) (Hendry, 1984). The former believed that the uncertainty principle was intrinsic to nature, while the latter believed that nature was deterministic despite the experimental limitation of the uncertainty principle.

Internally the photon—wavicle is a classical electromagnetic field whose domain is defined by the relativistic requirement for causality. However, the finite size of the photon limits the accuracy of measurements, for it is an indivisible entity, and for a momentum $p=h/\lambda$ its position can only be determined to within the length of its ellipsoidal domain = λ. Hence in interactions (measurements) the product of momentum and positional uncertainty is at least the minimum action h in accordance with the Heisenberg uncertainty principle. The wavicle's internal coordinates may be regarded as hidden variables (Belinfante, 1973).

The wavicle model suggests an intimate relationship between quantum mechanics and the theory of relativity, for the length of the wavicle is the distance that it travels (moving at the velocity of light) in one period of its oscillation, and this finite length leads to the uncertainty principle as explained above. This suggests that the finite value of Planck's constant h is related to the finite value of the velocity of light. David Bohm has alluded to this idea; that the limitation imposed upon causality by the finite velocity of light is related to the uncertainty principle (Bohm, 1965). If this conjecture is correct, it should lead to a quantitative relationship between h and c.

Einstein regarded the relation $E=h\nu$ as an enigma. This enigma is partially revealed by the wavicle model of the photon, for there is a simple explanation for the fact that all photons carry the same action h, and that all physical processes (in which photons are absorbed or emitted) involve integral multiples of this quantum of action; the action of a photon—wavicle is a Lorentz—invariant property. Energy and frequency transform in the same way under a Lorentz transformation (Einstein, 1905), and hence the ratio of energy to frequency (i.e. action) is invariant under a Lorentz transformation. In this sense all photons, from gamma rays to radio waves, are essentially the same particle, just as all electrons are regarded as identical even though they may have different energies (and hence different De Broglie wavelengths) in different experiments.

A corollary of the Lorentz invariance of action, is that Planck's constant h must have a finite value, for if h were zero then the amplitude of the photon—wavicle's electromagnetic field would have to be zero in order to make its action zero, for its action is the product of its momentum and its length, and its length is finite. Its finite length is deduced from Maxwell's equations and the relativistic requirement for causality, regardless of the field's amplitude. Hence if the quantum of action h, were zero, light would not exist. The photon would be a null particle; i.e. its field amplitude would be zero. While this explains why the de facto existence of light implies that Planck's constant is not zero, it does not account for the specific, observed value of h.

It should be noted that our quantisation of the photon's energy and momenta (linear and angular) was achieved by introducing the De Broglie relationship (1) between momentum p and wavelength λ into the classical electromagnetic field defined by (2). Thus our wavicle model is based upon the quantum axiom (1) in addition to the principles of classical, relativistic electromagnetism.

Our finite—field model of the photon has a similarity to the three—wave model of elementary particles proposed by Kostro (1985), and the soliton model proposed by MacKinnon (1981).

In a subsequent publication we will present the general theory of electromagnetic wavicles that have a spin angular momentum of any integral, or half—integral multiple of $h/2\pi$; the spin=1/2 wavicles, in particular, are thought to be neutrinos. We also hope to derive a similar model of non—zero rest—mass particles such as the electron. The present work should be extended with detailed consideration of, and computations on multiphoton decay, photon—photon interference, and absorption and emission in interactions with matter.

Acknowledgments

We are grateful to Professor Ian Walker of York University for financial support. Fritz Engler's skillful experimental work will be presented in detail in a subsequent publication.

References

Andrews, C. L., *Optics of the Electromagnetic Spectrum*, p.328
    (Prentice—Hall, Englewood Cliffs, N.J., 1960).
Bandyopadhyay, P. and P. Raychaudhuri, Phys. Rev. **3**, 1378 (1971).
Belinfante, F. J., *A Survey of Hidden Variable Theories*
    (Pergamon Press, Oxford, 1973).
Bethe, H. A., *Intermediate Quantum Mechanics*,
    (Benjamin, New York, 1964).
Bohm, D., *The Special Theory of Relativity*, Ch.28 and pp.177—178
    (Benjamin, New York, 1965).
Buonomano, V., 1986 (These Proceedings).
Chin, S. L. and P. Lambropoulos (Editors) *Multiphoton Ionisation*
    of Atoms (Academic Press, New York, 1984).
Coulson, C. A., *Waves*, 7th Ed., p.103 (Oliver & Boyd, London, 1955).
Craig, D. P. and T. Thirunamachandran, *Molecular Quantum*
    Electrodynamics, p.25 (Academic Press, London, 1984).
Dewdney, C., A. Garuccio, A. Kyprianidis and J. P. Vigier,
    Phys. Lett. **105A**, 15 (1984).
Diner, S., D. Fargue, G. Lochak and F.Selleri (Editors), *The Wave—*
    Particle Dualism (D. Reidel Company, Dordrecht, Holland, 1984).

Dirac, P. A. M., *The Principles of Quantum Mechanics*, 4th Ed. p.9
    (Oxford University Press, 1967).
Dontsov, P., and A. I. Baz, Sov. Phys JETP, **25**, 1 (1967).
Einstein, A., Ann. Phys. **17**, 1905; reprinted in English translation
    in *The Principle of Relativity*, pp.56—58 (Dover Publications, 1952).
Fewkes, J. H. and J. Yarwood, *Electricity and Magnetism*, pp.509—513
    (University Tutorial Press, London, 1956).
Grangier, P., G. Roget and A. Aspect, Europhysics Letters, Feb. 1986.
Green, H. S., and E. Wolf, Proc. Phys. Soc. **66**, 1129 (1953).
Held, B., G. Mainfray, C. Manus, and J. Morellec, Phys. Lett.
    35A, 257 (1971).
Hendry, J., *The Creation of Quantum Mechanics and the Bohr—Pauli
    Dialogue* (Reidel Pub. Co., Dordrecht, Holland, 1984).
Honig, W., Found. Phys. **4**, 367 (1974).
Hunter, G., and H. O. Pritchard, J. Chem. Phys, **46**, 2146 (1967).
Hunter, G., 1986, *The Covariant Wave Equation* (to be published).
Jammer, M., *Concepts of Mass*, Harvard University Press,
    Cambridge, Massachusetts, U.S.A. (1961).
Jeffers, S., (These Proceedings).
King, R. W. P., and T. T. Wu, *The Scattering and Diffraction of Waves*,
    Harvard University Press, Cambridge, Massachusetts, U. S. A. (1959).
Knight, D. L., and L. Allen, *Concepts of Quantum Optics*, p.174
    (Permagon Press, Oxford, 1982).
Kostro, L., Physics Letters, **107A**, 429 (1985).
Kraus, J. D., *Antennas*, p.178 (McGraw—Hill, New York, 1950).
Lawden, D. F., *An Introduction to Tensor Calculus and Relativity*,
    pp.17—20 (Methuen, London, 1962).
Levitt, S., Lett. Nuovo Cimento **21**, Ser. 2, 222 (1978).
Lewis, G. N., Nature, **118**, 874 (1926).
Loudon, R., *The Quantum Theory of Light*, 2nd Edition, pp.226—228
    (Clarendon Press, Oxford, 1983);
Love, A. E. H., Proc. London Math. Soc., Ser.2, **1**, 37 (1903).
MacKinnon, L., Lett. Nuovo Cimento, **32**, 311 (1981).
McQuarrie, D. A., *Quantum Chemistry*, p.78 and p.83—87
    (University Science Books, Mill Valley, California, 1983).
Nishiyama, Y., J. Phys. Soc. Japan **43**, 228 (1977).
Panarella, E., 1974, Found. Phys. **4**, 227 (1974).
Panarella, E., 1977, Phys Rev. **A16**, 677 (1977).
Panarella, E., and T. E. Phipps Jr., Spec. Sci. Tech. **5**, 509 (1982).
Panarella, E., 1985, Spec. Sci. Tech. **8**, 35 (1985).
Paul, H., Rev. Mod. Phys, **58**, 209 (1986).
Pipkin, F. M., in Advances in Atomic and Molecular Physics
    (Edited by D. R. Bates), **14**, 294 (1978).
Pfleegor, R. L. and L. Mandel, Phys Rev., **159**, 1084 (1967).
Stratton, J. A., P. M. Morse, L. J. Chu, J. D. C. Little, and F. J. Corbato,
    Spheroidal Wavefunctions, M. I. T. Press (1956).
Surdin, M., 1986 (These Proceedings).
Swartz, C. E., *The Fundamental Particles*, p.94
    (Addison—Wesley, Reading, Massachusetts, 1965).
Taylor, G. I., Proc. Camb. Phil. Soc. **15**, 114 (1909).
Thomson, J. J., Phil. Mag. Ser. 6, **48**, 737 (1924).
Thomson, J. J., Phil. Mag. Ser. 6, **50**, 1181 (1925).
Thomson, J. J., Nature **137**, 232 (1936).
Wadlinger, R. L. P., J. Chem. Educ. **60**, 943 (1983).

THE THEORETICAL NATURE OF THE PHOTON IN A LATTICE VACUUM

Harold Aspden

Department of Electrical Engineering
University of Southampton
Southampton SO9 5NH, England

INTRODUCTION

The empirical formulations of quantum theory do not afford an adequate insight into the physical processes which generate the photon. Experimentally confirmed photon correlations, unretarded by light speed propagation, are paradoxical from the viewpoint of relativity. Resolution of the resulting dilemma can only come from a better understanding of the physical nature of the photon. Just as quantum theory is, in the main, confirmed by the successful deciphering of quantitative data provided by optical spectra, photon theory, to be viable and complete, should account for the quantitative relationship between h, c and e, which collectively define α, the dimensionless fine-structure constant.

The photon theory to be described meets this requirement, but raises one open issue which has bearing upon experimental investigation and points towards another potential paradox. The theory gives reason for suspecting that, though h is invariant in the formula $E = h\nu$, the energy quantum shed by an atom in producing the photon radiation may have a very small amount of its energy deployed into priming the perturbation of the photon spin at the frequency ν. The result should be that h, as measured from analysis of optical spectra, will appear to be larger by a few parts in ten million compared with the value exhibited, for example, in Josephson and quantized Hall resistance experiments. Precision measurements of the fine-structure constant by different techniques may eventually clarify this issue.

PRIOR DEVELOPMENT OF THE THEORY

The author has evolved the theory progressively over the past 30 years. It relies upon instantaneous charge interaction to assure a harmonious cyclic motion component shared by all charge. This is seen as the underlying basis of Heisenberg's Uncertainty Principle. Given action-at-a-distance and assigning electric properties to the vacuum itself, the ordering of charge motion and lattice formation become vacuum features favoured on energy grounds. Retardation effects associated with energy transfer only occur when charge is in relative motion, a characteristic of the matter state but not of the undisturbed steady state condition of free space. Thus a sudden change in charge motion at A can be immediately sensed at B and energy exchanges at B between matter and the vacuum can occur before the retarded transfer processes restore the energy balance.

345

A theory built on these lines leads to a fine-structure constant of approximately 1/137 with a point charge approximation. A fairly precise value emerged once it was realized that a virtual muon pair has a special association with each lattice cell and asserts an equilibrium regulating the finite size of the lattice charges. The theory was put on firmer ground in 1972, thanks to independent computer analysis by Dr. D. M. Eagles and Dr. C. H. Burton of the CSIRO National Measurement Laboratory in Australia. Their calculations paved the way for determining the effect of a resonance involving the electron and the lattice charge so as to control the physical size of the lattice charge whilst as nearly as possible satisfying a condition that the lattice charges were at positions of zero potential. This collaboration resulted in calculation of α^{-1} as 137.0359148, a value within one part per million of that measured[1]. This compares very favourably with other theoretical proposals, as shown recently in a review by Petley[2]. However, the advance discussed in this article concerns the rather obvious step, which has somehow eluded the author until recently, of replacing the electron resonance with a muon resonance, a step which brings α into even closer accord with its measured value.

The basic theory, which has been of record since 1960[3] in its point charge approximation and since 1966[4] in its virtual muon pair regulation form, has attracted little interest, no doubt because its action-at-a-distance features are incompatible with relativistic doctrine. This departure from mainstream physical theory is reflected by the title 'Physics without Einstein' of the 1969 update published by the author[5].

Now that Aspect et al[6,7,8] have provided experimental confirmation that photon correlations do occur more rapidly than actions communicated at the speed of light, the proposition that there is some hidden coupling asserting action-at-a-distance is no longer hypothetical. Relativistic constraints assume less importance.

The author's photon theory has also a complementary feature giving it considerable support. This dates from 1974[9] and is the role which a lattice cell energy quantum (equal to that of two muons) can play in a minimal energy relationship involving the proton. This led in 1975, also in collaboration with Dr. D. M. Eagles, to theoretical evaluation of the proton-electron mass ratio[10]. The value obtained agrees with the latest experimental value to within one part in ten million. Van Dyck et al[11], whose precision measurement has a standard deviation at the 41 part per billion level, note that the author's theory gives a theoretical value that is *'remarkably close'* and that *'this is even more curious when one notes that the result was published several years before direct precision measurements of this ratio had begun'*. Also, several recent advances show that other fundamental constants are calculable with a typical accuracy of one part per million[12]. However, in this work attention is concentrated on the structure of the vacuum and the photon process. Recently, Sinha, Sudarshan and Vigier[13] have written about the superfluid vacuum and the need for it to contain superfluid states and convey real Einstein-de Broglie waves. This viewpoint is only partially supported, because here the author makes no attempt to qualify the analysis by relativistic considerations. The approach relies on there being an underlying physical reality, which owes nothing to the special role of the relativistic observer.

ENERGY TRANSFER WITH ACTION-AT-A-DISTANCE

It is necessary to put aside the relativistic proposition that magnetic fields are merely the effects of electric fields witnessed by an observer moving relatively to the source charges. Instead, attention must focus upon energy transfer. Also, the photon should not be seen as an energy quantum that somehow moves from A to B, but rather as an event that occurs at A as

energy is released by matter at A, followed by an event at B where energy
sourced in the vacuum itself transfers to matter. A wave set up at A and
radiating from A can trigger the event at B. This process requires the
vacuum to have specific properties and contain energy. It has a universal
uniformity and is invisible so long as it is not disturbed. Our analysis
concerns its structure and its primary disturbance condition in which a
small unit of its lattice acquires a state of spin to set up the wave
disturbance or resonate in response to existing waves, with energy quanta
being exchanged in the process. In restoring its equilibrium state the
vacuum transfers the surplus energy at A to compensate for the deficit at
B. Thus there is an apparent transfer of energy quanta at the wave speed,
but in reality the energy transfer involves a displacement of what we can
regard as a sea of energy permeating all space. This action is best considered as involving electric particles, the statistical presence of which
represents the wave and the energy.

Electric charge sets up an action-at-a-distance and asserts a potential
that is unretarded. This proposition differs from that of Wheeler and
Feynman[14], who accepted action-at-a-distance and the non-radiation of energy
by accelerated charge. They introduced reaction fields set up by the response of other charges, assuming that these were a combination of half of a
retarded and half of an advanced Lienard-Wiechert solution of Maxwell's
equations. Bearing in mind that electromagnetic waves are usually set up
by the collective acceleration of numerous charges, where the energy of
the mutual interaction is prevalent, we have no real evidence to suggest
that any charge can radiate its intrinsic electric field energy or set up
a reaction on itself that is not conservative of energy.

The Larmor formula for radiation by a discrete accelerated charge
depends for its derivation upon the assumption that its electric field is
distorted by a wave propagating outwards at the speed c and gives a continuous rate of energy radiation inversely proportional to the cube of this
speed. There is clearly conflict here with quantum theory, but the conflict
vanishes if the electric field of the isolated charge does not admit any
distortion and effectively requires the speed to be infinite, consistent
with action-at-a-distance. There is then no Larmor radiation. However,
to avoid infinities in the self-energy of the charge, it must have finite
form and one can consider finite speed of electric field disturbance within
the body of the charge. Rigorous analysis[15,16], fully allowing for the
interaction with an electric field of other charge producing the acceleration, then shows that the no-radiation condition applies within the body of
charge only if there is an inertial property characterized by a mass M equal
to E/c^2, where E is the electric energy and c the disturbance speed.

One charge can, therefore, accelerate another instantaneously without
first demanding energy transfer. The resulting motion will upset the mutual
potential energy. Its magnitude will change as the two particles come
closer or move further apart. There must then be energy transfer. Now,
though a charge cannot radiate its intrinsic electric energy continuously,
it can become involved in quantum phenomena. It may change state spontaneously, eg. the decay of a muon into an electron, or it may be created or
annihilated along with a similar charge of opposite polarity. Thus the
destiny of energy shed by the mutual potential change is the statistical
creation of charge pairs, involving vacuum energy fluctuations. The kinetic
energy of a moving charge is seen as that of charge pairs closely associated
with the primary charge. This model has been used elsewhere to show why
particle lifetime relates to energy as a function of speed[17,18], otherwise
attributed to 'time dilation'. Magnetic energy is simply energy in transit
from the mutual electric potential state to the local kinetic form of either
interacting charge. Such energy travels at a finite speed and may, in fact,
materialize transiently en route, so that all energy is really electric in

character, being wholly constituted by particles of electric charge and a lepton field. Such action-at-a-distance theory with retarded transfer of energy to replenish vacuum energy fluctuations not only leads to the Lorentz force but also leads directly to a unified force law corresponding to Einstein's gravitational equation by which anomalous perihelion motion has been explained[19].

In a uniform magnetic field the active transfer of energy becomes debatable but it does seem that free charge in motion in the field will be mutually neutralizing in part and the remaining part will become ordered to the extent that it develops a half-cancelling reacting field[20,21]. This requires the deployment of that number of reacting charges into the ordered state that has a kinetic energy density equal to the magnetic field energy density. This can account for the gyromagnetic factor of 2, without recourse to Dirac's theory, a result used below in the discussion of the quantized Hall effect.

Energy in transit will travel at the limiting speed c, measured in the local frame set by the vacuum lattice, seen also as the frame in which charge pair creation is referenced. Waves are disturbances of the lattice. A free wave not affected by other overlapping waves moving in the opposite direction will propagate at the same speed as the energy, but wave velocity can be subject to other criteria, especially in standing wave conditions. For example, in a standing wave set up by reflection between mirrors, the nodes of the electric and magnetic components are displaced relative to one another along the beam axis. This makes it probable that, in addition to being transported through space with the mirror system, there is a constant cyclic exchange of energy forwards and backwards along the beam between the two energy forms. Wave propagation in the two directions cannot follow this motion, which occurs through unit distance in a time $2/c$, as measured in the frame of the apparatus. All we can be sure about is that in making a round trip the waves will keep in step with the energy oscillations. Thus $2/c$ becomes equal to $1/u + 1/u'$, where u and u' are the wave velocities in the respective opposite directions. We could suppose that these velocities adopt the vacuum lattice as their reference, as does the energy. Indeed, this has been the author's assumption until hearing of a new experimental constraint (see discussion of Silvertooth experiment below). The alternative is to suppose that the wave velocities become referenced on the muon field, which assumes the role of a preferred frame associated with the isotropic cosmic background radiation. On this hypothesis the wave speed will be denoted c', applicable in either direction along the beam, a value which is referenced on a frame through which the mirror system moves at velocity v in the beam direction. This causes c'+v and c'-v to be u and u', respectively, and we then obtain the relationship:

$$c = c'(1 - v^2/c^2) \tag{1}$$

if we ignore terms in $(v/c)^4$.

EXPERIMENTAL EVIDENCE

The Michelson-Morley Experiment

The null of this experiment has been interpreted as support for the relativity hypothesis, assuming that wave velocity is isotropic either in the preferred frame or in the observer's frame. However, as there are standing waves in the compared optical paths of this experiment we might apply equation (1) to find also that the null result is fully compatible with wave velocity in the standing wave paths being referenced on a preferred frame. The experiment is inconclusive in view of this possibility.

The Sagnac Experiment

The positive result of this experiment demonstrated that standing waves set up in a rotating mirror system need not share the rotation of the apparatus. This allows rotation to be sensed by interferometer techniques as if wave velocities are referenced on the non-rotating frame. This experiment has long been regarded as consistent with ether theory and so it is consistent with the possibility that the vacuum has a lattice system which need not rotate with the apparatus. The relativity hypothesis is, however, in trouble in explaining this experiment. This has been a subject of critical review by Turner and Hazelett[22], who write in this connection: *'According to a leading relativistic treatment, even the formerly sacrosanct constancy of c blurs into any of a wide range of values, depending upon circumstances.'*

The Silvertooth Experiment

This is a recent experiment using a standing wave sensor[23] specially designed to scan along a standing wave set up by light rays of the same frequency travelling in opposite directions through the sensor. If these waves travel at the same velocity in both directions and relative to the apparatus, then the amplitude of the standing wave should be unmodulated along its length. Otherwise, a discrepancy in the two speeds should cause some modulation, such as a phase variation of the standing wave oscillations along the beam. The position of the nodes is also shifted as a function of any variation in the velocity discrepancy. As in the Michelson-Morley experiment, the standing wave system does, of course, move bodily with the apparatus in its translational motion through space. This is known already by direct test from the Wiener experiment of 1890[24].

Silvertooth[25] has measured the nodal shift in a standing wave for different orientations of the beam direction and has achieved a detection of the Earth's motion through space that is first-order in v/c. His initial experimental results correspond to motion of the Earth in a line drawn to constellation Leo, the measured velocity being 378 km/s. When these results are confirmed they will have invalidated Einstein's theory of relativity. Meanwhile, it seems prudent to develop the photon theory without reliance upon any relativistic qualifications. Clearly, Silvertooth's researches will have extensive impact upon those aspects of quantum theory which build on a connection with relativistic formalism that goes beyond analysis using the verified formulae for mass-energy and its variation with speed.

The Charge-Mass Ratio of Atomic Nuclei

There is some indirect evidence of the existence of a vacuum lattice having the form of a cubic array of charge sites in a neutralizing background and sharing the motion of matter. This is revealed in the charge-mass ratio of the atomic nuclei.

Berezin[26] has shown that minimal energy criteria can favour unexpected charge distributions about a central nucleating charge, all based upon electric interaction. Though this has been challenged on stability grounds by appeal to Earnshaw's theorem, as noted by Scott[27], this theorem fails if applied to conditions in a uniform charge continuum, a point discussed at length by the author[28] and recently briefly mentioned in relation to the Berezin proposition[29].

A three-dimensional array of A charges -e, nucleated by a central positive charge Ze, though conventionally deemed unstable when A is greater than Z can be stable if two conditions are met. Firstly, if the A charges meld into the vacuum background by replacing lattice charges, then the system will exhibit only the Ze core charge as sensed by matter. Secondly, the array of A charges will tend to cluster together under the influence of the nucleating charge provided A is not too large in relation to Z and, for the specific array, breaches minimal electric energy criteria. For a uniform symmetrical distribution of the A charges over a spherical volume A must be less than 2.5Z. If the A charges are uniformly distributed over a spherical surface, with Ze at the centre, then A must not exceed 2Z. To verify this, for example in the latter case, compare the Coulomb interaction energy between Ze and $-Ae$ at radius R, namely $-ZAe^2/R$, with the Coulomb self-interaction energy of the A charges, namely $A^2e^2/2R$. The total energy is negative, so favouring reduction of R and containment, only when A is less than 2Z.

This has direct relevance to atomic nuclei, if we can regard the proton and each bound neutron as a charge A and remember that when a neutron is driven out its charge becomes neutralized as the lattice site is filled by an opposite charge to that neutralizing the neutron. A becomes the atomic mass number and, for a multi-shell nucleus, the approximation by which A cannot exceed 2.5Z should hold. Note then that nuclei become unstable just above Bismuth in the periodic table, that is just above Z = 83 and, for the abundant isotope of Bismuth, A = 209. This is good accord. Also, over the lower range of atoms, for the more abundant isotopes from He^4, we know that the ratio A/Z progresses steadily from the value 2 upwards as A increases. It is as if the vacuum itself plays a role in determining the atomic stability conditions, by virtue of its lattice dimensions. Indeed, on this atomic evidence, the lattice spacing can be estimated and found to accord with that deduced from the photon theory below.

The Quantized Hall Effect

The action-at-a-distance effects which reproduce a synchronous electric lattice are in evidence in the quantized Hall effect and particularly in the fractional features of this effect. We suppose that the charges in a semi-conductor can form into a cubic lattice array and that current can be passed through the semi-conductor by the bodily displacement of this lattice, a process that occurs without energy exchange and so without radiation losses.

The relevant equations in the CGS system of units are:

$$Heu/c = Ee \qquad Neu/c = i \qquad\qquad (2)$$

relating magnetic field intensity H, charge e, translational speed of charge u, speed of light c, transverse electric field intensity E, density of charge carriers N and current i. The material is supposed to have unit permeability and must be homogeneous, so that the measured Hall resistance R_H, which is E/i, is given by:

$$R_H = H/Ne \qquad\qquad (3)$$

Whereas the small translational drift velocity is denoted u, the charges of mass m also have a much higher velocity component v directed lateral to H. This causes them to be deflected into orbital motion of radius r, in reacting to H according to the equation:

$$Hev/c = mv^2/r \qquad\qquad (4)$$

Such reacting charges will also move so that in their collective action they describe orbits which are linked by zero magnetic flux. This means that the primary field acting over the area πr^2 will cancel the effect of a magnetic moment set up by the reacting charge. This is equivalent to writing:

350

$$(S)H\pi r^2 = 4\pi(I)\pi r^2(N) \tag{5}$$

where I is the elemental current $ev/2\pi rc$ of each individual orbital charge and S is the number of such charges in unit area of each layer of the lattice grouping, the plane of each layer being perpendicular to H. N is determined in terms of the kinetic energy of each reacting charge:

$$N(\tfrac{1}{2}mv^2) = H^2/8\pi \tag{6}$$

as shown below. From (4), (5) and (6), it then follows that:

$$2\pi r^2(S) = 1 \tag{7}$$

From experiment involving superconductors it is known that the basic reaction flux quantum is $hc/2e$. Thus each unit $H\pi r^2$ must equal $nhc/2e$, where n is an integer. From (4), this corresponds to the classical Bohr atom quantization:

$$mvr = nh/2\pi \tag{8}$$

Equations (3), (4), (7) and (8) afford the equation:

$$R_H = (nS/N)hc/e^2 \tag{9}$$

where N/S is merely the number of layers of charges in the test specimen and contributing to the Hall resistance.

The value of n, being a number of flux quanta, can only be odd and must be the same for all charges forming the ordered lattice. The reason for this is that, to create a reacting flux quantum with n = 3, for example, two n = 1 reacting charges have to neutralize as a pair by one reversing its magnetic flux direction. This involves the transfer of double flux quanta to units with n = 1, so that n can increase to 3, 5, 7 etc. In short, formula (9) holds only provided n is odd, but N/S can be an odd or even integer. This is confirmed by the experimental data on the fractional Hall effect.

Equation (6) is justified because the reaction magnetic field is proportional to $2\pi(evr/c)N$, which from (4) is proportional to $\tfrac{1}{2}mv^2(N/H)$. This is further proportional to H'-H, where H' is the active effect of the applied field that is offset by the reaction field to leave H as the residual field. Taking the energy density term $N(\tfrac{1}{2}mv^2)$ as the variable tending to a maximum, meaning that from all the charges that are present only a limited number set by this condition can become regimented to provide a lattice and develop the ordered reaction, we then find that the differential with respect to H is zero when H' = 2H. Thus the effective field H always arises from a double action effect offset by a halving reaction. This accounts for the gyromagnetic factor of g = 2 concerning the specific magnetic moment in relation to angular momentum of orbital charge found in the ferromagnet. It further results in the equation (6), because $(g)2\pi(evr/c)N$ is $4\pi gN(\tfrac{1}{2}mv^2)H$, which equals H.

Complementary with equation (6) we obtain from (4) and (8):

$$\tfrac{1}{2}mv^2 = (He/2mc)(nh/2\pi) \tag{10}$$

which shows that the field strength H determines the reacting kinetic energy of each charge included in N, leaving equation (6) then to determine N. Such analysis shows that it is feasible to analyze electric lattices by direct application of simple classical style physical principles, which gives support to our adoption of similar methods in analysis of the vacuum.

THE NATURE OF THE VACUUM LATTICE

The lattice state of the vacuum comprises discrete charges e^- in a uniform continuum of opposite charge density e^+ per unit cell volume d^3, where d is the lattice spacing. The lattice structure is determined by minimum energy, but unlike structures in the matter state, where relative potentials can be negative, charges in the vacuum itself are governed by absolute criteria and minimum energy cannot be negative. This means that the lattice has a simple cubic structure. Locally the vacuum lattice is locked into the shell structure of atomic nuclei and forms microcosmic domains coextensive with the whole atom. Each lattice charge is subject to a restoring force resisting displacement relative to the background continuum charge. This makes the lattice a two-dimensional linear oscillator. It has a natural frequency $\nu' = mc^2/h$, which determines the mass m of the electron. This leads us into the calculation of h in terms of e and c, but before performing this calculation we need to consider the essential role of a virtual muon pair denoted μ^+, μ^-, which has association with each cell unit of the vacuum state.

In Fig. 1 a cubic cell of vacuum space in shown to include a spherical zone centred at P. This zone is displaced from the centre of a cell and is at a position of minimal positive electric potential as determined by the state B condition. In state A there are two muons located outside the zone P and in state B the zone P is occupied by a lattice charge e^-, with the charge e^+ deemed to be uniformly distributed over the remainder of the cell, shown shaded.

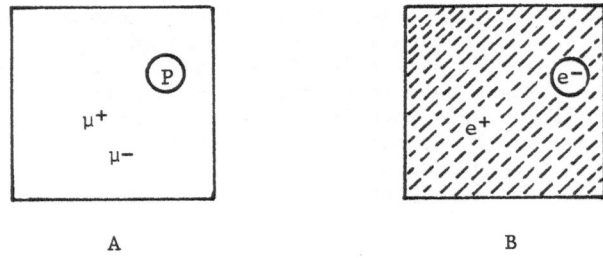

Fig. 1. Charge constituents of the vacuum lattice cell.

The virtual muon pair is deemed to migrate at random and to do this by successive mutual annihilation and creation at new positions, this action occurring at the frequency ν'. Thus, if the muon pair annihilates and is recreated outside zone P, state A prevails. If, however, the muon pair is created in a zone P not already occupied by charge in its degenerate state, then the lattice charge and continuum condition of state B will be set up in company with a release of energy as a vacuum fluctuation. Such unoccupied regions exist at boundaries where lattice domains move apart, but there is always a boundary elsewhere where lattice domains come together and it is here that the reverse process can occur. Energy is absorbed by the degenerate charge form, for which there is no longer available space, and the muon pair is reconstituted. This explains the quantum electrodynamic process by which lattice structure can exist in a continuum and have its several parts in relative motion without exhibiting linear momentum properties. Normally, however, the muon pair entering a zone P will find that it is filled with charge. It is then driven out to sustain a state A. This really amounts to saying that the vacuum is a sea of leptons (virtual muons) which has a propensity to become partially degenerate, setting up a lattice which shares translational motion of local matter, without exhibiting linear momentum, but which exhibits angular momentum and forces matter to adopt its cyclic jitter motion.

The overall effect of this sequence of events is that for equilibrium (apportionment of mass energy in proportion to space occupied by charge), the energy of the degenerate state B is 2μk, where k is the ratio of the volume of the zone P to the cell volume d^3 and μ here denotes the muon rest mass energy. The mean energy per cell is always 2μ, but this takes the form of the muon pair (state A) for 1-k of any unit period of time, with state B applying continuously. All space has become a charge plenum, with space itself being conserved in regulating transitions between charge forms.

A remaining consideration is that the charge e^-, being at a position of positive but minimal electric potential at P, is attracted towards the centre of the cell. It can only be 'stable', with the electric force balanced by centrifugal force, if it describes the circular orbit of equipotential about this centre in company with the same synchronous motion of all other such lattice charges in surrounding cells. All parts of the lattice therefore describe circular orbits of radius denoted r and at a universal frequency which we take to be $\nu' = mc^2/h$. A particle of matter moving with this frame (the electromagnetic reference frame) is at rest electrodynamically but it has an ever-changing position and an ever-changing momentum. The product of these quantities in the case of an electron is the product of two uncertainties (2r) and (mc) or h/2π, because it is only if the speed in orbit is c/2, giving uncertainty of momentum over a range mc, that the frequency in orbit satisfies the mc^2/h condition. Furthermore this also gives:

$$r = h/4\pi mc \tag{11}$$

THE PHOTON PROCESS

If the lattice is disturbed by the transport of energy, it keeps its synchronism regardless of the microdomain boundaries and merely expands its orbits. As with any two-dimensional linear oscillator, the energy stored is given by:

$$E = H(2\pi\nu') \tag{12}$$

where H is the accompanying angular momentum added to and so transported through the lattice.

At the source or at a position where this energy quantum is absorbed in exchanges involving matter, we need to assure a basis for balancing this angular momentum H. Something has to spin. We only recognize the lattice itself as a structure that can spin with a variable quantum and so we take a very small cubic unit of the vacuum lattice as the seat of the photon event involved. In spinning, this unit disturbs surrounding lattice or is resonantly receptive of existing disturbances in a receiving mode. The spin is at one quarter of the disturbance frequency ν, because the cube corners nudge the surrounding lattice four times per cycle. This gives the spin angular momentum as I(2πν)/4, where I is now the moment of inertia of the cubic photon lattice unit. The angular momentum balance gives, from equation (12):

$$E = (\pi^2 I\nu')\nu \tag{13}$$

which means that Planck's constant h is $\pi^2 I\nu'$ and, since ν' is mc^2/h:

$$h = \pi c\sqrt{(Im)} \tag{14}$$

Note, from equation (12), that when ν = ν' the photon spin angular momentum is h/2π. Also note that we have neglected the energy W needed to prime the spin of the photon unit. This is $\frac{1}{2}I(2\pi\nu/4)^2$, which from (13) gives:

$$W/E = \nu/8\nu' \tag{15}$$

It should also be noted that, in setting up the initial spin state, the two energy quanta E and W separate, because W remains with the cubic spin unit and E disperses. Thus these two quanta are separately conserved. Then, since angular momentum is conserved and W, both I and ν must remain constant throughout the period of spin. It follows that we can regard the value of I as that of the cubic unit at the initial onset of spin and since a cubic lattice can spin about any central axis with the same moment of inertia, the task of calculating h from (14) simplifies considerably.

The task of determining Planck's constant has become simply that of determining I. It will now be shown that the value of I is set by a 3X3X3 cubic lattice group fitting neatly within the spherical field cavity of diameter equal to the Compton wavelength of the electron.

Given that the spin unit has a lattice charge at its centre, is of simple cubic form and has a minimal moment of inertia, the 3X3X3 array is the only choice. There are 12 lattice charges at distance d and 12 at distance $\sqrt{d}$ from an axis drawn centrally through the other 3 lattice charges and this gives the unit a radius of gyration squared of $36d^2$. Let s denote the effective mass of each such charge in electron mass units m. Then:

$$I = 36smd \tag{16}$$

and, from (14), this gives:

$$h = 6\pi mcd\sqrt{s} \tag{17}$$

The theory then advances by noting that the displacement of the lattice charges in setting up the centrifugal balance in a fully dynamical system requires:

$$2(4\pi e^2/d^3) = sm(2\pi\nu')^2 \tag{18}$$

because the total mass of the e^+, e^- system of the state B in Fig. 1 is effectively divided in juxtaposition in the inertial frame to set up the balance. The charges are separated effectively by 2r, explaining the initial 2 in equation (18). Also, we will expect 2sm to equal 2μk.

From equations (11), (17) and (18), together with $\nu' = mc^2/h$, it is a matter of algebra to deduce:

$$hc/2\pi e^2 = 216\pi\sqrt{s} = 144\pi r/d \tag{19}$$

and the equation:

$$d = 72\pi e^2/mc^2 \tag{20}$$

Bearing in mind, from (11), that the Compton wavelength of the electron h/mc is $4\pi r$, we see that knowledge of r/d allows us to deduce the fine-structure constant and verify whether the photon unit is housed within the cavity of diameter equal to this Compton wavelength.

The task of calculating r/d is relatively simple, being merely that of computing the position of the zero potential of the zone P in Fig. 1, for, of course, an approximation to an infinite lattice. There is a marginal dependence upon the finite size of this zone P. A rigorous study of this problem in the context of the derivation of $E = Mc^2$ by the inertial method mentioned earlier has shown that the radius occupied by charge is given by the J. J. Thomson expression $2e^2/3Mc^2$. Computer calculation[1] gives the zero potential value of r/d as 0.30287 and so, from (19) a purely theoretical derivation of the reciprocal of the fine-structure constant as 137.017. This also gives an electron Compton wavelength of 3.8 d, which is just right for this diameter to embrace the photon unit of cube edge 2d.

However, in order to calculate the true value of the fine-structure constant, it is the term involving s in equation (19) that matters because this sets a resonance condition at a potential very slightly above the zero level. In the author's earlier version of the theory it was thought that the resonance was between the lattice charges and the electron-positron field. The resonance criterion is that the space volume corresponding to the Thomson charge formula should be conserved in particle transitions. Thus the energy $2smc^2$ occupies the space that can accommodate $(1/2s)^3$ electrons and positrons, the number of which must be odd. The least energy resonance of this kind is that for which this number is 1843, giving, from (19) the value of $hc/2\pi e^2$ of 137.0359148.

Whereas the intrinsic mass of the lattice charge is 2sm, the charge exhibits a mass density equal to that of the enveloping muon population and this means that, as in the hydrodynamic situation, the mass of the spherical charge form is effectively halved owing to the motion through the equally dense medium. Hence the effective mass sm.

THE MUON RESONANCE

Now, it has been shown above that sm = μk, where k is a measure of the ratio of the volume of the sphere centred at P in Fig. 1 in relation to the lattice cell volume. In terms of units of electron radius cubed, the volume of the lattice charge is $(1/2s)^3$ and, from the Thomson formula, the cell volume d^3 is $(108\pi)^3$ such units. This tells us that:

$$1/k = (3/4\pi)(108\pi)^3(2s)^3 \qquad (21)$$

and since we know that s is $(2r/3d)^2$ from (19) and that r/d is approximately 0.30287 the value of μ/m is calculable and is found to be just a little greater than 206. The mass of the muon emerges from the theory.

As a result of theoretical developments concerning the nature of the real muon which has a mass 206.7683 times that of the electron, the author has found that the real muon is continuously changing state between a form of 207 and 205 electron mass units, as it puts energy into electron-positron pair creation and annihilation. The exact muon mass has been calculated from a theory involving field cavity resonance[30],[31],[32]. For this reason it has occurred to the author that the virtual muon pairs in the vacuum lattice system might involve a proportion of these two types of muon and that it is the lattice theory which explains why this mix does not include 203 and 209 electron-mass muon types. The theoretical value of just over 206 and the need for an odd integer multiple of electron mass combine to determine this possibility. It was then realized that the electron resonance with the lattice particle could only occur under rare and very energetic conditions, because it involves an energy fluctuation very slightly in excess of that of the proton mass-energy. It could then be supplemented by an over-riding resonance state, where s in equation (21) is determined by the equation:

$$(p + q)s/k = 205p + 207q \qquad (22)$$

and p and q are small integers giving a value of s which assures the near zero potential condition in the lattice. The surprising result of this study was that with p = 1 and q = 2 equations (21) and (22) gave a value of s, which, upon insertion in (19), gives:

$$hc/2\pi e^2 = 137.035950 \tag{23}$$

This is extremely close to the measured value, being marginally less by a few parts in ten million. Indeed, it is representative of some values that have been reported from tests involving quantized Hall resistance and Josephson measurements. Generally, it seems that other methods involving analysis of optical spectra indicate a slightly higher value. Such tests are based on theory which concerns the energy levels of electrons in atoms and assume that all the energy released goes into the photon energy E. If, however, a small energy quantum W of equation (15) is deployed into priming the spin of the photon unit, we would expect the photon frequency to be very slightly smaller than normal theory predicts. Conceivably, this could be interpreted as indicating a higher value of h and so a higher value of $hc/2\pi e^2$. The magnitude of this effect is small, being, from (15), of the order of parts in ten million at frequencies of the order of 10^{15} Hz. Evenso, as research improves measurement of the fine-structure constant, this possibility should be resolved and could provide confirmation experimentally of the photon theory described in this work.

PROTON CREATION IN RELATION TO PHOTON THEORY

It has been argued that the vacuum lattice provides the physical basis for the photon process and that an energy quantum equal to that of two muons is associated with each cell of the vacuum and yet passes unseen except for occasions when there are violent disturbances of the lattice equilibrium. It is inevitable, therefore, that such a muon energy quantum plays a role in the creation of heavier particles, whereas the electron and positron are directly developed as the energy of the photons when in full resonance with the natural frequency of the vacuum lattice.

Proton creation involves the dimuon, a particle of charge e and mass 2μ. A muon pair can merge to be transformed into a compact unit formed by a muon and a dimuon of opposite charge. The dimuon has the same energy as two muons because, in this compact unit, the negative mutual electric energy balances exactly the rest-mass energy of the muon. However, the unit, as a whole, requires 9/8 times the space available from the muon charge pair. For balance involving muon pair annihilation a group of 9 muon pairs is needed to create 8 muon-dimuon systems. Thus, a concentration of muons will develop natural groups occupying the characteristic space of 18 muons or 144 dimuons. The proton develops when very high energy particles of near point form are also available to trigger the creation of a proton in company with enough protons and antiprotons so as to occupy exactly this characteristic space.

The relevant equations determining this proton mass are:

$$M_p = (\sqrt{3}/\sqrt{2} - 1)^{-1} 2\mu \tag{24}$$

as derived elsewhere[33]. This is an approximation based on a minimal energy relationship between the proton form and the dimuon. The controlling effect is the resonance between the proton and dimuon group, which involves space conservation and is formulated thus:

$$(M_p/2\mu)^3 = N/144 \tag{25}$$

where N is an odd integer. These two equations determine the value of N as 12,685 uniquely. Then, because we know from (22) that 3μ is precisely 619m, we can further check the photon theory by deducing an exact value of the proton-electron mass ratio using (25). The result is:

$$M_p/m = 1836.152\ 365 \tag{26}$$

This value is about 30 parts per billion higher than the value originally deduced[10] based on the electron resonance but closer to the experimental value measured by Van Dyck et al[11]. As with the fine-structure constant, it requires further precision experiment to really test the theory proposed.

CONCLUDING REMARKS

Whatever paradoxical problems emerge in quantum theory, these can but indicate that we need to probe the physical basis of the photon process. Implicit in this is the common property evident throughout space which ensures that Planck's constant is always the same. The structural property of a lattice system of some kind provides the only likely route to our understanding of the dimensionless fine-structure constant which embodies the Planck constant. Thus Eddington is famous for his prediction that:

$$hc/2\pi e^2 = (16^2-16)/2 + 16 + 1 = 137 \qquad (27)$$

based on the number of independent elements in a symmetrical matrix in 16-dimensional space, where 16 = 4X4 (4 being the number of dimensions in Minkowski's world). The comparison of such theories, including an updated version by Eddington and the author's theory, was the subject of a review by Eagles[34] published in 1976.

Theories such as that of Eddington, which are based on the abstract metric of relativity, run into their own problems, because relativity itself is prone to paradox. It too lacks that full physical basis which is an essential element to anything approaching a full understanding of how Nature is working. Evenso, scientists persevere in trying to force out the answers from relativistically-based lattice metrics, whatever the complexity of the computation. The most notable active example of this is the project undertaken by IBM to harness computing power so powerful that it can perform in only one year a calculation that would take hundreds of years on other supercomputer systems. Beetem et al[35] explain how IBM has set out to calculate the proton-electron mass ratio by methods that will test quantum-chromodynamics. Space and time are approximated by a 4-dimensional hyper-cubic lattice of points. The calculation proceeds by estimating transition probabilities and expecting quite accurate predictions by extending the computations over a 16X16X16X16 lattice. It seems to the author that such a protracted calculation is unlikely to give an answer matching that given above in equation (26), which is based upon transition probabilities of a virtual muon pair occupying a lattice site in a 3-dimensional space and an infinite lattice. It was long ago realized that, even if this approach was only an approximation to the real world, there was a resonance condition that could fix the values of the fine-structure constant and the proton-electron mass ratio.

Most important, however, in this debate is the ultimate need to accept that electric charge asserts an action-at-a-distance upon other electric charge. Our supposition that such actions are retarded is based upon the fact that electro-magnetic waves travel at the finite speed we associate with the symbol c. Yet such waves are attenuated in amplitude in inverse proport-ion with distance from source and the direct electric field varies according to the inverse square of this distance. Here was the warning that there could be a fundamental difference in action. It was this that caused the author to avoid the use of a time dimension in formulating the vacuum lattice theory and it comes as no surprise to the author that Aspect's experiments on quantum correlations confirm what has been termed 'Spooky action-at-a-distance'[36]. It remains to be seen whether independent experimental work will confirm Silvertooth's experimental claim to have measured the Earth's motion through space[25]. Such invalidation of relativity would undoubtedly lead to revival of a three-dimensional space governing the photon process.

REFERENCES

1. H. Aspden and D. M. Eagles, Aether Theory and the Fine-Structure Constant, Physics Letters, 41A: 423 (1972).
2. B. W. Petley, "The Fundamental Constants and the Frontier of Measurement", Adam Hilger, Boston, p. 161 (1985).
3. H. Aspden, "The Theory of Gravitation", Sabberton, Southampton (1960).
4. H. Aspden, "The Theory of Gravitation", 2nd. Ed., Sabberton, Southampton (1966).
5. H. Aspden, "Physics without Einstein", Sabberton, Southampton (1969).
6. A. Aspect, P. Grangier and G. Roger, Experimental Tests of Realistic Local Theories via Bell's Theorem", Phys. Rev. Lett., 47: 460 (1981).
7. A. Aspect, P. Grangier and G. Roger, Experimental Realization of Einstein-Podolsky-Rosen-Bohm Gedankenexperiment: A New Violation of Bell's Inequalities, Phys. Rev. Lett., 49: 91 (1982).
8. A. Aspect, J. Dalibard and G. Roger, Experimental Test of Bell's Inequalities using Time-Varying Analyzers, Phys. Rev. Lett., 49: 1804 (1982).
9. H. Aspden, "The Chain Structure of the Atomic Nucleus", Sabberton, Southampton (1974).
10. H. Aspden and D. M. Eagles, Calculation of Proton Mass in a Lattice Model of the Aether, Nuovo Cimento, 30A: 235 (1975).
11. R. S. Van Dyck Jr., F. L. Moore, D. L. Farnham and P. B. Schwinberg, New Measurement of the Proton-Electron Mass Ratio, Int. Jour. Mass Spectrometry and Ion Proc., 66: 327 (1985).
12. H. Aspden, The Theoretical Nature of the Neutron and the Deuteron, (to be published).
13. K. P. Sinha, E. C. G. Sudarshan and J. P. Vigier, Superfluid Vacuum Carrying Real Einstein-de Broglie Waves, Physics Letters, 114A: 298 (1986).
14. J. A. Wheeler and R. P. Feynman, Interaction with the Absorber as the Mechanism of Radiation, Rev. Mod. Phys., 17: 157 (1945).
15. H. Aspden, Inertia of a Non-Radiating Particle, Int. Jou. Theor. Phys., 15: 631 (1976).
16. H. Aspden, "Physics Unified", Sabberton, Southampton, p. 80 (1980).
17. H. Aspden, Theoretical Resonances for Particle-Antiparticle Collisions Based on the Thomson Electron Model, Lett. Nuovo Cimento, 37:307 (1983).
18. H. Aspden, Meson Lifetime Dilation as a Test for Special Relativity, Lett. Nuovo Cimento, 38: 206 (1983).
19. H. Aspden, Unification of Gravitational and Electrodynamic Potential Based on a Classical Action-at-a-Distance Theory, Lett. Nuovo Cimento, 44, 689 (1985).
20. H. Aspden, Crystal Symmetry and Ferromagnetism, Spec. Sc. Tech., 1:281 (1978).
21. H. Aspden, Electromagnetic Reaction Paradox, Lett. Nuovo Cimento, 39: 247 (1984).
22. R. Hazelett and D. Turner, "The Einstein Myth and the Ives Papers", Devin-Adair, Old Greenwich, p. 249 (1979).
23. E. W. Silvertooth and S. F. Jacobs, Standing Wave Sensor, Appl. Optics, 22: 1274 (1983).
24. O. Wiener, Ann. Phys., 40: 203 (1890).
25. E. W. Silvertooth, Experimental Detection of the Ether, (to be published).
26. A. A. Berezin, An Unexpected Result in Classical Electrostatics, Nature, 315: 104 (1985).
27. W. T. Scott, "The Physics of Electricity and Magnetism", Wiley, New York, p. 43 (1966).
28. H. Aspden, "Modern Aether Science", Sabberton, Southampton, p. 87 (1972).
29. H. Aspden, Earnshaw's Theorem, Nature, 319: 8 (1986).
30. H. Aspden, The Nature of the Muon, Lett. Nuovo Cimento, 37: 210 (1983).
31. H. Aspden, The Mass of the Muon, Lett. Nuovo Cimento, 38: 342 (1983).

32. H. Aspden, The Muon g-Factor by Cavity Resonance Theory, Lett. Nuovo Cimento, 39: 271 (1984).
33. H. Aspden, Don't Forget Thomson, Physics Today, 37: 15 (Nov. 1984).
34. D. M. Eagles, A Comparison of Results of Various Theories for Four Fundamental Constants of Physics, Int. Jour. Theor. Phys., 15: 265 (1976).
35. J. Beetem, M. Denneau and D. Weingarten, The GF11 Supercomputer, in: "The 12th Annual International Symposium on Computer Architecture", Conference Proceedings, IEEE Computer Soc. No. 634, p. 108 (1985).
36. N. D. Mermin, Is the Moon there when Nobody Looks? Reality and the Quantum Theory, Physics Today, 38: 38 (April 1985).

IS THE PHOTON CONCEPT REALLY NECESSARY ?

Maurice Surdin

Centre des Faibles Radioactivités
Laboratoire Mixte CNRS/CEA - Boîte Postale n° 1
91190 Gif-sur-Yvette, France

ABSTRACT

A definition of a normalized fundamental impulse, a wave packet of e.m. radiation of positive energy and momentum is given. A complementary impulse of negative energy and momentum is also defined. These wave packets may be depicted as a wave above the "sea level" of the Zero Point Field (of Stochastic Electrodynamics) and a complementary wave, or hole, beneath the "sea level", they propagate as a combined pair.

Using these notions several optical experiments are interpreted without the intervention of the photon concept.

1 - INTRODUCTION

Einstein's main motivations to consider the corpuscular nature of radiation[1,2] were the following :

- the behavior, in the thermodynamic sense, of monochromatic radiation of low density (within the validity of Wien's radiation law) is the same as if radiation consisted of mutually independant quanta of energy of magnitude $h\nu$.

- the emergence of the explanation of the photoelectric effect. Thus, the maximum kinetic energy of an ejected electron is given by

$$E_{max} = h\nu - W \qquad (1.1)$$

where $h\nu$ is the energy of each incoming light quantum[3] and W is the work function, i.e. the work required to remove the electron from the metal. Eq. (1.1) is now known as Einstein's photoelectric equation.

   – considering the energy fluctuations of blackbody radiation in the frequency range (ν, $\nu + d\nu$) and the cavity subregion of volume V, Einstein established the following relation

$$\overline{(\Delta E_\nu)^2} = \left[h\nu \rho(\nu) + \frac{c^3}{8\pi \nu^2} \rho^2(\nu) \right] V. d\nu \qquad (1.2)$$

where,

$$\rho(\nu) = \frac{8\pi h \nu^3}{c^3} \left(e^{h\nu/kT} - 1 \right)^{-1} \qquad (1.3)$$

is the Planck spectral density of blackbody radiation. In the domain of frequencies and temperatures where the Rayleigh–Jeans formula applies, eq. (1.3) becomes

$$\rho(\nu) = \frac{8\pi \nu^2}{c^3} kT \qquad (1.4)$$

and eq. (1.2) is now

$$\overline{(\Delta E_\nu)^2}_{waves} = \frac{c^3}{8\pi . \nu^2} \rho^2(\nu) V d\nu. \qquad (1.5)$$

This term may be computed on the basis of classical Maxwell's wave theory of light.

   The first term in the r.h.s. of eq. (1.2), which is inexplicable from the wave theory point of view, is accounted for on the basis of Einstein's quanta of light. Thus

$$\overline{(\Delta E_\nu)^2}_{particles} = h\nu \rho(\nu). V d\nu \qquad (1.6)$$

Finally, eq. (1.2) which is known as Einstein's fluctuation formula for blackbody radiation, may also be written as

$$\overline{(\Delta E_\nu)^2} = \overline{(\Delta E_\nu)^2}_{particles} + \overline{(\Delta E_\nu)^2}_{waves} \qquad (1.7)$$

   Here is also the first time that the so called wave-particle duality was introduced.

   In a subsequent paper Einstein[4] confirmed and elaborated on the corpuscular nature of light quanta and attributed to a quantum of energy hν

362

a momentum whose magnitude is $h\nu/c$, which has a definite direction. This concept was well verified by the now well known experiment of A.H. Compton[5].

Soon after Schrödinger published his first paper on wave mechanics in 1926, several papers appeared where, on the one hand G. Wenzel and G. Beck (for references see Jammer[6]) gave a wave-mechanical derivation of Einstein's photo-electric equation and, on the other hand, G. Wentzel[6] and Schrödinger[7] provided a wave mechanical description of Compton's effect.

More recently, Lamb and Scully[8] and Franken[9] used the semi-classical theory of interaction atom-radiation field, where atoms obey the laws of Quantum Mechanics (QM), and the radiation field is treated classically, according to Maxwell's equations, i.e. without the use of the photon concept. They have been successful in deriving Einstein's photo-electric equation.

Moreover, the results of various interference experiments, such as those of Young's slits or Michelson's interferometer by Janossy and Naray[10] at very low source intensities are quite naturally accounted for by considering a wave packet the length of which is comparable with the coherence length of the spectral line used. The intensity of the light source is set to such a low level as to ascertain that at any instant only one such wave packet is present in the instrument. Interference patterns are observed provided the coherence length is larger than the optical path difference, for example, in Michelson's interferometer.

Conversely, a group of several experiments was performed purporting to show the indispensability of the photon concept. Thus, the wave theory of light cannot explain these experimental results, whereas the photon theory of light does so, adequately.

Typical of such experiments, also one of the most sophisticated, is that of Clauser's[11], who measured diverse twofold coincidence rates between four photomultipliers viewing two cascade photons on opposite sides of two beam-splitters.

For the purpose of the present paper, the essence of such experiments may be presented as follows: a beam of light of a low intensity source is directed on a beam-splitter. Two photomultipliers view the beams on both sides of the beam-splitter. The coincidence rate between the two photomultipliers is measured. According to the wave theory of light, a wave packet, depicting the emission from an atom of the source, is split into two parts by the beam-splitter and one should observe coincidences between the two photomultipliers. Whereas, if one considers the photon theory, a photon arriving at the beam-splitter is either reflected or it is transmitted, since it is not possible to split a photon into two parts, no coincidences should be observed. Experiments disprove the wave theory and are in agreement with the photon concept of light.

Another simple experiment may be considered as proof of the photonic nature of light. Consider a source of low intensity whose beam is directed on a photomultiplier. One measures the rate of arrival of quanta of light on the photomultiplier. Then one inserts an absorber between the source and the photomultiplier. One measures the new rate of arrival of quanta of light. In the classical theory of light, as a consequence of the inter-position of the absorber, the amplitude of the field decreases, whereas the photon theory predicts that the number of photons arriving, per unit time, on the photomultiplier decreases. The latter prediction is confirmed by the experiment.

Summing up, one may conclude that the ensemble of the experiments considered above fall into two groups: the first group comprises those experiments which may be explained either by the wave theory or the photon theory of light. Whereas the experiments of the second group are explained only by the photon theory.

The following proposition was considered: Is it possible to advance a theory, albeit a heuristic one, based on the classical wave theory which does not necessitate the photon concept and which accounts for the results of the second group of experiments ?

The rest of the paper will be devoted to the presentation of such an endeavor.

2 - PRELIMINARY CONSIDERATIONS

2.1 The motivation. The above proposition will be elaborated in the framework of Stochastic Electrodynamics (SED). SED, as it is well known (see also the next paragraph), is a classical theory augmented with a fundamental postulate, viz: the existence at the absolute zero of temperature of a universal fluctuating electro-magnetic field, the Zero Point Field (ZPF).

During the last two decades, this theory was developed and applied to various physical phenomena with interesting results. These results, in particular the Planck spectrum for blackbody radiation[12], were obtained without the invocation of the photon concept. It appears, thus, quite natural that an attempt, within the framework of SED, should be made at the elaboration of the above proposition.

The development and the application of SED is an interesting line of investigation _per se_ and should not be considered as a snub to Q.M. On the contrary, quantum mechanical results were often considered as the standards to which SED results were compared. Since the approach to physical phenomena in the framework of SED is different from that of Q.M., on several occasions SED results contributed to shed a novel light on the comprehension of the phenomena considered.

2.2 _Stochastic Electrodynamics_. Developed, essentially during the last two decades, SED is a classical theory augmented with a fundamental postulate of the existence of the ZPF.

Applying thermodynamical considerations, such as were used by Wien to establish his famous law, an expression giving the energy spectral density of the ZPF is derived. For a one-dimensional case, one has

$$\mathcal{E}(\omega)\, d\omega = \frac{K\omega^3 d\omega}{3\pi c^3} \qquad\qquad (2.1)$$

where ω is the circular frequency, c the velocity of light and K is a constant having the dimension of action. For comparison of results obtained in SED with those of Q.M. one sets $K = \hbar$. The application of the Lorentz invariance to the spectrum of the ZPF yields the same result. However, neither derivation furnishes the numerical value of K.

Recent surveys of SED may be found by Boyer[13], and Surdin[14] where further references are given.

The relevant facts to the pursual of the present paper are :
- the results obtained in SED were derived using strictly the classical wave theory of radiation; the photon concept was not introduced.

- in the framework of SED it appears as if all the laws of physics stem from a single concept, the ZPF.

Thus, it was shown[15] that Einstein's spontaneous emission may be considered as the emission induced by the ZPF. It was also shown[16] for the hydrogen atom that the higher energy levels and, hence, their energy difference are the consequences of the action of the ZPF. The emitted energy is in the form of an elementary coherent wave-packet, the energy of which is extracted from the ZPF. It will be accompanied by a "complementary" coherent wave-packet, say an "anti-wave packet", a "hole" of energy in the ZPF.

The concepts of coherence, elementarity and complementarity of a wave packet are clarified hereafter.

2.3 <u>The elementary wave-packet</u> as considered hereafter is the wave-packet emitted by an atom. Usually it is reckoned that emission by an atom is effected in the form of a monochromatic radiation of energy $h\nu$. However, due to the ZPF, the emitted line is not pure monochromatic but has a "natural line-width". This is due to the fact that the energy of an atom for an orbit of order n (n=1,2,3..) does not have a sharply defined value, but possesses a natural line-width. For a spherically symmetric hydrogen atom, the relative line-width for an orbit of order n is given by[17]

$$\Delta\nu_0/\nu_0 = \frac{8}{3} \cdot \frac{1}{n^2} \left(e^2/Kc\right)^3 \tag{2.2}$$

for the ground state n=1, one obtains

$$\Delta\nu_0/\nu_0 = \frac{8}{3} \left(e^2/Kc\right)^3 \tag{2.2a}$$

The value given by eq. (2.2a) may be considered as the minimum relative line-width of the emitted radiation, resulting from a transition between orbit of order n and the ground state. In practice the line is further broadened due to thermal or Doppler effects. However, as $\Delta\nu_0/\nu_0$ is generally small, say 10^{-5}, it will be considered hereafter that the energy of the emitted line is well approximated by $h\nu_0$, or in the SED notation by $K\omega_0$, where ν_0 or ω_0 is the nominal frequency of the emitted line.

An atom may be considered as a resonant system of central frequency ν_0 and of relative band-width given by eq. (2.2a). The relaxation time of the resonant system is given by

$$\tau_o = \frac{1}{2\,\Delta\nu_o} \qquad (2.3)$$

2.4 <u>The natural coherence length</u> is defined as

$$L_o = \frac{c}{\Delta\nu_o} = \lambda_o\,\frac{\nu_o}{\Delta\nu_o} = \lambda_o\,\frac{\lambda_o}{\Delta\lambda_o} \qquad (2.4)$$

λ_o is the nominal wave-length of the wave-packet and $\Delta\lambda_o$ is the width corresponding to $\Delta\nu_o$.

For a transition to the ground state of the hydrogen atom the minimum value of $\Delta\nu_o/\nu_o$ is given by eq. (2.2a), it comes

$$L_o = \lambda_o \left(\frac{3}{8}\right)\cdot\left(\frac{Kc}{e^2}\right)^3 \qquad (2.5)$$

L_o is the natural coherence length ; in practice, however, because of the broadening of the emitted line due to thermal or Doppler effects, the coherence length is smaller than L_o.

2.5 <u>The energy content of a wave packet</u>[14] at the nominal frequency ν_o and of line width $\Delta\nu$ is obtained as follows: the coherence length is $L = c/\Delta\nu$ and the volume of the "coherence sphere" is

$$\upsilon = \frac{4\pi}{3.8}\,c^3\big/(\Delta\nu)^3 = \frac{\pi}{6}\,c^3\big/(\Delta\nu)^3$$

The energy density in the packet is then

$$\Delta_{\nu_o} = \frac{12\,K\nu_o\,(\Delta\nu)^3}{c^3} \qquad (2.6)$$

It is interesting to compute the energy density of the ZPF centered on the same frequency ν_o and having the same band-width $\Delta\nu$ as the wave-packet. Considering the contributions of both the electric and magnetic components of the (three dimensional case) ZPF, one has

$$\Delta_z = \frac{4K(2\pi)^3\,\nu_o^3\,d\nu}{c^3} \qquad (2.7)$$

The ratio of the two energy densities is then

$$\frac{\Delta_z}{\Delta_{\nu_o}} = \frac{32\,\pi^3\,\nu_o^2}{12\,(\Delta\nu)^2} \cong 8\pi^2\,\nu_o^2\big/(\Delta\nu)^2 \qquad (2.8)$$

This ratio depends on the relative band-width of the radiation considered.

The corresponding electric fields are in the ratio

$$\frac{E_z}{E_{y_o}} \cong 3\pi \, \nu_o/_{\Delta\nu} \tag{2.9}$$

For optical frequencies, taking $\lambda_o = 5.000$ Å, $\Delta\lambda = 0.25$ Å, one has

$$\nu_o/_{\Delta\nu} = \lambda_o/_{\Delta\lambda} = 2 \times 10^4 \quad \text{and} \quad \frac{E_z}{E_{y_o}} \cong 1.8 \times 10^5$$

2.6 The complementarity. A wave-packet, such as considered above, may be Fourier analyzed into its monochromatic components. Each component's propagation is governed by the electromagnetic wave equation. For a source-free space the equation is

$$c^2 \nabla^2 \vec{F} = \frac{d^2\vec{F}}{dt^2} \tag{2.10}$$

where ∇^2 is the Laplacian and $\vec{F}$ stands for either the electric field vector $\vec{E}$ or for the magnetic vector $\vec{B}$ of the wave. Eq. (2.10) is second order in space and in time and it has two independent space and two independent time solutions.

Consider a one-dimensional case where eq. (2.10) describes the propagation of the wave along the x-axis and where $\vec{E}$ is along the y-axis. Then the two independent solutions are

$$\vec{E}_{\pm}(x,t) = \hat{y} \, E_o \sin\left[2\pi\left(\frac{x}{\lambda} \pm \nu t\right)\right] \tag{2.11}$$

and

$$\vec{B}_{\pm}(x,t) = \vec{B}_o \sin\left[2\pi\left(\frac{x}{\lambda} \pm \nu t\right)\right] \tag{2.12}$$

the alternating signs in eqs. (2.11) and (2.12) correspond to the two independent time solutions.

Let the source of radiation, localized at the origin, be emitting in the +x direction; then, in common practice, the sign − corresponds to the usual retarded potential and the sign + corresponds to the advanced potential solution. In what follows the concept of the advanced potential solution is relinquished. Here one contemplates the possibility of both solutions as propagating in the +x direction for t>0. Eq. (2.11) for the usual retarded solution is then

$$\vec{E}_- \left(x>0, t>0 \right) = \hat{y}\, E_0 \sin\left[2\pi \left(\frac{x}{\lambda} - \nu t \right) \right] \qquad (2.11a)$$

and for the other solution, the complementary solution, one has

$$\overline{E}_+ \left(x>0, t>0 \right) = \hat{y}\, E_0 \sin\left[2\pi \left(\frac{x}{\lambda} - \{-\nu t\} \right) \right]$$

$$= \hat{y}\, E_0 \sin\left[2\pi \left(\frac{x}{\lambda} - \overline{\nu} t \right) \right] \qquad (2.11b)$$

Eq. (2.11b) corresponds to a retarded wave propagating in the +x direction and having a negative frequency $\overline{\nu} = -\nu$ as well as a negative energy and negative momentum.

Similar equations may be obtained for the magnetic field vector, viz

$$\vec{B}_- \left(x>0, t>0 \right) = \vec{B}_0 \sin\left[2\pi \left(\frac{x}{\lambda} - \nu t \right) \right] \qquad (2.12a)$$

$$\vec{B}_+ \left(x>0, t>0 \right) = \vec{B}_0 \sin\left[2\pi \left(\frac{x}{\lambda} - \overline{\nu} t \right) \right] \qquad (2.12b)$$

The energy and momentum flow of the wave governed by eqs. (2.11b) and (2.12b) is now considered. Using Maxwell's equations, one finds for the Poynting vector S_+ that

$$\vec{S}_+ = \frac{1}{\mu_0} \left(\vec{E}_+ \times \vec{B}_+ \right) = \left(\hat{y} \cdot \hat{z} \right) \frac{(-E_0^2)}{\mu_0 c} = \frac{\hat{x}\, (-E_0^2)}{\mu_0 c} \qquad (2.13)$$

where $\hat{x}$, $\hat{y}$ and $\hat{z}$ are unit vectors along the Cartesian axes and $\lambda \nu$ = c. Eq. (2.13) indicates that the flow of negative energy and negative momentum of the complementary wave is along the +x direction.

Thus eqs. (2.11a) and (2.12a), on the one hand, describe the usual retarded solution of the wave of frequency ν propagating along the +x direction and, on the other hand, eqs. (2.11b) and (2.12b) describe a "complementary" wave of frequency $\overline{\nu} = -\nu$ and of negative energy and momentum propagating along the +x direction and accompanying the former wave.

The complementary wave has some unusual properties. In particular the reflection and the transmission properties of the wave at the surface of separation between two optical media of indices n_1 and n_2 are of interest. In Annex I the case of normal incidence is examined.

Correlatively, the inverse Fourier transform of the complementary wave corresponds to a complementary wave-packet, the anti-wave packet considered above. The pair of them, the wave-packet and its complementary wave-packet, propagate conjointly along the +x direction, the one transporting positive energy ($2\pi K \gamma_o$) and positive momentum, the other transporting negative energy ($2\pi K \overline{\gamma_o} = -2\pi K \gamma_o$) and negative momentum. It should be stressed here that <u>this state of affairs is possible only because of the existence of the ZPF</u>.

In SED the energy of a wave-packet emitted by an atom is supplied directly, or through intervening processes, by the ZPF. Thus, to a wave-packet of positive energy which is abstracted from the ZPF corresponds a wave-packet of negative energy, a "hole" of energy in the ZPF.

There is a similarity of behavior of the usual wave above the "sea-level" of the ZPF and the complementary wave beneath the "sea-level". The "zero reference" level for energy is the ZPF. For conciseness the elementary wave-packet will be dubbed ADNI (<u>A</u>rchetypal <u>D</u>efinite <u>N</u>ormalized <u>I</u>mpulse) and will be designated by $\mathcal{A}$, its complementary component, the ANTI-ADNI will be $\overline{\mathcal{A}}$.

Adnis and the complementary anti-adnis propagate as a pair without interaction. Interaction occurs only in the presence of matter which acts as a "irreversible" element.

The adni + anti-adni behavior conforms to the following group of rules (R):
    - the adni is an indivisible entity, the same applies to an anti-adni,
    - the spectral density of ZPF, on the average, is given by eq. (2.1).
    To the positive energy of an adni, extracted from the ZPF, corresponds a negative energy of the companion anti-adni.
    - in free space an adni and its companion anti-adni propagate as a pair without interaction.
    - interaction occurs in the presence of matter, which acts as a "irreversible" element.

370

- during interaction each Fourier component of an adni may be cancelled by the corresponding component of the companion anti-adni. Interferences may be obtained only with adnis and anti-adnis which were previously separated.

- the usual effects, such as the photo-electric effect, are due to adnis. The possible anti-adni effect on matter is considered in Annex III.

3 - INTERPRETATION OF EXPERIMENTS

3.11 <u>Absorption</u>. Consider a flow of adnis $\mathcal{R}$, accompanied by their companion anti-adnis $\overline{\mathcal{R}}$, incident on a slab of absorbing medium of thickness ℓ. The adni loses its kinetic energy to an electron of one of the atoms of the medium, whereas the anti-adni does so to a "hole". The energy of the excited electron and hole is then dissipated by collisions with the crystalline lattice or, which is equivalent from the present point of view, by the recombination of the excited electron and hole.

Let $N(x)$ be the number of adnis present at the depth x from the front face of the slab, $dN(x)$ the number of adnis lost along the distance dx in the absorber, then if

$\dfrac{dN}{dx} \ll N/\ell$, one may write

$$\frac{dN(x)}{dx} = -\alpha N(x) \tag{3.1}$$

where α is a proportionality constant (having the dimension of the inverse of length). Integrating eq. (3.1) gives for $x = \ell$

$$N_\ell = N_o \, e^{-\alpha \ell} \tag{3.2}$$

where N_ℓ denotes the number of outcoming adnis from the slab of thickness ℓ for N_o incoming adnis. As photo-multipliers count only adnis, the counting rate will be proportional to N_ℓ. Suppose now that one inserts between the light source and the absorbing slab a beam-splitter. As will be seen (Annex II) the beam-splitter acts as a non-linear device which separates adnis from their companion anti-adnis. Thus, the average rate of incoming adnis and anti-adnis on the absorbing slab is now one half of the preceding rate. However, due to the beam-splitter, now adnis and anti-adnis are no more simultaneous. Considering the energy gained by the crystalline lattice of the absorber, the same considerations as above apply. Nevertheless, the counting rate of the photo-multiplier is now halved.

One may characterize the absorbing medium by the formal relation

$$n\mathcal{A} + n\overline{\mathcal{A}} = 0 \tag{3.3}$$

where n is an integer (n=1,2,3..)

3.2 <u>Fringes obtained with a plane parallel plate</u>. Henceforward, the sources of the experiments considered are of very low intensity, so that at any instant only one pair adni+anti-adni is present in the instrument.

An adni is an indivisible entity which can exist only in whole numbers. The same concept applies to anti-adnis. When an adni arrives at the frontal face of the plane parallel plate, it is either reflected or transmitted as a whole. Its companion anti-adni is then transmitted or reflected. The occurrence of reflection or transmission is a random event. As in the case of photons, the frontal face has a non linear behavior.

Consider, first, the case where an adni is reflected from the frontal face, then its companion anti-adni is transmitted into the plate, traverses it and is reflected by the rear face. The other case occurs when an adni is transmitted, traverses the plate and is reflected by the rear face, the companion anti-adni is then reflected by the frontal face. In both cases the Fourier components of the adni and the anti-adni undergo on reflection a phase-shift, of either 0 or π , at the boundary surfaces, as shown in Annex I (for normal incidence).

The combined adni+anti-adni Fourier component at frequency ν arriving at the frontal face of the plate is the solution of eq. (2.10), and may be written as

$$\vec{A} = \vec{A_0} \left\{ \sin\left[2\pi\left(\frac{x}{\lambda} - \nu t\right)\right] + \sin\left[2\pi\left(\frac{x}{\lambda} - \overline{\nu}t\right)\right] \right\} \tag{3.4}$$

Fourier components of different frequencies ν behave all in the same manner at the boundary surfaces as their parent adnis or anti-adnis.

Due to the fact that an adni and its companion anti-adni are separated at the frontal face, they travel different optical paths. For the corresponding Fourier components a path difference results.

Bearing in mind the rules (R) above, set forth for the behavior of adnis and their componion anti-adnis, one may reproduce the calculations of Born and Wolf[18] (p. 282, their Fig. 7.21). Here, however, one has to take into consideration that at the surfaces of separation (front face and rear face) of the plate, the phase shifts on reflection of the adni and anti-adni are the same. Using Born and Wolf's notations, one obtains bright fringes when :

$$2 n'h \cos \theta' = m' \lambda_0 \quad , \quad m' = 1/2 , \; 3/2 , \; 5/2 , \cdots \quad (3.5)$$

and dark fringes when

$$2 n'h \cos \theta' = m' \lambda_0 \quad , \quad m' = 0, 1, 2, \cdots \quad (3.6)$$

or for bright fringes

$$2 n'h \cos \theta' \pm \frac{\lambda_0}{2} = m \lambda_0 \quad , \quad m = 0, 1, 2, \cdots \quad (3.5a)$$

and for dark fringes

$$2 n'h \cos \theta' \pm \frac{\lambda_0}{2} = m \lambda_0 \quad , \quad m = 1/2, \; 3/2, \; 5/2, \cdots \quad (3.6a)$$

Eqs. (3.5a) and (3.6a) are the same as those (eqs (8a) and (8b)) of Born and Wolf, obtained using the classical Maxwell theory of light. Similar arguments may be advanced to explain the more complicated pattern of Newton's rings. In particular, when a liquid of refractive index differing from those of the plate or the lens is interposed between them.

3.3 Interference with Young's slits. An adni being an indivisible entity should pass through a slit in its entirety. It is considered here that its companion anti-adni passes then through the other slit. The passage through the slits is best understood by considering Babinet's Principle[19].

This principle relates the diffraction of an electromagnetic field by a screen to the diffraction by a "complementary screen". A phase difference of $\pi/2$ between the two diffracted fields occurs.

The "complementary screen" of a slit is a perfectly conducting strip of the same geometrical form. At this screen the behavior of an adni and anti-adni are the same, they are reflected and undergo the same phase shift on reflection. At the slit, whereas each Fourier component of an adni undergoes a phase shift of $\pi/2$ each Fourier component of the companion anti-adni undergoes a complementary phase-shift of $-\pi/2$. Thus, the emerging adnis and their companion anti-adnis have now a total phase shift of π .

Applying the usual interference considerations and bearing in mind the rules (R), one obtains the classical expression for the interference pattern.

To justify the above statement that while an adni passes through one slit, its companion anti-adni passes through the other slit, one considers the following argument which avoids the introduction of information travelling at velocities higher than the velocity of light. Tetrode[20] and Wheeler and Feynman[21] have considered the relation between a source and an absorber. Their findings may be summarized concisely as follows "The absorber is an essential element in the mechanism of radiation".

This concept was applied within the framework of SED[22] to experiments of the EPR-type. It was considered there that the information about the absorber (the entirety of the receiver in these experiments) is furnished to the source of radiation by the ZPF. In the case of Young's slits, the information about these is conveyed to the source by the ZPF.

3.4 <u>The Clauser type experiment</u>. Considering the essence of such experiments, as presented in the <u>Introduction</u> above, their explanation in terms of adnis+anti-adnis is quite simple. At the beam-splitter an adni is reflected or transmitted. The event is random, the probability being 1/2 for reflection and 1/2 for transmission. The companion anti-adni is then either transmitted or reflected. The adni, since it carries positive energy, can extract an electron from the photo-cathode of the photo-multiplier and, thus can be detected ; whereas the anti-adni received on the photo-cathode of the other photo-multiplier, since it carries a negative energy, cannot extract electrons and is not detected. Hence the non-observation of coincidences between the two photo-multipliers.

3.4 <u>A combined experiment</u>. Consider now an experiment which combines those described in   3.2, 3.3 and 3.4 as depicted in Figure 1.

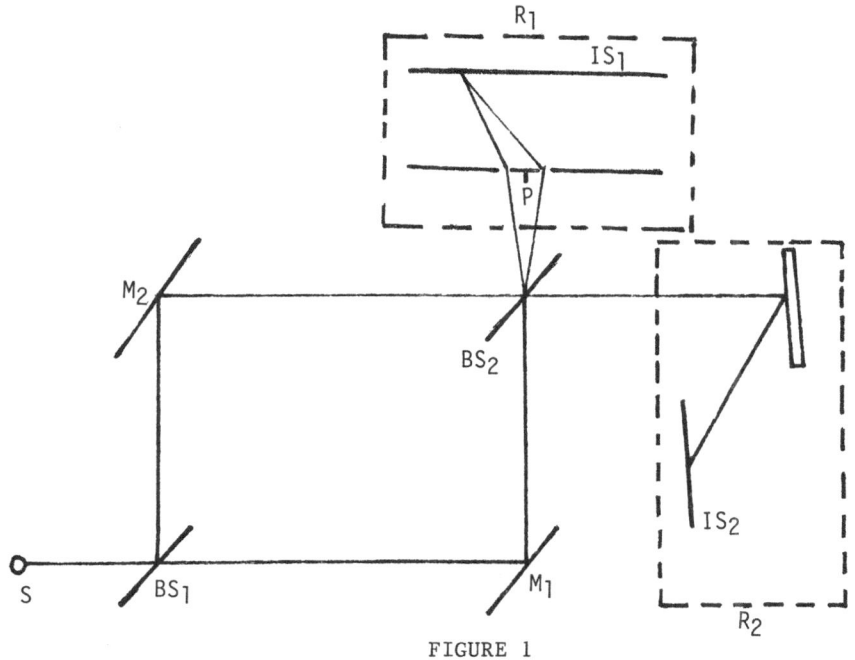

FIGURE 1

The source S emits individual quanta of light, so that at any time only one such quantum is present in the instrument. BS_1 and BS_2 are 50-50% beam splitters. M_1 and M_2 are mirrors. The two "receivers" are R_1 —a Young slits interference arrangement and R_2 is a plane parallel plate interference set-up. The Young slits arrangement has an extra plate p which ensures that beam N°1 passes through slit N°1 and, correspondingly, beam N°2 passes through slit N°2. The optical paths of both beams are equal.

The analysis of this experiment within the frameworks of the three theories, viz: the photon picture, the classical e.m. wave theory and the adni+anti-adni picture, yield the same result, i.e. interference patterns should be observed on both screens IS_1 and IS_2.

4 - DISCUSSION

In the framework of SED it appears quite appropriate to envisage the concept of an adni and its companion anti-adni. Since in this theory the energy of a wave packet emitted by an atom is supplied directly, or through intervening processes, by the ZPF. Thus, to a wave-packet of positive energy, which is ultimately abstracted from the ZPF - an adni - corresponds a wave-packet of negative energy - the anti-adni - a "hole" in the energy of the ZPF.

Using the adni+anti-adni concept the results of a variety of optical experiments can be explained without the use of the photon concept. In particular, experiments of the Clauser type, which were designed to demonstrate the indispensability of the photon concept, are also accounted for.

Moreover, the adni+anti-adni concept sheds new light on the mechanism of operation of such objects as the beam-splitter.

The analysis of the combined experiment of 3.5 shows that the direct detection of an anti-adni is not feasible when using the usual optical phenomena, such as beam splitting and interferences.

The direct detection of an anti-adni necessitates the use of a new "detector". In Annex III, the essentials of an experiment, using an as yet unknown detector, is proposed.

Recently, several papers appeared where "ghost waves" were considered and experiments for their detection were proposed[23]. Here, a somewhat different point of view is adopted. Roughly speaking, the "ghost waves" considered in Ref.[23] would be identified here with the anti-adni wave-packet, although their detection is somewhat different.

Finally, the questions put in the title of this paper may be answered by the negative, at least for the experiments considered.

Note

After this paper was completed the attention of the author was called to the paper by P. Grangier, G. Roger and A. Aspect, Europhysics Lett. 1(4) 173 (1986) where the results of a most beautiful experiment on photon anti-correlation effect on a beam splitter are given. These results confirm and amplify those of Clauser's.

ANNEX I

Consider the classical case where a continuous incident complementary wave EH in the optical medium n_1 is failing on interface x0y of a second medium n_2. The transmitted wave is E_1H_1 and the reflected wave is E_2H_2, as depicted in Figure A1.

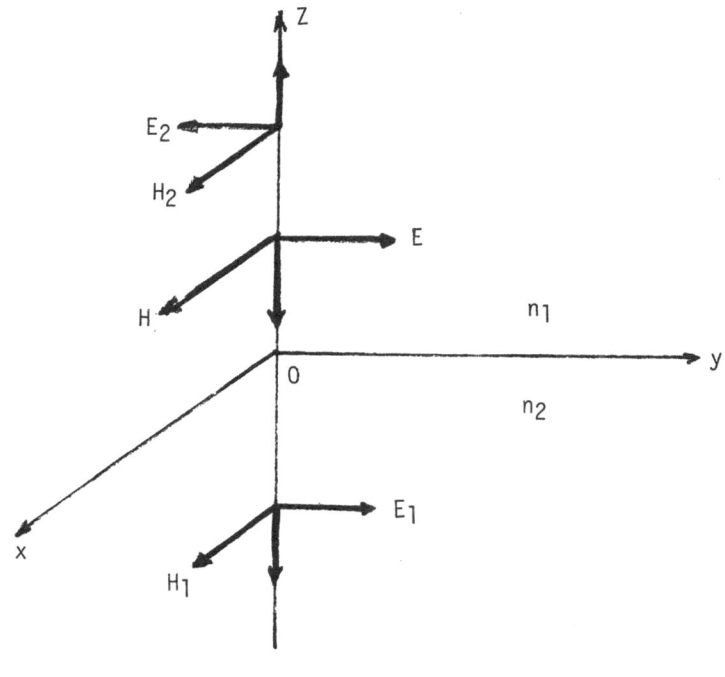

FIGURE A1

Both media have permeabilities $\mu_1 = \mu_2 = 1$. The continuity of the tangential component of E at the interface requires

$$E-E_2 = E_1 \qquad (A1.1)$$

at any given time and at any given point on the surface. Similarly, the continuity of the tangential component of H requires that

$$n_1E + n_1E_2 = n_2E_1 \qquad (A1.2)$$

from eqs. (A1.1) and (A1.2) one obtains

$$E_1 = \frac{2n_1}{n_1+n_2} E \qquad (A1.3)$$

$$E_2 = \frac{n_2 - n_1}{n_1 + n_2} \, E \qquad\qquad (A1.4)$$

and

$$n_1 E^2 = n_1 E_2^2 + n_2 E_1^2 \qquad\qquad (A1.5)$$

eq. (A1.5) is the energy conservation equation.
Let $n_2/n_1 = n$; eq. (A1.3) and (A1.4) become

$$E_1 = \frac{2E}{n+1} \qquad , \qquad E_2 = \frac{(n-1)E}{n+1} \qquad\qquad (A1.6)$$

The reflectivity R is then

$$R = E_2^2/E^2 = \frac{(n-1)^2}{(n+1)^2} \qquad\qquad (A1.7)$$

and the transmissivity T is

$$T = nE_1^2/E^2 = \frac{4n}{(n+1)^2} \qquad\qquad (A1.8)$$

If one replaces n by 1/n, the reflectivity R of eq (A1.7) remains unchanged, i.e. the ratio of the reflected energy to the incident energy is the same whether the incident wave crosses the interface from n_1 to n_2 or from n_2 to n_1. It is also the same as for the usual retarded wave (this is also true for T of eq. (A1.8)). A notable difference between the complementary and the usual retarded wave is that the phase change which occurs on reflection at the interface is 0 or π for the complementary wave, whereas it is π or 0 for the usual retarded wave, when $n_2 > n_1$ or $n_1 > n_2$.

Using the above considerations and the ensemble of rules (R), one can obtain the interference patterns of the various cases considered.

ANNEX II

A beam-splitter has the same effect on adnis, anti-adnis as on photons, viz: adnis are either reflected or transmitted: the reflection or the transmission are random events with respective probabilities 1/2. Moreover, a beam splitter dissociates the pair adni+anti-adni, so that an adni is reflected while its companion anti-adni is transmitted, and vice-versa.

The purpose of this Annex is to propose a model accounting for this non linear behavior of a beam splitter.

Various types of beam-splitters are used ; consider those made of a thin metallic film deposited on a transparent substratum. If the light is to be transmitted through the film its thickness should be smaller than the corresponding skin-depth. If the thin film had the conductivity of bulk metal, say for aluminium, the skin-depth would be of the order of 30 Å. In general, films of such thicknesses, if not deposited according to a definite procedure or not submitted to a special heat treatment, do not possess the properties of bulk metal. In fact, the thickness of films used as beam-splitters is of an order of magnitude higher than that cited above and have properties more akin to thse of semi-conductors than to those of metals.

For such thin films, experiments show[24] that large deviations from Ohm's law occur when an electric field higher than a few volts.cm^{-1} is applied parallel to the surface of the film. One finds that the longitudinal conductivity of the film decreases with increasing applied field.

Considering eq. (2.7), one may write that the electric field E_z due to the ZPF applied along the surface of the thin film at some instant t is

$$\frac{E_z^2}{8\pi} = \frac{2K(2\pi)^3 \nu_o^3 \, d\nu}{c^3} \qquad (A2.1)$$

This value of E_z will be considered in conjunction with E_{ν_o} , the value of the electric field of an adni of frequency ν_o and band-width $\Delta\nu$.

For example $\nu_o = 6.10^{14}$ s^{-1} and $\Delta\nu = 3.10^{10}$s^{-1} (corresponding to $\lambda_o =$ 5.000 Å and $\Delta\lambda = 0.25$ Å), and considering the same band-width for the ZPF at ν_o , one obtains $E_z = 700$ Vcm^{-1}. At such high fields, one may assume that complete "saturation" occurs, i.e. the resultant electric current remains practically constant when a small extra applied field is added or substracted from this value E_z. This extra applied field E_{ν_o} , due to the above adni, according to eq. (2.9) is 1.8×10^5 times smaller than E_z, hence

$$E_{y_0} \cong 4 \times 10^{-3} \ V.cm.^{-1}$$

At any point on the thin film, since the ZPF is random, the orientation of $\vec{E}_z$ will be random in 2π around this point. In view of the large ratio E_z / E_{y_0} the component $\vec{E}_{zy_0}$ of $\vec{E}_z$ along the direction of $\vec{E}_{y_0}$ will be large as compared to $\vec{E}_{y_0}$, excepting for two small angles lying symmetrically around the directions of $+\pi/2$ and $-\pi/2$ with respect to the direction of $\vec{E}_{y_0}$.

Summarizing, one may conclude that $\vec{E}_{y_0}$ is either parallel or anti-parallel to a large component $\vec{E}_{zy_0}$ of the ZPF. In view of the saturation effect considered above, when $\vec{E}_{y_0}$ is parallel to $\vec{E}_{zy_0}$ the addition of the small quantity E_{y_0} to E_{zy_0} has no noticeable effect on the existing conditions of the thin film and the adni, depicted by $\vec{E}_{y_0}$, is transmitted. When $\vec{E}_{y_0}$ is anti-parallel to $\vec{E}_{zy_0}$ the (adni's) electric field is cancelled. However, the subtraction of a small quantity E_{y_0} from E_{zy_0} has no noticeable effect on the conditions of the thin film. The fact that $\vec{E}_{y_0}$ was cancelled on the thin film is tantamount to the reflection of the adni. Since the direction $\vec{E}_z$ is random, reflection and transmission of adnis are random events with respective probabilities of 1/2.

For anti-adnis, since on the thin film the direction of the corresponding $\vec{E}_{y_0}$ is opposite of that of the adni, they will be transmitted when adnis are reflected and reflected when adnis are transmitted.

This is the description of the ideal 50-50% beam-splitter for normal incidence. It requires a perfect isotropy, electrical conduction-wise, of the thin metallic film as well as the requested non linearity of its conductivity. Under these conditions the beam-splitter has a reflectivity and transmissivity of 1/2 and it dissociates all adnis from their companion anti-adnis.

If, however, the condition of isotropy was not satisfied, as for example for thinner films, the transmissivity along the directions of high electrical resistivity will be higher and adnis as well as their companion anti-adnis will be preferably transmitted. Likewise, higher conductivity, eventually along preferred directions, will improve the reflectivity of the beam-splitter for both adnis and their companion anti-adnis. For such beam-splitters the separability of adnis from their anti-adnis is not complete.

The mechanism of beam splitting described above applies only when $E_{\nu_0} \ll E_z$, it applies in the case of a unique pair adni+anti-adni. However, for the usual optical source intensities the conditions $E_{\nu_0} \ll E_z$ is not satisfied. Thus, for the usual source intensities in the Mach-Zehnder[18] interferometer the zeroeth fringe is bright, whereas for a unique (at any time in the instrument) adni+anti-adni this fringe is dark.

ANNEX III

Once the concept of adni+anti-adni is accepted it becomes quite natural to try and show their existence experimentally, particularly that of the anti-adni. The essentials of a simple layout of a proposed experiment are given in Figure A.3

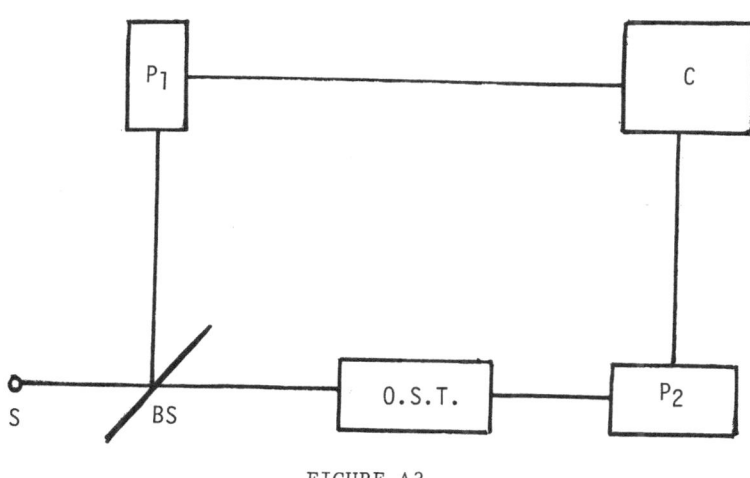

FIGURE A3

One recognizes the schematics of the Clauser type experiment, where in the transmit branch a special device, unknown yet, the OST (Oscillator Stop Tube) is inserted. The OST oscillates permanently unless it receives at its input an anti-adni, when it stops oscillating and sends an optical signal to P_2.

The succession of events is then as follows: when an adni is reflected by a beam-splitter BS it is received by photo-multiplier P_1 while the anti-adni is transmitted and is received by the OST, as a consequence an optical signal is received by photo-multiplier P_2. One observes a coincidence between the two photo-multipliers P_1 and P_2. In the case of the complementary event, viz: when the anti-adni is reflected at BS and the adni is transmitted, no signals are delivered by P_1 and P_2.

Another possibility would be to replace the OST by an "anti-adni multiplier", which is a photo-multiplier where the potential of the first dynode is negative with respect of the cathode. All the other dynodes have the same potentials as the usual photo-multiplier's. Probably, the nature of the cathode and of the first dynode should be chosen for the best efficiency.

Based on the remarks of Annex II and those of 3.3 a less straightforward proof of the existence of the couple adni+anti-adni may be obtained with the following experiment.

A Young slits arrangement is set along the reflected or the transmitted beam of a beam-splitter. If the separation, by the beam-splitter, of adnis and their corresponding anti-adnis were complete, no interference fringes behind the slits should be observed. However, due to fluctuations interference fringes will occur but their visibility will necessitate a longer exposure time than that required when the beam-splitter is removed.

REFERENCES

1. Einstein, A., Ann. Physik, 4th Sec. 17:132 (1905)
2. Einstein, A., Phys. Z., 10:817 (1909)
3. Lewis, G.N., Nature (London) 118:874 (1924), where for the first time the term"photon" was used.
4. Einstein, A., Phys. Z., 18:121 (1917)
5. Compton, A.H., Bull. Nat. Res. Council (USA) 4, Part2:1 (1922)
6. Jammer, M., The Conceptual Development of Quantum Mechanics (Mc Graw Hill, New York) (1966)
7. Schrodinger, E., Ann. Physik 82:257 (1927)
8. Lamb, W.E. and Scully, M.O., Polarization, Matter and Radiation. In: Jubilee volume in honor of Alfred Kastler (Presses Universitaire de France, Paris) (1969)
9. Franken, P.A., Collision of Light with Atoms. In: Atomic Physics, Bederson, Cohen and Pichanick Eds. (Plenum Press) (1969)
10. Janossy, L. and Naray, Z., Supp. Nuovo Cimento 9:588 (1958)
11. Clauser, J.F., Phys. Rev. D9:863 (1974)

12. Surdin M., Braffort, P. and Taroni, A., Nature (London) $\underline{210}$:405 (1966)
13. Boyer, T.H., A brief survey of Stochastic Electrodynamics. In: Foundations of Radiation Theory and Quantum Mechanics, Barut Ed. (Plenum Press) (1981)
14. Surdin, M., Stochastic Electrodynamics, on overview. In: Old and new questions in Physics, Cosmology, Philosophy and Theoretical Biology. Essays in honor of Wolfgang Yougrau, Van der Merwe Ed. (Plenum Press, New York) (1983)
15. Surdin, M., Lett. Nuovo Cimento $\underline{39}$:86 (1984)
16. Surdin, M., Int. J. Theoret. Phys. $\underline{14}$:207 (1975), Ann. Fond. Louis-de-Broglie $\underline{10}$:125 (1985)
17. Surdin, M., Found Phys. $\underline{12}$:873 (1982)
18. Born, M. and Wolf, E., Principles of Optics, 5th Ed. (Pergamon Press, Oxford) (1975)
19. Jackson, J.D., Classical Electrodynamics (John Wiley and Sons, New York) (1975)
20. Tetrode, H., Z. Phys. $\underline{10}$:317 (1922)
21. Wheeler, J.A. and Feynman, R.F., Rev. Mod. Phys. $\underline{17}$:157 (1945)
22. Surdin, M., Lett. Nuovo Cimento $\underline{42}$:153 (1985)
23. Selleri, F., In: The wave-particle dualism, S. Diner et al. Eds. (D. Reidel Pub. Co.) (1984)
24. Bernamont, J., Ann. Physique $\underline{7}$:71 (1937).

BEYOND QUANTUM MECHANICS USING THE SUBVAC

A.B. Datzeff

Faculty of Physics
University of Sofia
Bulgaria

ABSTRACT

Quantum Mechanics (QM), despite its well known many successes, meets with difficulties, some of which are insurmountable. This requires that QM be reconsidered and rebuilt on a new physical basis. For this purpose we start from the hypothesis of a material carrier of the electromagnetic field, which we name Subvac. Owing to its fluctuations, a microparticle (an electron for instance) will need a probabilistic description by a function $w \geqslant 0$. We find that the function $F(x,y,z,t)$, with $|F|^2 = w$, satisfies the Schrödinger equation. In this way all of the mathematical formalism of QM is retrieved. The analysis further shows that QM must be generalized as a nonlinear theory. Hence, we propose and discuss a generalized nonlinear Schrödinger equation, which is given here in the most general form for many-particle problems.

INTRODUCTION

I would like to express my satisfaction that, in my opinion, this very important Conference is not dedicated to general questions of which quantum mecanics (QM) is abundant, but to the basic question: Are there experimental violations of the QM predictions or there are no such ones? In the second case no problem exists. In the first case, QM must be revised, and such a revision can only be a fundamental one since QM is a fundamental theory.

The basic criterion here would be, naturally, the agreement with the corresponding experiment. The QM development from its beginning has imposed upon physicists the conviction that it always explains the experiments correctly. But recently it has appeared that, while in one category of phenomena concerning ensembles of microparticles with small densities ρ and high velocities v, the QM description is correct, it seems that for large ρ and low v a clear divergence exists between theory and experiment. Let us consider as an example the known case of the electron microscope: a homogenous beam of electrons with the same velocity v and space charge ρ, when dispersed on some object in a corresponding electromagnetic field, gives on a screen an image of the object that is similar to it and magnified. The theory used here is in fact the classical theory of the optical microscope in the geometrical optic approximation. Otherwise, the wave function Ψ, which is a solution of the one-particle Schrödinger equation (SE), is the same (semiclassical) approximation. But if now, one keeps the accelerating potential constant, i.e. the same V, and if one lets ρ grow, then the image will be deformed; with increasing ρ it will no longer resemble the object.

This deformation is not interesting, even boring for the experimentalist, but in reality it exists and it may become interesting at times (in principle, it is always interesting for a physicist, who must know how to describe it). But does it follow from the known SE? Not at all. It is connected with the complicated many-particle problem for the electrons in the beam, and one has to take into account the interaction between them, which was neglected thus far. Nor does this involve the many-particles SE. It appears here as a nonlinear problem and, when taken into account, changes the SE to a nonlinear one, i.e. it leads to the basic modification in QM. As we shall show below, the nonlinearity is imposed in each case on particles treated by the corresponding linear Schrödinger equation (LSE).

I.   Linear Quantum Mechanics

QM, as well as its mathematical formalism, are well known to all of us. But it is known as well that there is no unanimous understanding of some of its basic statements, and this leads to many difficulties[1]. For this reason, allow me to restate some of its results that are necessary for the following exposition, naturally, from my own point of view.

The historical onset of QM is based on two great ideas: (1) a wave associated to every microparticle, and (2) a matrix associated with every parameter of the microparticle. So far there is no generally accepted opinion on, or some model for, the physical sense of these two notions, but there is agreement that they give a correct probabilistic description of the behaviour of the particle. It turns out that the lines of thought following the two initial ideas produce an identical mathematical apparatus: the known linear mathematical formalism of QM. The last one was formulated by von Neumann as a strict system of hypotheses, axioms and a scheme for calculations, for which one prescribes absolute mathematical rigour. In general, its physical interpretation follows the Copenhagen interpretation of QM. It contains as well the known von Neumann theorem forbidding the introduction of "hidden parameters" in QM.

A central feature in the mathematical apparatus of QM is the Schrödinger equation (SE), or the linear Schrödinger equation (LSE), for the wave function Ψ. For the one-particle problem $\left(\Psi = \Psi\,(x,y,z,t)\right)$ one has:

$$H\Psi = i\hbar(\partial\Psi/\partial t) \tag{1}$$

The continuity equations for the probability density ρ and the current density $\vec{J}$ are:

$$\frac{\partial\rho}{\partial t} + \text{div}\ \vec{J} = 0, \quad \rho = \Psi*\Psi, \quad \vec{J} = \frac{\hbar}{2im}\,(\Psi*\nabla\Psi - \Psi\nabla\Psi*) \tag{2}$$

If the potential is constant, the solution of (1) is the plane wave:

$$\Psi' = C\ \exp\ i(\vec{k}\cdot\vec{r} - \omega t). \tag{3}$$

This is the simplest case of a continuous spectrum. As in all cases of a continuous spectrum, here and for the many-particle SE, it produces insurmountable difficulties. So, normalizing Ψ' to unity in some volume $v_0 = b^{-2}$, it follows that the wave Ψ' corresponds to particles with a constant average density $\rho = V_0^{-1}$, namely to an infinite number of particles of mass m in all space, which is obviously absurd. This is an expression of the fact that the function ψ' is not normalizable.

Let us take as a model of a many-particle SE, an atom A with n electrons, $M_i(X_i,Y_i,Z_i)(i = 1,2,\ldots,n)$ and nucleus $M_0(X_0,Y_0,Z_0)$. The corresponding SE for n+1 particles with wave function $\Psi(x_1,\ldots,z_n,t)$ can be suitably written using the Jacobi coordinates ξ, η, ζ. Let us denote by $\rho_1(\xi_1,\eta_1,\zeta_1)$ the vector $M_1 M_2$, by $\rho_2(\xi_2, \eta_2, \zeta_2)$ the vector $G_1 M_3$ (G_1 is the mass center of points M_1, M_2), and so on until ρ_{n-1}, where $G_{n-1} = G_{n-1}(X,Y,Z)$. Here

$$\xi_k = \sum_{}^{k} m_i x_i / \sum_{}^{k} m_i - x_{k+1}, \quad \eta_k = \ldots, \quad \zeta_k = \ldots, \quad X = \sum_{}^{n} m_i x_i / \sum_{}^{n} m_i, \quad k = 1,\ldots,n-1 \quad (4)$$

The variables are separated with (4) and one has

$$\Psi = \phi(X,Y,Z,t) X(\xi_1, \ldots , \zeta_n, t), \quad \phi = C \exp i(k_x X + k_y Y + k_z Z - \omega t). \quad (5)$$

The plane wave (5) describes a uniform movement of the m.c.G. After the interpretation given above the function ϕ describes an infinite number of points as G in all the space (which is an evident paradox) or in some corresponding closed final domain with volume V, containing N' particles as G, with constant average space density $\bar{\rho} = |C|^2 = N'/V$, average constant velocity $\bar{v}$, and average specific volume $\bar{V}_0 = 1/\rho_0 = V/N'$. Consider the abstract picture, where the randomly distributed points G with density $\bar{\rho}$ are replaced with regularly arranged points G at the sites of a cubic crystal lattice, of the same density $\bar{\rho}$. Then the edge of the elementary cube (the distance between one point G and its nearest neighbours) will be $\ell = \sqrt[3]{\bar{V}_0} = \sqrt[3]{V/N'}$. We shall use this abstract scheme in III.

It is known that QM does not define its object, say an electron, from the very beginning. Although it must be so, the strict mathematical formalism is applied to it as to some known object. But as for the natural question as to whether or not the object is localizable in the space-time, as for a classical particle, the answer of QM seems to be "no". But if a categorical answer with "yes" or "no" is required, the answer of the followers of the Copenhagen interpretation is in general "I do not know". So honestly it is given for instance by Feynmann[2].

On the other hand, it is claimed that, because of the Correspondence Principle, QM in the classical approximation becomes classical mechanics. But there is no strict mathematical proof of it (for instance, letting Planck's constant $h \to 0$, or letting the mass $m \to \infty$). (I consider that there is no such transition.) However, QM does not impose any limitation on the mass of its objects; hence, even for $m \to \infty$, QM will not become a classical theory, although the mass will then be that of a macro-object.

These and other difficulties clearly show that changes must be introduced in QM, and certainly some limitations must be imposed on its mathematical formalism as well. However QM, when the physical aspect is not clear and its mathematical formalism is a unique system, will not support some isolated corrections. The necessary changes will certainly turn out to be profound. So, it will not contradict the constraint contained in the von Neumann theorem (even if it agrees with the author's view for such a case). Likewise, this will not affect at all the great merits of today's QM. On the contrary, it will be thus raised to a higher stage. This is the conception on the problem which I would like to expose here.

II. A new viewpoint

My fundamental idea in reconsidering QM is based on two basic hypotheses:

1. All the objects in nature, in the micro and in the macro-world, can be described in space-time, i.e. one can use space-time models for them. (This is practically the opinion of many physicists.)

2. The electromagnetic field has a material carrier which we have called the subvac (substance of the vacuum). A similar notion (the ether) has played some role in the history of physics and, no doubt, a positive role during the formation of the Faraday-Maxwell conceptions. Later, Special Relativity (SR) rejected it, replacing it by the notion of field. The field is an important notion in today's physics and has well-defined mathematical and formal physical properties. In previous works[3-7] we have shown that such a material carrier is quite compatible with SR and especially with the Lorentz transformations, if one introduces some changes in its physical interpretation. The subvac hypothesis turns out to be fruitful elsewhere as well: for instance in the study of electron structure[6] and in the notion of Planck's quantum[7].

We assume that the subvac has a discrete structure, namely that it is composed of sub-particles AS (Atoms of the Subvac, a conditional name). The AS have a discrete structure generally unknown until now. They possess a micro-electromagnetic field (EF) represented by two vectors $\vec{e}_o, \vec{h}_o$ which have a random distribution in each finite domain. If they are arranged by some external EF, a macro EF will be created. The subvac possesses a fluctuation as well. Because of this fluctuation, the interaction of each charged particle (for example an electron) with the subvac will have both classical and random behaviour, which could be described by some probabilistic function $w(r,t) \geqslant 0$. In a step-by-step study one shows that the function F, where $|F|^2 = w$, satisfies a differential equation very similar to the LSE; to note the difference we shall name it a Probability SE (PSE). The function F describes the electron macrodistribution in the space, w being proportional to the relative electron time of residence $\Delta t/T$ in each volume ΔV which is a part of the volume V. But, if one considers an ensemble of N electrons in volume V, having a similar movement, then for their relative average number in $\Delta \tau$, namely $\Delta N/N$, one admits the equality $w = \Delta t/T = \Delta N/N$, (reminiscent of the Ergodic Hypothesis in classical statistical physics). Hence, the probability equation (evidently the LSE as well) will describe both the stochastic movement of every electron and the statistical distribution of the same ensemble of electrons.

About the electron, we accept that there is in it, thanks to its inner structure, some oscillation process of frequency ν. Because of the electron-subvac interaction, an oscillation process will be created in some domain σ around the electron of frequency ν', depending on its velocity v. These two processes will certainly become weaker with increasing distance from the electron. This phenomenon could be schematically written as a stationary wave $\phi(\psi_+, \psi_-)$, where the waves ψ_+, ψ_- propagate radially in opposite directions (outward and inward). When the electron passes in the proximity of some other object and the domain σ touches the last one, the wave ϕ in it will be influenced and changed. This will give a supplementary force which acts on the electron, and affects its movement. Hence, the electron, thanks to the wave ϕ which accompanies it, will "know" of the existence of other objects in its vicinity and will "take them into account". This will have some reflection on the final probability position distribution in space. This model simply explains one of the basic difficulties in QM interpretation: the Young experiment, in which one observes electron beam interference through one or two slits. Each electron considered as noninteracting with its neighbours is, by itself, responsible for the probability picture corresponding to the function w (a stochastic process). Of course, every real experiment is always connected with a multitude of particles (an electron beam), considered without mutual interactions. Thus, the statistical distribution N(r) on a screen will present the same picture as the stochastic distribution w(r).

One can extract from the last hypothesis, and the picture which follows from it, many general qualitative consequences conserving QM, where there exists an analogy between the probability function F and the wave function Ψ. Some examples are: reflection of a particle on a wall, passage through a potential barrier, diffraction and interference phenomena, and so on[3]. Similarly, if the particle of mass m grows and becomes macroscopic, then the effect of the vacuum fluctuations on it will be insignificant, and its movement passes from quantum to classical. (Passage from QM to classical mechanics).

After these qualitative deductions from the new viewpoint, let us move to quantitative ones. After specifying all the conditions which must be imposed on the one-particle probability function F, for which $|F|^2 = w$, the considerations lead to the result that F satisfies the linear differential equation:

$$HF = i\hbar(\partial F/\partial t), \quad Hf = Ef, \quad (F = f \exp(-iEt/\hbar) \tag{6}$$

where the second equation is the amplitude probability equation. At first sight, the first equation coincides with SE, without being identical to it, because some limitations are imposed on it according to the physical ideas used in its formulation. These are:

1) The equation (6) is always defined in a finite space domain of volume V, an arbitrary but limited one. This makes the spectrum of the amplitude equation in (6) always discrete. Besides, it is divided into two domains: in one of them the spectral lines have a relatively large density in comparison with the other. When the volume V grows and occupies all the space, equation (6) becomes identical to LSE and the two domains in question will coincide in accordance with the domains of the continuous and the discrete spectra of SE.

2) One imposes a limitation on the electron velocity by the inequality $v > v_0 > 0$. This excludes from consideration a Brownian electron movement.

3) The case of macroscopic mass of the particle is excluded, i.e. equation (6) is valid for microparticles only. Outside of these limitations equation (6) is identical to SE.

The above results are generalized in a natural way to the n-particle probability equation and to the entire mathematical formalism of QM.

Here we shall make more precise one question that is important for QM, but not clearly defined and thereby leading to many misunderstandings namely the particle-ensemble relation. There is a widespread understanding that, when we speak of the one-particle SE, what we really mean is one unique particle; for instance one electron described by this equation, and similarly by the n-particle SE (such as an atom with n-1 electrons). This understanding comes from classical mechanics, where the equation of motion for n-particles really describes n-particles. In fact, the name of n-particle SE stands for the existence of a large ensemble of N particles ($N = \infty$ in QM). This ensemble can be considered as a multitude of identical systems, S, of particles with the same average space density, where one takes into account the interactions in each system but neglects the interactions between the different systems. (This corresponds to the tacit assumption that the average distance between the systems is large in comparison with their average sizes). Hence, the designation of n-particle SE ($n = 1, 2, \ldots \ll N$) is only a conditional designation. Of course, the limit $N \to \infty$ does not have a physical sense (in reality N is always finite, but this requirement cannot be introduced in the actual QM without a fundamental change in it). Hence, it must enter in it by its formulation simultaneously with some other limitations mentioned above.

The above considerations explain the problem for particle-system of particles-ensemble, connected with LSE. If one does not take into account all the interactions in the ensemble, then the above considerations will be incomplete, i.e. they will explain the experiment only in an approximate way. The introduction of the interactions mentioned above in the probability equation (i.e. in LSE) extends these equations to a nonlinear form, which we shall make further on.

Starting from the last explanation one can divide the phenomena treated by the many-particle SE in two parts according to whether or not one takes into account the interactions between the systems in the ensemble of particles (without a strict limit between them):

1. The LSE explains the experiment in the following cases.

a). If in a limited volume V one has an ensemble of N particles (N_1 systems (atoms) containing n particles where $N = nN_1$), then, for small densities ρ, the linear atomic spectra are well described by QM as independent of ρ.

b) If an average $N = nN_1$, n-particles with small ρ and large velocities v, for instance an electron beam enter an aperture and go out of the volume V (a stationary case), then the interference picture is well explained as independent of ρ.

2. This is also the case as in the preceding examples when the interactions between the systems cannot be neglected. In case b) with an electron beam, $n = 1$, if the velocity v is small and ρ is large, then the beam dilates and deforms in general, but according to LSE it is invariable here. If the system is made up of hydrogen atoms ($n = 2$) in case (b) the spectral atom lines dilate, which does not follow from LSE. In case of a solid body (or crystal) one usually starts from LSE written for all the particles N in the crystal (often putting $N = \infty$). Then the estimation of all the interactions here is not made in a correct way, and the corresponding LSE is treated by improper operations such as: omitting many members, localizing the nuclei, following an arbitrary scheme of arrangements of them etc. It is known that some good results are

found this way, but the logic of LSE is certainly violated. Here it is necessary to take into account nonlinear effects. The nonlinear generalized SE proposed below treats the last problems with more logic and better grounded approximations, but the calculations are naturally more complicated.

I must admit that, when thinking over the proposed schemes, I am astonished that people who are interested in fundamental problems of QM only direct their efforts to the cases of point 1, where SE holds; hence, it is difficult to find weak points there, while under point 2 (with examples from the many-particle problem), SE is not sufficient and gives reason for a revision of QM. I think that it is correct and necessary to focus the attention of physicists on the cases under point 2.

III. Nonlinear Generalization

Let us return to the problem of generalizing the SE. Let us examine the case of a dense beam of electrons moving with uniform velocity v. Let it be incident upon a nontransparent unlimited plane screen, A_2 with a circular aperture σ, and let it be absorbed on a second unlimited plane screen B_1 parallel to A_1. The diffraction effects are well described by classical wave optics as well as by QM. Obviously, this picture does not depend on the density of the incident electron beam, since SE is linear. Indeed, if the latter has a solution Ψ, then the function $\Psi_1 = A\Psi$, in which A is an arbitrary constant, is also a solution. This should be true for arbitrary large values of $|A|$. (The zeros of the function Ψ_1 do not depend on $|A|$, and neither does the relative density of the electrons that have fallen on the screen B_1). Experiments show, however, that the diameter of the electron beam increases behind screen A_1 and gets larger as B_1 is moved further from A_1.

Let F(X,Y,Z,t) be the probability function for an electron P(x,y,z) moving in an external potential field $U^\circ(x,y,z)$. The effect of all the other electrons on P can be taken into account by means of a supplementary term U(x,y,z,t), which expresses the average Coulomb potential energy caused by the other electrons. We admit that in this case of free movement of the electrons, the function F will satisfy the following integro-differential nonlinear probability equation, which generalizes the linear Schrödinger one-particle equation[3-5], namely:

$$\Delta F - \frac{2m}{\hbar^2}(U^\circ + \lambda U)F = -\frac{2im}{\hbar}\frac{\partial \Psi}{\partial t}$$

$$U(r,t) = \int_V \frac{|F(r^1,t)|^2 \, d\tau'}{|\vec{r} - \vec{r}'|}, \quad \lambda = e^2 N_0, \quad N = \int_V |F|^2 d\tau \tag{7}$$

The general idea of a nonlinear SE is considered in Ref. 10 and specifically for the one dimensional case in Ref. 11. (In other words, Eq (7) takes into account the interaction of the representative electron P with all the other electrons in the ensemble; this interaction is not present in the usual SE for the same case). The integrals in U and in N are carried out over the volume V between the two screens and they are always convergent. The continuity equation $\partial \rho/\partial t + \text{div}\vec{J} = 0$ is immediately obtained from (7).

For the stationary case ($\partial U^\circ/\partial t = 0$), eq. (7) has an integral of the type F = f(x,y,z) exp($-itE/\hbar$), whence we obtain the amplitude probability equation

$$\Delta f + \frac{2m}{\hbar^2}(E - U^\circ - \lambda U)f = 0, \quad U(r) = \int_V \frac{|f(r')|^2 \, d\tau'}{|\vec{r} - \vec{r}'|}, \quad N = \int_V |f|^2 d\tau. \tag{8}$$

In Eqs. (7) and (8), the parameter $\lambda = e^2 N_o$, where e is the electron charge, N_o is the average electron density ρ in the falling beam, and N is the average number of electrons in the volume V.

Let the incident beam be characterized by the function $\bar{f} = A' \cdot \exp i(kr - \omega t)$, $|A'|^2 = N_o$. Then N is given by the approximate formula $N = \rho V \sigma_o (\ell/V) = \rho \sigma_o \ell$, where ℓ is the distance between screens A_1, B_1. If the origin, 0, is in σ, then the axis $OZ \perp A_1$, and the boundary conditions are:

$$f(P) = \bar{f} \; (P \quad \sigma), \; f(P) = 0 \; \big(P \; (A_1 - \sigma)\big), \; (Z = 0) \tag{9}$$

Here one considers the inner sides of the boundary of the domain τ as absolutely absorbant for the incident electrons, i.e. the values of the solution are not given there. This is an emission problem. So, the boundary condition (9) completely determines the solution of equation (8) in volume V. Equation (8) will determine an eigenvalue problem if the volume V is enclosed, for instance, by completely reflective walls (F = 0 at the walls) and described by the initial condition $F = F_o(X,Y,Z)$ for t = 0.

The nonlinear equation (7) can be written in the usual form of the linear SE with a nonlinear Hamiltonian,

$$HF = i\hbar \, (\partial F/\partial t), \; H = H^\circ + \lambda U \tag{10}$$

$$U = \int K(r, \, r') L(r, \, r') F(\tau) \, d\tau', \; K = |\vec{r} - \vec{r}'|^{-1}, \; LF(r) = F^X(r')F(r')F(r)$$

The corresponding amplitude equation will be Hf = Ef (F = f exp(-iEt/ℏ). The generalization to the two-particle equation (hydrogen atom) can be made by using the relative coordinate $\vec{r}_\tau$ and the coordinate $\vec{t}_g$ of the mass center G. The motion of G will be represented by a plane wave, determining an ensemble of points as G in all the space, as above.

The many-particle problem for n + 1 particles, as shown by (5), can be schematized by an atom $\bar{A}$ with electrons M_i (X_i, Y_i, Z_i) (i = 1,...,n) and a nucleon A_o (X_o, Y_o, Z_o). The corresponding LSE for n + 1 particles can be treated in Jacobi coordinates as in eq. (4). According to eq. (5), its solution splits as $\Psi = \phi$. The plane wave ϕ describes an ensemble of N' atoms A_p, p = 1,2,...N' >> n, or systems S_p in uniform motion. They are represented by their mass enter G_p. Each particle (electron, nucleus) is defined by its vector $\vec{r}$, $\vec{R}$, starting from a fixed point 0. Thus, the nucleus M_o and the electrons $M_{o1},...M_{on}$ of atom A_o will be given by $\vec{R}_o$, $\vec{r}_{oi}$ (i = 1,2,...,n), and correspondingly by $\vec{R}_p$, $\vec{r}_{pi}$ for A_p. Using the distribution scheme for points G_p at the vertices of a cubic lattice, the distance between the nearest vertices will be $l = \sqrt[3]{V_o} = \sqrt[3]{V/N'}$. The function itself can, in principle, be found by solving the corresponding LSE, for instance by the method of self-consistent fields.

Let T°, U° be the corresponding kinetic and potential (inner) energy of the atom A_o:

$$T^\circ = -\frac{\hbar^2}{2m_0}\Delta R_0 - \frac{\hbar^2}{2m}\sum_{i=1}^{n}\Delta r_{oi}$$

$$\overline{e}^2 U = \sum_i |\vec{R}_o - \vec{r}_{oi}|^{-1} + \frac{1}{2}\sum_{i\neq k}|\vec{r}_{oi} - \vec{r}_{ok}|^{-1} \qquad (11)$$

with Hamiltonian $H^\circ = T^\circ_o + U^\circ$ ($i = 1,\ldots,n$; $k = 1,\ldots,N'$).

Let [r] be the set of all vectors $\vec{r}$, and similarly for $[\vec{R}]$. The total potential energy U of atom A_o will depend on the mutual interactions of its own electrons and nucleus, as well as on the interactions between the electrons and nucleus of A_o with the electrons and nuclei of the other atoms. If they are in an integral, then the $\vec{r}$, $\vec{R}$ will be denoted by $\vec{r}'$, $\vec{R}'$. U° will be represented as

$$e^{-2}U = \sum_{p\neq o} |\vec{P}_{io} - \vec{P}_{ip}|^{-1} - \sum_{ip}|\vec{r}_{oi} - \vec{R}_p|^{-1} + \frac{1}{2}\sum_{ip\tau}|\vec{r}_{oi} - \vec{r}_{pq}|^{-1}. \qquad (12)$$

Here $i = 1,\ldots,n$; $p = 1,\ldots,N'$, $nN' = N$. L will be the nonlinear operator

$$LF(r) = F*(r')\,F(r')\,F(r). \qquad (13)$$

Let $F(r,t) = \Psi(r,t)$ be the wave function referring to atom A_o. Using (11), (12), and (13) we shall write the generalized nonlinear probability n-particle equation (generalized nonlinear n-particle SE) as

$$\left\{H^\circ + \lambda \int U([\vec{r},\vec{R}']\,Ld\tau'\right\} F(r,t) = i\hbar\frac{\partial F}{\partial t} \qquad (14)$$

For the sake of convenience we introduced the parameter λ (here $\lambda = 1$) as a factor of the integral in (14). Obviously for $\lambda = 0$ (or otherwise rejecting the nonlinear term), equation (14) will pass into the usual LSE in its usual form, for atom A_o.

Equation (14) is obviously not easy to solve. A basic method, although formal in the beginning, is to expand the function Ψ in a power series of λ. But, although equation (14) for atom A_o contains all of the A_p atoms numbering N', the principal contribution will be given by its nearest neighbours, as the general Coulomb potential created by each atom A_p will decrease exponentially with distance. So we shall have the greatest simplification for equation (14) if we retain the sum (14) only atom A_1, nearest to A_o. Then, using the total sum, as above, equation (14) will mainly describe the variation in the A_o spectrum, namely its expansion, as a function of the atom density ρ (an experimental fact).

Here the following question arises in a natural way. Consider for simplicity only atom A_o and its nearest neighbour. The latter acts on A_o as a perturbation, as the potential created by its averaged, and it can provoke in A_o some deformation, but not a fundamental change. But, to create a molecule A_oA_1 from atoms A_o, A_1 (which is not excluded theoretically), the electrons and the nuclei of A_o and A_1 must play the same role in the molecule (A_o, A_1). Hence, here we must directly treat a molecule made from 2 nuclei and n electrons. Following the preceding logic, we shall write for the molecule A_oA_1 an equation like (14) with

2+2n particles, where the molecules that are neighbours of A_oA_1 will act on it as a perturbation, as by the atoms in equation (14).

An objection raised to the generalized nonlinear SE (GSE) is that it is essentially equivalent to the Hartree equations resulting from the self-consistent fields. This is, however, only an apparent resemblance because of the similarity of the integral terms. These are, however, completely different things. The GSE (9) (n = 1) without the integral term is identical to the one-particle SE, but in this case, a corresponding Hartree equation does not exist. In the general case, the Hartree equations for n particles are given as an approximation to the n-particle LSE (although this is not precise), while the same LSE is generalized by supplementary terms to become GSE (14). Moreover the physical ideas contained in the Hartree equation and GSE are completely different.

A second and more fundamental objection was raised; the GSE is not a novelty in QM, and it certainly follows from its mathematical apparatus, probably by some averaging over the n-particle SE. This, however, is not true. Indeed, let us assume that the GSE (14) follows from LSE for some fixed number of particles N' found in some way. But, according to such an equation, and because of the current formula (2), the number of particles (electrons) falling on the second screen B_1, for time t is $N_t = \alpha't$ (α' is a constant); if the number N_t increases without limit with t, then $N_t \gg N$. So, one cannot find the necessary number N" to write the necessary LSE.

Finally, we state that the GSE proposed here is not some consequence from the apparatus of QM, but it is a <u>new postulate</u>. Therefore, every experiment for a precise check on this equation may become a <u>crucial experiment</u>.

CONCLUSION

The problem of the physical interpretation of QM is obviously of greatest importance, although it has been, in fact, neglected for many years. The widespread point of view here were the ideas of the Copenhagen school. They place the responsibility of existence of the unusual new regularities in the microworld on the observer's intervention and his experiment. Many physicists, including some of the creators of QM, provided new explanations, but they remained isolated in general. Besides, for a long period of time, proposals to revise QM were met coldly, even negatively. On the other hand, the mathematical apparatus of QM was supported by experiments in an enormous number of cases, and this consolidated among physicists the conviction that this will always take place. So they lost interest in the problem of the physical essence hidden behind the formulae. But, with time, on the one hand the weak logical points in QM are pressing to be explained and, on the other hand, experiments clearly show that, in many cases, the actual interpretation of QM is powerless to deal with them.

We are convinced that a basic reinterpretation of QM is necessary and unavoidable. This is important, not only for a clarification of its weak points, but also for Quantum Field Theory, whose physical basis is QM itself. Perhaps, then its famous divergences will be attacked succesfully. We think it is an important task for the physicists - experimentalists and theorists - to direct their efforts to surmount the disagreement between experiment and theory, especially for many cases of the many-particle problem. Besides, the finding that the observer's

role, so strongly underlined by the one-particle problem in the beginning
of QM, fades here and disappears before the complicated objective
regularities of the many-particle problem. On the other hand, the notion
of experiment itself, whose sense is rather undefined in QM, acquires
here an unusual reality independent of the observer.

The above considerations provided for these questions show that it
is necessary to take a realistic, firm stand on the fundamental notions
of QM, and to expand the nonlinear apparatus of QM while recovering the
known linear apparatus as a limiting case. From here it is of great
importance to plan and carry out many experiments where LSE ceases to be
true. We think that the significance of such experiments goes beyond the
limits of QM; thus it is most desirable that they should be realized and
well-interpreted.

REFERENCES

1. M. Jammer – The Philosophy of Quantum Mechanics, Wiley, New York,
    1974.
2. R.P. Feynmann – The Character of Physical Law, London, 1965.
3. A.B. Datzeff – Mécanique quantique et réalité physique, Ed. Bulg.
    Ac. Sci., Sofia, 1969.
4. A.B. Datzeff. Phys. Lett., $\underline{59A}$, No. 3, 185 (1976); Bulg. J. Phys.,
    $\underline{6}$, 3, 225; $\underline{4}$, 394 (1979).
5. A.B. Datzeff – Nuovo Cimento, $\underline{29B}$, No. 1, 105 (1975); "Open
    Questions in Quantum Physics", G. Tarozzi and A. van der Merwe
    (eds.), Reidel, (1985).
6. A.B. Datzeff – Phys. Lett., $\underline{80A}$, 6 (1980); Intern. J. Quantum Chem.
    $\underline{23}$, 81 (1983).
7. A.B. Datzeff – Phys. Lett., $\underline{100A}$, 71 (1984).
8. L. de Broglie. Une tentative d'interprétation causale et
    non-linéaire de la mécanique quantique, Gauthier-Villars, Paris,
    1956.
9. L. de Broglie – La théorie de la mesure en mécanique ondulatoire,
    Gauthier-Villars, Paris, 1957.
10. Bialiniski-Birula and J. Mysialksi; Ann. Phys. (N.Y.) $\underline{100}$, 621
    (1976).

DISCUSSION IV.

Compiled and interpreted by D.W. Kraft and E. Panarella

Panarella. What rotation velocity does Phipps envisage as necessary for detecting the experimental effect he described?

Phipps. My numerical example employed 10^{-6} c; you can't get too many orders of magnitude higher than that without centrifugal force pulling the device appart.

Panarella. What about the vacuum requirement?

Phipps. If one tries to work in a dispersive medium such as air, one quickly gets into conceptual problems about signal vs. energy transport vs. phase velocity; all of this is merely for phase velocity, so I wanted a good vacuum to relieve the conceptual problems in that area.

Panarella. Will phase shifts be detected by fringe shifts?

Phipps. No. I was envisioning a time resolution, i.e. one had to see a very sharp distinction in the response time of the two photodetectors. In the example I considered, I took 10^{-10} sec, and that called for a long vacuum chamber. If you can reduce it to 10^{-11} sec, then you can shorten the vacuum chamber by a factor of 10. I think the photodetector response time is more limiting than the flash modulation of the source; I think we now have source modulations way below that.

Edmonds. You have a mirror spinning at very high speed. What about motion at its center?

Phipps. There is motion of second order, v^2/c^2. The degree of the tilt and therefore the movement of the reflection is of order v^2/c^2. For many purposes you can regard it as practically stationary.

Edmonds. I also see problems with the mirror having nonuniform density (because of the high rotational speed), so that as you move away from the center, you are bouncing light off regions of greater density. Perhaps you don't accept special relativity, but you could use it to analyze the light bouncing off the mirror. I see this and other problems contributing subtle effects that could obscure the effect you are looking for.

Phipps. Well, there is always a great gap between the actual experiment and the idealized thinking, the Gedanken experiment, that precedes it. Although the points you raise certanly should be considered, the basic question is whether it is worthwhile to attempt something of this sort; for that purpose, my rough order of magnitude calculations are sufficient to convince me that the answer is in the affirmative.

Edmonds. The detectors are flying around the edge of this wheel very fast, catching the photon. How do you get the information out of the fast-moving detectors and into the laboratory equipment? Do you have transmitters on there and are you concerned about the phase shifts of the radio waves that are sent to the lab?

Phipps. I envision wires of equal length leading to the center and using slip rings where things are slower moving.

It would be delightful to do the experiment without moving detectors. I considered treating a moving mirror as an absorber/re-emitter, but that introduces a conceptual problem. With a photodetector, you have a local process and there is a real completion of the quantum process. But with an extended surface such as a mirror, you are possibly dealing with a nonlocal interaction with the whole surface. That may be so different from the true final severing of the phase connection that occurs in a genuine localized photodetection that such attempts as doing the whole thing on a moving platform with mirrors may be actually missing the whole point. I am, in fact, undecided as to exactly what a quantum detector is in the context of testing neo-Hertzian theory. I have pictured the field detector as a black box, but, with radiation detection, when is the quantum process actually completed? We speak of absorption and re-emission of a mirror but if there are nonlocalized processes, then it is probably physically different from a true photodetector detection. Which am I speaking of? I don't know. If the experiment were done with actual photodetectors and obtained a negative result, that would falsify the whole line of thought that I have provided here. But, if it were done with mirrors, and with photodetectors stationary in the lab, then I would consider the matter undecided. To me the crucial experiment is of the general type that I described, where you actually go through the agony of moving a true photodetector.

Selleri. In your reformulation of electromagnetic theory, à la Hertz, essentially you say that there is complete agreement with Maxwell theory as far as interactions between moving charges and the electromagnetic field is concerned. Now when you move your detector, I would think that the detector is a set of charges, for it must be to detect electromagnetic radiation. Therefore I cannot see why you should expect a difference.

Phipps. What difference is that?

Selleri. The Maxwell and Hertz theories make the same predictions about the interaction of moving charges and electromagnetic fields.

Phipps. The Maxwell and Hertz equations have field quantities as their dependent variables. I have to imagine that there are detectors giving operational meaning to these quantities in the equations. You are speaking of what amounts to test charges moving in external fields; to speak of what happens to these test charges, I have to speak of a force law. In order to state what the $\vec{B}$ is that appears in the Lorentz form of the force law, we have to imagine a $\vec{B}$-meter having some general state of motion in our laboratory. In the particular case when the $\vec{B}$-meter is at rest, we get the customary form of the Lorentz force law, but if it moves, as it would in the view of a differently traveling inertial observer, then we obtain a force law in which the $\vec{v}$ in the $\vec{v} \times \vec{B}$ term is actually the relative velocity of the test charge and the field meter. The latter is a bit of matter that has a definite state of motion with respect to the test charge. This is a conceptual difference between the Hertz and Maxwell formulations, because the Maxwell-Lorentz theory has

398

this type of velocity which involves the relative motion of the test charge in a frame of reference; this is referred to by O'Rahilly in his book on electromagnetic theory as a "schesic" velocity. He devotes much discussion to the troubles of having to have such a velocity in the formulation in the laws of physics. In the Hertzian reformulation, one gets away from that; that is, the only velocity appearing in a force law is the velocity between the field meter, one bit of matter, and another bit of matter which is the test electrical charge. Of course the two are identical in the special case that the field meter is at rest in the laboratory, for then the meter's state of motion is the same as that of the laboratory itself - this brings you back to Maxwell-Lorentz case.

Buonomano. I would like to comment on Santos' remarks relating to memory, by summarizing some results that I published a few years ago.

Bell's inequality is usually derived under the assumption that consecutive photon pairs are independent. In a real experiment, photon pairs follow each other and one could validly imagine that there is some sort of memory effect. For example, one can imagine emission events with consecutive emissions depending on one another, or imagine a memory in the polarizer so that successive photons see a different polarizer. We may imagine similar memory effects in the detector. If you take an average over a single experimental run, i.e. many particles consecutively following each other, then you can have agreement with quantum mechanics. Actually if you use just four experimental runs, then one can invent theories that agree with quantum mechanics and disagree with Bell. But if you average over all possible experimental runs, then even these theories must agree with Bell's inequality. If you average over all possible time averages, to disagree with Bell, you still must imagine that the polarizers affect the source or each other. I don't know how this relates to real experiments; I imagine that in real experiments you do small numbers of time averages.

Edmonds. If you let the system sit before you turn it on again, would the results after one week be different from those after one day?

Buonomano. It depends on what is independent. What is repreparation of a state? Repreparation of a state is just the quantum mechanics way of saying that you have independent systems. To know whether you have independent situations presupposes that you understand the physics of the situation.

Selleri. With this mechanism you can concretely show that Bell's inequality can be violated.

Buonomano. Sure. You can do it with four individual, independent experimental runs. If you do it over an ensemble of independent runs, you must agree with Bell's inequality. Think of having many labs, each one doing an experimental run; if you average over all of these, then you have to agree with Bell's inequality.

Panarella. I have a question for Kyprianidis in relation to the probability for effective photons to appear in low intensity laser beams. According to Quantum Potential Theory, such probability is $\sim 10^{-30}$ for the parameters of my experiment. Since you asked me for more data concerning ionization phenomena, I have here a paper by de Brito (Can. J. Phys. 62, 1010, 1984), where such data can be found. What strikes me, and indeed this is the question that I am going to put to you, is that sometimes de Brito, with a much larger probability of ionization than yours from his theory, is unable to get ionization. I wonder then if your own probability of 10^{-30}, which justified the negative result of my experiment, will not lead to no ionization at all for all gases?

Kyprianidis. The calculations that I presented referred to condition of focussed beams. For the values of the parameters of your experiment, I get a probability of photoelectric emission of ~10^{-30} or less. The result is intensity-independent. In other words if, with the same experimental apparatus that you have, you would increase the power density, the same result would be obtained. However, if one were able to change the gaussian parameter of the beam by $\sqrt{10} \simeq 3$, by better focussing the laser beam, then the probability would dramatically change to 10^{-3}.

Selleri. I would like to direct a comment to Santos. I agree completely with much of what Santos said, but there is one point on which I do not quite agree. Santos pointed out that twenty years have passed since the discovery of Bell's inequality, and in this interval no violation of quantum mechanics or of local realism have been found, and his conjecture is then that perhaps quantum theory is compatible with Einstein locality and we have still to understand how that is possible. I do not think that I agree with this. Also, from a historical point of view, it is quite true that it is twenty years since the discovery of Bell's inequality, but if we look at the real history of this problem, we find that it has always been surrounded by strong ideological pressures, a sort of smoke screen which has not allowed one to see clearly what the real problem was. This is a residue of the Copenhagen interpretation and that is why twenty years have passed, and perhaps another ten shall pass before we have a real solution of the problem.

Pertinent experiments have been performed only one at a time, with years between one experiment and the next. Contrast this with elementary particle physics where an interesting problem would have been clarified within only three years. Instead, here it is difficult to find individuals willing to undertake such experiments, and the money required to support them. Then also there are some strange results which have not been investigated enough. For example, there are some strange features about circular polarization, and there is the problem of radiation trapping. If radiation trapping exists, and if it has the effect of lowering the correlation, then why is there agreement of experiment and theory? I don't understand how a wrong experiment which contains spurious effects can agree within forty standard deviations of the Bell limit, as reported by Aspect. But we know that radiation trapping is present in that experiment, so I don't understand it. Problems of this importance should be investigated in all possible ways, and some nice proposals have been advanced. For instance, the decay of the pi-meson into K^o, $\bar{K}^o$, and theoretical proposals on Λ, $\bar{\Lambda}$, and the Lo-Shimony proposal. This is a rich field, conceptually, but surrounded by thick clouds that prevent one from seeing what is going on.

Wadlinger. If the single wave model is correct, then no experimental claim for singe photon diffraction is valid, for no one can claim to chop a single wave from a wave train. Therefore until you know the makeup of the photon, you may not claim single photon diffraction from probability calculations. Secondly, if the Hunter-Wadlinger model is correct, it is a quantum violation, for Heisenberg said that we cannot know the nature of the photon as being one wavelength long, for we cannot make simultaneous accurate measurements of energy and time. We cannot know what we cannot measure.

Phipps. I would like to ask Hunter about his cigar shaped model for the photon. Are the size and shape preserved in time, and if so, what about de Broglie's theorem that a linear mathematical system describes only things that come apart in time?

Hunter. Our model is a nondispersive, soliton type of model. The de Broglie theorem applies to a wave packet containing a distribution of frequencies, whereas we are dealing with a single frequency. Thus even though we are dealing with the solution of a linear equation, it is not dispersive.

I would like to reply briefly and to take the opposite viewpoint of my colleague Wadlinger. The key piece of evidence in regard to our model is the question of low intensity interference. The evidence that supports our model is basically the same as that for Panarella's model, which predicts that diffractive intereference is a multiphoton effect. Thus in a double slit experiment, some photons go through one slit, some go through the other slit, they diffract at different angles, and interfere when they reach the plane of observation. It's the same as Panarella's photon clumps. The photons within such a clump, or bunch are coherent. Panarella's experimental work and that of Dontsov and Baz support this view. However all the other experimental evidence, including that of our own work with Jeffers which was undertaken to confirm Panarella's results, does not confirm this view. The key question is whether one can get single photons, that is, however low the intensity, can you be sure that you have only one photon in the diffractometer at any one time? Now the experiments of Grangier seem to support single photon interference rather than photon - photon interference, and therefore I must admit that the experimental evidence is against our model.

Vigier. We have here two conflicting views. Quantum mechanics is based on Dirac's statement that a particle interferes with itself, and I believe that this has been shown experimentally by Grangier and by the neutron experiments. This does not mean that the photon cannot be described by some extended structure which you call a wavicle and I call a soliton. The problem is that if you want to interpret single photon interference from slits separated by more than one wavelength, then in this model you must say that the extended soliton moves in the direction of the Poynting vector. This is perfectly possible for you can introduce a nonlinear Maxwellian term and the total solution of the nonlinear equation will consist of a solution for the linear part plus the soliton solution; the latter, of the form sin kR/kR, is a singularity and follows the drift lines of the surrounding regular Maxwell field. The two solutions are geared together and this is the reasoning whereby Einstein, Infeld, Hoffmann and others figured the test particle as an R^{-1} singularity in the surrounding regular $g_{\mu\nu}$ field, and this forced the test particle to follow geodetics. I like that argument for now I can understand why you can make single photon interference; in a sense the singularity locates the extended R^{-1} structure which is the photon and which follows the drift trajectories of the Maxwell equations. So, I don't think your model is finished; it is a first step and it can be very well fitted into the picture I described and there is no contradiction between the model and the essential predictions of quantum mechanics. I know that efforts are under way to improve on the Grangier experiment, and I am not at all convinced by your results. I believe that Dirac's statement about particles interfering with themselves is fundamentally correct.

Greenberger. What happens if you introduce your photon into a dispersive medium and have it come out again? Has it preserved its original size?

Hunter. I don't know what happens to it inside the medium, but when it comes out again, I believe it will resume its original shape. The essence of our model is just a photon propagating in free space.

Phipps. I am delighted that Datzeff has focused on the many-body problem, for not too much ingenuity has been devoted to it. There may be some more mileage in linear mathematics, for an unsatisfactory feature of the quantum many-body problem is that one speaks of a 4n-dimensional space, so the decision to include or not include an extra particle in the discussion alters the dimension of the descriptive space, which seems a bit artificial. I have found that by introducing idempotent operators in the classical many-body problem, one can interpret the Hamiltonian in the Hamilton-Jacobi theory as a linear operator. The idempotent operators can be thought of as n × n matrices, all of whose entries are zero except for one at the i-th position on the main diagonal. These things can be treated as tags on the energy of the i-th particle. The collective energy is just the sum of the individual tag energies, and you can obtain the main aspects of the classical relativistic many-body problems out of this approach. It amounts to interpreting what has always been treated as a scalar quantity, the collective energy of the system, as an operator. But that is useful also for the purpose of linearizing the classical Hamiltonian in the sense that Dirac did, and for that technique you have to introduce operators. So operators are not foreign to the classical mechanical problem and, when introduced, permit a direct correspondence transition to the quantum description. So there is much to be gained for getting the mechanics formally similar on the two sides of the correspondence transition.

Well I am hopeful that there may be exciting new physics lurking here and it is possible that even biology is concealed in a better understanding of the many-body problem.

Panarella. (to Hunter) Your model of the photon is really many photons.

Hunter. Our model of the photon is a ellipsoidal volume, associated with a single frequency, but it can contain multiples of hν of energy. The normal single photon in normal, low intensity visible light, would consist entirely, ordinarily, of single photons. In stimulated emission, what initially emerges is a photon of 2 hν energy and in a focused laser beam, what one is producing are these same volume, same frequency photons with many hν of energy in the same volume.

Panarella. You measured the transmission of microwave power through slits of varying width. For very small slit size, how does your model differ from the predictions of classical diffraction theory?

Hunter. The classical electromagnetic theory for this situation was done by Andrejewski and is recorded in the book by King and Wu. If you extrapolate their calculated classical curve to the region of interest, you will find it very difficult to distinguish experimentally between the classical curve and our billiard ball-like model, which says that if the slit is large enough to let the "cigar" go through, then it will go through and otherwise, it won't. With respect to diffraction, as the slit approaches the critical width, the proportion of the radiation diffracted increases and becomes very large. Our experiment was designed not to collect the diffracted light, but only that which went straight through.

Belinfante. Did you use just Maxwell's equations without any non-linearities?

402

Hunter. Yes.

Belinfante. So, if two of these billiard balls approach each other, they would just go through each other without disturbing each other. Then they could not create interference just by interacting with each other because there is no interaction.

Hunter. For free states, you are right, but what if two of these things come together at the observation screen?

Belinfante. That's a different case. What I mean is that fringes are not places where billiard balls interfere but are places where the billiard balls arrive.

Hunter. Our model said that the interference is between two of these billiard balls.

Belinfante. That is only within the size of the billiard balls, but those fringes are macroscopic compared to the billiard ball.

Hunter. I don't think the theory of the interference is substantially different from what it would be in classical optics because the model is that, as we go through a diffraction slit, they are bent at different angles depending on the impact parameter of the cigar with the slit, and then the actual interference, as in the classical theory, depends upon path differences. The theory of the interference is basically the same as in classical optics.

Belinfante. In that case it would be necessary to catch the constructive or destructive interference on the screen and it would be necessary that one billiard ball coming from one slit and one coming from the other slit just happened to arrive at exactly the same time and at the same spot on the screen. This can also explain what happens at low intensities.

Hunter. Yes. The model seems to be the same as Panarella's for interference at low intensities. The model depends on clumps of coherent photons, so one predicts for very low intensities where the clumps disappear, the interference disappears.

Wadlinger. But the clumps may simply be a wavetrain. If n waves are in the train, then n waves are interacting with each other.

Hunter. I believe our next step should be to study this more thoroughly.

Panarella. In my model there is an interaction relation between the photons and they position themselves in such a way that the clump carries with it at any one time all the characteristic features of interference. These features are destroyed when the clumps are broken.

Selleri. I wish to direct some observations to Panarella. I recall an experiment containing a source in which only one atom at a time is visible. How can one atom emit a clump photons?

Panarella. Now we are going into the physical model for the formation of clumps. This is not something which occurs within the source. The photons are emitted freely out of the source; once they are out of the source, the interaction relation guides them so they reposition themselves.

Selleri. But there isn't enough time for this. Atoms move slowly and light goes with the velocity of light.

Panarella. Once two photons leave the source and travel at the velocity of light, then the interaction relation is such that these two photons catch one another.

Selleri. I repeat, there is only atom at a time.

Panarella. Then we come back to the question how did you detect one photon emitted from one atom. I must stress again that there is some extrapolation in the concept of measurement, and that all detectors have a limited quantum efficiency, and therefore if one photon is detected, we have no knowledge of the other photons that were neglected by the detector.

Phipps. I would like to ask about the concept of photon length. Originally, I associated this with the coherence length which had to do with the $\Delta E \cdot \Delta t$ product of the photon emission process. At this conference, we have heard of models in which the photon length is very much shorter than this, namely the Hunter-Wadlinger model of the photon as one wavelength long. In addition, there are also length parameters inferrable from the response time of the detectors. For instance, the case of a photographic plate was pictured as behaving like a coincidence counter with a certain time interval within which several photons must arrive to effect the grain response. Now one doesn't measure the lengths of objects moving at relativisitic speeds, much less of light; what we do measure are time intervals at a given place in the laboratory. What is your view of photon length and what do length and time mean in your conception?

Panarella. My conception of the photon derives from the expression for the photon-photon interaction which results in the formation of clumps. Within these clumps, there is a minimum longitudinal separation between photons; this minimum separation is equal to λ and no more than one photon can occupy a volume of $\simeq \lambda^3$.

Buonomano. I spoke to members of Rauch's group who did the low intensity neutron interferometry experiments, in which they could see the detection of individual neutrons. They were separated by large time intervals of the order of, on average, one second. They talked of neutron detectors of very high efficiency.

Panarella. Since we want to be sure that there is only one neutron within the interferometer, we should consider the details of the timing mechanism and demonstrate that the separation between successive neutrons exceeds the length of the interformeter.

Selleri. This has already been done.

Greenberger. There are many experiments that one can do. For example, the insertion of a piece of aluminum on one side of the beam slows up that side because the index of refraction of aluminum is slightly different from one. By using polarized neutrons and by applying an external magnetic field, they can detect the angular deviation in the amount of precession, so that you have a very accurate timing mechanism. Now these neutrons are moving very slowly, of the order of 10^5 cm/sec. Then it takes them 10^{-5} sec to go through a 10 cm device. Now they see only about one coherent neutron per second, so that the chances of two coherent neutrons being in the device are 10^{-5}. As Rauch puts it, when you count one neutron, the next one to come through hasn't been born yet.

Panarella. This is also said of photons.

Greenberger. In the neutron experiments, standard quantum mechanics explains the results very easily. It's a linear theory and all the results of superposition work. This is not to ignore alternative formulations, such as that of Bohm and Vigier, which can reproduce the results of the standard quantum mechanics; the result is that there happens to be more than one way of describing quantum mechanics. Nonetheless, it is the standard quantum mechanics which facilitates calculations and predicts very easily the results of delayed choice type of experiments. I grant that the interpretation poses problems, but it is not the case that quantum mechanics is wrong or even not natural.

Kyprianidis. It is clear that there is no conceptual difference between neutron and photon interferometry. That's a critical point because the technical differences between the two permit one to see more clearly what is happening in neutron interferometry. It is quite enlightening to compare the results of the static absorption with the time-dependent absorption results obtained by chopping. There is a big difference between these results - you have a much lower intensity with the time-dependent absorption. When the chopper frequency approaches the order of the coherence length of the neutron packet, then you arrive at the static case again.

Selleri. I agree with Greenberger that quantum mechanics is a very natural theory when one deals with neutron or photon interferometry. But I do not think the quantum mechanics is natural when it comes to the EPR paradox; there it is natural to reason in terms of local realism and quantum mechanics is incompatible with local realism. I think that this subject should be taken more seriously and that experiments should be performed under conditions in which quantum mechanics is really incompatible with local realism. This has never been done.

Greenberger. I agree with you that as an arena for testing quantum mechanics, EPR is much more to the point. Quantum mechanics is always non-intuitive; here is a case where it violates not only our normal intuition, but violates the condition that many people would put on the experiment for it to have any interpretation whatsoever.

Wadlinger. Phipps asked earlier about the length of the photon, and Panarella answered about one lambda. I would like to add some more evidence. On page 1192 of their book, Misner, Thorne and Wheeler indicated that from fluctuation phenomena, they estimate the volume of a photon of energy $h\nu$ to be $\sim\lambda^3$. This is from quantum geometrodynamics considerations.

Greenberger. I can't believe that you can get any independent evidence on this from gravity. There is an enormous coincidence in nature, the Magic Numbers, which can be interpreted in many ways. Whether this is coincidence or meaningful is hard to know.

Santos. I should like to comment on whether quantum mechanics is natural or not. When I use stochastic optics to try to explain phenomena, I am not for throwing quantum mechanics away. Rather I am trying to reinterpret quantum mechanics. For example, the principal novelty of stochastic optics is the assumption of a zero point background radiation, which is also present in quantum field theory; we merely reinterpret this as a real rather than as a virtual backgrouned radiation. We also reinterpret creation and annihilation operators as a means of representing random variables. Thus, to some extent, we are trying to go as close to quantum mechanics as possible. Even if we cannot obtain very accurate results at the moment, we are trying to gain a qualitative understanding. Our final intention is to go as close as possible to the

quantum formalism and to reinterpret it. Perhaps there must be some
small modifications to avoid the conflict with local realism.

Aspden. I would like to comment on the connection between the photon and
gravitation. First of all, I do not think that the photon and
gravitation have any immediate, direct connection, but there could be an
indirect connection. I presented a picture of a lattice in the vacuum.
That lattice must have formed from chaos. My picture of how this lattice
was formed is analogous to the formation of ferromagnetic structures
which come into being upon cooling below a critical temperature. I found
these models to be useful when applied to the vacuum to deduce the nature
of photon birth. I would suggest that if we find eventually that there
is a good explanation for the photon mechanism that relates to a lattice
structure for the vacuum, we may infer that at the instant when that
chaos in the vacuum turned into order, gravitation appeared. That is,
when the Universe cooled below its "Curie Temperature" and formed the
lattice, photons and gravitation appeared and all the matter in the
Universe began to coalesce. I suggest that quantum mechanics, once it is
sorted out, might help solve some cosmological questions.

DOES QUANTUM ELECTRODYNAMICS EXPLAIN THE OBSERVED LAMB SHIFT?

Frederik J. Belinfante

Department of Physics, Purdue University, Lafayette, IN 47907
Present address: P O Box 901, Gresham, OR 97030

INTRODUCTION

While so far the success of nonrelativistic quantum theory for systems with finite numbers of degrees of freedom has been astounding, and relativistic quantum theory of electrons has made predictions about spin and about positrons that have well been confirmed by observations, applications of quantum theory to systems with infinite numbers of degrees of freedom suggest that quantum theory in its present form is not yet completely correct. Predictions made by quantum field theory, in particular by quantum electrodynamics, are obtained only by what a mathematician would call "swindles," like treating divergent renormalizations as small corrections. Physicists so far have accepted these swindles, because they consider it reasonable that at very small distances the present theory would yet require corrections, which, they hope, will in the future lead to finite results that would lie close to the results presently obtained by treating logarithmically divergent integrals as finite quantities.

I could live with this trust in the future theory, if I were convinced that in other respects there are no blemishes in the calculations which we are told to show beautiful agreement between quantum-electrodynamical predictions and experimental observations. A critical look at the conventional calculation of the Lamb shift by second-order perturbation theory, however, reveals the need for a revision of part of these calculations, in particular of the contribution to the Lamb shift from intermediate states with virtual photons of moderate energies from $0.003 \, mc^2$ up, where m is the electron mass. For these intermediate states, it is conventional to use Bethe's nonrelativistic treatment of the Lamb shift,[1] and its later refinements,[2] even when relativistic methods are used for calculating the contributions from intermediate states with virtual photons of energies around mc^2 and higher.[3] Some of the approximations made by Bethe are not justified, and we have started a program for correcting these errors. Until this work has been completed, we do not have the right to claim agreement between quantum electrodynamics and the observed facts.

CONVENTIONAL DERIVATION OF THE LAMB SHIFT

The first to complete the relativistic calculation of the Lamb shift were French and Weisskopf.[3] They showed how formally the shift of hydrogen atom energy levels caused by interaction between electrons and the electromagnetic field may be written as the sum of "exchange terms" and "nonexchange terms," which both, in the radiation gauge, are sums of "dynamic" terms due to interaction with the transversely polarized electromagnetic field, and "static" terms due to interaction through the instantaneous Coulomb field. The latter may, in the Lorentz gauge, also be written as the sum of terms due to interaction with the longitudinally polarized vector potential and terms due to interaction with the retarded Coulomb potential.

These formal expressions diverge. However, similar perturbations would have been present also if the electron of which we calculate the energy had been a free electron instead of an electron in the Coulomb field of the proton. In particular, the exchange part of this perturbation of a free electron would be similar to the effect of a change of the electron mass. Therefore, in the "mass renormalization" procedure it is assumed that this (divergent) part of the perturbation is already included by using the experimentally observed mass in the unperturbed electron Hamiltonian, so that only the difference between the perturbation effects on a bound and on a free electron would cause the observed shift of the hydrogen energy levels.

Similarly, the nonexchange part of the perturbation is a sum of several contributions. The perturbed energy contains terms proportional to the Coulomb potential at the position of the electron. This may be regarded as describing a change of charge of the electron. In the "charge renormalization" procedure, we assume that this part of the perturbation has already been taken into account by using in the Coulomb term of the unperturbed hydrogen Hamiltonian the experimentally observed charge of the electron. There are also terms that look like an electromagnetic interaction between the charge of the electron in its hydrogen state, and the charge of "vacuum polarization" (presence of virtual electron pairs) around the proton. The static part of these terms cause the Uehling shift of the energy levels.[4] Here, we will not further discuss it. The dynamic part of these nonexchange terms is negligible for a hydrogen atom here assumed to be at rest.

Our present discussion will be confined to consideration of the exchange terms in the shift of energy of the electron in its hydrogen state. This is now the difference between the perturbations calculated once while taking into account the influence of the proton's Coulomb field upon the electrons, and once assuming the electrons were free. As this is the difference between two infinite integrals, we should take the difference between the integrands before performing the integration, but the result would remain ambiguous, unless we use the correct way of combining the two integrals to a single one. As explained by French and Weisskopf, for this purpose one should use the Lorentz gauge. Yet, the result obtained in this way may be written as the sum of its dynamic part and its static part, the former with the intermediate e.m. field transversely polarized.

Properly combining the two integrals to a single one amounts to properly establishing a one-to-one correspondence between intermediate states of the electron in the proton's field, and of the electron that is free. The sum and integral over intermediate electron states is a sum and integral over a complete set of electron states. The trick here is to find a valid approximation in which the electron field both with and without the proton's Coulomb field may be described by the same complete set of functions. This may be done, but differently for intermediate states in which the intermediate virtual photon is hard or is soft. Here, hard photons have energies $> mc^2\delta$, where δ is a numerical constant somewhere between $\alpha = 1/137.0390$ and 1. Soft photons have momentum $< mc\delta$.

The first attempt at calculating the Lamb shift of the hydrogen energy levels, by Bethe[1] in 1947, took into account merely the nonrelativistic dynamic interaction with soft photons, assuming that for $\delta = 1$ the hard-photon interaction would be negligible. The first calculation taking into account the relativistic perturbation was by French and Weisskopf,[3] who considered in detail the effect of hard photons in the intermediate states, assuming $\alpha < \delta < 1$. For the (much larger) contribution from soft-photon intermediate states, they simply added Bethe's results adjusted for $\alpha < \delta < 1$ instead of Bethe's $\delta = 1$. They found that δ would cancel out in this sum of soft- and hard-photon contributions.

Later calculations using Schwinger's utilization of Tomonaga's interaction picture or using Feynman diagrams differ from French and Weisskopf's work by more obvious relativistic covariance and by some of the mathematical techniques used, but from the point of view of physics they are essentially identical. Higher-order calculations later made by others add some refinement, but will here not be considered.

While thus the hard-photon case has been discussed extensively in the literature, this is not so for the soft-photon case. There have been some papers[5,6] improving on the numerical calculation of the soft-photon contribution predicted by the formulas of Bethe, who expressed the soft-photon shift of the 2s level by

$$\delta E_{2s} = (\alpha^3/3\pi) \; \text{Ry} \; \{ (\textstyle\sum g_j) \; \ln \delta + (\textstyle\sum g_j) \; \ln (2/\alpha^2) - (\textstyle\sum g_j \ln v_j) \} \qquad (1)$$

in terms of quantities g related by $g = 3 f v^2/2$ to the oscillator strengths $f(2s,n)$ and $df(2s,n)$, where $v = 1/4 - 1/n^2$ for the discrete spectrum of intermediate states, and $v = 1/4 + 1/n^2$ for the continuous spectrum. We write $\textstyle\sum g_j$ for $\Sigma \, g(2s,n) + \int dg(2s,n)$; $\ln (2/\alpha^2) = 10.533678$ and $\text{Ry} = mc^2\alpha^2/2$, so that $\alpha^3 \text{Ry}/3\pi = h v_0$ for a frequency $v_0 = 135.63485$

megacycles. According to French and Weisskopf, hard-photon states would add to (1) the amount

$$(\alpha^3/3\pi) \; \text{Ry} \; \{ \tfrac{5}{6} - \tfrac{1}{5} - \ln 2 - \ln \delta \} , \qquad (2)$$

where $\tfrac{5}{6} - \tfrac{1}{5} - \ln 2 = -0.059\,813\,847\,226\,612$. From the sum rule

$$- \text{Ry}^3 \textstyle\sum v_j^3 \; | \int \psi_o^* \vec{x} \, \psi_j \, d^3 x |^2 = \textstyle\sum (\int \psi_o^* [H_{op}, \vec{x}] \, \psi_j \, d^3 x) (\int \psi_j^* [H_{op}, [H_{op}, \vec{x}]] \, \psi_o \, d^3 x)$$

$$= (e^2 \hbar^4/m^2) \int \psi_o^* \{ \psi_o \vec{\nabla}^2 \tfrac{1}{r} + (\vec{\nabla} \tfrac{1}{r}) \cdot (\vec{\nabla} \psi_o) \} \, d^3 x \qquad (3a)$$

it is known that

$$\textstyle\sum g_j = 1 , \qquad (3b)$$

but $\textstyle\sum g_j \ln v_j$ requires numerical computation. With present-day computational tools, this is easily calculated with high precision. For the sport of it, I have done it myself with ten-digit accuracy and better, yielding

$$\left. \begin{array}{ll} \Sigma g(2s,n) = 0.026433608787 , & \Sigma g(2s,n) \; \ln v_n = - 0.046467908953 , \\ \int dg(2s,n) = 0.9735663912 , & \int dg(2s,n) \; \ln v_n = 2.85823780207 , \\ \text{so,} \quad \textstyle\sum g_j = 1.0000000000 , & \textstyle\sum g_j \ln v = 2.81176989312 . \end{array} \right\} \qquad (4)$$

This accuracy is, however, totally spurious, not only because of the lack of accuracy in $(\alpha^3/3\pi)$, but much more so because Bethe's formulas are based upon "approximations" that are not valid at all for $\delta \geq \alpha$, so that the soft-photon contribution to the Lamb shift needs recalculation. The present paper outlines where these nonrelativistic calculations can be improved and discusses the results of the first step in these new calculations.

CHOICE OF THE COMPLETE SET OF ELECTRON WAVE FUNCTIONS USED AS INTERMEDIATE
STATES

If for an intermediate electron state of energy E_j and initial unperturbed energy level E_o we put $\nu_j = (E_j - E_o)/Ry$ and $D_j = \alpha \nu_j/2$, and if K is the momentum or wave vector of the intermediate photon in Hartree units, then the exchange contribution to the level shift δE_o is

$$\delta E_o = \frac{2\alpha^2}{\pi} Ry \int_0^\infty K \, dK \oint \frac{\Sigma \left| \int \psi_o^* e^{i\vec{K}\cdot\vec{x}} V \psi_j \, d^3x \right|^2}{-(K + D_j)} , \qquad (5)$$

where V is the operator for a component of the electron velocity (like x and ψ in Hartree units), and where Σ is a sum over the components of V of which we calculate the contribution to the level shift (these components correspond to the polarization directions considered for the intermediate electromagnetic field, so they are two components perpendicular to $\vec{K}$ for the dynamic contribution to δE_o), while Σ also sums over all intermediate states ψ_j belonging to the same energy level E_j. If we express ψ_o in terms of plane waves corresponding to electron momenta $\vec{P}$ in Hartree units, then mass renormalization adds to (5) a similar expression with the denominator

$$-(K+D_j) \text{ replaced by } K + [1/\alpha^2 + (\vec{P} - \vec{K})^2]^{\frac{1}{2}} - [1/\alpha^2 + \vec{P}^2]^{\frac{1}{2}} , \qquad (6)$$

and where in $\oint$ we average over the angle between $\vec{P}$ and $\vec{K}$.

French and Weisskopf treat the intermediate electrons in the proton's Coulomb field as nearly free, so that in zeroth approximation they use free-electron wave functions as the complete set of intermediate electron states. These states form a continuum; there is here in $\oint$ no sum over discrete energy levels. For the bound intermediate states in (5), the effect of the Coulomb field on the ψ_j is then calculated by perturbation theory. This treatment is satisfactory for intermediate states with hard virtual photons, of energy $> mc^2\delta$ with $\delta \geq \alpha$. They then found the result (2) above for the shift of the 2s level caused by these hard-photon intermediate states or by nonexchange interaction. (The term $-\frac{1}{5}$ in (2) stands for the Uehling shift.) The averaging over angles between vectors $\vec{P}$ and $\vec{K}$ was performed by French and Weisskopf after first expanding in powers of P and $(\vec{P}\cdot\vec{K})$, neglecting higher powers of $\vec{P}$ than the second. When (2) is added to Bethe's contribution (1) with (3b) and (4) for the contribution from soft-photon intermediate states, (2) is merely a small correction, bringing down the total shift δE_{2s} from $7.72190084 \times 135.63485 = 1047.359$ megacycles to $7.66208699 \times 135.63485 = 1039.246$ megacycles.

When the virtual photon has small energy, one should use for a complete set of states the unperturbed hydrogen wave functions, of which some belong to the continuous spectrum corresponding classically to hyperbolic orbits, but some belong to the discrete energy levels of the atom. It is reasonable to assume with Bethe that we may use nonrelativistic approximations in calculating (5) and (6) for these intermediate states, although we should take into account the electron spin, as there will be nondipole virtual transitions in which the spin flips. We therefore will use the Pauli approximation to the Dirac wave functions, in which one assumes that in the Pauli representation (with Dirac matrices $\vec{\alpha} = \rho_1 \vec{\sigma}$ and $\beta = \rho_3$) the ψ_j will for positive energies have two large components $\Psi = \psi_{1,2}$ satisfying the Schrödinger equation, and two small components obtained from the large components by $\psi_{3,4} = -(i\alpha/2) \vec{\sigma}\cdot\vec{\nabla} \Psi$ (in Hartree units). Excluding from ψ_j

the negative-energy states, the states Ψ then form a complete set (part continuum and part discrete). (The negative-energy states supposedly contribute little to (5) or (6) for soft photons. For hard photons, one *subtracts* the sum over negative-energy states.[3])

As the averaging over angles between $\vec{P}$ and $\vec{K}$ complicates the calculation, we now with Bethe assume that P is entirely negligible, and we replace the mass renormalization terms for photons of energy $< mc^2\delta$ (with $K < \delta/\alpha = 137.039\,\delta$) by

$$\frac{2\alpha^2}{\pi}\,\text{Ry}\int_0^{\delta/\alpha} dK \oint \frac{\Sigma\,|\int \psi_0^*\,e^{i\vec{K}\cdot\vec{x}}\,V\psi_j\,d^3x|^2}{1 + [(1/\alpha K)^2 + 1]^{\frac{1}{2}} - (1/\alpha K)}\ , \tag{7}$$

where V now is a component of $\vec{\alpha}/\alpha = 137.039\,\vec{\alpha}$.

In Eqs. (5) – (7), according to French and Weisskopf, Σ should for a z-axis parallel to $\vec{K}$ still include a sum

$$\Sigma\,|\psi_0\,V\psi_j|^2 = \alpha^{-2}\{|\psi_0^*\,\alpha_x\,\psi_j|^2 + |\psi_0^*\,\alpha_y\,\psi_j|^2 + |\psi_0^*\,\alpha_z\,\psi_j|^2 - |\psi_0^*\,\psi_j|^2\}\ , \tag{8}$$

where the last two terms correspond to the static exchange interaction. French and Weisskopf, however, agree with Bethe that the static contribution would for soft photons be negligible compared to the dynamic contributions from the first two terms in (8). This claim may require verification by actual calculation of the static terms; so far, I have not yet found the time for doing so, and will in the following merely discuss the transverse dynamic terms.

DUBIOUS APPROXIMATIONS MADE IN BETHE'S CALCULATIONS

Bethe makes the following approximations, which are satisfactory for very small K (< 0.4, corresponding to photon energy $< 0.003\,mc^2$), but which certainly break down long before K reaches δ/α:

(1) He neglects the phase factor $\exp(i\vec{K}\cdot\vec{x})$ in (5) and (7);

(2) In (7), he neglects the recoil energy of the electron, replacing the denominator by 1.

These two approximations certainly greatly simplify the calculations, but they distort the numerical results, and therefore we are forced to redo the calculations, avoiding these dubious approximations.

It would be even better to take $\vec{P}$ in (6) also for $K < \delta/\alpha$ into account at least up to terms quadratic in $\vec{P}$, like French and Weisskopf did for hard photons. As the calculations become already rather complicated without introducing this additional complication, I have not yet done so.

WORK TO BE DONE

We discuss here merely the 2s level. For obtaining the Lamb shift, similar calculations should be made for the 2p level, and the difference should be taken.

Our tasks then are the following: (We interchange the summation and integrations $\oint$ over E_j or n, and the integration over K:)

(A) Calculate the matrix elements, and perform the summations Σ of their absolute squares at fixed E_j or n. This will give us

$$T_n = \alpha^{-2} \sum_{l,m,s} \sum_{i=x,y} |\int \psi_{2s}^* e^{iKz} \alpha_i \psi_{nlms} d^3x|^2 \qquad (9)$$

for any fixed values of n and K. (Note that (9) is independent of α, as the α_i have matrix elements only between the large and the small components of ψ, and the latter contain a factor α.)

(B) We perform the integrations over K up to various maxima K between 1 and 137.039 that might serve as choices for δ/α. This gives us the quantities

$$F_n = \int_0^K dK \left\{ \frac{T_n}{1 + [(1/\alpha K)^2 + 1]^{\frac{1}{2}} - (1/\alpha K)} - \frac{T_n}{(\alpha v_n/2K) + 1} \right\} \qquad (10)$$

For the discrete hydrogen levels, we will find the two terms in (10) for $K \to \infty$ to give two convergent integrals, contrary to Bethe's result of linear divergence of the integrals of the single terms, and logarithmic divergence of their difference. However, when similar integrals for the continuous spectrum are added, for instance the dynamic contribution to the mass renormalization terms in (7) will add up to

$$\frac{2}{\pi} Ry \int_0^K dK \frac{\int \psi_0^* e^{iKz} (\alpha_x^2 + \alpha_y^2) e^{-iKz} \psi_0 d^3x}{1 + [(1/\alpha K)^2 + 1]^{\frac{1}{2}} - (1/\alpha K)} , \qquad (11a)$$

where the phase factors cancel out, so that we obtain

$$\frac{1}{\alpha\pi} Ry \{2\alpha K - \alpha^2 K^2 + \alpha K [1 + \alpha^2 K^2]^{\frac{1}{2}} + \ln (\alpha K + [1 + \alpha^2 K^2]^{\frac{1}{2}})\} , \qquad (11b)$$

which for $K \to \infty$ diverges like $\frac{1}{\alpha\pi} Ry \{\frac{1}{2} + 2\alpha K + \ln (2\alpha K)\}$. Therefore, our only hope for convergence for $K \to \infty$ remains taking the difference as in (10), before the integration over K is performed. For the soft-photon case, however, the perturbation term and the mass renormalizaion term may be integrated separately up to the cutoff $K = \delta/\alpha$ chosen for a fit to the result (2).

(C) Finally, we have to sum the F_n obtained for the discrete spectrum, or to integrate over n for the continous spectrum as in (17) below.

So far, I have completed this program for the 2s state only for the discrete spectrum.

SIMPLIFICATION BUT DISTORTION OF RESULTS BY BETHE'S APPROXIMATIONS

When the factors e^{iKz} are omitted, we can calculate the matrix elements of $\vec{\alpha}$ by

$$\vec{\alpha} = [\vec{x} H_{op} - H_{op} \vec{x}]/i\hbar c^2 \qquad (12)$$

in cgs units, and integrate the $H_{op} \vec{x}$ term by parts for making the $H_{op}^\dagger$ operate upon the ψ^* on the left. Also, Bethe's approximation T_n^B to our T_n becomes independent of K. Moreover, for the 2s state, there is now in Σ no sum over l, as only $l = 1$ will give a nonvanishing $\int \psi_{2s}^* x_i \psi_{nlms} d^3x$. The result then is

$$T_n^B = \frac{4096}{3 n^3} \frac{(1 - 2/n)^{2n-4} (1 - n^{-2})}{(1 + 2/n)^{2n+4}} , \qquad (13)$$

412

so that, for $n \to \infty$,

$$T_n^B \to \frac{4096}{e^8 n^3}(1 + \frac{11}{3n^2}) . \tag{14}$$

Also, T_n^B is now by (12) related to the oscillator strengths $f(2s,np)$ mentioned below Eq. (1), by

$$T_n^B = \nu_n f(2s,np)/2 = g_n/3\nu_n . \tag{15}$$

Replacing T_n in (10) by (13), and replacing the first denominator in (10) by 1, Bethe then finds for F_n the "approximation"

$$F_n^B = T_n^B D_n \ln (K/D_n + 1) , \tag{16}$$

in which he finally neglects the "+ 1", as small compared to $K/D_n = 2\delta/\alpha^2 \nu_n$. For the continuous spectrum, the only difference in (16) is a value for T_n^B different from (13). By

$$\delta E_{2s} = \frac{2\alpha^2}{\pi} \text{Ry} \oint F_j , \tag{17}$$

(16) then gives Bethe's result, Eq. (1) above.

From (13) we see that in Bethe's approximation there are in the sum over discrete states in (17) no terms for n = 1 or 2, while by the corrected formulas (10) with (9) we find that n = 1 and n = 2 give the largest contributions to the sum over discrete states. (These contributions come mostly from transitions in which the virtual photon gets its angular momentum from a flip of spin of the electron, a process which is entirely overlooked by Bethe's calculations.)

From (14) we see that the sum over n in (17), like $\Sigma \, 1/n^3$, converges very slowly. Therefore, when we stop the summation, say at N = 50, there is a large remainder $\sum_{n=N+1}^{\infty} F_n$ that should be replaced by an integral. The same will be true in the revised calculation. The only difference is that the integrand in that integration over large n can be obtained analytically for Bethe's "approximation," while in the revised calculation we first calculate individual F_n for a number of large sample values of n, and then find empirically a function of n which satisfactorily describes the results for F_n found in these sample cases .

THE ACTUAL MATRIX ELEMENTS

As e^{iKz} does not commute with H_{op}, we cannot use (12), and we must use in (9) the Dirac matrices with the Pauli approximation for the four-component ψ. Resolving the electromagnetic field into circularly rather than linearly polarized waves, we use $(\alpha_x + i\alpha_y)/\sqrt{2}$ and $(\alpha_x - i\alpha_y)/\sqrt{2}$ for the α_i in (9). We resolve the initial 2s state as well as any intermediate states $\psi_{n\ell ms}$ into states with spin up or spin down in the large components, by using for the latter

$$\psi_{n\ell m\uparrow} = \begin{pmatrix} \Psi_{n\ell m} \\ 0 \end{pmatrix}, \quad \psi_{n\ell m\downarrow} = \begin{pmatrix} 0 \\ \Psi_{n\ell m} \end{pmatrix}, \quad \Psi_{n\ell m} = R_{n\ell} Y_{\ell m} \tag{18}$$

with

$$R_{n\ell} = \frac{2}{n^2} \left[\frac{(n+\ell)!}{(n-\ell-1)!} \right]^{1/2} (\frac{2r}{n})^{\ell} e^{-r/n} \sum_{h=0}^{n-\ell-1} (\frac{-2r}{n})^h \frac{(n-\ell-1)!}{h! \, (n-\ell-1-h)! \, (2\ell+1+h)!} . \tag{19}$$

413

We then find , for instance for the 2s↑ state, the matrix elements

$$\langle 2,0\uparrow | \; \alpha^{-1} e^{iKz} \alpha_{x+iy}/\sqrt{2} \; |n\mathit{l}m\downarrow\rangle = (K/\sqrt{2}) \int \Psi_{2,0}^{\dagger} e^{iKz} \Psi_{n\mathit{l}m} \, d^3x \; , \qquad (20a)$$

$$\langle 2,0\uparrow | \; \alpha^{-1} e^{iKz} \alpha_{x-iy}/\sqrt{2} \; |n\mathit{l}m\downarrow\rangle = 0 \; , \qquad (20b)$$

$$\langle 2,0\uparrow | \; \alpha^{-1} e^{iKz} \alpha_{x+iy}/\sqrt{2} \; |n\mathit{l}m\uparrow\rangle = (i/\sqrt{3}) \int D_{2,0}^{\dagger} Y_{1,-1}^{*} e^{iKz} \Psi_{n\mathit{l}m} \, d^3x \; , \qquad (20c)$$

$$\langle 2,0\uparrow | \; \alpha^{-1} e^{iKz} \alpha_{x-iy}/\sqrt{2} \; |n\mathit{l}m\uparrow\rangle = (-i/\sqrt{3}) \int D_{2,0}^{\dagger} Y_{1,1}^{*} e^{iKz} \Psi_{n\mathit{l}m} \, d^3x \; , \qquad (20d)$$

where $D_{n\mathit{l}} = dR_{n\mathit{l}}/dr$ (in Hartree units). Thence, for fixed n and l,

$$T_{n\mathit{l}} = T_{n\mathit{l}}^{s} + T_{n\mathit{l}}^{o} = \frac{K^2}{2} |\int_0^\infty R_{20} R_{n\mathit{l}} I_{00} r^2 \, dr|^2 + \frac{2}{3} |\int_0^\infty D_{20} R_{n\mathit{l}} I_{11} r^2 \, dr|^2 \qquad (21)$$

with

$$I_{nm} = \int_0^{2\pi} d\varphi \int_{-1}^{+1} du \; e^{iKru} Y_{nm}^{*} Y_{\mathit{l}m} \; , \qquad (22)$$

and, for fixed n,

$$T_n = T_n^{s} + T_n^{o} = \sum_{\mathit{l}=0}^{n-1} T_{n\mathit{l}} \; . \qquad (23)$$

Equations (20a-d) actually are the matrix elements for the transitions back from the intermediate state to the unperturbed 2s↑ state under the reabsorption of the virtual photon. In (20c-d) we performed an integration by parts, which altered Ψ_{20} into $D_{20}^{\dagger} Y_{1,\pm 1}^{*}$.

We see from (20a-d) that for virtual emission of circularly polarized photons in the z-direction, under transition to a state of specified quantum numbers n and l, there are for an electron with spin up three options:

(1) The photon may take an angular momentum $\hbar$ in the z-direction, by a flip of spin of the electron from spin up to spin down. This fixes the direction of circular polarization of the photon wave.

(2) and (3) A photon circularly polarized in either direction may be virtually emitted under a corresponding decrease or increase of the orbital azimuthal quantum number m by 1.

We will call option (1) a spin-flip emission of the virtual photon, and will call (2) and (3) orbital emissions of the photon. We indicate these two cases by superscripts s and o. The two terms in (21) correspond to these two cases. Options (2) and (3) do not exist in a transition from an s state to an s state, and therefore do not exist for transitions from the 2s state to any state with n = 1. That is, for our 2s state, we have

$$T_{n0}^{o} = 0 \quad \text{and} \quad T_1^{o} = 0.$$

In Fig. 1, we have plotted for n from 1 to 4 the dependence of T_n^{s} and of T_n^{o} upon K between 0 and 3. The solid lines are for the spinflip cases, and the broken lines are for the orbital cases. We find a strong dependence on K. Bethe assumed for all values of K the same value which we find at K = 0. Therefore, he did not consider the spinflip cases at all, and therefore he did not find the matrix elements toward states with n < 3.

In Fig. 2, we first have added T_n^{s} and T_n^{o} to give the T_n that we need for use in our Eq. (10). Next, we have divided this T_n by its value at K = 0. That is, in Fig. 2 we have plotted the ratio $y_n = T_n/T_n^{B}$ as a function of K. Bethe's approximation is therefore here shown by the line $y_n = 1$.

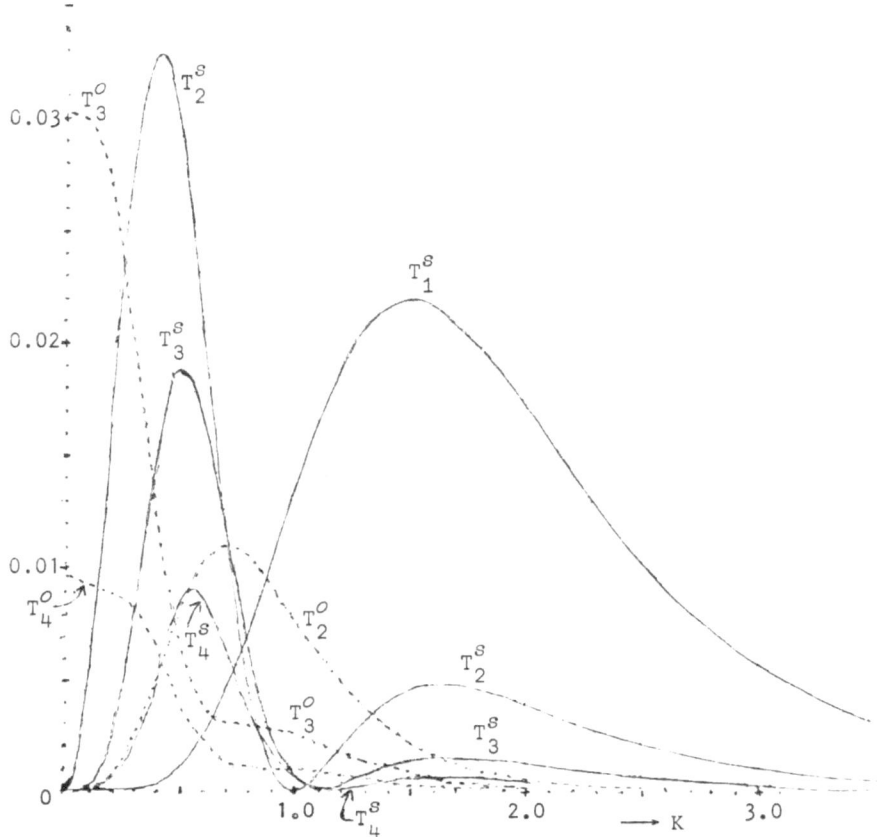

Figure 1. As functions of K (the photon momentum in Hartree units, we plot here by solid lines the sums T_n^s of absolute squares of matrix elements for transitions from the 2s state by spin flip to any states with a given value of n, and by broken lines similar sums T_n^o for "orbital" transitions without flip of spin. Predominant are the spinflip transitions from the 2s state to the 1s state, which were completely overlooked in Bethe's calculation.

We show in Fig. 2 curves for n = 3, 4, 6, and 25. For small K, T_n for n ≥ 3 soon starts to decrease parabolically from T_n^B. For n ≥ 4, T_n near K = 0.2 starts to rise again, and climbs to a steep maximum near K = 0.6. T_n then keeps dropping steeply toward a value much lower than predicted by Bethe at K = 0 (photon energy $mc^2/137$). This is followed by a rather flat second maximum. Finally, $T_n \to 0$ like K^{-6}, which makes the integrals in (10) convergent for $K \to \infty$. We noted already that there cannot be such a convergence for transitions of the electron to the states of the continuous spectrum.

For large n, the sum over l in (23) considerably slows down the numerical computation, which is already slow because for fixed n and l we express T_{nl} as a triple sum. (There is in (19) a sum over h; we represent Y_{l0} and

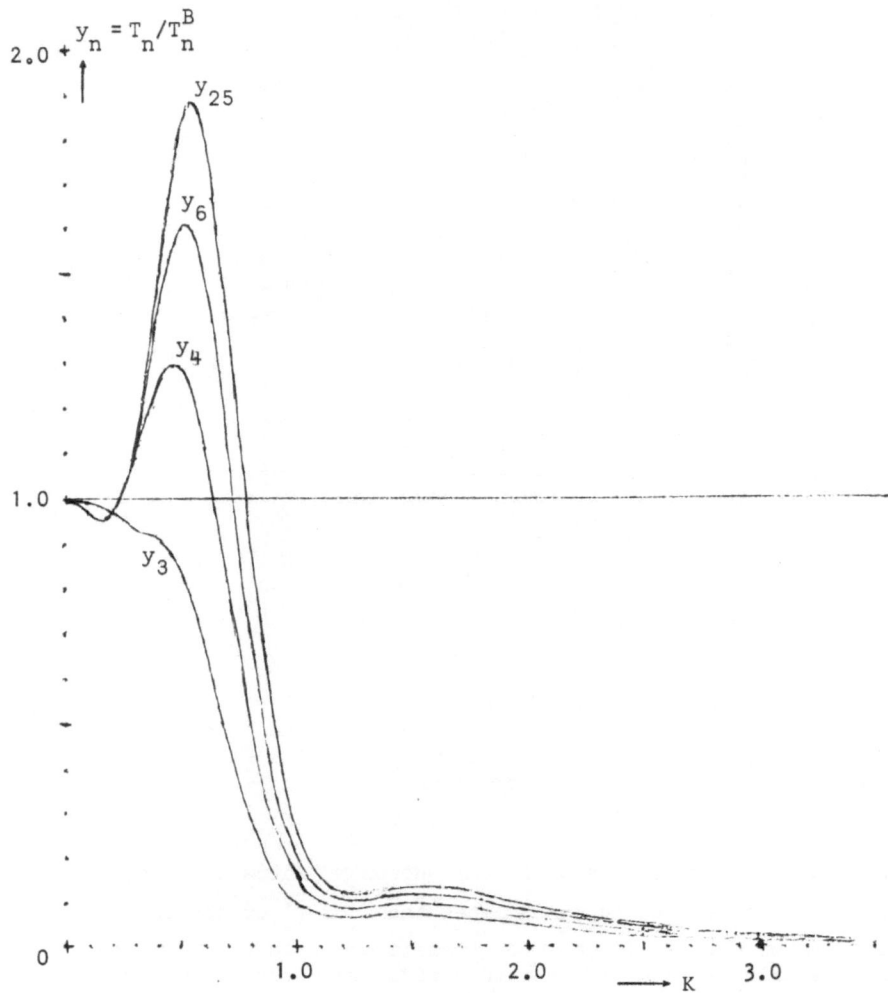

Figure 2. Ratios $y_n = T_n/T_n^B$ as functions of K. Here, T_n is
the sum of absolute squares of matrix elements for transitions
from the 2s state to any discrete states with principal quantum number n; T_n^B $(= T_n$ at K = 0) is Bethe's value for T_n .

Y_{l1} by sums over i, and the integration over r introduces a sum over j.)
For large n, the T_{nl} become smaller and smaller as l increases toward n-1.
However, the triple sum by which each T_{nl} is calculated has terms of alternating sign that become larger and larger. This happens even faster
when K is very small, and soon, for increasing l, though the sums are
finite, we obtain for the T_{nl} by rounding errors very large numerical
outputs, though we know that they must add up to a T_n value close to T_n^B .
This occurs when the individual terms in the summation become so large
that their significant small difference does no longer show up in the
number of digits which the computer can handle.

It is therefore not only useful for saving time, but also necessary for preventing this explosion of the summations, to cut off the sum over l in (23) as soon as we reach an l (>2) for which T_{nl} is small enough.

For obtaining in this way any precision at all, one must use a computer language that works with floating-point numbers of many digits. On a 512K MACINTOSH, I used for this purpose the computer language Rascal^{TM} developed in 1985 at Reed College (Portland, Oregon) by G. Stein, S. Gillespie, and R. Crandall. It performs all calculations with approximately 19-digit accuracy, using 10-byte floating-point numbers. With this accuracy of the computer, and avoiding calculations for $K < 0.005$, explosion of summations could be avoided by a cutoff as soon as $T_{nl} < 10^{-16}$ at $l \geq 3$. For most calculations, we avoided $K < 0.03125$ and cut off at 10^{-15}.

THE INTEGRATION OVER K

The region $0 < K < \frac{1}{32}$ is rather critical, as in this region the $\alpha \nu_n / 2K$ in the energy denominators become important, so that Simpson integration in this region would require very many small steps. On the other hand, we wanted to avoid calculating T_n in this region as much as we could.

Therefore, after establishing how T_n behaves in this region, by a number of sample calculations in the region $0.005 < K < 0.03125$, we replaced each T_n for $K < \frac{1}{32}$ by a function of K which well reproduced the sample values calculated. For large n, a function of K and n was used. For good accuracy, this general function was replaced by individual functions of K for individual n values, for n up to 19.

These functions of K allowed us to perform the integrals in (10) analytically for K up to 0.03125, with the results expressed as expansions.

From there up to $K = 1$, we integrated by Simpson's rule, using 31×2^n halfsteps dK. For best results, we used here $n = 6$ for $n > 2$, and $n = 5$ for $n \leq 2$. At $n = 11$, and again at $n = 27$, we actually halved the number of steps, but then by sample calculations found a general formula for correcting for the difference between numerical results from Simpson integration with a smaller or a larger number of steps. (The accuracy becomes 16 times better, when the number of steps is doubled, and, for large n, the errors are proportional to n^{-3}.)

From $K = 1$ up to $K = e^4$, we changed the integrals over K into integrals over $\ln K$. The integral was performed piecemeal, with printouts of integrals up to in-between points, in case we would want to use one of these K values later as $K = \delta / \alpha$. For accuracy in these integrations, it turned out to be necessary to use for $n \leq 2$ twice as many Simpson steps between $\ln K = \frac{7}{4}$ and $\ln K = \frac{7}{2}$, as were sufficient for $n > 2$.

From $K = e^4$ up to $K = 137.039$, and (for the fun of it) from there up to infinite, the integrations were again performed analytically, replacing individually calculated T_n values by a function of K and n that was found to reproduce satisfactorily sample values of $T_n(K)$ calculated for $K > 40$ for any n. We used here

$$T_n \text{(at } K > 40) = \frac{16}{n^3 K^6} - \frac{109 + 106/n^2}{n^3 K^8} . \qquad (24)$$

Here like for $K < \frac{1}{32}$, analytical integrations lead to further sums.

417

As n increases, the time required for calculating by (10) an individual F_n value increases considerably, and could take more than 24 hours of computing time on the Macintosh. On the other hand, with F_n crudely proportional to n^{-3}, the sum over n converges exceedingly slowly.

For an accurate calculation of the remaining $\sum_{n=N+1}^{\infty} F_n$ after having obtained $\sum_{n=1}^{N} F_n$ by summing individually calculated F_n values, we first replace F_n for $n > N$ by a function $F(n)$ that will sufficiently accurately reproduce the individually calculated F_n for $\frac{1}{2}N \le n \le N$. Next, we calculate a few sample F_n for n a lot larger than N, for verifying that those are also well reproduced by $F(n)$.

We then use the formula

$$\sum_{n=N+1}^{\infty} F(n) = \int_{N+\frac{1}{2}}^{\infty} dn \sum_{j=0}^{\infty} a_j \, d^{2j} F(n)/dn^{2j} \tag{25}$$

with the a_j chosen so that, for *any* function $F(x)$,

$$\sum_{j=0}^{\infty} a_j \int_{-\frac{1}{2}}^{\frac{1}{2}} dx \, d^{2j} F(x)/dx^{2j} = F(0) . \tag{26}$$

Inserting here Taylor expansion of $f(x) = d^{2j} F(x)/dx^{2j}$, we obtain for a_j the recursion formula

$$a_k = -\sum_{j=1}^{k} \frac{a_{k-j}}{4^j \, (2j+1)!} \tag{27}$$

with $a_0 = 1$, so that

$$a_1 = -\frac{1}{24}, \qquad a_2 = \frac{7}{5760}, \qquad a_3 = -\frac{31}{967680}, \qquad a_4 = \frac{127}{154828800}, \quad \text{etc.} \tag{28}$$

We then find that in (25) terms with $j > 3$ are for $N = 50$ negligible.

RESULTS

We performed the calculations with redundant accuracy, using three choices for the cutoff factor δ: (1) $\delta = 1$, as suggested by Bethe,[1,2] corresponding to $K = 137.039$ and photon energy mc^2. (2) $\delta = \alpha$, as suggested by French and Weisskopf,[3] corresponding to $K = 1$ and photon energy αmc^2. (3) The in-between value $\delta = \alpha \exp 3.5 = 0.2416498$, corresponding to $K = 33.11545$. We find

Case	δ	$\sum_{n=1}^{50} F_n$	$\sum_{n=51}^{\infty} F_n$	$\sum_{n=1}^{\infty} F_n$
(1)	1	$-0.000\,330\,910\,952$	$0.000\,000\,306\,719$	$-0.000\,330\,604\,232$
(2)	α	$+0.000\,108\,251\,785$	$0.000\,000\,420\,012$	$+0.000\,108\,671\,797$
(3)	0.24...	$-0.000\,330\,898\,752$	$0.000\,000\,306\,721$	$-0.000\,330\,592\,031$

From this, contributions to δE_{2s} are obtained by Eq. (17). Comparison of Eqs. (1) and (17) shows that multiplication of the last column above by a factor $(6/\alpha) = 822.234$ gives the quantities comparable to Bethe's quantities $\{(\Sigma g_n) \ln \delta + (\Sigma g_n) \ln (2/\alpha^2) - (\Sigma g_n \ln \nu_n\}$, which were part of Eq. (1). Inserting in this the top line of Eqs. (4), we find the following comparison between Bethe's results for the discrete spectrum, and our results:

Case	δ	Bethe's $\Sigma g_n \ln (2\delta/\alpha^2\nu_n)$	Our $(6/\alpha) \Sigma F_n$
(1)	1	+ 0.324 911 040	− 0.271 834 040
(2)	α	+ 0.194 850 665	+ 0.089 353 646
(3)	0.2416498	+ 0.287 368 296	− 0.271 824 080

We see that Bethe's results do not agree at all with the results of a more careful calculation. While Bethe's above results for increasing δ *increase* by 0.026 433 608 787 ln δ, we find our results to *decrease* for increasing δ. Before concluding from this anything about the fit between the soft-photon calculation and French and Weisskopf's hard-photon calculation, we first have to add the results of a revised calculation of the soft-photon transitions to the continuous spectrum of hydrogen states. (Also in the calculations for those transitions, Bethe, Brown, and Stehn left out the exponential phase factor and the recoil of the free electron in the mass renormalization term.)

CONCLUSION

Our results show that there is no justification for leaving out the photon phase factor or the recoil of free electrons from soft-photon calculations that are extended beyond photon energies of $mc^2/400$, while there are no reasons for believing that the approximations made in the conventional relativistic calculations would be valid down to that low photon energies. The in-between range of photon energies may possibly be covered by calculations using nonrelativistic Pauli wave functions for the hydrogen atom, but this will require taking into account the phase factor and the recoil energies neglected in the conventional calculations.

For the 2s state we have calculated the contribution to the Lamb shift from transitions to the states of the discrete spectrum under virtual emission of a photon of energy below somewhere between mc^2 and $mc^2/137$, but to this should be added the much larger contribution from soft-photon transitions to the continuous spectrum. It is therefore necessary to perform similar calculations for those transitions, using a computer language allowing many-digit calculations, and preferably on a faster computer than I had available for the present work. Then, similar calculations should be made also for the 2p state. Also, it would be wise to verify whether really the static contributions to the soft-photon exchange case are negligible, and perhaps $\vec{P}$ in Eq. (6) should not be entirely neglected.

Until all of that work has been completed, nobody has the right to claim that we would know that quantum electrodynamics with renormalizations yields results that agree with the observed facts.

REFERENCES

1. H. A. Bethe, The Electromagnetic Shift of Energy Levels, Phys. Rev. 72: 339 (1947).
2. H. A. Bethe and L. M. Brown and J. R. Stehn, Numerical Value of the Lamb Shift, Phys. Rev. 77: 370 (1950).
3. J. B. French and V. F. Weisskopf, The Electromagnetic Shift of Energy Levels, Phys. Rev. 75: 1240 (1949).

4. E. A. Uehling, Polarization Effects in the Positron Theory, <u>Phys. Rev.</u> 48: 55 (1935).
5. John M. Harriman, Numerical Values for Hydrogen Fine Structure, <u>Phys. Rev.</u> 101: 594 (1956).
6. Charles Schwartz and J. J. Tiemann, New Calculation of the Numerical Value of the Lamb Shift, <u>Ann. Physics</u> 2: 178 (1959).

THE QUANTUM AND REST FRAME VACUUM - ABSOLUTE MOTION WITH LORENTZ

CONVARIANCE

James D. Edmonds, Jr.

Physics Department
McNeese State University
Lake Charles, LA 70609 USA

INTRODUCTION

In reading a translation of Lorentz's original relativity paper, which preceded Einstein's by about five years, I was startled to see how much of Special Relativity he developed. His foundation was the idea that Maxwell's equation of electromagnetism is more fundamental than Newton's equation of motion. He explored the "symmetry" of Maxwell's equation, discovering the famous Lorentz transformations. He also then modified Newton's motion equation to confirm to this symmetry and thus hit upon the correct Newton-Einstein equation of motion.

Einstein simply reinvented the wheel starting from a different perspective. He guessed that no absolute motion exists, so $c = c'$, and Lorentz covariance is derived. Everyone seemed to like this better and it is the way relativity is introduced in most textbooks today. Einstein did recognize time dilation as quite physical, but his real contribution of genius was his curved space-time gravity equation. The equivalence pseudomorphism helped his imagination grasp bending light, thus curved space, though his belief in real equivalence of gravity and acceleration was quite mistaken. We no longer need such motivation. His gravity equation has withstood every test to three or four digit accuracy so far. It is incompatible with quantum electrodynamics, itself tested to seven or eight digit accuracy, so something has to give in the future here.

The expanding or contracting universe follows naturally from Einstein's gravity equation (though he tried to avoid this instability with a "patch" added to the natural equation). In the 1940s, Gamow and associates theorized that the big bang should leave radiation reaching us today from all directions. This "Gamow" radiation was forgotten, then accidentally discovered in the mid 1960s. By now it has been used to determine that the earth is moving "absolutely" at about (2/1000)c toward the galaxy Hydra. (About half of this speed is accounted for by considering the sun's motion around our galaxy.) There is no better operational definition of absolute motion than with respect to a rest frame stretching to perhaps 18 billion light years in all directions. So it appears Lorentz was more correct than Einstein in his approach to Special Relativity. Relatively few other physicists seem to believe this these days, but I have confidence it will prove correct in the long run of history. Moving clocks run slow and moving atoms contract. The vacuum itself must "produce" these effects.

The modern electroweak theory and the "standard model" involve a uniform vacuum Higgs field. Similarily, the old quantum electrodynamic theory involved a dynamic vacuum of virtual pairs. The older version had the negative energy states all filled - a vacuum with infinite energy density; this is not compatible with gravitation. All of this points to problems with isotropy when one moves, absolutely, through the dynamic vacuum. The effects might be small but, I believe, they must be real and, in principle, could be detectible. Perhaps only subtle, low energy effects, such as the Lamb shift, would be able to show them. We must look for anisotropy effects where the <u>virtual particles</u> in the vacuum play their <u>most</u> significant part. The Michelson-Morley experiment failed due to equipment length contraction. Since the whole lab and earth are length-contracted, it may be difficult to see anisotropic, <u>quantum</u> effects.

I believe, and have for over a decade, that quantum relativity is founded on Lorentz symmetry for some purely mathematical reasons, independent of Einstein's metaphysics of moving observers and what they might see. This perspective gives us a new window on the basis equation and, I hope, a new inspiration for someone to successfully generalize them and merge them with quantum gravity. The supersymmetry efforts are somewhat along these lines, since group symmetries are guessed at. These have nothing to do with moving <u>human</u> observers. Yet, supersymmetry proponents overstress the foundation of 4-space, I believe.

Let me now show why Lorentz symmetry fits the world of relativistic quantum physics and how moving observers are irrelevant at the fundamental level. The quantum equations and the ether dictate what moving observers will find, when they exist and move. They are not really important.

QUATERNIONS

Hamilton invented quaternions in the mid 1840s because the world is three- and not two-dimensional. They were "overthrown", about 1890, by Gibbs and Heaviside who invented vectors from them. Mathematicians showed in the early 1900s that quaternions are unique among <u>all</u> hypercomplex numbers that one can invent. This is "why" the universe has a 3 + 1 space-time structure! I have no doubt about this, though it is conjecture and must always be such.

Mathematics links to physics in ways the Pythagoreans would be delighted to know about, were there any alive today. Our countably-infinite rational number system describes the world amazingly well. To get distributive and associative laws we must break the symmetry of the number line. We define

$$(-1)(-1) = +1.$$

There is a natural bias toward one half of the line. Since the negative numbers naturally relate to antimatter (which in turn naturally relates to backward time travel in quantum electrodynamics), the second law, our mental time bias, and the scarcity of antimatter in the universe probably all result from this natural asymmetry in the number line. Most physicists deny an asymmetry of the number line when I point this out to them. It is quite necessary, however, and very profound.

422

To achieve a sufficiently subtle and complex world, the basic mathematical substructure cannot be too simple. Otherwise, the big bang might produce only an expanding universe of gas and no people to contemplate its physics. There is a dynamic tension in the design -- keep it simple and elegant yet not so simple that DNA cannot evolve and self-replicating beings of high intelligence come to exist, "fleetingly", before the second law reaps its due.

We see in the mathematics that complex quaternions are needed. These eight-element hypercomplex numbers are isomorphic to the 2-by-2 complex matrices. There are two natural conjugations, the complex and the quaternion, or symplectic. These conjugations give two natural Lie groups within this system. One is the Lorentz group, SL(2,C) with six parameters, and the other is SU(2) X U(1) with four parameters. They have the common subgroup SU(2) which "covers" the rotation group R(3) in 3-space. Herein lies the why of Lorentz covariance, not in moving fleets of humans peering at experiments going by them. The dimensions of space-time and its basic symmetries are all tied to the simplest, elegant number system concepts - or so it would seem to me.

But there is more. The complex quaternions only can deal with a quantum relativistic world of massless quanta such as photons, gravitons, and even neutrinos. That could have been all that our universe contained, in the design, and then we wouldn't be here to wonder about it all. To introduce rest mass is no easy task! A drastic generalization of the number foundation is needed. Dirac stumbled onto this in the 1920s for the wrong reason. He thought Schrödinger's relativistic wave equation was inconsistent because of the second order time derivative. He hit upon the Diract algebra of 4-by-4 complex matrices. This algebra is nothing more than a dirct product of the two subsystems in the complex quaternion algebra. Thus, from our number mysticism perspective, nature "wanted" rest mass and perhaps intelligent beings, so the number system was enlarged in a natural and elegant way.

I spent a decade analyzing physics from this radical perspective. I discovered that, as a hypercomplex number system, the Dirac algebra has four natural conjugations instead of two. [This is sort of like two spin-1/2 subsystems giving rise to a spin-0 and a spin-1 (triplet) superstructure.] Sure enough, these four conjugations give rise to only two natural groups. One conjugation gives SU(2) X SU(2), with six parameters, and the other three, isomorphically, give the covering group of de Sitter's group of rotations in 5-space with 10 paramters. [There are 10 planes in 5-space (coordinate pairs for rotation)]. Mathematical details have been published and a summary series of articles is being published in Speculations in Science and Technology as space allows, beginning with 7, 289 (1984).

FUTURE DIRECTIONS

Within this elegant number theory context, we see the Dirac equation as an incomplete fragment. Space-time is naturally 6-dimensional, or possibly 10-dimensional or "both". We get quantum electrodynamics by setting one of these 6-dimensions to zero, setting to other to the empirical mass value of the quantum with spin-1/2, and also reducing the basic symmetry group to its principal subgroup, rotations in 4-space, i.e., Lorentz symmetry.

Although the four conjugations naturally lead to a group covering rotations in 5-space, they just as naturally lead, minimally, to 6-space for the quantum world! Thus, the two extra dimensions are different from each other in a fundamental way! Perhaps this relates to hadrons and leptons or such. It certainly relates to something fundamental if my number mysticism is on the right track.

This is as far as my insights have taken me. Action-at-a-distance, nonlocality, and subsequent collapse of the wavepacket, may all relate to these extra dimensions of space. Quantum theory is a stopgap theory, just as curved space 4-dimensional classical gravity is. There is at least one more physics revolution yet to come. (I hope we live to see it, but then that is a greedy attitude. We have had more than our share of physics revolutions in our lifetime). The best brains have struggled over this since the mid 40s with no substantial improvement in QED except perhpas for the electroweak force. So far, charmodynamics is unproved one way or the other. The same is true of 9- or 11-dimensional supersymmetry theories. We must keep trying all of these ideas, and others as expounded at this conference. I hope I can win a few bright converts to number mysticism, to use the number structures as if they were lab data to suggest new approaches to this, the biggest question of science: "What's it all made of?" This is second only to the philosophical question, "What's is all for, anyway?" Whatever the reason, we _are_ here (I think) so we can play with problems of how it all works. There is no greater scientific challenge, unless perhaps it is how the brain functions. I doubt that these 6-dimensional, subtle action-at-a-distance quantum concepts play an important role in the mind. Neuro-transmitters at junctions in 3-space or just chemistry may be all that is required for thought. The wiring circuitry is incredibly complex and poorly diagrammed so far; that is why I abandoned that field of study. I think the brain will prove to be a basically classical machine, and duplicatable (in principle) with each each neuron being stimulated by one microcomputer. This would produce an incredibly large, but classical array of interconnections and exchanges of information. Each microcomputer ("neuron") would be connected to thousands of others in the massive array of 10^{12} computers. The individual computers would simulate the subtle chemical biases and feedbacks that represent real neuron functions.

Until quantum theorists can tell us why the muon is 200 times heavier than the electron, instead of say 2 times or 2 million times heavier, and until there is successful theory of quantum gravity, I won't believe we know much more than how to predict the mass of the hydrogen atom, _given_ the masses of a proton and an electron. In other words, we understand chemistry now but still not physics. I hope that supercolliders of the future will clearly show whether the theoretical ideas abounding today in particle physics have much validity.

REFERENCES

Edmonds, J.D., Jr., 1984, "The Mystifying Mathephysics of Microreality," Part 1, Speculations in Science and Techology., 7, 289. (Part 2 in 1986).
Edmonds, J.D., Jr., 1974, "Quaternion Quantum Theory," Am. J. Physics., 42; 220.

APPENDIX

The basic mathematical structure of nature is as follows. A Dirac algebra element, with 16 parts, consists of complex coefficients c^μ and

d^μ associated with hypercomplex number basis elements e_μ and f_μ, as follows (sum on $\mu = 0,1,2,3$):

$$D = C_R{}^\mu (e_\mu) + c_I{}^\mu (ie_\mu) + d_R{}^\mu (f_\mu) + d_I{}^\mu (if_\mu).$$

There is a 16-by-16 multiplication table for the e_μ and f_μ which is easy to memorize, [Edmonds, (1986-see above)]. The four basic conjugations $D^\dagger$, D, $D^\downarrow$, and D^V change various signs among the 16 elements in such a way that the antiautomorphic property holds: $(D_1 D_2) = D_2 D_1$, etc. The $D^\dagger$ conjugation is the usual Hermitian conjugation of matrix algebra. The complex quaternion subsystem is described by the (e_μ), $\leftrightarrow i\sigma_\mu$, elements. These are closed under multiplication. The σ_μ are the Pauli matrices (quaternions essentially). The $f_\mu \leftrightarrow \gamma_\mu$, where γ_μ are the Dirac matrices in the common notation handed down from Dirac's original work. I have argued for a new, systematic notation, evolving from the quaternion substructure.

The four natural groups are generated by $D\,D \equiv 1(e_o)$, etc. By examining their "infinitesimal" elements and Lie group commutator structures, we find that $D^\downarrow D$ has 6 parameters, while the other three, $D^\dagger D$, etc., all have 10 parameters and are equivalent. These describe the covering group of the de Sitter group, just as SU(2) covers the 3-space rotation group.

We make physics contact by examining $P = P^\dagger$. In the massless world this gives a 4-vectors: $P = P_4^\mu(e_\mu) = -i\hbar\partial^\mu(e_\mu)$. In the Diract algebra this becomes $P = P^\mu(e_\mu) + P^4(f_o) + P^5(if_o) = P^\dagger$. The Dirac equation, $P\psi = 0$, then follows from $P^4 \equiv mc$, $P^5 \equiv 0$ or from $P^4 \equiv 0$, $P^5 \equiv mc$. The Maxwell equation for massless photons, A, comes from the form $[P\,P + (P\,P)\,]A - Px\,[(P\,A) + (P\,A)\,] = J = J^\mu(e_\mu)$, with $P^4 \equiv 0$ and $P^5 \equiv 0$. In the more general case where $P^4 \neq 0$ or $P^5 \neq 0$, then $P\,P$ is more complex in structure and gives a generalization of the Klein-Gordon equation. Rich subtlety abounds here in 6-space. There is much to be explored.

Curved space-time can be added by making the e_μ and f_μ dependent upon space-time position (x^μ, x^4, x^5). I used this to write Einstein's gravity law in this language, in 4-space. It does not come out looking very elegant. It is not at all as pretty or as natural as the Dirac and Maxwell equations. This should suggest, the way to improve on Einstein's gravity equation so as to make it fit within the natural quantum algebra. It may well be that no successful quantum gravity is possible at the quantum level in 4-space. Mass is crucial in gravity and it may well dissolve into new, subnuclear cyclic coordinates.

This has been too brief a sketch of the mathematical backbone of my approach. The continuing series in Speculations in Science and Technology will eventually display all the details.

THE DOUBLE SLIT EXPERIMENT AND THE NON-ERGODIC

INTERPRETATION OF QUANTUM MECHANICS

V. Buonomano

Instituto de Matematica
Universidade Estadual de Campinas
Campinas, São Paulo, Brasil

ABSTRACT

The Non-Ergodic Interpretation of Quantum Mechanics is a local realistic particle attempt to confront the difficulties in the Foundations of Quantum Mechanics. It is described in this work using the double slit experiment which motivated it. It explains how a photon passing through one slit knows if the other slit is open or closed by assuming particles interact (interfere) with each other indirectly via an hypothesized medium with certain memory type properties. It is assumed that when a sufficient number of photons with sufficiently identical properties consecutively pass through a region of the hypothesized medium they affect the average properties of the medium which in turn affect other photons that later pass through the same region. Crucial to this view is to imagine a medium which can be forced into stable "modes" by many similarly prepared photons consecutively passing it.

A feasible experimental test is described. The polarization correlation experiments are discussed in relationship to this view. The joint probability question and the object-apparatus interface are commented on.

I. INTRODUCTION

The Non-Ergodic Interpretation of Quantum Mechanics is a local realistic particle attempt to confront the difficulties in the Foundations of Quantum Mechanics. It is reviewed in this work using the double slit experiment which motivated it. This is done in Section 2. In Section 3 the interpretation is formally defined and several comments are made. Section 4 describes a feasible experimental test. The polarization correlation experiments are discussed in relationship to this view in Section 5, where comments are also made on the joint probability question and the object-apparatus interface. It's relationship to Stochastic Quantum Mechanics is given in Buonomano and Prado. This work follows closely Buonomano (1985c).

II. THE BASIC PHYSICAL IDEA

Consider the double slit experiment. The basic problem in trying to understand the physics of this experiment from a local realistic particle point of view is to explain how an individual photon passing through one slit "knows" if the other slit is open or closed. The Non-Ergodic Interpretation is based on the physical idea of photons interacting (interfering) with each other indirectly via a hypothesized medium with certain memory type properties. We now proceed to describe what we mean by this.

We will assume that there exists a medium in empty space and that our laboratory is stationary in this medium. It is supposed to have certain (average) properties which affect the path of photons passing through it and vice versa. That is, we imagine that many consecutive photons (prepared in a sufficiently similar way) passing through a given small region, R, of space, affect the properties of that region, which in turn affects the paths of other photons that later pass through R. Here photons may interact (i.e., interfere) with other photons, but only indirectly via this medium with memory effects. In other words, a photon passing through one slit knows if the other slit is open or closed from the local properties of the medium (in the common path) which will contain this information from other photons actually having previously passed the other slit or not.

Logically one might make an analogy of two professors who are never simultaneously in a room but communicate with each other by leaving messages on the blackboard. The experiments that almost never have two photons in the apparatus simultaneously, only eliminates the possibility that photons interfere directly (i.e., interact by "touching" so to speak).

What is the Medium?

We would like to emphasize that we are not presenting a concrete physical theory. We are only talking about a physical idea of a medium with some sort of memory effects. One may try to justify this medium as a stochastic medium or as a fluid medium. One may wish to remain abstract and refer to it as a field, whose properties depend on what passed through it. One could think about vacuum states. One might talk about an index of refraction of the medium or field which depends on its past history, etc.

Any future physical theory would have to somehow justify a medium of some type which can be forced into stable modes by many similarly prepared photons which consecutively pass through it. The Lorentzian invariance of this memory would also have to be dealt with. It should also be emphasized that despite the qualitative physical nature of the Non-Ergodic Interpretation, it is completely testable and gives a version of quantum mechanics which is as well defined as either the Copenhagen or Statistical Interpretations. This will become clear in the next section.

Memory Development and Decay

The existing interpretations of quantum mechanics predict that interference must exist independent of the light intensity. This forces us to consider a more specific type of memory. We must imagine that we can obtain the same memory buildup in the case of intense light, with relatively few photons and in the case of weak light, with many photons. That is, in this view, as one reduces the light intensity one must compensate for this by using many more photons (i.e., much longer time averages) in order to obtain the same level of interference. In the limit of infinitely low intensity light one must use infinitely many photons to condition the medium.

428

For example, one might choose to imagine the following. Consider an interference experiment (with both paths open) in which the apparatus is already aligned. Assume the photon intensity is I. The initial photons that pass will not show any interference pattern until a sufficient number of them pass to condition the medium. That is, interference will not be seen until the medium arrives in an equilibrium condition with the photons (i.e., with the state preparation). The number, N, of photons that will be necessary to do this will depend on the intensity, the bandwidth and perhaps the frequency. We will call N the memory development parameter.

Note that once the medium is conditioned then we have a stable situation and will continue to see interference until we stop photons from entering the system or change the intensity. When we stop photons from entering the system, then the memory will decay with some characteristic decay time, τ. Let ∂ be the average time separation between consecutive photons at the given intensity. An important observation is that this τ must somehow indirectly depend on ∂ in order to be consistent with quantum mechanics. Another way of saying this is that for low intensities the memory buildup takes longer, but on the other hand the memory decay must be larger also. We are thinking of this memory as resulting from a dynamic equilibrium (or cooperative phenomena) between the intrinsic properties of the medium and the properties of the photons as a group (i.e., their state preparation).

III. THE FORMAL VIEWPOINT

Now we formally define The Non-Ergotic Interpretation of Quantum Mechanics. It assumes the same Hilbert Space Formalism used in both the Copenhagen and Statistical Interpretations (Ballentine) except for the following. In these interpretations one associates the mathematical object, $<A> = <\Psi|A|\Psi>$, representing the average of the observable A in the state Ψ, with the laboratory procedure of taking an ensemble average. Instead the Non-Ergotic Interpretation identifies this same object with the laboratory procedure of taking a time average. This is the *only* difference in formalism. Here we are making a different association between *mathematical objects* of the Hilbert Space Formalism and *laboratory procedures* than either the Copenhagen or Statistical Interpretations. The Non-Ergodic Interpretation always makes the same numeric predictions as the usual interpretations but it makes them only for time averages and not an ensemble averages. This point will be elaborated on in the next section.

An additional point should be noted here. The Non-Ergotic Interpretation identifies $<\Psi|A|\Psi>$ with an average over laboratory time averages. That is, we must take our average over an ensemble of time averages. Each ensemble element is an entire experimental run. This interpretation is a statistical interpretation in the sense of Ballentine.

IV. TESTING THE NON-ERGODIC INTERPRETATION

Ensemble Versus Time Averages

The Non-Ergodic Interpretation clearly gives different experimental predictions than the usual interpretations of quantum mechanics since it makes its predictions for time averages and not ensemble averages. For a true ensemble average it cannot agree with the usual interpretations in general. For example, in the double slit experiment, the conceptually correct laboratory average that ideally should be made according to the usual interpretations is the following. One should have many distinct independent identical apparatus with exactly one photon in each all prepared in the same state Ψ. The superposed positions of all the photons in all the apparatus must then show the appropriate interference pattern. The Non-Ergodic Interpretation clearly predicts no interference whatsoever, since there can be no possibility of photons indirectly communicating with each other here. There can be interference only for time averages.

Of course, an ideal ensemble average is not practical in both economic terms and due to our inability to exactly duplicate state preparations. All existing quantum mechanical experiments are time averages out of necessity. They must implicitly or explicitly make an ergodic assumption that the time averages are equal to the ensemble averages.* Also in the spirit of the empiricist claims of the Copenhagen Interpretation, one might remark that strictly speaking the Non-Ergodic Interpretation is more consistent with the experimental evidence than either the Copenhagen or Statistical Interpretations.

An Experimental Test

We now describe a feasible concrete experiment (Mermin) which avoids the necessity of taking a true ensemble average. Various other less interesting experiments have been described in Buonomano (1985c, 1980). Consider an interferometer arrangement with an ideal laser light source. Assume we have only two counters which we take to be 100 percent efficient. They are to be positioned on the collecting screen at the expected minimum and maximum of the interference pattern. Also we may assume that equal number of photons will arrive at the two counters when one or the other of the interferometer paths are blocked.

Imagine that we are doing an experiment with both paths open and are seeing a good stable interference pattern. For the sake of concreteness, say we are seeing a pattern with a .8 visibility. That is, 90 percent of the photons are going into one of the counters and 10 percent into the other. (For now we ignore the statistical fluctuations in the counting rates. Also, we only consider the total number of photons entering both of the counters when we mention a percentage.) Now quickly close one of the paths. The Non-Ergodic Interpretation predicts that 90 percent of the immediately following photons, all of which passed the unblocked path, will continue to go into the counter situated at the maximum. Further as more and more photons pass, the percentages will *continuously* change to 50 percent as the medium becomes reconditioned to the single slit situation. The immediately following photons are fooled, so to speak, by the medium, into acting as if both slits are still open. Of course, the Copenhagen and Statistical Interpretations predict that the counting rate must abruptly change to 50 percent in each counter when a path is closed, even for the immediately following photons.

The opposite experiment may then be performed. That is, let one path be blocked and assume that we are reliably obtaining 50 percent of the photons in each of the counters. Then quickly (i.e., in a time small compared with the average time separation of the photons) unblock the blocked path. The Non-Ergodic Interpretation would predict that 50 percent of the immediately following photons would continue to go in each of the counter. As more and more photons passed, this percentage would continuously change

---

*It is sometimes said that a time average may be treated as an ensemble average if the system is reprepared between consecutive measurements. For example, in the double slit one could maintain that waiting a certain time is a repreparation of state (e.g., that time for which there is almost never more than one photon in the apparatus at a time). Of course, this just changes one's assumption from being explicitly ergodic to one about what constitutes a re-preparation of the system.

until it stabilized at 90 and 10 percent respectively, in the counters. Again the usual interpretations would predict an abrupt change upon un-blocking the path. In particular, the latter percentages would be obtained for the immediately following photons.

Some Statistical Comments on the Experiment

We make some further remarks on the above experiment. Let N represent the number of photons that are necessary to recondition the medium (at the given intensity level) from one configuration to the other (one path open to both paths open or vice versa). We referred to N as the memory development parameter in Section II. It would only have some rough statistical significance. A problem with the above experiment might then be that the statistical fluctuations in the counting could be large in comparison to N. This would hide the effect that we are looking for and therefore invalidate our experiment. Let us consider some concrete numbers.

First we recall some facts. Interference is, of course, a statistical effect. One cannot see or measure interference with one, two or just several photons. One needs a certain number of photons to see interference. More accurately one needs a certain minimum number of photons to claim that the measured visibility of the pattern is representative of the true visibility within a specified precision and probability. Quantum mechanics not only predicts interference, but it also predicts the number of photons necessary to see it (with a specified precision and probability). For example, one may calculate (Buonomano, 1985b) that 1600 photons (for an ideal laser light source) is sufficient to guarantee a measurement of the visibility with a precision of .05 and a probability greater than .95.

In the above experiment, let us take the light intensity to be 1600/sec on the average arriving in both the counters together when both paths are open. Then when one of the paths is blocked, 400 photons on the average would be detected in each of the counters by our assumptions. Since laser light is Poissonian distributed, the standard deviation of the counting rate for a given counter is the square root of the average counting rate. This means that most of the time between 1440+/-40 and 160+/-13 would be detected in the two counters respectively with both paths open. With one path closed the figure would be 400+/-20. Therefore, the fluctuations in the counting rate can only hide the effect we are looking for if N is roughly smaller than 40. That is, if the memory developed with just 40 photons. We do not consider this a limitation.

Panarella's Experiment

The low intensity interference experiments (e.g., Reynolds and Spartalian) do not attempt to analyze the statistics of interference, that is, the number of photons necessary to obtain interference. Panarella's Experiment (Panarella) is an exception to this. Quantum mechanics (in the existing interpretations) predicts for laser light that if n photons are sufficient to guarantee good interference at a given intensity level, then the same number serves at any intensity level (Buonomano, 1985a). Panarella's experiment does not agree with this. He reports non-linear effects in the sense that the number of photons necessary to see interference is not independent of the light intensity. More accurately he performs an experiment at a certain intensity level and obtains a good interference pattern collecting a certain number of photons. He then dramatically reduces the intensity, but collects roughly the same number of photons by increasing the collection time. Interference is not seen. The Non-Ergodic Interpretation is consistent with these results, but it would predict that the interference must return if one collected sufficiently many

photons. More photons would be needed at the lower intensity level in order
to obtain the same quality of interference.*

V. BELL'S INEQUALITY AND SOME COMMENTS

In this section we discuss Bell's Inequality and the polarization correlation experiments in relationship to The Non-Ergodic Interpretation.
We first recall that Bell's Inequality is only derivable for an ensemble
average and not a time average. That is, it validly characterizes that experimental situation in which one has many identical apparatus and exactly
one photon pair in each of the apparatus. Real laboratory averages are, of
course, time averages and an ergodic type assumption must be made in order
to interpret them as ensemble averages. Thus, it must be explicitly assumed that photons which consecutively pass through the apparatus are absolutely independent of each other (i.e., no memory effects exist). This,
of course, may be true but there is no experimental evidence for it that
we are aware of.

It has been shown (Buonomano 1978 and Caser) that there exists non-
ergodic local realistic theories which agree with Quantum Mechanics in the
polarization correlation experiments. One may invent, for example, a theory
which has the following properties. Consider the usual experimental arrangement in these experiments. That is, we have two distant polarizers
with the photon pair source located between them. A first photon passing
through a polarizer leaves the medium in the immediate vicinity, R1, of the
polarizer in a state which depends on the state of the photon and the polarizer. A second photon passing through R1 leaves a neighboring region,
R2, (somewhat closer to the source) in a state which depends on the state of
the photon and R1, and therefore, on the states of the two photons and the
polarizer, etc. Thus, after many photons one may justify that the source
depends on the state of the polarizer, and similarly for the other side
(see Buonomano 1978). This is a valid local realistic theory and certainly
may agree with quantum mechanics since the states of the polarizers and
source are interdependent. But any future concrete physical theory would
have to justify this type of feedback over time from the polarizers to the
source. This example shows it is possible to invent a formal theory which
would be consistent with all the polarization experiments except Aspect's
last experiment (Aspect 1982).

Note that if one increased the distance between the polarizers and the
source, one may still agree with Quantum Mechanics but only by using many
more photons to allow an equilibrium to set in. In this view the two
polarizers, the source and the medium are one dynamical system which must
reach a mutual equilibrium in order for the quantum mechanical predictions
to be obtained. The above explanation is clearly invalid for an ensemble
average.

Another comment is that the Non-Ergodic Interpretation would point to
a possible violation in Maltus's Law for the initial photons of an experimental run. Only for a sufficiently long time average (i.e., for a sufficiently
large number of photons) would one have to obtain the cosine squared law.
The same comments could be made about the reflection and refraction laws
(Scalera). It would be interesting to know if there was any data available
that may give us the statistical development of the time average in these
experiments.

------

*One should take care here to note that it is only meaningful to count the
number of photons in an interference pattern from the beginning of a physical experimental run, and not from the beginning of a data collection period.

Aspect's Experiment

We now discuss Aspect's last experiment (Aspect). If in this experiment the polarizers were randomly switched between parameter values then the Non-Ergodic Interpretation could not agree with Quantum Mechanics in this experiment. This is because even if there is a mechanism for information flow between a polarizer and the source, the information must arrive at the source at a delayed time and therefore must be uncorrelated with the actual state of the polarizer. In Aspect's experiment the switching between parameter values is not random but periodic. One might imagine theories in which the polarization information about the parameter value of a polarizer is modulated by the commutator. In other words, one may justify saying that the source accumulates memory such that it knows when the photons leave the source what the parameter settings of the polarizers will be.

A necessary condition for this to make sense is that the autocorrelation time of the commutator switching be large in comparison to the time of flight of a photon between the source and the polarizers, or conversely the autocorrelation should be small in order to test the type of theory we are describing. Aspect (1976) intended to have this condition satisfied in his experimental proposal. In the actual experiment (Aspect 1982) this condition was not satisfied. Aspect has made this limitation clear, in particular, in the preprint of his above quoted article.

A Comment on the Joint Probability Question

In The Non-Ergodic Interpretation, there is a logical mechanism for feedback between the measuring part of the apparatus and the state preparation part. They are not independent. The measurement you choose to make affects the state preparation. The logical mechanism is simply our time average. (We have offered no physical mechanism for this feedback.) For a true ensemble average this would not make classical sense. In our view the phase distribution of the properties of our particles will, in general, depend on the observable (but only for a sufficiently long time average). This means that the marginality requirement (Cohen) is not valid here. Therefore, the conclusions about the non-existence of classical joint probability distributions do not apply to The Non-Ergodic Interpretation (Mugur-Schachter, Buonomano 1980).

This same comment may be applied to the question of understanding the object apparatus interface in quantum mechanics. In the Non-Ergodic Interpretation the properties of the object are not independent of the properties of measuring apparatus and vice versa for sufficiently long time averages. It says when you first begin an experimental run, they (the states of the object and apparatus) will be independent, but after a sufficient number of particles pass through the apparatus they will not. One may easily think of examples of this type of behavior from both the physical and social sciences.

VI. REFERENCES

Aspect, A., "Experimental Test of Bell's Inequality Using Time-Varying Analyzers," Physical Review Letters, 49(25), 1804 (1982).
Ballentine, L., "The Statistical Interpretation of Quantum Mechanics," Reviews of Modern Physics, 42, 358 (1979).
Buonomano, V., "A Limitation on Bell's Inequality," Ann. Inst. Henri Poincare, Sect. A, 29, 379 (1978).
Buonomano, V., "Quantum Mechanics as a Non-Ergodic Classical Statistical Theory," Il Nuovo Cimento, 57B, 146 (1980).

Buonomano, V., "The Number of Counting Measurements, the Visibility and the Intensity," Lettere al Nuovo Cimento, 43(2), 69 (1985a).

Buonomano, V., and Bartmann, "Testing the Ergodic Assumption in the Low Intensity Interference Experiments," submitted (1985b).

Buonomano, V., "The Non-Ergodic Interpretation of Quantum Mechanics," to be published in The Proceedings of the International Conference on Microphysical Reality and Quantum Formalism, Urbino, Italy 1985 (G. Tarozzi, ed.) (1985c).

Caser, S., "On a Local Model that Violates Bell's Inequality," Physics Letters.

Cohen, L., "Can Quantum Mechanics be Formulated as a Classical Theory," Philosophy of Science, 33, 317 (1966).

Mermin, D., Personal communication.

Mugur-Schachter, M., "Elucidation of the Probability Structure of Quantum Mechanics and the Definition of a Compatible Joint Probability," Foundations of Physics, 13, 419 (1983).

Nelson, E., "Derivation of the Schrodinger Equation from Newtonian Mechanics," The Physical Review, 150(4), 1079 (1966).

Panarella, E., "Diffraction of Light," Annales de la Fondations Louis DeBroglie, 10, 1 (1985).

Prado, F., "Particulas que Interagim Indiretamente e a Interpretação Estocástica de Mecânica Quântica," Ph.D. Thesis, Instituto de Matemática, Universidade Estadual de Campinas, Campinas, S. P., Brasil. Also Buonomano and Prado 1985, "Indirectly Interacting Particle Processes and the Stochastic Interpretation of Quantum Mechanics," Foundations of Physics, in revision.

Reynolds, G. T., and Spartalian, K., "Interference Effects Produced by Single Photons, Il Nuovo Cimento, LXI(B2), 355 (1969).

Scalera, G., personal communication.

EINSTEIN'S CONCEPTION OF THE ETHER

AND ITS UP-TO-DATE APPLICATIONS IN THE RELATIVISTIC WAVE MECHANICS

Ludwik Kostro

Institute of Experimental Physics
University of Gdańsk
Wita Stwosza 57, Poland

INTRODUCTION

Einstein's conception of the ether (ECE) is almost unknown by physicists and philosophers. The majority of them is even convinced that Einstein has removed the notion of the ether from physics for ever. In such a situation, a more detailed presentation of ECE is necessary to understand the applications which it has found nowadays in the relativistic wave mechanics.

A SHORT HISTORICAL OUTLINE OF EINSTEIN'S IDEAS CONCERNING ETHER

In 1894 (or 1895) Einstein, being 15 (or 16) years old, wrote his first "scientific" paper (which he never published) entitled "Über die Untersuchung des Ätherzustandes im magnetischen Felde"[1]. At that time Einstein believed in the existence of a stationary quasi-rigid luminiferous ether. He regarded it as an elastic medium and wondered in particular[2] how "the three components of elasticity affect the velocity of an ether wave"[1] which is generated when the electric current is turned on.

As ETH student, Einstein wanted to construct an apparatus which would accurately measure the earth's movement against the ether[2]. In 1901 he wrote a letter to his friend Grossman in which he told him: "A new and considerable simpler method for investigating the motion of matter relative to the light-ether has accurred to me"[2]. In his speech at Kyoto University Einstein informs us about this method:

"I tried to find the clear experimental evidence for the flow of the ether in the literature of physics, but in vain. Then I myself wanted to verify the flow of the ether with respect to the earth, in other words, the motion of the earth. When I first thought about this problem, I did not doubt the existence of the ether or the motion of the earth through it. I thought of the following experiment using two thermocouples: Set up mirrors so that the light from a single source is to be reflected in two different directions, one parallel to the motion of the earth and the other antiparallel. If we assume that there is an energy difference between the two reflected beams, we can measure the difference in the generated heat using two thermocouples. Although the idea of this experiment is very similar to that of Michelson, I did not put this experiment to a test. While I was thinking of this problem in my student years, I came to know the strange result of Michelson's experiment. Soon I came to the conclusion that our idea about the motion of the earth with respect to the ether is incorrect, if we admit Michelson's null result as a fact. This was the first path which led me to the special theory of relativity".[3]

In 1905, having formulated the special relativity theory, Einstein began to deny the existence of the stationary luminiferous ether. He considered it as "superfluous"[4] and wholly useless[5-7] because according to the relativity principle an absolute space at absolute rest does not exist and because the electromagnetic fields have to be regarded as independent realities which are not states of a medium. Einstein maintained even that:"the ether in the old sense does not exist"[8] and propagated this opinion not only in the scientific reviews but also in newpapers e.g. in the Vossische Zeitung.[9]

The history of the new (relativistic) ether conception began in 1916 i.e. after the definitive formulation of the general relativity theory. The introduction of the new conception was provoked, in a certain sense, by H. A. Lorentz and Ph. Lenard.

Lorentz wrote a letter to Einstein in which he maintained that the general theory of relativity admits of a stationary ether hypothesis. In reply Einstein introduced a new definition of the ether:

$$\text{"state } g_{\mu\nu} = \text{Aether"}$$

He wrote to Lorentz the 17 June 1916:

"I agree with you that the general relativity theory admits of an ether hypothesis as does the special relativity theory. But this new ether theory would not violate the principle of relativity. The reason is that the state $g_{\mu\nu}$ = Aether is not that of a rigid body in an independent state of motion, but a state of motion which is a function of position determined through the material phenomena"[10].

As we see the physical space (connected closely with time) described by the symmetrical tensor $g_{\mu\nu}$ ($g_{\mu\nu} = g_{\nu\mu}$) was considered by Einstein as a relativistic ether. Einstein did not publish his new idea either in 1916 or 1917. The first appearance in print of the new conception was provoked by Ph. Lenard.In 1917 Lenard published a paper against Einstein's relativity theory entitled "Über Relativitätsprinzip, Äther, Gravitation"[11]. In this paper he maintained that in the general relativity the disqualified ether (disqualified by the relativity theory) came back under a changed name "Space". In reply Einstein wrote an essay entitled "Dialog über Einwände gegen die Relativitätstheorie"[8] in which he published the above presented new definition of the ether. This definition will be called by Einstein in the famous Einstein - Lenard discussion concerning ether and relativity theory in Bad Nauheim(1920): "Eine neuartige Definition für den Begriff Äther"[12].

Einstein introduced three new models of the ether:
(1) The first one is that of the special relativity theory. In the mathematical description of this ether the 10 components of the metrical tensor are constant $g_{\mu\nu} = \eta_{\mu\nu}$ ($g_{11} = g_{22} = g_{33} = 1$; $g_{44} = -1$ and the other 6 components = 0). The ether of the special theory of relativity is rigid and to a certain extent four-dimensional. It is infinite and flat. Its metric is pseudo-euclidian.
(2) The second one is that of the general relativity theory. In the mathematical description of this ether the 10 components of the $g_{\mu\nu}$ tensor are no longer constant. The space states described by the tensor $g_{\mu\nu}$ can change not only from place to place, but also in time. The ether of the general relativity theory is no longer rigid and flat. Its metric is pseudo-Riemannian
(3) The third one is that of the unitary relativistic field theory. In the mathematical description of this ether the symmetrical tensor $g_{\mu\nu}$ does no longer describe the ether in the complete way because the geometrical structure of it is more than riemanian. New structure elements have to be introduced for a complete description of the ether because it has to determine

not only the inertio-gravitational phenomena, but also the electromagnetic ones.

Summarizing we can say that since 1916 Einstein's physics of space-time became a physics of a new ether. Nevertheless, we must mention that after 1934 Einstein began to use the word "ether" less and less often, although he wrote still in 1938: "This word ether has changed its meaning many times in the development of science... Its story by no means finished is continued by relativity theory"[13], and though he indicated still in 1954 that e.g. the: "rigid four-dimensional space of the special theory of relativity is to some extent a four-dimensional analogue of H.A. Lorentz's rigid three-dimentional ether"[14].

THE REAL PHYSICAL SPACE CONSTITUES A RELATIVISTIC ETHER

According to Albert Einstein:

"There is an important argument in favor of the hypothesis of the ether. To deny the existence of the ether means, in the last analysis, denying all physical properties to empty space"[15]

In Einstein's theory of relativity the three physical notions: "space" "ether" and "field" have found their complete unification through consequent identification.

"Physical space and the ether, are only different terms for the same things; fields are physical states of space"[16]

ECE presents Einstein's original interpretation of the models of physical space constructed in his special and general relativity and in his unitary field theory. It constitues a gradual conceptual activation, dynamization and materialization of the physical space. According to ECE, in its most developed form, the physical space closely connected with time is not a passive and static container of events and not physically indifferent or neutral arena of physical phenomena but an active and dynamic field which determines the inertio-gravitational, electromagnetic and other processes and produces even elementary particles. The real physical space, as an active field of this kind, possesses energy and therefore mass as well and that is why it is material. It constitues an active matter sui generis for which the term "ether" is the best name.

The activation of the physical space

It has "seemed utterly absurd to the physicists of the nineteenth century to attribute physical functions or states to space itself"[16]. It is not so in Einstein's theory of relativity, the physical space plays there a real active part in physical processes. When Einstein speaks about ether:

"We are dealing with those things thought as physically real which alongside the ponderable matter composed of elementary particles, play an important part in the causal nexus studied in physics"[17]

The "physically real things" mentioned here are the "real qualities of space" and that is why Einstein continues in the same paper:

"Instead of speaking about ether, somebody might just as well speak about the 'physical qualities of space'"[17]

According to Einstein's new conception, it is impossible to formulate a complete physical theory without an (at least latent) ether hypothesis, because every complete physical theory must take into consideration the real properties of the physical space i.e. the "Milieu-Einflüsse"[17]. Somebody might not use the word "ether" but has to recognize that the physical space has real physical properties which play an active part in physical happening and therfore Einstein maintains:

"The ether hypothesis was bound always to play a part even if it was mostly a latent one at first in the thinking of physicists"[15]

According to ECE the absolute (i.e. independent from time and matter) space of Newton, because of its active "inertia-determining function"[16] constitues one of the models of the ether[17]. In this model:

"Space was conceived as absolute in other sense also; its inertia determining effect was conceived as autonomous i.e. not to be influenced by any physical circumstances whatever; it affected masses but nothing affected it"[16]

Einstein's special and general relativity and his unitary field theory, as it was mentioned already, have their own models of ether identified with the physical space.

(a) In Einstein' special relativity model, where the ether became "to a certain extent four-dimensional"[17] (because of the relativity of simultaneity) the physical space accomplishes the active function "determining the inertial behavior of a test body introduced into it"[14] and has the physical property of transmitting electromagnetic waves"[13] but "it no longer stands for a medium built up of particles"[14] or points[17] and is no longer regarded as an immobile or stationary medium as it was supposed in the Newtonian model of the physical space and in Lorentz's conception of the ether.

"The special principle of relativity forbids us to regard the ether as composed of particles the movement of which can be followed out through time, but the theory is not incompatible with the ether hypothesis as such. Only we must take care not to ascribe a state of motion to the ether"[15]

"The whole difference the special theory of relativity made in our conception of the ether lay in this, that it divested the /Lorentz's/ ether of its last mechanical quality namely immobility"[15]

"According to the special relativity, the ether remains still absolute because its influence on the inertia of bodies and on the propagation of light is conceived as independent of every kind of physical influence"[17]

(b) The ether of Einstein's general relativity is no longer absolute in the above mentioned sense because "it not only conditions the behavior of inert masses but is also conditioned, as regarded its state by them"[15].
    Einstein's general relativity is incomprehensible without an active ether.
    "According to the general relativity space is endowed with physical qualities; in this sense, therefore, an ether exists. In accordance with the general theory of relativity space without an ether is inconceivable. For in such a space there would not only be no propagation of light, but no possibility of existence of scales and clocks, and therefore no spatio-temporal distances in the physical sense. But this ether must not be thought of as endowed with the properties characteristic of ponderable media, as composed of particles the motion of which can be followed; nor may the concept of motion be applied to it"[15]

The general relativity ether manifests its activity through its function determining the inertio-gravitational behavior of the bodies and through the creation of elementary particles. A test body or particle which is only under the influence of the physical space is at rest or follows a geodetic (curved or straight) respectively in curved or locally flat spaces of reference.

Einstein has at first occasionally noted the possibility that material particles might be considered as singularities of the material field but subsequently he arrived to the conviction that this point of view could not be accepted at all. "For a singularity brings so much arbitrariness into the theory that it actually nullifies its laws"[18]. He made therefore attempts to find solutions of general relativity field equations free of singularities which might "be interpreted as presenting corpuscules"[19]. Together with Rosen, he found such solutions of the centrally symmetrical gravitational field equations for both the neutral and for the electrical particles. Having found them he repeated his opinion expressed in 1924[17] that:"The neutral, as well the electrical particle is a portion of space"[18], material space of corse.

(c) In Einstein's general relativity (as in his special relativity) the electromagnetic field appears still as something which "fills space"[14] i.e. as something which does not belong to the structure of the physical space described by the metrical tensor $g_{\mu\nu}$. Since the real physical space was regarded by Einstein as the "fundamental" or "total field"[20] of all physical actions and not only of the inertio-gravitational one, he began to look for "a theory of the continuum in which a new structural element appears side by side with the metric such that it forms a single whole together with the metric"[16]. Thus the formulation of an unitary field theory became the main aim of Einstein's research programme.

He often emphasized that the pseudoriemanian space-time described by the tensor $g_{\mu\nu}$ does not constitue a complete description of the physical space connected with time. He made several attempts to generalize it e.g. through enriching "Riemannian space by adding the relation of direction or parallelism"[16]. He was even convinced that he "found the most natural form for this generalisation"[14] in his "theory of unsymmetrical field"[21] (which he considered as his longtime sought unitary field theory) which unifies in his opinion the gravitational and electromagnetic interactions.

The activity of the ether described by Einstein's unitary field theory is richer than that described by Einstein's general relativity because it includes also the electromagnetic interactions, but today Einstein's unitary field theory is considered as unsatisfactory.

Dynamization of the physical space

In the Newtonian physics the physical space was regarded by physicists as a changeless reality."Space was still for them, a rigid homogeneous something incapable of changing or assuming various states"[16].In Einstein's theory of relativity the physical space is no longer an immutable physically indifferent container entirely foreign to modifications but a dynamic changing in time medium.

(a) In Einstein's special relativity however, the ether is still "rigid",

("The fourdimensional space of special theory of relativity
is just as rigid and absolute as Newton's space"[16])

but the fusion of space and time in Einstein's special relativity: "has to be characterized as dynamization of space"[22]as it has been indicated e.g. by M. Capek, because the physical space is no longer timeless. In Newtonian physics:

"The true reality of space is timeless, change and succesion
belong merely to the physical processes, not to the space as such"[22]

439

The fusion of time and space means an "incorporation of space into the physical becoming"[22].

(b and c) In Einstein's general relativity and especially in his unitary field theory we are no longer dealing with the traditional distinction between an immutable and static spatial container and its concrete and changing content.

"Space as opposed to 'what fills space'... has no separate existence"[14]

For instance, in general relativity it is meaningless to speak about the gravitational field as being located in space when the whole reality of this field is reduced to the modifications of the noneuclidian spatio-temporal medium. The pseudoriemanian space-time with its curvature varing not only from place to place, but even in time, and in particular the idea of expanding and contracting space whose radius of curvature is continuesly changing and also the real vibrating and waving of the mentioned spatio-temporal medium show the "nonrigidity" and the dynamic nature of Einstein's relativistic ether.

The idea of "nonrigid" and active physical space has been already introduced by Rieman[23]. According to Einstein's relation, we owe to Riemann:

"A new conception of space in which space was deprived of its rigidity and in which its power to take part in physical events was recognized as possible"[16]

Materialization of the physical space

On the basis of the principle of equivalence of energy and mass (formulated already in the special relativity) Einstein arrived at the following conclusions:

(a) The real physical space (even though it was empty) as an active field possessing energy (and therefore mass as well) constitues an active matter sui generis i.e. an ether.

(b) There is no a qualitative difference between the material physical space and the ponderable matter composed of particles.

(c) The formulation of a consequent unitary field theory, where the material physical space constitues the primary matter producing the secondary one i.e. the elementary particles, must be possible.

"The division into matter and field is after the recognition of equivalence of mass and energy something artificial...Matter is where the concentration of energy is great, field where the concentration of energy is small. But if this is the case, then the difference between matter and field is a quantitative rather than a qualitative one. There is no sense in regarding matter and field as two qualities different from each other. There would be no place in our new physics for both field and matter field being the only reality"[13]

The mentioned "new physics" is the unitary field theory the formulation of which became Einstein's main research programme. According to this programme, the elementary particles have to be regarded as born in field and from field or in ether and from ether or also in space and from space. For, as we know, in Einstein's theory of relativity "field", "ether" and "space" are synonyms and they have to be conceived as the primary reality.

"The strange conclusion to which we have come is this - that now it appears that space will have to be regarded as a primary thing and that matter is derived from it, so to speak, as a secondary result. Space is now turning around and eating up matter. We have always regarded matter as a primary thing and space as a sencondary result. Space is now having its revenge, so to speak, and is eating up matter. But that is still a pious wish"[24]

440

As we see, in 1930, the formulation of a unitary field theory was Einstein's pious wish. In an other paper, written also in 1930, Einstein emphasized that the material physical space became for him the unique carrier of reality ("alleiniger Träger der Realität")[25].

"The real is conceived as a four-dimensional continuum with a unitary structure of a definite kind (metric and direction). The laws are differential equations, which the structure mentioned satisfies, namely, the fields which appear as gravitation and electromagnetism. The material particles are positions of high density without singularity.
We may summarize in symbolical language. Space, brought to light by the corporeal object, made a physical reality by Newton, has in the last few decades swallowed ether and time and seems about to swallow also field and the corpuscles, so that it remains as the sole carrier of reality"[25]

EINSTEIN'S RELATIVISTIC ETHER CONSTITUES AN ULTRA-REFERENTIAL FUNDAMENTAL

REALITY

Einstein does not identify ether with the "reference spaces" (the number of which is infinite) composed of points and being at rest or motion with respect to each other. He identifies it with the "physical space as such" which is one and unique, not composed of points and to which the notion of motion in the mechanical sense cannot be applied at all. Einstein's relativistic ether ERE i.e. the physical space as such is something ultra-referential. It does not constitue a reference frame and has not a proper reference frame. If ERE had had a proper reference frame it would have been at rest in it. ERE however is not a stationary ether.
The ultra-referential physical space cannot be conceived as composed of immobile points because an immobile point constitues something totaly relative. An immobile point of a reference space constitues a set of collocal (or isotopic) events in this reference space. Since in Einstein's theory of relativity collocality is something totaly relative therefore the ultra-referential physical space is inconceivable as composed of immobile points. The Newton's absolute space is conceived as composed of immobile points, but not the ultra-referential Einstein's physical space as such.
Every point in the four-dimensional world has its world-line and therefore an extended entity composed of points (such as e.g. a reference space) can be presented in such a world as a set of world-lines. The extended ERE, of course, cannot be presented in such a manner.

"In the language of Minkowski this is expressed as follows. Not every extended entity in the four-dimensional world can be regarded as composed of world-lines "[15]

The physical space as such is closely connected with time as such. It is important to note that the time as such is also an ultra-referential reality. There are infinite reference times intimately connected with their proper reference spaces but there is only one and unique ultra-referential time as such. The ultra-referential time is not composed of moments like the ultra-referential space is not composed of points. A moment constitues a set of simultaneous events which belong to it. Since in the theory of relativity the simultaneity is a strictly relative thing the ultra-referential time cannot be composed of moments. Nevertheless, the ultra-referential time is something "extended" composed of past, present and future. With respect to a freely chosen event considered as present there exists always a set of events which are absolutely past, and a set of events which are absolutely future. Every reference time is one of the possible orientations in the ultra-referential time. The ultra-referential time rends possible an infinite set of reference spaces.

The ultra-referential physical space is with respect to the reference spaces a more fundamental reality. The reference spaces are quasi-objects which move with respect to each other in the ultra-referential physical space but not with respect to it. The ultra-referential physical space rends possible the existence and motion of the reference spaces but it does not move at all in the mechanical sense.

On the other hand, the ultra-referential space is never passive or quiet. Einstein considers the nonatomically and nonmechanically conceived ether as the fundamental source of every physiacal activity, the creation of particles included. His presentation of this activity, (except the inertio-gravitational one), cannot be considered today as satisfactory. In this point Einstein's research programme cannot be regarded as accomplished in a definitive way.

Novadays this programme, as it has been shown by Faddeev[26], is continued in those hypothesis in which the elementary particles are presented as solitons on top of an active field. One of the reasons of Einstein's ill-succes was the lack of the introduction of the constant of Planck into the description of ether activity. In the creation of the elementary particles however, the elementary quantum of action must play a fundamental part.

EINSTEIN'S CONCEPTION OF THE ETHER UP-TO-DATE APPLICATIONS IN THE RELATI-

VISTIC WAVE MECHANICS

In 1923[27,28] and 1924[29] L. de Broglie having introduced Planck's constant into Einstein's special relativity through the identy $mc^2 = h\nu$ which constitues the most basic assumption of his relativistic wave mechanics, discovered the relativistic waves called "waves of matter". This discovery, in our opinion, proves the real existence of ERE active excitation describable in the reference frames by wave functions. L. de Broglie however, formulating his wave mechanics, did not use the notion of the ether at all[30], but later, as his collaborator J.-P. Vigier testifies[31] took into consideration the possibility of an introduction of such a notion. He taked e.g. about the "deeper background of space"[31].

J.-P. Vigier, F. Halbwachs, F. Piperno, A. Kyprianidis, D. Sardelis et al.[32-34] developing de Broglie relativistic wave mechanics in the framework of so-called Stochastic Interpretation of Quantum Mechanics (SIQM) opposed to the Copenhagen Interpretation use Einstein's conception of the ether. In SIQM this conception became however completed by Dirac's conception of the ether[35]. According to J.-P. Vigier et al. Einstein's relativistic ether i.e. the material $g_{\mu\nu}$-field is filled with Dirac's covariant ether-like vacuum[34] which constitues a mixture of endowed with spin $J = 0$, $J = 1/2$ and $J = 1$ extended particles and antiparticles. Such a covariant mixture constitues according to J.-P. Vigier et al. a background sea at absolute zero temperature on which the de Broglie real waves travel. Every particle (considered in SIQM as an extended entity) is surrounded by a real de Broglie wave. Since the Dirac's non empty vacuum constitues a mixture of particles and antiparticles a de Broglie "pilot" quantum wave has to be regarded as a superluminal phase like collective drift and random motion on top of this non empty vacuum which implies subquantal fluctuations or jumps at velocity of light.

J.-P. Vigier emphasizes that Einstein's relativity theory is perfectly compatible with such an underlying relativistic stochastic ether model and that inherent to this model is Einstein's idea that quantum statistics reflects a real subquantal physical vacuum alive with fluctuations and randomness. The concept of a non empty vacuum has been revived not only to yield a foundation to the SIQM but also to explain causally possible nonlocal superluminal interactions resulting from Einstein-Podolski-Rosen paradox.[32]

442

J.-P. Vigier in his paper entitled "Non-Locality, Causality and Aether in Quantum Mechanics"[36] revisits Einstein's conception of the ether presented by Einstein in the essay "Über den Äther"[17] in the light of recent development in SIQM. He adds in this article to the usual $g_{\mu\nu}$ terms stochastic $\delta g_{\mu\nu}$ terms and describes space-time as a real subquantal covariant random medium which implies subquantal fluctuations. Thus the material space-time is considered by him as a fluctuating $\delta g_{\mu\nu}$-field.

Einstein's relativistic ether and the "three-waves hypothesis"

Einstein's conception of the ether is also used in the "three-waves hypothesis" (TWH) proposed by the author in 1978[37-39] also in the framework of de Broglie relativistic wave mechanics. The TWH constitues an attempt to develop some ideas of Einstein's research programme concerning the elementary particles. In Einstein's research programme the elementary particles are conceived as "fields of particular kind" ("Felder besonderer Art"[17]) which constitue "particular states of space" ("besondere Raum-Zustände" [17]) Remaining in the framework of Einstein's programme and using de Broglie concept of "wave field" ("champ ondulatoire"[40,41]) the TWH presents the elementary particles as particular threefold wave fields (TwFs) which constitue particular states of the material physical space i.e. of Einstein's relativistic ether.

The TwFs can be observed from infinite reference frames. In the TWH they are studied, for the time being, only in the locally inertial reference frames i.e. where in the mathematical description, the components of the $g_{\mu\nu}$ tensor describing the gravitational potentials of the real physical space are constant and where the Christoffel symbols vanish i.e. where in the physical space the state of weightlessness governs. In such reference frames the physical quantities of the TwFs are varying according to the linear Lorentz transformation law and therefore the mathematical formalism of special relativity can be used.

A relativistic material TwF constitues an extended vibrating field with a central point at rest in its proper reference frame. In such a reference frame it has a proper period T_o, frequency ν_o and energy $E_o = h \nu_o$ concentrated around the central point. Having energy the TwF has also mass m_o concentrated around the central point as well. The central point of the TwF constitues its center of mass (CM). The TwF has also an incessantly vibrating center of energy (matter) density (CED). The CED vibrates in the circumambiency of the CM. The CED as distinct from the CM has been introduced (by means of a hydrodynamic model) into the relativistic wave mechanics by Bohm and Vigier[42].

The frequency of the CED vibration is equal to that of the TwF and is in phase with it where the CED vibrates. The CED vibration as a CED vibration of a wave field is wave-like i.e. its frequency transforms according to the eq. $\nu = \nu_o (1 - v^2/c^2)^{-1/2}$ as opposed to the frequency of a clock-like vibration which transforms according to the eq. $\nu = \nu_o (1 - v^2/c^2)^{1/2}$. There is no a reference frame of the central point of the TwF in which the CED does not vibrate. Also in this sense Einstein's relativistic ether is never quiet. The CED as an active oscylating point "produces" in Einstein's relativistic ether two wave fields. One propagating at superluminal velocities (from ∞ to c) and another propagating at subluminal velocities (from 0 to c).

(1) The superluminal wave field constitues the first component of the TwF. The CM and the CED are surrounded first of all by de Broglie wave field (BwF) the waves (B-waves) of which are described by the well known function:

$$\psi_B (x,y,z,t) = a \exp\left[2\pi i\nu(t-x/u)\right]$$

(with well determined amplitude a) and characterized by the physical quantities: phase velocity $u = c^2/v > c$, and wavelength $\lambda_B = h/mv = (h/E) u$ (where c is the velocity of light and v the velocity of CM).

According to the TWH, the BwF constitues a particular kind of superluminal radiation which does not transport energy but transports a special kind of momentum $\vec{p}_B = (h/c^2)\vec{v}_B\vec{u}$ (momentum of Einstein's relativistic ether wave excitation)[39].

The BwF penetrates the whole empty space (i.e. the unoccupied Einstein's relativistic ether). In the proper reference frame of the TwF, the BwF is characterized by an infinite wavelength of its waves and propagates at infinite phase velocity in all directions begining from the central point. If in a locally inertial reference frame (which constitues our laboratory frame) the CM moves at constant velocity v e.g. in the +x direction along the x axe, then the BwF appears as propagating from the central point at different superluminal velocities in different directions: from the infinite velocity in the direction parallel to the y,z plane to the least one $u_{+x} = c^2/v_{+x} > c$ in the +x direction (where v_{+x} is the velocity of CM). The wavelengths of the B-waves (of the BwF propagating in this way), diminish from the infinite wavelength in the directions parallel to the y,z plane to the shortest one $\lambda_{B_{+x}} = h/mv_{+x} = (h/E)u_{+x}$ in the +x direction.

In all directions which are not parallel to the y,z plane and not parallel to the +x axe the BwF propagates at velocities smaller than infinite but greater than u_{+x} and its B-waves have wavelengths shorter than infinite but longer than $\lambda_{B_{+x}}$.

If we place a set of observers (stationary with respect to the laboratory frame) on a plane parallel to the y,z plane in a certain distance from the y,z plane in the +x direction, then the CM (moving along the x axe) moves only in the direction of one observer A_o which is placed where the mentioned plane intersects with the x axe. The CM can move only in a unique direction but it approaches other observers of the plane as well at varying velocity smaller then v_{+x}. The shortest distance of approach is equal $A_o A_n$ when the CM meets A_o. At that moment the velocity of approach is equal to zero. The CM does not meet other observers but the BwF arrives at all of them and it is important to note that it happens at the same time. This relativistic effect can be presented by means of geometrical diagrams. We will note here only that this effect is a simple consequence of de Broglie relation $c^2 = vu$. The slower the CM approaches an observer the faster the BwF propagates in his direction and therefore the propagating BwF meets all the observers even the most distant ones at the same time. The B-waves surfaces of the BwF appear to them as planes which approach at velocity u equal to the phase velocity u_{+x} of the B-wave which meets the observer A_o.

(2) The BwF, if observed from different reference frames has different relativistic images in every of them. These images if observed from the laboratory frame constitue a particular superimposition of B-waves. L. Mackinnon who is the first who indicated this relativistic effect has also shown that it constitues a nondispersive wave-packet having properties of a soliton[43-45] Mackinnon's soliton is characterized by a Compton transforming wavelength $\lambda_C = \lambda_{o_C}(1 - v^2/c^2)^{1/2}$ and an intrinsic phase velocity c. It is desribed in our laboratory frame by the function:

$$\psi_C(r,x,t) = \left[\sin(kr)/kr\right]\exp\left[i(\omega t - k_o x)\right]$$

with $k = m_o c/\hbar$, $r = \left[(x-vt)^2/(1 - v^2/c^2) + y^2 + z^2\right]^{1/2}$, $\omega = mc^2/\hbar$, $k_o = mv/\hbar$

The solitary C-wave constitues the second component of the TwF. Its formation can be presented by means of space-time diagrams[43]. Mackinnon's solitaon constitues an extended material microobject in the proper sense. The energy and the inertia of the TwF are closely connected with it. The nondispersive wavepacket forms itself where the B-waves are in phase and where therefore the amplitude of the packet is the greatest. The energy of the TwF is therefore concentrated in the solitary C-wave. Hence the CED is located inside the Mackinnon soliton and the inertia of the TwF is related to the amplitude terms of the solitary C-wave[44].

(3) The mentioned above subluminal wave field (introduced in 1978 by the author[37]) the waves of which are described by the function:[38,39]

$$\psi_D(x,y,z,t) = a \exp\left[-2\pi i \nu (t-x/v)\right]$$

constitues the third component of the threefold wave field (TwF). Its properties are in a certain sense opposite to those of the BwF and therefore it can be named as dual to the de Broglie wave field (DwF). Its waves (D-waves)* are characterized by the phase velocity $v \langle c$, wavelength $\lambda_D = h/mu = (h/E)v$ and momentum $\vec{p}_D = (h/c^2)\nu_D\vec{v}$.

The DwF, if observed from the proper reference frame of the TwF does not propagate at all. Its velocity and wavelength of propagation are equal to zero in all directions begining from the central point. In the proper reference frame the DwF manifests itself only through the CED vibration as a merely local periodic phenomenon of frequency ν_0. If observed from our laboratory frame, the DwF propagates in different directions at different subluminal velocities: from the velocity equal to zero in the directions parallel to the y,z plane to the greatest one in the +x direction equal to the CM velocity. The wavelengths of DwF propagation increase from zero in the directions parallel to the y,z plane to the longest one in the +x direction

$$\lambda_{D_{+x}} = h/mu_{+x} = (h/E)v_{+x}$$

In all directions which are not parallel to the y,z plane and to the +x axe the DwF propagates at velocities greater than zero but smaller than v_{+x} and has wavelengths longer than zero and shorter than λ_{D+x}.

The DwF propagates like expanding sphere the diameter of which increases in the direction +x. If we single out three points AOB of this diameter, then A does not move, O moves at velocity $(1/2)v_{+x}$ and B at velocity v_{+x} equal to the CM velocity. The DwF follows the CM and propels it. The DwF front does not arrive at all our laboratory frame observers at the same time.(It reaches together with the CM the A_0 observer the first). This relativistic effect is a simple consequence of the TWH relation[37,38]

$$\lambda_C^2 = \lambda_B \lambda_D$$

based on the mentioned de Broglie relation. The TWH relation can be presented as follows

$$c^2 T^2_C = (uT_B)(vT_D)$$

(where $T_C = T_B = T_D$ /equal also to $T_{CED} = T_0(1 - v^2/c^2)^{1/2}$/, because conditions of local metrical homogeneity govern in our laboratory frame). The faster the BwF approaches an observer the slower the DwF propagates in his direction.

In our laboratory frame, the trajectory of the CM will be a straigthline. The trajectory of the vibrating CED will have in a certain sence a wave-like form. The wavelength of such a wave-like trajectory is equal to the wavelength of the D-wave propagating in the +x direction

$$\lambda_{CED_{traj}} = \lambda_{D_{+x}} = v_{+x}T_D$$

---

*In my unpublished paper written in 1978[37] the D-waves are named by me V-waves because of their subluminal velocity v. The name D-wave (dual to the de Broglie wave) has been introduced by R. Horodecki who on the basis of my unpublished paper (presented to him for an estimation) has formulated his own version of the TWH. In his works (Phys. Lett: 87 A:95 (1981); Phys. Lett.91 A: 269 (1982); Phys. Lett. 96 A:175 (1983); Lett. Nuovo Cimento 36:509 (1983) R. Horodecki propagates, develops and modifies my TWH. He thanks me for the basis provided for his works in Phys. Lett. 87 A:95 (1981), see p.97.

Thus the D-wave$_{+x}$ manifests itself, in a certain sense, through the CED vibration.

In our laboratory frame, The DwF carries the C-wave soliton on its wavefront at the point which which propagates the fastest i.e. where we find the wavefront of the longest D-wave $\lambda_{D_{+x}}$ and in the direction indicated by the wave vector:

$$k_{D_{+x}} = 2\pi / \lambda_{D_{+x}}$$

CONCLUSION

The conclusion of this paper is the following. An elementary paricle can be presented as a threefold wave field (TwF) on top of Einstein's relativistic ether (ERE). In such a TwF the C-wave soliton constitues an extended microobject in the proper sense. Such a microobject stores up the whole energy of the TwF in its intrinsic C-wave vibration, has inertia properties and is characterized by a transforming Compton wavelength. The Compton wavelength of the intrinsic C-wave vibration belongs to the internal structure of the microobject.

An elementary particle however, is not only a microobject but also an extended widespread wave field composed of the BwF and the DwF. The superluminal BwF precedes the soliton-microobject preparing the way for it among different obstacles[37-39]. Other solitons-microobjects are obstacles for the BwF and the DwF. The BwF is responsible for all reflection, dyfraction, interference and suprluminal correlation phenomena[37-39]. The DwF follows the soliton-microobject and propels it in the space-time where the BwF has prepared the way. It is responsible for all energy exchange phenomena because carring the soliton-microobject it carries also its energy and inertia[38,39].

All the three wave fields are relativistic wave fields on top of Einstein's relativistic ether. Their physical quantities are intimately interconnected and correlated[37-39]. Their interconnection andcorrelation find an expression e.g. in the following equations:

$$\lambda_C^2 = \lambda_B \lambda_D \, , \qquad k_C^2 = \vec{k}_B \vec{k}_D \, , \qquad p_C^2 = \vec{p}_B \vec{p}_D$$

(where k_C is the wave number and p_C the intrinsic momentum of the solitary C-wave; $\vec{k}_B$ and $\vec{k}_D$, $\vec{p}_B$ and $\vec{p}_D$ the respective wave vectors and momenta of the B-waves and of the D-waves).

Summarizing we can say. The physical space (closely connected with time) conceived nonatomically and nonmechanically (i.e. ERE)constitues a material active subquantal medium the activity of which manifests itself, among other things, through the creation of the elementary particles. We are able to descibe this creation if we use de Broglie introduction of Planck's constant into relativity theory.

ACKNOWLEDGMENTS

The author would like to express his sincer thanks to the VW-Stiftung (i.e. to the Volkswagen Company) for financial support which made this research possible. The author would also like to thank Deutsches Museum, Dr.Jürgen Teichmann and all of the individual people who have been so kind to be of assistence.

REFERENCES

1. A. Einstein, Über die Untersuchung des Ätherzustandes im magnetischen Felde, Phys. Blätter, 27:390 (1971).
2. A. Pais, 'Subtle is the Lord...', The Science and the Life of Albert Einstein, ClarendonPress, Oxford University Press, Oxford, New York, (1982) p. 130.
3. A. Einstein, Speech at Kyoto University, December 14, 1922, NTM-Schriftenr. Gesch. Naturwiss., Technik, Med., Leipzig, 20:25 (1983).
4. A. Einstein, Zur Elektrodynamik bewegter Körper, Ann. d. Phys., 17:891 (1905).
5. A. Einstein, Relativitätsprinzip und die aus demselben gezogenen Folgerungen, Jahrb. d. Radioaktivität, 4:411 (1907).
6. A. Einstein, Über die Entwicklung unserer Anschauungen über das Wesen und die Konstitution der Strahlung, Phys. Zeitschrift. 10:817 (1909).
7. A. Einstein, Le principe de la relativitè et ses conséquences dans la physique moderne, Arch. Sci. Phys. et Nat. 19:5 and 125 (1910).
8. A. Einstein, Dialog über Einwände gegen die Relativitätstheorie, Naturwissenschaften. 6:697 (1918).
9. A. Einstein, Vom Relativitätsprinzip, Vossische Zeitung, Nr. 209:1, 26 April 1914.
10. A. Miller, Imagery in Scientific Thought Creating 20th Century Physics, Birkhäuser, Boston, Stuttgart (1984) p. 55.
11. Ph. Lenard, Über Relativitätsprinzip, Äther, Gravitation, Jahrb. d. Radioakt. u. Elektronik, 15:117 (1918).
12. A. Einstein - Ph. Lenard, Allgemeine Diskussion über Relativitätstheorie, Phys. Zeitschr. 21:666 (1920)..
13. A. Einstein, L. Infeld, The Evolution of Physics, Univ. Press., Cambrige (1947).
14. A. Einstein, Relativity and the Problem of Space, in: Ideas and Opinions, Crown Publishers. Inc. New York, Fifth Printing, (1960) pp. 360-377.
15. A. Einstein, Äther und Relativitätstheorie, Springer, Berlin (1920).
16. A. Einstein, Da; Raum-, Äther- und Feld-Problem der Physik, in: Mein Weltbild, Querido Verlag, Amsterdam (1934).
17. A. Einstein, Über den Äther, Schweiz. Naturforsch. Gesellsch. Verhandl. 105:85 (1924).
18. A. Einstein, N. Rosen, The particle problem in the general theory of relativity, Phys. Rev. 48:73 (1935).
19. A. Einstein, Physics and Reality, J. Franklin Inst. 221:313 (1936)
20. A. Einstein, On the Generalized Theory of gravitation, Scientific American, 182, No. 4. April, 1950, pp.13-17.
21. A. Einstein, Apendix II, in: The Meaning of Relativity, Fifth Ed., Princeton Univ. Press. Princeton (1955).
22. M. Capek, Relativity and the Status of Space, The Review of Metaphysics, (December 1955) pp. 167-195.
23. B. Riemann, Über die Hypothesen, welche der Geometrie zu Grunde liegen (1854), Springer, Berlin (1921).
24. A. Einstein, Address at the University of Nottingham, Science, 71:608 (1930).
25. A. Einstein, Raum, Äther und Feld in der Physik, Forum Philosophicum, 1:173 (1930).
26. L.D. Faddeev, Le vedute di Einstein e talune vedute contemporanee sulle particelle elementari, in: Astrofisica e Cosmologia, Gravitazione, Quanti e Relatività negli sviluppi del pensiero scientifico di A. Einstein. "Centenario di Einstein 1879-1979", Giunti Barbèra, Firenze (1979) pp. 765-793.
27. L. de Broglie, Ondes et quanta, C. R. Acad. Sci., 117:507 (1923).
28. L. de Broglie, Quanta de lumière, diffraction et interférence, C. R. Acad. Sci., 117:548 (1923).
29. L. de Broglie, Recherches sur la théorie des quanta, Univ. Press, Paris (1924).

30. E. Whittaker, A History of the Theories of Aether and Electricity, Harper and Brothers, New York (1960) Vol. II. p.268.

31. J.-P. Vigier, Louis de Broglie - Physicist and Thinker, Found. Phys. 12:932 (1982).

32. J.-P. Vigier, De Broglie Waves on Dirac Aether; A Testable Experimental Assumption, Lett. Nuovo Cimento, 29:467 (1980).

33. F. Halbwachs, F. Piperno, J. -P. Vigier, Relativistic Hamiltonian Description of the Classical Photon Behavior: A Basis to Interpret Aspect's Experiments, Lett. Nuovo Cimento, 33:311 (1982).

34. A. Kyprianidis, D. Sardelis, A H-Theorem in the Causal Stochastic Interpretation of Quantum Mechanics. Lett. Nuovo Cimento, 39:337 (1984)

35. P. A. M. Dirac, Is there an Aether ?, Nature, 168:906 (1956).

36. J.-P. Vigier, Non-locality, Causality and Aether in Quantum Mechanics, Astron. Nachr. 303:55 (1982).

37. L. Kostro, A wave model of the elementary particle, A three-waves hypothesis (1978) unpublished paper.

38. L. Kostro, A Three-wave Model of the Elementary Particle, Phys. Lett. 107 A:429 (1985).

39. L. Kostro, Planck's constant and the three waves of Einstein's covariant ether, Phys. Lett. 112 A:283 (1985).

40. L. de Broglie, Nouvelles perspectives en microphysique, Ed. A. Michel, Paris(1956).

41. L. de Broglie, Recherches d'un demi-siècle, Ed. A. Michel, Paris (1976).

42. D. Bohm, J.-P. Vigier, Phys. Rev. 109:1882 (1958).

43. L. Mackinnon, A nondispersive de Broglie Wave Packet, Found. Phys. 8:157 (1978).

44. L. Mackinnon, Particle Rest Mass and the de Broglie Wave Packet, Lett. Nuovo Cimento, 31:37 (1981).

45. L. Mackinnon, A Fundamental Equation in Quantum Mechanics?, Lett. Nuovo Cimento, 32:311 (1981).

APPENDIX

Einstein presented and developed his ether conception in the following papers:

1916 Letter to H.A. Lorentz (June 17, 1916)

1918 Dialog über Einwände gegen die Relativitätstheorie Naturwissenschaften 6:697

1920 A. Einstein-Ph. Lenard, Allgemeine Diskussion über Relativitätstheorie, Phys, Zeitscher, 21:666

1920 Äther und Relativitätstheorie, Springer, Berlin.

1924 Über den Äther, Schweiz, Naturforsch, Gesellsch, Verhandl, 105:85.

1929 Raum, Äther und Feld in der Physik, Forum Philosophicum, 1:173, (1930.

1930 Das Raum-, Feld- und Äther-Problem in der Physik, Second World Power Conference, Transaction, 19:1.

1930 Address at the University of Nottingham, Science, 71:608.

1930 Das Raum, Feld- und Äther- Problem in der Physik, Die Koralle, 5/11/:486.

1934 Das Raum, Äther- und Feld-Problem der Physik, in: Mein Weltbild, Querido Verlag, Amsterdam, (1934).

1938 The Evolution of Physics, Univ. Press, Cambridge, (with L. Infeld) (1947).

1954 Relativity and the Problem of Space, in: Ideas and Opinions, Crown Publ. Inc., New York, Fifth Printing, (1960), 360-3777.

DISCUSSION V.

Compiled and interpreted by D.W. Kraft and E. Panarella

Edmonds. To how many digits will the theoretical Lamb shift agree with experiment in the worst case? I saw a paper eight or ten years ago which claimed agreement to seven or eight places.

Belinfante. The fine structure constant is known to six digits, and the calculation is based on the fine structure constant. I don't trust the seventh and eigth digits because the α to a power comes in.

Edmonds. Why have the people who wrote that paper missed all the things you say are wrong?

Belinfante. Many of those who worked in this field were fascinated by Feynman diagrams which are appropriate for the high energy part of the calculation. This is nice from a mathematical point of view. Essentially, these are higher order calculations and some errors were later found. But in the lowest order, the calculation is nothing more than taking French-Weisskopf and putting it in fancy form so you can understand better why it's covariant.

Edmonds. You didn't give a number. Will the theory and experiment agree to at least three or four digits?

Belinfante. Right now I know nothing; I have to convince myself that if you don't make crazy assumptions, you still get the right answer. Then I will say that there is perfect agreement, and I hope there is because I like quantum theory.

Aspden. Do you know of alternative explanations of the Lamb shift? I recall for example seeing some proposals based on stochastic electrodynamics.

Belinfante. I don't know that work. Weisskopf wrote an early paper, based on ideas of Welton, that considered vacuum fluctuations but provided only a qualitative understanding.

Greenberger. A number of textbooks include quick and dirty calculations based on vacuum fluctuations that provide order of magnitude agreement.

Marshall. I'm sure you can't improve on the first few digits, because stochastic electrodynamics is not capable of giving the hydrogen spectrum; that is, it does not give you even the zero-th order correctly, so, in my opinion, you can't do any better that the Welton calculation.

Selleri. I think the approach outlined by Edmonds can be described as neo-Platonist. Your ontology is really made of mathematics and your work contains mathematical simplicity. It is well to remember that Feynman has a similar approach to physics which has led him to great successes. Now we start understanding that an approach of this type is not really defensible, although it is an important part of modern science. We know that mathematics works beautifully in many cases, and we know that we do not understand why it works so well. Einstein wrote this several times. I want to stress that not all of reality, and certainly not all of science, can be based on mathematics. Two examples of this are Kepler's failed attempt at construction of a model of the Solar System, and Darwin's Origin of Species which contains not a single equation, but whose great lines of the development of life have been confirmed by modern molecular biology and genetics.

Also, I believe that ideas of space-time were given up too easily by the Copenhagen School. If it is proven that we must give up these ideas, so be it, but let that be because additional research compels us to do so. One beautiful feature of this Conference has been the attempt by several individuals to build models in space, time, and causal terms.

Edmonds. I think we all have a natural desire to picture the world neatly in our imagery. But in order to start with gases and end with people, I suspect that the laws would have to be so weird and abstract that they may be beyond human comprehension.

Secondly, with regard to evolution, that discovery was an important one because it helped to discredit the idea that miracles must be included to explain the world as we see it. But in modern terms, the basic paradigm of science makes Darwin's work trivial, for if we start with hydrogen gas and end up people and no miracles in between, then there had to be evolution; it is merely a question of working out the details. It is essentially quantum physics operating over a massive scale of many, many particles and many, many years. Eventually, those details should, in principle, be worked out.

Honig. Mathematics is a tremendously fertile field at the present time and is undergoing a period of flowering. The field of nonstandard numbers and in particular, the transfinite numbers, are concepts into which one can pack ideas, such as those we have been talking about, extremely neatly, and then pull them out and manipulate them in ways that give explanations and a clearer overall picture of the inconsistencies and paradoxes.

Greenberger. A number of years ago I worked out a variable mass formalism based on scale transformations. From that point of view, the mass gets built into the theory which then becomes a five-dimensional theory. I also found that the Dirac equation, when written in such a five-dimensional space, has the sort of structure that you talked about. I would like to point out then that just beyond the Dirac equation, there are structures that can include your approach.

Marshall. I would like to support Selleri's remarks that advances on fundamental questions are made by using our notions of space and time and not by the use of mathematics. Edmonds appears to prefer quaternions to vectors, because the quaternions are more bizarre; I would say that we should stick to vectors.

Edmonds. Responding first to Honig, I don't think that higher infinities and that sort of mathematics will be applicable to the real worlds. I think the dense set of numbers and the rational numbers will suffice to describe the world, and theories based on these numbers will be the only ones that human beings will be able to test. With respect to

452

Greenberger's comment, I am glad to see someone thinking about higher spaces for I think this is where the future lies and mass has to be replaced by a deeper level. Quantum physics is quite elegant when written in quaternion language, while Einstein's equation is not. That suggests to me that Einstein's equation is in need of improvement so as to get it to fit this language - there is a hint there that Einstein's gravity equation is not complete, which may be why they have not yet been able to quantize it. As for bizareness, it, like beauty, is in the eye of the beholder. I take the position that if you examine the progression from nonrelativistic classical physics to relativistic quantum physics, you see an enlargement of the number systems which elegantly and naturally describe that progression. That evolution has successfully provided deeper levels of understanding in terms of comparison with experiment. This should lead us to a generalization of that trend into something more sophisticated and to a deeper understanding of the world.

Greenberger. I know of two experiments that bear on Buonomano's ideas. One is an electron interference experiment done at Tubingen, in which interference patterns from various numbers of electrons were obtained. Even when there are just a few electrons, you can see that they seem to know where to go, or not go, in the interference pattern. The other was a neutron interference experiment done by Rauch. He compared the case of blocking the neutrons with an absorber that passes, say, 10% of the neutrons, versus the case of blocking 100% of the neutrons, 90% of the time, so that 10% of the time they go through unaffected. He found different interference patterns for these two cases.

Panarella. But this is identical to what I find with photons. Given the same number of photons at different photon fluxes, I do not obtain the same diffraction patterns. Such nonlinearity, which I believed was peculiar to photons, seems to be applying to neutrons also. There seems to be problems therefore in proving the wave-particle duality for neutrons, because a single particle phenomenon has to be linear.

Buonomano. I don't think that electron experiment applies to what I said, because of the way they collect the electrons. It is difficult for me to say whether the neutron experiment is applicable. I should point out the effects I described are transient effects and, although I would like to analyze existing experiments in terms of them, if the experimenter isn't paying attention to what I am interested in, it's very difficult to analyze the data.

Surdin. Not long ago, I did just what you explained here. The zero point field goes through the polarizer, stimulates the emission of the source, and when the emission returns to the polarizer, it has the same polarization and there is a cosine squared dependence on the angle. I have performed certain modifications of this scheme by modulating the polarizer, with interesting results which I can show you.

Phipps. I gather you've made your peace with quantum mechanics, but it seems to me you still have to make your peace with relativity principles. You are imputing to the medium itself the physical mechanism of interference; this would suggest to me that there must be something there which must have a state of motion. This then introduces the ether wind, and if we are traveling toward Leo at 400 km/sec, then every microsecond you are conditioning a medium that's being swept away 40 cm.

Buonomano. Absolutely right. My apparatus is stationary in the medium and one would have to physically justify the medium and worry about Lorentz invariance.

Edmonds.   Your model must also deal with the phenomenon of a particle
being in two slits at once and by being able to tunnel through a barrier;
this means that it has to sense the barrier thickness before it gets
there in order to know the probability of tunneling through. Imagine if
each of 10,000 people took their finger and hit it against the next one;
then there is a finite nonzero probability that someone's finger would
tunnel through. If we each did this only once, then according to your
theory no fingers would go through because the medium isn't conditioned
yet. Consider instead, if just I hit my finger 10,000 times, each time
my finger comes down, it conditions the medium differently, so that,
eventually one finger will be able to tunnel through because the medium
has been altered by enough fingers approaching it prior to that.

Another thought is, why can't you put 10,000 pairs of slits around a
single source with two photomultipliers behind each one, flash this once
so the pairs catch on the average one photon in each pair, and do an
ensemble average that way, and add up all those individual clicks and
then see if you can put back together the interference fringes or not?
Such an experiment could show whether your model is really valid.

Buonomano.   I'd have to think about that.

Marshall.   I found Kostro's cultural and historical introduction
wonderful and fascinating. I also found absolutely in tune with my own
ideas of the ether what I could perhaps describe as the Kostro filtering
of the Einstein view of the ether; presumably we cannot get an overall
picture of the Einstein view because he is not here, but I think the
Kostro filtering was extremely interestng. It fits with my idea of the
SED description of the ether and I would say that within this
persepctive, we have some recent verifications of this picture. The SED
ether, as developed by Planck and Nernst, was precisely of the same kind
as Kostro described for the Einstein ether, i.e. an ether in which it is
not meaningful to speak about the motion of the ether, or motion relative
to the ether. The basic property of the spectrum which is given by SED
is the same for all inertial observers. Recently, accelerations with
respect to this frame have been studied. Whether one uses the quantum or
the semiclassical description of the zero point field, it was found that
the ether can be detected because it warms up. There is here some
physical justification of an ether which can be observed only when you
are in accelerated motion with respect to it. Furthermore, you say that
what is missing from the Einstein description of the ether up to 1924 is
Planck's constant. Well, in my view of the zero point field as a real
physical field, h is precisely a measure of the strength of the zero
point field. And so we find that the Planck theory, which was developed
in rivalry with Einstein's outlook, contains many of the features which
are required in the Einstein view of the ether.

Edmonds.   The basic symmetry of the laws of Dirac and Maxwell have
Lorentz covariant structures which has nothing to do with what moving
observers see. The mathematics dictates the symmetry and the symmetry is
the Lorentz symmetry, and so the laws have a Lorentz symmetric structure.
Once you have those laws, they dictate what happens, and so we get the
phenomena of length contraction and time dilation; within those kind of
measurements, using light, you can't demonstrate absolute motion. But
this still leaves totally open the question of the vacuum. If there is a
preferred reference frame dictated by the background radiation showing
that we are moving at 0.002c toward Hydra, and if the vacuum is filled
with Higgs bosons or quasi-particle pairs, then if we do an experiment
that is very sensitive to the virtual pairs that are part of the vacuum,
then we should see some asymmetry in experiments that are of a quantum
nature using the vacuum; these asymmetries could not be detected when

454

using light because of length contraction and time dilation. Whether or not the vacuum is Lorentz covariant doesn't change the fact that the laws are Lorentz covariant. Thus it reopens the question of a preferred reference frame and it is no longer tied to Einstein's equality of observers. Thus I believe the Lorentz point of view is more correct than Einstein's as to why the laws are Lorentz covariant.

I would also like to ask Kostro about tachyons. Your theory contains a superluminal component. Does that then lead to an instantaneous action at a distance component to the nature of reality?

Kostro. No. That would require an infinite propagation velocity.

Phipps. Kostro's paper beautifully dramatizes the two opposed views of what fields might mean. Einstein was quoted as saying that "fields are physical states of space," which I characterize as the ectoplasmic view of the field. The definition of field that I spoke of in my paper referred to the reading on a black box, and space comes into it as the theater in which one can locate the field point. The point can be anywhere, but the field is, in a sense, created locally by the detection at the point where you place this black box; all the other places are uninterrogated and purely methaphysical. This is what I call the operational view where you demand the operations to be performed to make the field appear. I wonder if Bridgeman who founded the operational point of view noticed this difference between himself and Einstein.

Jeffers. I would like to ask Kostrow if his theory of the ether makes any predictions that can be observationally or experimentally tested.

Kostro. I don't know. I don't see that such experiments can be done at this time.

Buonomano. One of the foundational questions of special relativity is whether we can perform a terrestrial experimental test that can distinguish between Einstein's special theory and neo-Lorentzian theory. The kind of nonergodic work that I described can be a test of the medium and so could give rise to a possible terrestrial type test.

Edmonds. Motion through the vacuum, which is full of virtual particles, relative to the background radiation in the Universe might be detected by an experiment sensitive to quantum fluctuations. Since the Solar System is moving at $0.002c$, an experiment performed for different orientations of the apparatus that produces a wind-in-the-face effect could reveal that the quantum is not isotropic. I urge that experiments to test for this be considered.

Aspden. An experiment that revealed light speed anisotropy was reported by Silvertooth who set up standing waves built up from the same laser in a circuital system. He finds that the spacing of the nodes changes with orientation, resulting in a velocity of 378 km/sec in the direction of Leo. However, these experiments still require independent verification. A 1983 paper in Applied Optics by Silvertooth and Jacobs described the development of the standing wave sensor which scans along the standing wave path without obstructing the signal.

Edmonds. The reference frame for that measurement of our speed is with respect to matter moving in all directions over 15 billion light years, and therefore, this experiment unequivocally determined the absolute motion of the Earth. Thus, if you are going to look for any kind of effect, you can work backward from this speed to calculate the magnitude of the effect your apparatus must detect.

CONCLUDING REMARKS

G. Hunter

Department of Chemistry, York University
4700 Keele Street
North York (Toronto), Ontario, Canada M3J 1P3

I would like to give you my perspective of what I heard at this conference and I ask your forgiveness if I neglect to mention things that were most important to you.

It is our continuing unhappiness with the state of quantum theory, particularly its counter-intuitive character, that brought us together. One of the most fundamental problems is the question of the meaning of space and time, and how they are measured. This is related to the concept of the ether, for which we heard several conceptions, and for which there does not appear to be any reconciliation. Several papers at this conference touched upon this problem.

Although we all accept the predictive ability of quantum theory, we seek a deterministic interpretation, something other than the conventional Copenhagen view that Nature is intrinsically statistical. The work of Vigier and his collaborators emphasized the wave-particle duality and, in terms of the quantum potential theory, are able to explain trajectories along with waves. Their formalism is entirely equivalent to the nonrelativistic Schrödinger equation and so they are providing quantum theory with a deterministic interpretation.

Other work that has been presented here was concerned with the question of whether quantum theory is right. That is, even in the nonrelativistic domain, can one create experimental situations that show that it is not sufficient? The really hard thing in this field is the interpretation of the experiments and we had some quite heated debate at times concerned with just that.

The nature of interference in neutron and light scattering experiments was a subject that surfaced on several occasions. We have to accept that the experimental evidence to me strongly suggests that a particle can interfere with itself. In particular, the neutron experiments seem to support the view that a neutron in a diffraction apparatus somehow goes along two channels at the same time and interferes with itself. While this is counter-intuitive, it is difficult to devise an alternative explanation. However, there were at least two papers that attempted to do just that: Buonomano stated that previous events imprint themselves on the apparatus to influence subsequent events, and in this way explained interference without having the particle go through two channels at the same time. Surdin introduced two kinds of waves to provide a deterministic explanation of these intereference effects, especially the recent experiments of Grangier, Roger and Aspect, whose results seem to support single photon interference.

This conference has been a nice blend of the presentation of the experimental evidence and a variety of new ideas which have the potential for providing deterministic interpretations that are alternatives to the conventional statistical view point.

Belinfante addressed the continuing problems of divergences and noncovariance in QED, and brought home the idea that theoretical physicists often make approximations that are mathematically convenient, but lacking in physical justification. His attempt to calculate the Lamb shift accurately is a laudable pursuit and we look forward to those results.

I was especially impressed with the work of the Italian School of quantum theory where both experiments and theory are done in the same laboratory. In particular, I think back to the presentations of Selleri and of De Martini, each of which contained a lovely combination of experimental evidence and theoretical interpretation. While they didn't necessarily provide all the answers, that kind of work is the epitome of what we all ought to be doing.

CONCLUDING REMARKS

F. Selleri

Dipartimento oh Fisica dell'Universita
INFN - Sezione di Bari
Via Amendola 172, I-70126 - Bari, Italy

I wish to start my summary with many thanks for all the work that has been done for making this Conference possible. We all enjoyed it very much and are very grateful to D. Kraft, W. Konig and E. Panarella, who solved all our problems. We shall remember with pleasure this Conference of Bridgeport.

The history of physics is sometimes complicated rather than linear and simple. An illuminating event was a conference entitled "New Theories in Physics," held in Warsaw in 1938, at which von Neumann presented his theorem stating that the casual deterministic completion of quantum mechanics was not possible. At the same conference, Bohr expressed his admiration for von Neumann's theorem and said that it came to the same conclusion that one of his then-recent papers had obtained. I believe that paper was the 1935 paper on complementarity, refuting Einstein, Podolsky and Rosen. Thus, there is a close logical link between von Neumann's Theorem and the Complementarity Principle, although they may be very different philosophically. Heisenberg, in one of his books, also tried to prove the impossibility of the causal completion of quantum theory with a qualitative argument on the scattering of alpha-particles from a grating. Shortly afterwards, Pauli and Jordan along with Bohr, Born and others began to stress that the causal completion of quantum theory could not be obtained. This should be recalled because it has been a gigantic historical mistake; the pressure exerted by these important individuals was so strong that even such people as Einstein and de Broglie became completely isolated. But now, thanks to the work of a few courageous people, we have overcome all that. From a logical point of view the overcoming of von Neumann's Theorem and of the Complementarity Principle dates from 1952 with the papers by Bohm, who simply did what had been considered impossible: he found a causal completion of quantum theory. After that, de Broglie resumed work and then, in 1964, came Bell's paper in which he found the weak point in von Neumann's Theorem: the third action. The discovery of Bell's Inequality signalled a change of mood, resulting in a growth of interest in this field and in this approach. I believe that, today, we have every reason to say that space-time and causality should be defended as much as possible. By this I don't mean that they are necessarily true, for they may be falsified. But this has not yet happened, and they were abandoned only because of cultural and philosphical fashions, not because of scientific reasons. We should now seek scientific reasons and have experiments to prove that space-time and causality cannot be applied or, as I believe to be more likely, that they can continue to be applied.

In this conference we have had many contributions in that direction. I enjoyed hearing Honig's photex concept, and the work of Vigier, Kyprianidis, Dewdney and Cufaro-Petroni, which goes in the direction of a space-time and causal description of quantum phenomena. I have learned much from Aspden on his lattice in space and the possibility of calculating the value of α and of the mu-meson mass, and I enjoyed the paper by Kostrow on the three-wave model.

But, there have also been other contributions: Surdin talked about stochastic processes, Hunter and Wadlinger about a physical model for the photon, and Marshall on stochastic electrodynamics. I don't have the capability to comment on these, but they were all very interesting papers.

The contributions of Buonomano, Santos and Garuccio brought out the problems of locality and of the EPR paradox. I am a little unhappy with the position of Vigier and his collaborators on locality because, from a historical point of view, it is not correct to think that de Broglie held to a nonlocal position; it was mixed. In a 1976 paper, de Broglie states that quantum mechanics has been misinterpreted for no state vectors of the second type exist and therefore no EPR paradox exists. Well, that is a way of saying that quantum theory is local. Also, when I had the opportunity to meet personally with de Broglie, he stressed that he agreed completely with a paper of mine in which I defended locality. Of course, all ideas should be investigated, and the idea of nonlocality is perfectly possible from a logical point of view. In fact, all of the solutions proposed for the EPR paradox are far-fetched. I can recall the following ones:

1. Quantum mechanics is wrong.
2. There are superluminal influences, which seems to me impossible from a relativistic point of view, but one never knows.
3. The future can influence the past.
4. Negative probabilities as proposed by Feynman and by Dirac.
5. The standard Copenhagen point of view that there is nothing real in the world except human activities, human experiments, but not an external reality.
6. Nonergodic processes as proposed by Buonomano.

There may be other solutions, also far-fetched, and that is why the EPR paradox is so interesting.

In this conference we have not had the perfect mix of the experimental and theoretical levels, in the sense that perhaps not all known experiments have been taken seriously enough. There are many experiments that have been done from which we can learn much and I think we should take them more seriously. There have been some very interesting experiments presented at this conference. We had experiments presented by De Martini, by Panarella, by Matteucci, and by Jeffers, and they were all very exciting and stimulating.

I would like to mention the very interesting theoretical proposals put forward for new relationships between theory and experiment, from different points of view, by Datzeff, by Belinfante, and by Phipps. I also recall the stimulating idea of a truly mathematical ontology presented by Edmonds, Raychaudhuri's ideas on the composite nature of the photon, and the solution of long standing conceptual paradoxes proposed by de Haan. We all enjoyed the stimulating remarks contributed by Boyer and by Greenberger.

The growing interest in this type of activity provides us with cause for optimism. Now, it is true that science is not always logical; rather, it is a complicated activity in which different ideas, prejudices, and mistakes converge. However, logic has much to do with science, in fact more than with any other human activity. So we may say there is a growing logical space for a renewal of the foundations of modern physics; moreover, in the last 10 years, we can say there is also a growing sociological space for this change. Thus I believe we are slowly heading towards a complete recasting of physics. Of course, it will take time; it will not be easy and it will not be short. But perhaps we can prepare for the best.

COMMENTS

H. Rymer

Department of Physics
University of Bridgeport
Bridgeport, Connecticut - U.S.A. 06601

I hope that the comments that I will make here will not be viewed as criticism of a colleague's work, especially if these remarks are made by one outside the field of quantum physics.

In my opinion, the meeting that was held at the University of Bridgeport concerned the most important topic that is before the community of scientists today. There is no doubt in my mind, what so ever, that the questions which were brought to the floor of the meeting affect the entire field of physical science. No part of science is immune from the values, and concepts, which rise from the investigations in this area of fundamental physics. The questions which have been asked of quantum physics have long since departed from the narrow confines of "quantum mechanics." We are now in the domain of the foundations of physics.

It is also quite clear that the answers to the probing questions which have been asked will have to come from the experimentalist. Until some concerns have been laid to rest, the detailing of theory upon theory will not, I believe, result in the movement toward the understanding of the fundamental processes of physics. Quantum physics is in much the same position that the study of cosmology was in the 1940's or 50's. It was not until a system of cosmological standards and coordinate systems, in the most general sense, was established that order was achieved in this field of study. This task was accomplished by the observational astronomer and the development of a new generation of instrumentation.

I would like to suggest that this new generation of instrumentation for use in quantum physics experimentation has been developed. We can, at this time, pulse a wave train of between five and ten lambda in length, and create an optical chirp of uncontested quality. We can measure an absolute distance with an accuracy of lambda/400 with the possibility of lambda/1000 in the near future. Point sources with a separation of .0009 arc seconds can be resolved and the computer furnishes us with an unbiased observer.

All of these techniques are expensive and require some form of permanent laboratory facilities. These are, however, necessary tools which are needed to establish the base of modern physics.

We have, at this time, a plethora of experiments and experimental procedures, many of which are indicative of the needs, direction and skill of the experimentalist. Many of the experiments require considerable effort and expense to duplicate. Because of this cost, verification of results has been difficult by other workers. However, the keynote of experimental procedure is that work can be duplicated by an unbiased and perhaps unschooled, observer.

I would, therefore, like to suggest the following for consideration and critique:

1. I think that a program similar to the international study that was conducted to achieve uniformity of fundamental astronomical tables should be considered for quantum physics. This means that a comprehensive survey of all experiments pertaining to quantum physics at its most fundamental level should be initiated. The purpose of this study would be to assess the experimental work needed to establish a base for future directions of laboratory work to guide the development of instrumentation and to analyze the internal agreement between present experiments to measure fundamental quantities. This end result would be a knowledge of what is needed in the way of experimental evidence in order to establish a firm foundation for quantum physics. Such a study would indicate the future trends of experimental work in this field.

2. In those experiments where linear distance measures are of extreme importance I do not think that interference methods, in the normal sense of the word, are what is needed. These methods are not dynamic, but static, and it is important that a continued baseline be established during experimental runs. This would indicate that methods of heterodyne interferometry would be more accurate. With this method one can measure, and maintain, distances to lambda/400. In this technique the distance measures are replaced with time measures which are much better to work with.

3. Electromagnetic events should be sensed, whenever possible, with the CCD array rather then with any other type of detector. Response times are much better in these intruments than in any other type of detector. This device also imposes an implied coordinate system into the experiment and brings the computer along as the monitor. This is also a dynamic system in that continuous samples of the optical image can maintain a diffraction-limited system throughout an experimental run.

4. I think it is a bad mistake to allow mechanical chopping of an optical signal, except for the most gross of purposes. Mechanical choppers, even very good ones, have been the source of much trouble in accurate optical experiments, especially when the experiment involves scattered radiation. Whenever possible, optical chopping should be used. This is not a trivial job as these instruments are quite exotic and difficult to use. However, with proper care, these instruments do the best possible job.

5. Optical components in an experiment should be mounted using adaptive optical techniques. In this way an experiment can maintain a diffraction-limited optical system and still retain the bore sighting of the optical path. These systems do require the use of a computer, but one can maintain a system with

eighteen to twenty degrees of freedom without a great deal of
trouble. In addition, the quality of the image obtained more
than justifies the effort.

6. I think that time measure should be based on an in-situ
vibrational mode standard rather than the International second.

7. Laser beam management standards must be established with respect
to what can be tolerated in an experiment. During the course of
an experimental run the laser beam characteristics must be
maintained with a beam spectrometer. This is extremely important
when the beam frequency is in the IR. A beam with known quality
must be delivered to the experimental platform.

8. If any form of adaptive optical system is used, it must be made
certain that no optical component is in motion during the
measurement mode of the system. This is particularly true of
moving filters, and of any other component, during the course of
an experimental run.

9. Coordinate platforms for the laboratory reference for any
experiment should be maintained, for stability, by ring laser
techniques. In fact, it would be a good idea to have all
experiments referred to some system of standard coordinates.

I am sure that many more problems can be discussed and some form of
correction procedure be established in the interest of standardization
and reproducibility. In any case, what I have suggested is going to be
expensive; I do not think that any one group can afford such a facility.
Therefore I suggest that international cooperation is needed to establish such a laboratory. It would not, by any means, be as expensive
as high energy physics. In fact, by comparison, it would be cheap,
especially when one considers the possible returns for pure science.

I think that another meeting should be called to discuss experimental standards and procedures for quantum physics experiments. I hope
that such a meeting would result in a recommendation for the establishment of an international laboratory dedicated to quantum physics
experiments.

466

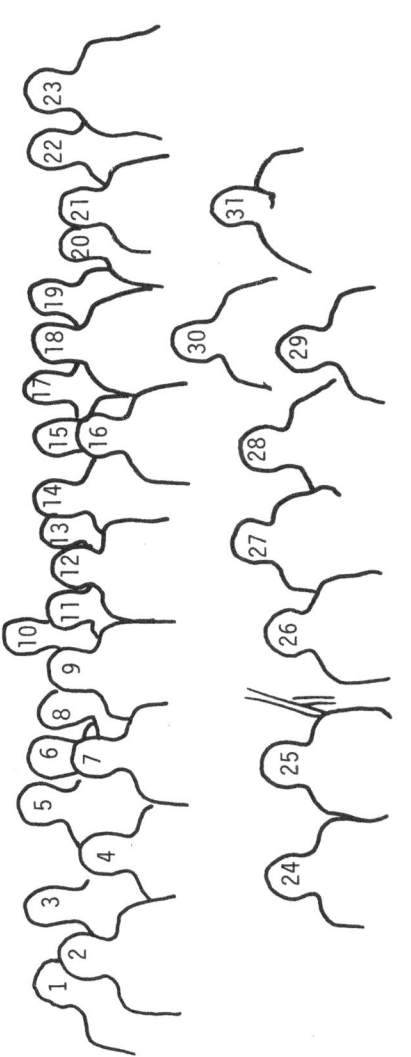

QUANTUM UNCERTAINTIES - PARTICIPANTS IDENTIFICATION

1. F. De Martini
2. D.W. Kraft, Associate Director
3. T. Boyer
4. E. Santos
5. W.M. Honig, Director
6. A. Kyprianidis
7. E. Panarella, Associate Director
8. C. Dewdney
9. F. Selleri
10. R. Wadlinger
11. M. Surdin

12. F. Belinfante
13. T.E. Phipps, Jr.
14. G. Hunter
15. S. Jeffers
16. A.B. Datzeff
17. L. Kostro
18. H. Aspden
19. W. Ward
20. H. Rymer
21. M. Browne

22. J. Tucci
23. J.P. Vigier
24. P. Raychaudhuri
25. A. Garuccio
26. V. Buonomano
27. G. Matteucci
28. T. Marshall
29. J. Edmonds
30. N. Cufaro-Petroni
31. M. De Haan

PARTICIPANTS AND OBSERVERS

Participants

H. Aspden
University of Southampton
Southampton SO9 5NH
England

V. Buonomano
Universidade Estadual de Campinas
13100 Campinas SP
Brazil

A.B. Datzeff
University of Sofia
Sofia 1000
Bulgaria

F. De Martini
Università di Roma
Rome 00185
Italy

J.D. Edmonds, Jr.
McNeese State University
Lake Charles, LA 20609
U.S.A.

W. M. Honig, Director
Curtin University
Perth, S. Bentley 6102
Western Australia

L. Kostro
University of Gdansk
80-952 Gdansk
Poland

A. Kyprianidis
Institut Henri Poincaré
75005 Paris
France

T.W. Marshall
36 Victoria Ave.
Manchester M20 8RA
England

F.J. Belinfante
P.O Box 901
Gresham, OR 97030
U.S.A.

N. Cufaro-Petroni
Università di Bari
70126 Bari
Italy

M. De Haan
Université Libre de Bruxelles
1050 Brussels
Belgium

C. Dewdney
Portsmouth Polytechnic
Portsmouth PO1 2DZ
England

A. Garuccio
Università di Bari
70126 Bari
Italy

G. Hunter
York University
Toronto, Ontario M3J 1P3
Canada

D. W. Kraft, Associate Director
University of Bridgeport
Bridgeport, CT 06601
U.S.A.

G. Matteucci
Università di Bologna
40126 Bologna
Italy

E. Panarella, Associate Director
National Research Council
Ottawa K1A 0R6
Canada

T.E. Phipps, Jr.
908 South Busey Ave.
Urbana, IL 61801
U.S.A.

E. Santos
Universidad de Santander
Santander
Spain

M. Surdin
Centre des Faibles Radioactivités
91190 Gif-sur-Yvette
France

R.L.P. Wadlinger
York University
Toronto, Ontario M3J 1P3
Canada

P. Raychaudhuri
Calcutta University
Calcutta 700 009
India

F. Selleri
Università di Bari
70126 Bari
Italy

J.P. Vigier
Institut Henri Poincaré
75005 Paris
France

Observers

T. Boyer
City College of New York
New York, NY 10031
U.S.A.

D. Greenberger
City College of New York
New York, NY 10031
U.S.A.

H. Rymer
University of Bridgeport
Bridgeport, CT 06601
U.S.A.

J. Tucci
University of Bridgeport
Bridgeport, CT 06601
U.S.A.

M. Browne
University of Bridgeport
Bridgeport, CT 06601
U.S.A.

S. Jeffers
York University
Toronto, Ontario M3J 1P3
Canada

L. Talkington
Science and Nature
Tappan, NY 10983
U.S.A.

W. Ward
York University
Toronto, Ontario M3J 1P3
Canada